PROGRESS IN
POLAROGRAPHY

VOLUME III

PROGRESS IN POLAROGRAPHY

Edited by

P. ZUMAN and L. MEITES

Department of Chemistry
Clarkson College of Technology
Potsdam, New York

with the collaboration of

I. M. KOLTHOFF

Department of Chemistry
University of Minnesota
Minneapolis

VOLUME III

1972

WILEY-INTERSCIENCE

a division of

John Wiley & Sons, Inc.   New York   London   Sydney   Toronto

Copyright © 1972, by John Wiley & Sons, Inc.

All rights reserved. Published simultaneously in Canada.

No part of this book may be reproduced by any means, nor transmitted, nor translated into a machine language without the written permission of the publisher.

Library of Congress Catalog Card Number: 61-16748

ISBN 0-471-98638-0

Printed in the United States of America.

10 9 8 7 6 5 4 3 2 1

To J. Heyrovský, R. Brdička, and B. Breyer
in gratitude for their roles in developing electrochemistry
to its present state

AUTHORS OF VOLUME III

B. KASTENING, Forschungsabteilung Angewandte Elektrochemie der Kernforschungsanlage Jülich, Jülich, Germany

MICHAIL KOZLOVSKY, The Kirov Kazakh State University, Alma-Ata, U.S.S.R.

GEN P. SATÔ, Sophia University, Kioi-cho, Chiyoda-ku, Tokyo, Japan

S. G. MAIRANOVSKII, N. D. Zelinskii Institute of Organic Chemistry, Academy of Sciences of U.S.S.R., Moscow, U.S.S.R.

REITA TAMAMUSHI, The Institute of Physical and Chemical Research, Yamato-machi, Kita-adachi-gun, Saitama, Japan

ALEXANDRA ZEBREVA, The Kirov Kazakh State University, Alma-Ata, U.S.S.R.

P. ZUMAN, Department of Chemistry, Clarkson College of Technology, Potsdam, New York

PREFACE

During the decade that has elapsed since the publication of the first two volumes of this series, polarography and the other electroanalytical techniques related to it have undergone expansion and development that have both deepened our theoretical knowledge and broadened the experimental basis for further advances. The progress in instrumentation that has been made in this decade led to the development of new, sophisticated, and often highly specialized techniques, which in some cases have provided more straightforward or accurate approaches to systems already well characterized by dc polarography, and in other cases have revealed phenomena difficult or even impossible to detect by classical techniques.

The present volume deals with a number of aspects of polarography that are of special interest in current research. Two contemporary electrochemical problems on which much attention is focused—the formation and reactions of radicals and radical ions, and the effects of the electrical double layer on electrode processes—are discussed by Doz. B. Kastening and Dr. S. G. Mairanovskii, respectively. In stripping techniques, which have assumed great importance in trace analysis, the compounds that deposited metals may form with each other and with the mercury into which they are deposited are of great importance, and these and the techniques by which they have been identified and studied are reviewed by Professor M. T. Kozlovsky and Mme. A. Zebreva. Interest in inorganic polarography has been divided between studies of labile and substitution-inert complexes; studies of the former group are surveyed by Professor R. Tamamushi and Dr. G. Satô. Finally, the question of how to interpret a given set of polarographic data, with special reference to the effects of varying the pH, is discussed by Professor P. Zuman.

This is by no means an exhaustive selection of topics of current interest. We believe it illustrates the breadth of modern polarography without unduly emphasizing any one narrow area, and we have chosen to restrict the number of fields included so that each could be discussed fully while keeping the size of the volume within reasonable bounds.

Since the first two volumes of this series were published, three of those who contributed to them have been lost to the polarographic family. The deaths of J. Heyrovský, R. Brdička, and B. Breyer have left us all poorer.

P. ZUMAN
L. MEITES
I. M. KOLTHOFF

Potsdam, New York
Potsdam, New York
Minneapolis, Minnesota
November 1971

CONTENTS—VOLUME III

	PAGE
Application of Polarography and Related Electrochemical Methods to the Study of Labile Complexes in Solution By Reita Tamamushi and Gen P. Satô	1
Effects of Acidity in the Elucidation of Organic Electrode Processes By P. Zuman	73
Intermetallic Compounds in Amalgams By Michail Kozlovsky and Alexandra Zebreva	157
Free Radicals in Organic Polarography By B. Kastening	195
Double-Layer and Related Effects in Classic Polarography By S. G. Mairanovskii	287
Author Index	370
Subject Index	386

CONTENTS—VOLUME I

Polarographic Literature
  By Marie Heyrovská

Die Gleichung für Polarographische Diffusionsströme und die Grenzen Ihrer Gültigkeit
  By Jaroslav Koutecký und Mark von Stackelberg

The Instantaneous Currents (i–t Curves) on Single Drops
  By Jaroslav Kůta and Ivan Smoler

Double-Layer Structure and Polarographic Theory
  By Paul Delahay

Adsorption in Polarography
  By Charles N. Reilley and Werner Stumm

Concentration Polarization and the Study of Electrode Reaction Kinetics
  By John E. B. Randles

General Theoretical Treatment of the Polarographic Kinetic Currents
  By Rudolf Brdička, Vladimír Hanuš, and Jaroslav Koutecký

Constant-Current Polarography and Chronopotentiometry at the Dropping Mercury Electrode
  By Taitiro Fujinaga

The Electroreduction of Anions
  By Alexander N. Frumkin and Nina Nikolaeva-Fedorovich

Recent Advances in Inorganic Polarography
  By George F. Reynolds

Mechanism of the Electrode Processes and Structure of Inorganic Complexes
  By Antonin A. Vlček

Polarography of Complex Compounds
  By Jiři Koryta

Trends in Organic Polarography
  By Petr Zuman and Stanley Wawzonek

Current Trends in the Study of the Influence of Structure on the Polarographic Behaviour of Organic Substances
  By Petr Zuman

Polarographie in Nichtwässrigen Lösungen
  By Kurt Schwabe

CONTENTS—VOLUME II

Recent Modifications of the Heyrovský Dropping Mercury Electrode
  By Izaak Kolthoff and Yutaka Okinaka

The Hanging Mercury Drop in Polarography
  By Jiři Říha

Chromato-Polarography
  By Wiktor Kemula

Square-Wave and Pulse Polarography
  By Geoffrey C. Barker

Single-Sweep Method
  By Jiři Vogel

Oscillographic Polarography with Alternating Current
  By Robert Kalvoda

Alternating Current Polarography and Tensammetry
  By Bruno Breyer

Solid Electrodes
  By Ralph N. Adams

Advances in Polarographic Instrumentation in the United States
  By Louis Meites

Polarographic Instrumentation in Japan
  By Nobuyuki Tanaka

Advances in Classical Polarographic Instrumentation in Europe
  By Lubomír Šerák

Cells for Polarographic Electrolysis
  By Zbigniew P. Zagórski

Continuous Polarographic Analyzers
  By Jiří V. A. Novák

Important Factors in Classical Polarography
  By Petr Zuman

Treatment of the Sample in Inorganic Polarography
  By George W. C. Milner

Organic Reagents in Inorganic Polarographic Analysis
  By Mutsuaki Shinagawa

Polarography in Organic Analysis
  By Philip J. Elving

Organic Reagents in Polarometric (Amperometric) Titrations
  By Jaroslav Zýka

Polarography in Metallurgy
  By Miloš Spálenka

Polarography in Medicine and Biochemistry
  By Miroslav Březina

Polarographic Analysis in Pharmacy
  By Petr Zuman and Miroslav Březina

Polarography in Mineralogy, Geology, and Archaeology
  By Danilo Cozzi

Polarography in the Chemical Industry
  By Isamu Tachi and Mitsugi Senda

Cumulative Author Index

Cumulative Subject Index

CHAPTER I

APPLICATION OF POLAROGRAPHY AND RELATED ELECTROCHEMICAL METHODS TO THE STUDY OF LABILE COMPLEXES IN SOLUTION

Reita Tamamushi

The Institute of Physical and Chemical Research, Wako-shi, Saitama, Japan

and

Gen P. Satô

Sophia University, Kioi-cho, Chiyoda-ku, Tokyo, Japan

Contents

I. Introduction	1
II. Theoretical Treatment of the Electrode Reaction of Labile Complexes	2
A. Electrode reaction and equilibrium potentials of labile complexes	2
B. Faradaic current	4
C. Surface concentration of reacting species	6
D. Potentiostatic current–time relationship	8
E. Galvanostatic transition time	12
F. Current–potential relationship in Dc polarography	13
G. Ac polarography and faradaic impedance	20
III. Equilibrium Studies of Labile Complexes	26
A. Determination of stability constants. Use of the reversible half-wave potential	26
B. Determination of stability constants. Use of the diffusion current	30
C. Study of ion association by analysis of the average diffusion coefficient	36
IV. Kinetic Studies of Labile Complexes	42
A. Mechanistic study of electrode reactions	42
B. Kinetics of complex formation and substitution reactions	51
References	68

I. Introduction

Historically, the polarographic study of inorganic complexes goes back to the pioneering work of von Stackelberg and von Freyhold (81) and of Lingane (46). The classic method of analysis is thoroughly treated by Kolthoff and Lingane in their famous monograph on polarography. Since then, both theory and methodology of polarography and related electrochemical techniques have made remarkable progress, and now these techniques enjoy wide application to the study of complexes in solution. Some of the general developments are presented and discussed by Crow and Westwood (14, 15) and by Crow (13) in their recent publications. Excellent reviews have also been published by Vlček (117, 118) on the relationship between the electrochemical reactivity and the structure of inert complexes and by Koryta (39) on the electrochemical kinetics of metal complexes.

This chapter concerns a theoretical treatment of the electrode reaction of labile complexes and its application to the study of equilibria and kinetics of labile complexes in solution. In addition to conventional polarography, some of the related techniques, such as potentiostatic, galvanostatic, and impedance measurements, are included as well. This article is not intended to be a comprehensive review of the given subjects; topics are chosen and treated on the basis of the authors' personal interests.

II. Theoretical Treatment of the Electrode Reaction of Labile Complexes

A. Electrode Reaction and Equilibrium Potentials of Labile Complexes

Labile complex species $M^{m+}L_i$ ($i = 0, 1, 2, \ldots, p, \ldots, \mu$) are generally subjected to the following stepwise equilibria in solution:

$$M^{m+}L_i + L \rightleftharpoons M^{m+}L_{i+1} \quad (i = 0, 1, 2, \ldots, p, \ldots, \mu - 1) \qquad (1a)$$

where $i = 0$ refers to a simple metal ion M^{m+} and μ is the maximum number of ligands L coordinated with M^{m+}. In this article the charge of each complex species and that of L are omitted in order to avoid complications in representing chemical species.

Let us assume that the complex species $M^{m+}L_p$ is involved in the charge transfer process as given by Eq. 1b:

$$M^{m+}L_p + ze \rightleftharpoons M^{(m-z)+}L_q + (p - q)L \qquad (1b)$$

If the reduced form of $M^{m+}L_p$ is also a labile complex species, another set of stepwise equilibria should be considered with respect to the reduced species:

$$M^{(m-z)+}L_j + L \rightleftharpoons M^{(m-z)+}L_{j+1} \quad (j = 0, 1, 2, \ldots, q, \ldots, \nu - 1) \qquad (1c)$$

ν being the maximum number of L coordinated with metal ion $M^{(m-z)+}$.

The overall electrode reaction of a labile complex species consists of these three processes (Eq. 1a–c), and it offers a typical example of the charge transfer process coupled with homogeneous chemical reactions. The contribution of heterogeneous coupled reactions is not discussed in this article.

The equilibrium potential E_e of the $M^{m+}L_i/M^{(m-z)+}L_j$ redox system is given by Eq. 2 after the Nernst equation:

$$E_e = E_p^{\ominus} + \frac{RT}{zF} \ln \frac{[M^{m+}L_p]}{[M^{(m-z)+}L_q][L]^{(p-q)}} \quad (2)$$

where $E_p^{\ominus}$ is the standard potential corresponding to the charge transfer process of Eq. 1b and [ ] represents the concentration of each species in the bulk of the solution. The effect of activity coefficients is neglected for the sake of simplicity.

Introducing the total concentration c_{ox} of the oxidized form and that of the reduced form c_{red},

$$\left. \begin{aligned} c_{ox} &= \sum_{i=0}^{\mu} [M^{m+}L_i] \\ c_{red} &= \sum_{j=0}^{\nu} [M^{(m-z)+}L_j] \end{aligned} \right\} \quad (3)$$

we obtain Eq. 4, representing the equilibrium potential in terms of c_{ox} and c_{red}:

$$E_e = (E^{\ominus})_B + \frac{RT}{zF} \ln \frac{c_{ox}}{c_{red}} + \frac{RT}{zF} \ln \frac{F_0^{ox}([L]=1)}{F_0^{red}([L]=1)} - \frac{RT}{zF} \ln \frac{F_0^{ox}([L])}{F_0^{red}([L])} \quad (4)$$

where $(E^{\ominus})_B$ is the standard potential corresponding to the standard state of the overall electrode reaction ($c_{ox} = c_{red} = [L]$ = unit concentration), and related to $E_p^{\ominus}$ by the relation

$$(E^{\ominus})_B = E_p^{\ominus} + \frac{RT}{zF} \left\{ \ln \frac{\beta_p^{ox}}{F_0^{ox}([L]=1)} - \ln \frac{\beta_q^{red}}{F_0^{red}([L]=1)} \right\} \quad (5)$$

The functions $F_0^{ox}([L])$ and $F_0^{red}([L])$ are represented by

$$F_0^{ox}([L]) \equiv 1 + \beta_1^{ox}[L] + \beta_2^{ox}[L]^2 + \cdots + \beta_{\mu}^{ox}[L]^{\mu} = \sum_{i=0}^{\mu} \beta_i^{ox}[L]^i \quad (6a)$$

$$F_0^{red}([L]) \equiv 1 + \beta_1^{red}[L] + \beta_2^{red}[L]^2 + \cdots + \beta_{\nu}^{red}[L]^{\nu} = \sum_{j=0}^{\nu} \beta_j^{red}[L]^j \quad (6b)$$

β_i^{ox} and β_j^{red} being the overall stability constants of $M^{m+}L_i$ and $M^{(m-z)+}L_j$, respectively, as defined by Eq. 7a and b:

$$\beta_0^{ox} = 1, \qquad \beta_i^{ox} = \frac{[M^{m+}L_i]}{[M^{m+}][L]^i} \ (i = 1, 2, \ldots, \mu) \quad (7a)$$

$$\beta_0^{red} = 1, \qquad \beta_j^{red} = \frac{[M^{(m-z)+}L_j]}{[M^{(m-z)+}][L]^j} \ (j = 1, 2, \ldots, \nu) \quad (7b)$$

The values of $F_0^{ox}([L] = 1)$ and $F_0^{red}([L] = 1)$ are numerically identical with

$$\sum_{i=0}^{\mu} \beta_i^{ox} \quad \text{and} \quad \sum_{j=0}^{v} \beta_j^{red}$$

respectively.

B. Faradaic Current

The faradaic current density I attributable to the charge transfer process of Eq. 1b is given by Eq. 8 when a positive sign is assigned to the cathodic current*:

$$\frac{I}{zF} = \tilde{k}_p^{\,c}\,[M^{m+}L_p]_s - \tilde{k}_p^{\,a}\,[M^{(m-z)+}L_q]_s\,[L]_s^{(p-q)} \tag{8}$$

where $[\;]_s$ denotes the concentration at the electrode surface and $\tilde{k}_p^{\,c}$ and $\tilde{k}_p^{\,a}$ are the electrochemical rate constants of the cathodic and the anodic processes, respectively, of the charge transfer process of Eq. 1b. The electrochemical rate constant can be expressed by Eq. 9 as a function of the electrode potential E referred to a given reference electrode:

$$\left.\begin{array}{l} \tilde{k}_p^{\,c} = k_p^{\,c} \exp\left\{-\dfrac{\alpha_p zF}{RT} E\right\} \\[6pt] \tilde{k}_p^{\,a} = k_p^{\,a} \exp\left\{\dfrac{(1-\alpha_p)zF}{RT} E\right\} \end{array}\right\} \tag{9}$$

where α_p is the cathodic transfer coefficient, $k_p^{\,c}$ and $k_p^{\,a}$ are the cathodic and anodic rate constants, respectively, at $E = 0$, and all the other symbols have their usual meanings. The double-layer effect on the electrode kinetics of metal complexes has been discussed by Koryta (39) and is ignored in this article.

At the dynamic equilibrium of the electrode reaction, which is defined as the state in which no net current is flowing, the electrode potential and the surface concentrations should be equal to E_e and the bulk concentrations, respectively, and the following relations must be satisfied:

$$k_p^{\,c}[M^{m+}L_p]\exp\left\{-\frac{\alpha_p zF}{RT} E_e\right\} = k_p^{\,a}[M^{(m-z)+}L_q][L]^{(p-q)}\exp\left\{\frac{(1-\alpha_p)zF}{RT} E_e\right\}$$

or
$$\tag{10}$$

$$k_p^{\,c} c_{ox}[L]^p \frac{\beta_p^{ox}}{F_0^{ox}([L])}\exp\left\{-\frac{\alpha_p zF}{RT} E_e\right\} = k_p^{\,a} c_{red}[L]^p \frac{\beta_q^{red}}{F_0^{red}([L])}\exp\left\{\frac{(1-\alpha_p)zF}{RT} E_e\right\}$$

$$\tag{11}$$

* We adopt this sign convention in this chapter in order to avoid confusion with the existing convention in polarography. Theoretically, however, the reverse convention may be preferable.

From Eqs. 10 and 11, we can define the electrochemical standard rate constants $\tilde{k}_p^\ominus$ and $(\tilde{k}_p^\ominus)_B$ of the charge transfer process of Eq. 1b at the potentials $E_p^\ominus$ and $(E^\ominus)_B$, respectively:

$$\tilde{k}_p^\ominus \equiv k_p^c \exp\left\{-\frac{\alpha_p zF}{RT} E_p^\ominus\right\} = k_p^a \exp\left\{\frac{(1-\alpha_p)zF}{RT} E_p^\ominus\right\} \quad (12)$$

$$(\tilde{k}_p^\ominus)_B \equiv k_p^c \frac{\beta_p^{ox}}{F_0^{ox}([L]=1)} \exp\left\{-\frac{\alpha_p zF}{RT}(E^\ominus)_B\right\}$$

$$= k_p^a \frac{\beta_q^{red}}{F_0^{red}([L]=1)} \exp\left\{\frac{(1-\alpha_p)zF}{RT}(E^\ominus)_B\right\} \quad (13)$$

where $\tilde{k}_p^\ominus$ refers to the exchange rate constant at the standard state of the charge transfer process ($[M^{m+}L_p] = [M^{(m-z)+}L_q] = [L] =$ unit concentration) and $(\tilde{k}_p^\ominus)_B$ to that at the standard state of the overall electrode reaction ($c_{ox} = c_{red} = [L] =$ unit concentration).

By using the relation

$$E_p^\ominus = \frac{RT}{zF} \ln \frac{k_p^c}{k_p^a} \quad (14)$$

the standard rate constant $\tilde{k}_p^\ominus$ can be expressed by Eq. 15 in terms of k_p^c, k_p^a, and α_p:

$$\tilde{k}_p^\ominus = (k_p^c)^{(1-\alpha_p)}(k_p^a)^{\alpha_p} \quad (15)$$

The standard rate constant $(\tilde{k}_p^\ominus)_B$ is related to $\tilde{k}_p^\ominus$ by the equation

$$(\tilde{k}_p^\ominus)_B = \tilde{k}_p^\ominus \left\{\frac{\beta_p^{ox}}{F_0^{ox}([L]=1)}\right\}^{(1-\alpha_p)} \left\{\frac{\beta_q^{red}}{F_0^{red}([L]=1)}\right\}^{\alpha_p} \quad (16)$$

Introducing these standard rate constants into Eq. 8, we obtain the following expressions representing the current density attributable to the electrode reaction of Eq. 1:

$$\frac{I}{zF} = \tilde{k}_p^\ominus \left([M^{m+}L_p]_s \exp\left\{-\frac{\alpha_p zF}{RT}(E - E_p^\ominus)\right\}\right.$$

$$\left. - [M^{(m-z)+}L_q]_s[L]_s^{(p-q)} \exp\left\{\frac{(1-\alpha_p)zF}{RT}(E - E_p^\ominus)\right\}\right) \quad (17)$$

or

$$\frac{I}{zF} = (\tilde{k}_p^\ominus)_B \left(\frac{F_0^{ox}([L]=1)}{\beta_p^{ox}} [M^{m+}L_p]_s \exp\left\{-\frac{\alpha_p zF}{RT}(E - (E^\ominus)_B)\right\}\right.$$

$$- \frac{F_0^{red}([L]=1)}{\beta_q^{red}} [M^{(m-z)+}L_q]_s[L]_s^{(p-q)}$$

$$\left. \times \exp\left\{\frac{(1-\alpha_p)zF}{RT}(E - (E^\ominus)_B)\right\}\right) \quad (18)$$

Another important kinetic parameter called the exchange current density I_p^0 is given by Eq. 19:

$$\frac{I_p^0}{zF} = k_p^c c_{ox}[L]^p \frac{\beta_p^{ox}}{F_0^{ox}([L])} \exp\left\{-\frac{\alpha_p zF}{RT} E_e\right\}$$

$$= k_p^a c_{red}[L]^p \frac{\beta_q^{red}}{F_0^{red}([L])} \exp\left\{\frac{(1-\alpha_p)zF}{RT} E_e\right\} \quad (19)$$

Useful information on the electrode kinetics is obtained from analysis of the exchange current density; for example, the transfer coefficient α_p can be determined according to the equation

$$\left\{\frac{\partial \ln I_p^0/c_{ox}}{\partial \ln c_{red}/c_{ox}}\right\}_{[L]} = \alpha_p \quad (20)$$

The method of analysis of I_p^0 has been discussed by Vetter (115) and by Delahay (18).

C. Surface Concentration of Reacting Species

The derivation of theoretical equations representing the faradaic current as a function of known or measurable quantities requires complete knowledge of the concentration at the electrode surface at which the charge transfer proceeds. Generally, the surface concentrations of reacting species are determined by the faradaic current and the mode of mass transfer of these species. The following treatment is restricted to a system in which the contribution of the electric migration can be neglected because of the presence of a large excess of supporting electrolyte in the solution.

Under the conditions mentioned above, the change in concentration with time t of reacting species X in solution is given by Eq. 21:

$$\frac{\partial [X]_{x,t}}{\partial t} = \left(\frac{\partial [X]_{x,t}}{\partial t}\right)_{diff} + \left(\frac{\partial [X]_{x,t}}{\partial t}\right)_{flow} + \left(\frac{\partial [X]_{x,t}}{\partial t}\right)_{chem} \quad (21)$$

where x is the distance from the electrode surface to the bulk of solution. On the right-hand side of Eq. 21: the first term, referring to diffusion, is determined by Fick's law; the second, referring to the convective flow, by the hydrodynamic characteristics of the electrode and solution; and the last, referring to the coupled chemical reactions, by their reaction rates.

In conventional measurements of electrode kinetics, the concentration of the reacting species involved in the charge transfer process satisfies the following conditions, for example, with respect to $M^{m+}L_p$:

$$t = 0; \quad [M^{m+}L_p]_{x,t=0} = [M^{m+}L_p] = \frac{\beta_p^{ox}}{F_0^{ox}([L])} c_{ox}[L]^p \quad (22a)$$

$$x \to \infty; \quad [M^{m+}L_p]_{x \to \infty, t} = [M^{m+}L_p] = \frac{\beta_p^{ox}}{F_0^{ox}([L])} c_{ox}[L]^p \quad (22b)$$

$$x = 0; \quad D_p \left(\frac{\partial [M^{m+}L_p]_{x,t}}{\partial x} \right)_{x=0} = \frac{I}{zF} \quad (22c)$$

where t is the time measured from the beginning of electrolysis and D_p the diffusion coefficient of $M^{m+}L_p$. For species not involved in the charge transfer process, Eq. 22c at $x = 0$ is replaced by

$$D_X \left(\frac{\partial [X]_{x,t}}{\partial x} \right)_{x=0} = 0 \quad (22d)$$

The surface concentration of reacting species is obtained by solving differential Eq. 21 for mass transfer under the initial and boundary conditions as given by Eq. 22a–d. A set of conditions similar to Eq. 22a–c should be given with respect to $M^{(m-z)+}L_q$. The mathematics involved in the derivation is relatively complicated, and only the results obtained are given. It is also assumed in this article that the surface concentration of ligand L is equal to the bulk concentration during the course of electrolysis. This assumption can be satisfied when the solution contains a large excess of L or is well buffered with respect to L.

At stationary electrodes, and when the time of electrolysis is so short ($t < 1$ min) that the effect of convective flow can be ignored, the surface concentrations of $M^{m+}L_p$ and $M^{(m-z)+}L_q$ are represented as (49, 54)

$$[M^{m+}L_p]_s = \frac{\beta_p^{ox}[L]^p}{F_0^{ox}([L])} \left\{ c_{ox} - \frac{I/(zF)}{\lambda_{ox}\sqrt{\bar{D}_{ox}}} - \frac{1}{\sqrt{\pi}\sqrt{\bar{D}_{ox}}} \int_0^t \frac{I/(zF)}{\sqrt{t-u}} du \right\} \quad (23)$$

$$[M^{(m-z)+}L_q]_s = \frac{\beta_q^{red}[L]^q}{F_0^{red}([L])} \left\{ c_{red} + \frac{I/(zF)}{\lambda_{red}\sqrt{\bar{D}_{red}}} + \frac{1}{\sqrt{\pi}\sqrt{\bar{D}_{red}}} \int_0^t \frac{I/(zF)}{\sqrt{t-u}} du \right\} \quad (24)$$

In these equations u is an arbitrary variable and $\bar{D}_{ox}$ and $\bar{D}_{red}$ are the average diffusion coefficients of the oxidized and reduced forms, respectively, weighted with regard to the stability constant of each complex species:

$$\bar{D}_{ox} = \frac{1}{F_0^{ox}([L])} \sum_{i=0}^{\mu} \beta_i^{ox}[L]^i D_i \quad (25)$$

$$\bar{D}_{red} = \frac{1}{F_0^{red}([L])} \sum_{j=0}^{\nu} \beta_j^{red}[L]^j D_j \quad (26)$$

The parameters λ are complicated functions of the rate constants of the coupled chemical reactions and the stability constants and diffusion coefficients of the complex species; the subscript "ox" refers to reaction 1a, and "red" to reaction 1c. The values of λ can be determined from the limiting currents, as discussed in Section II-D.

At a dropping mercury electrode (DME), the effect of convective flow resulting from the expansion of a mercury drop should be considered. Applying MacGillavry–Rideal's differential equation to the convective diffusion at a DME, we obtain the following equations for $[M^{m+}L_p]_s$ and $[M^{(m-z)+}L_q]_s$, provided that the effect of curvature of the electrode can be neglected (49, 54):

$$[M^{m+}L_p]_s = \frac{\beta_p^{ox}[L]^p}{F_0^{ox}([L])}\left\{c_{ox} - \frac{I/(zF)}{\lambda_{ox}\sqrt{\bar{D}_{ox}}}\right.$$

$$\left. - \sqrt{7/3\pi}\,\frac{1}{\sqrt{\bar{D}_{ox}}}\int_0^t \frac{I/(zF)}{\sqrt{t^{7/3} - u^{7/3}}}\,u^{2/3}\,du\right\} \quad (27)$$

$$[M^{(m-z)+}L_q]_s = \frac{\beta_q^{red}[L]^q}{F_0^{red}([L])}\left\{c_{red} + \frac{I/(zF)}{\lambda_{red}\sqrt{\bar{D}_{red}}}\right.$$

$$\left. + \sqrt{7/3\pi}\,\frac{1}{\sqrt{\bar{D}_{red}}}\int_0^t \frac{I/(zF)}{\sqrt{t^{7/3} - u^{7/3}}}\,u^{2/3}\,du\right\} \quad (28)$$

It must be remembered that Eqs. 23–28 hold only when the condition given by Eq. 29 is satisfied with respect to the first-order (or pseudo-first-order) rate constants $\vec{\rho}$ and $\tilde{\rho}$ of each step of the coupled chemical reactions of Eq. 1a and c:

$$(\vec{\rho} + \tilde{\rho})t \gg 1 \quad (29)$$

This condition means that the each one of the coupled chemical reaction proceeds at a sufficiently high rate so that the equilibria of the coupled chemical reactions are maintained except in a region very close to the electrode surface.

D. Potentiostatic Current–Time Relationship

Let us consider the current–time relationship at a given constant electrode potential. When electrolysis is carried out for a relatively short period of time (for example, $t < 1$ sec), Eqs. 23 and 24 for the surface concentration at stationary electrodes can be used with reasonable accuracy at a DME as well.

Solving the integral equation derived by introducing Eqs. 23 and 24 into Eq. 8, we obtain the following expression for the current density at a given constant potential:

$$\frac{I}{zF} = \{(\tilde{k}_p^{\,c})_B c_{ox} - (\tilde{k}_p^{\,a})_B c_{red}\} \exp(Q^2 t)\, \mathrm{erfc}\,(Q\sqrt{t}) \tag{30}$$

$$(\tilde{k}_p^{\,c})_B = \frac{\beta_p^{ox}[L]^p}{F_0^{ox}([L])} \tilde{k}_p^{\,c} \left\{ 1 + \frac{1}{\lambda_{ox}\sqrt{\bar{D}_{ox}}} \frac{\beta_p^{ox}}{F_0^{ox}([L])} \tilde{k}_p^{\,c} \right.$$

$$\left. + \frac{1}{\lambda_{red}\sqrt{\bar{D}_{red}}} \frac{\beta_q^{red}}{F_0^{red}([L])} \tilde{k}_p^{\,a} \right\}^{-1} \tag{31}$$

$$(\tilde{k}_p^{\,a})_B = \frac{\beta_q^{red}[L]^p}{F_0^{red}([L])} \tilde{k}_p^{\,a} \left\{ 1 + \frac{1}{\lambda_{ox}\sqrt{\bar{D}_{ox}}} \frac{\beta_p^{ox}}{F_0^{ox}([L])} \tilde{k}_p^{\,c} \right.$$

$$\left. + \frac{1}{\lambda_{red}\sqrt{\bar{D}_{red}}} \frac{\beta_q^{red}}{F_0^{red}([L])} \tilde{k}_p^{\,a} \right\}^{-1} \tag{32}$$

$$Q = \frac{(\tilde{k}_p^{\,c})_B}{\sqrt{\bar{D}_{ox}}} + \frac{(\tilde{k}_p^{\,a})_B}{\sqrt{\bar{D}_{red}}} \tag{33}$$

Equation 30 has a formal resemblance to that for a simple electrode reaction without coupled chemical reactions [cf. for example, references (17) and (60)], and the time dependence of current is represented by the function $\exp(u^2)\,\mathrm{erfc}\,(u)$. By using the overvoltage, $\eta = E - E_e$, Eqs. 30 and 33 can be written as (72):

$$\frac{I}{zF} = \frac{\sqrt{\bar{D}_{ox}}}{\sqrt{t}} c_{ox} \frac{1 - \exp\left(\frac{zF}{RT}\eta\right)}{1 + \sqrt{\bar{D}_{ox}/\bar{D}_{red}}(c_{ox}/c_{red})\exp\left(\frac{zF}{RT}\eta\right)} Q\sqrt{t}\exp(Q^2 t)\,\mathrm{erfc}\,(Q\sqrt{t}) \tag{34}$$

$$Q = \frac{(\tilde{k}_p^{\,c})_B}{\sqrt{\bar{D}_{ox}}} \left\{ 1 + \sqrt{\bar{D}_{ox}/\bar{D}_{red}}(c_{ox}/c_{red})\exp\left(\frac{zF}{RT}\eta\right) \right\}$$

$$= \frac{(\tilde{k}_p^{\,a})_B}{\sqrt{\bar{D}_{red}}} \left\{ 1 + \sqrt{\bar{D}_{red}/\bar{D}_{ox}}(c_{red}/c_{ox})\exp\left(-\frac{zF}{RT}\eta\right) \right\} \tag{35}$$

Parameter Q concerning the kinetics of the electrode reaction can be determined by the analysis of the potentiostatic current–time relationship. Equations 34 and 35 can be applied to the determination of Q, $(\tilde{k}_p^{\,c})_B$, and $(\tilde{k}_p^{\,a})_B$ when $\bar{D}_{ox}$, $\bar{D}_{red}$, and E_e are known (72).

Another useful method of analysis has been proposed by Tanaka and Yamada (112, 120). The current density $I_{t=\theta}$, at a given constant potential and at a sufficiently long time $t = \theta$ so that the condition $Q\sqrt{\theta} \gg 1$ is satisfied, is given by

$$\frac{I_{t=\theta}}{zF} = \frac{(\tilde{k}_p^c)_B c_{\text{ox}} - (\tilde{k}_p^a)_B c_{\text{red}}}{\sqrt{\pi}\sqrt{\theta}Q} \tag{36}$$

The introduction of Eq. 36 into Eq. 30 results in the relation

$$\frac{I}{I_{t=\theta}} \frac{\sqrt{t}}{\sqrt{\pi\theta}} = Q\sqrt{t} \exp(Q^2 t) \, \text{erfc}(Q\sqrt{t}) \tag{37}$$

Parameter Q can be obtained by means of Eq. 37 because the left-hand side of the equation is a measurable quantity. This method has the advantage that no information about the equilibrium potential is required in the determination of Q. Using the values of $I_{t=\theta}$ and Q, we can calculate the rate constants $(\tilde{k}_p^c)_B$ and $(\tilde{k}_p^a)_B$ according to the relation

$$\begin{aligned} I_{t=\theta} Q \frac{\sqrt{\pi\theta}}{zF} &= (\tilde{k}_p^c)_B c_{\text{ox}} - (\tilde{k}_p^a)_B c_{\text{red}} \\ &= (\tilde{k}_p^c)_B c_{\text{ox}} \left\{ 1 - \exp\left(\frac{zF}{RT}\eta\right) \right\} \\ &= (\tilde{k}_p^a)_B c_{\text{red}} \left\{ 1 - \exp\left(-\frac{zF}{RT}\eta\right) \right\} \end{aligned} \tag{38}$$

in which a knowledge of diffusion coefficients is not necessary.

The treatment of Tanaka and Yamada is particularly useful when applied to a system that does not contain either Ox or Red in the bulk phases. For example, in the case of $c_{\text{red}} = 0$, $I_{t=\theta}$ is represented by

$$\frac{\sqrt{\pi\theta}}{zF} I_{t=\theta} Q = (\tilde{k}_p^c)_B c_{\text{ox}} \tag{39}$$

from which $(\tilde{k}_p^c)_B$ is easily obtained. Anodic rate constant $(\tilde{k}_p^a)_B$ can be determined from Eq. 33 by using the values of Q and $(\tilde{k}_p^c)_B$. In addition, the standard potential $(E^{\ominus})_B$ of the overall electrode reaction can be determined according to the relations

$$\ln \frac{(\tilde{k}_p^c)_B}{(\tilde{k}_p^a)_B} = \frac{zF}{RT}(E^{\ominus})_B - \frac{zF}{RT}E + \ln \frac{F_0^{\text{ox}}([L]=1)}{F_0^{\text{ox}}([L])} - \ln \frac{F_0^{\text{red}}([L]=1)}{F_0^{\text{red}}([L])} \tag{40}$$

$$E\{(\tilde{k}_p^c)_B = (\tilde{k}_p^a)_B\} = (E^{\ominus})_B + \frac{RT}{zF} \ln \frac{F_0^{\text{ox}}([L]=1)}{F_0^{\text{ox}}([L])} - \frac{RT}{zF} \ln \frac{F_0^{\text{red}}([L]=1)}{F_0^{\text{red}}([L])} \tag{41}$$

The potential $E\{(\tilde{k}_p{}^c)_B = (\tilde{k}_p{}^a)_B\}$, at which $(\tilde{k}_p{}^c)_B = (\tilde{k}_p{}^a)_B$, is first obtained by plotting the left-hand side of Eq. 40 against E at a given constant value of [L]. The value of $(E^\ominus)_B$ may then be determined from the dependence of $E\{(k_p{}^c)_B = (k_p{}^a)_B\}$ on [L] according to Eq. 41. This treatment provides a new method of determining $(E^\ominus)_B$ from the kinetic measurement.

Limiting Current (9, 10, 17, 42, 43, 119). At sufficiently negative potentials, at which the conditions $(\tilde{k}_p{}^c)_B \gg (\tilde{k}_p{}^a)_B$ and $k_p{}^c \beta_p^{ox}/F_0^{ox}([L]) \gg \lambda_{ox}\sqrt{\bar{D}_{ox}}$ are satisfied, the cathodic current density reaches a limiting value I_l as represented by Eq. 42:

$$\frac{I_l}{zF} = \frac{\sqrt{\bar{D}_{ox}}}{\sqrt{t}} c_{ox} \lambda_{ox} \sqrt{t} \exp(\lambda_{ox}^2 t) \operatorname{erfc}(\lambda_{ox}\sqrt{t}) \qquad (42)$$

This relationship can be used for the determination of λ_{ox} from the measurement of I_l. The similarity between Eqs. 42 and 34 clarifies the physical meaning of parameter λ; parameter Q, which is related to the kinetic nature of the overall electrode process, determines the reversibility of the electrode reaction, while parameter λ, which is related to the kinetics of the supply and removal of the reacting species to and from the electrode surface, determines the kinetic nature of the limiting current.

If the rates of coupled chemical reactions are so large that the condition $\lambda_{ox}\sqrt{t} > 5$ is fulfilled, the limiting current is reduced to the diffusion current I_d which is determined only by diffusion:

$$\frac{I_l}{zF} = \frac{\sqrt{\bar{D}_{ox}}}{\sqrt{\pi t}} c_{ox} = \frac{I_d}{zF} \qquad (43)$$

On the contrary, if the rates of coupled chemical reactions are so small that the condition $\lambda_{ox}\sqrt{t} < 0.03$ is satisfied, the limiting current is totally controlled by the reaction rates and is represented by Eq. 44:

$$\frac{I_l}{zF} = \sqrt{\bar{D}_{ox}} \lambda_{ox} c_{ox} = \frac{I_k}{zF} \qquad (44)$$

I_k being the kinetic current.

Equations 43 and 44 suggest that the diffusion current density is proportional to $t^{-1/2}$, whereas the kinetic current density is independent of t. Therefore the time dependence of product $I_l\sqrt{t}$ (see Table I) provides a useful criterion for testing the possible contribution of coupled chemical reactions to the mass transfer process.

TABLE I. Time Dependence of the Product $I_l \sqrt{t}$ of Potentiostatic Limiting Current Density I_l and Square Root of Time t ($t = 0.01$–1 sec)

Parameter λ related to the rates of coupled chemical reactions	Limiting current controlled by	Time dependence of $I_l \sqrt{t}$
$\lambda > 50$ sec$^{-1/2}$	Diffusion	Independent of t
$50 > \lambda > 0.03$ sec$^{-1/2}$	Diffusion and chemical reaction	Proportional to $\sqrt{t} \exp(\lambda^2 t) \operatorname{erfc}(\lambda \sqrt{t})$
0.03 sec$^{-1/2} > \lambda$	Chemical reaction	Proportional to $\sqrt{t}$

E. Galvanostatic Transition Time

The galvanostatic (or chronopotentiometric) transition time τ of the electrode reaction at a given constant cathodic current density I_c is given by Eq. 45, which is derived from Eq. 23 when the rate constants of the coupled chemical reactions fulfil the condition of Eq. 29:

$$\sqrt{\tau} = \frac{zF\sqrt{\pi}\sqrt{\bar{D}_{ox}}}{2I_c} c_{ox} - \frac{\sqrt{\pi}}{2\lambda_{ox}} \tag{45}$$

The first term on the right-hand side of Eq. 45 refers to the contribution of diffusion, and the second to that of the coupled chemical reaction. A similar relation holds for the transition time at a given constant anodic current density.

When the rate of coupled chemical reaction is very large, the second term on the right-hand side of Eq. 45 may be neglected. In such a case the transition time is expected to be determined by diffusion only, and the product $\sqrt{\tau} I_c$ should be independent of I_c. The contribution of the coupled chemical reaction appears as a deviation of $\sqrt{\tau} I_c$ from the constancy with respect to I_c; the $\sqrt{\tau} I_c$ versus I_c relationship is expected to give a straight line with a slope of $-\sqrt{\pi}/2\lambda_{ox}$, from which the parameters λ_{ox} and $\bar{D}_{ox}$ can be determined.

The transition time of the reduction process accompanied by only one preceding chemical reaction,

$$Y \underset{\overleftarrow{\rho}}{\overset{\overrightarrow{\rho}}{\rightleftarrows}} O \tag{46a}$$

$$O + ze \rightleftarrows R \tag{46b}$$

has been analyzed by Delahay and Berzins (19) in a more rigorous way without assuming the conditions of Eq. 29. The result is given by Eq. 47:

$$\sqrt{\tau} = \frac{zF\sqrt{\pi}\sqrt{D}}{2I_c} c_{ox} - \frac{\sqrt{\pi}}{2K\sqrt{\vec{\rho}+\overleftarrow{\rho}}} \operatorname{erf}(\sqrt{\vec{\rho}+\overleftarrow{\rho}}\sqrt{\tau}) \qquad (47)$$

where c_{ox} is the total concentration of the oxidized form in the bulk of the solution (c_{ox} = [O] + [Y]), D is the common value of diffusion coefficients of O and Y ($D_O = D_Y = D$), and K is the equilibrium constant of reaction 46a as defined by

$$K = \frac{[\mathrm{O}]}{[\mathrm{Y}]} = \frac{\vec{\rho}}{\overleftarrow{\rho}} \qquad (48)$$

It is easily understood that Eq. 47 takes a form similar to Eq. 45 when $(\vec{\rho}+\overleftarrow{\rho})\tau > 2$. When the equilibrium of reaction 46a is very much in favor of Y, the condition $K \ll 1$ (and therefore $\overleftarrow{\rho} \gg \vec{\rho}$) is satisfied and Eq. 47 can be reduced to

$$\sqrt{\tau} = \frac{zF\sqrt{\pi}\sqrt{D}}{2I_c} c_{ox} - \frac{\sqrt{\pi}}{2\sqrt{K\vec{\rho}}} \operatorname{erf}(\sqrt{\vec{\rho}/K}\sqrt{\tau}) \qquad (49)$$

Under conventional experimental conditions, the effect of a coupled chemical reaction on the transition time is expected to be clearly observed when $\lambda \lesssim 500$ sec$^{-1/2}$, and the transition time measurement is better suited to the study of fast coupled chemical reactions than the analysis of polarographic limiting currents (cf. Section II-F-2).

F. Current-Potential Relationship in Dc Polarography

1. Equations of the Polarographic Reduction Wave. Kinetic analysis of the dc polarographic wave of complex species was first attempted by Tamamushi and Tanaka (84, 85) on the basis of the steady-state assumption and the concept of Nernst's diffusion layer. Their treatment has the advantage that it can be applied to relatively complicated electrode reactions without introducing mathematical difficulty. The results, however, are not accurate enough for quantitative analysis of the current–potential relationship at a DME.

A rigorous treatment of the polarographic reduction wave of complex species has been proposed by Matsuda and Ayabe (49, 54). Let us consider the following electrode reaction at the DME:

$$\mathrm{M}^{m+}\mathrm{L}_i + \mathrm{L} \rightleftharpoons \mathrm{M}^{m+}\mathrm{L}_{i+1} \quad (i = 0, 1, 2, \ldots, p, \ldots, \mu - 1) \qquad (50a)$$

$$\mathrm{M}^{m+}\mathrm{L}_p + ze + \mathrm{Hg} \rightleftharpoons \mathrm{M}(\mathrm{Hg}) + p\mathrm{L} \qquad (50b)$$

in which only one complex species $M^{m+}L_p$ is involved in the charge transfer process giving metal amalgam M(Hg), and the other complex species are subjected to dissociation–association reactions. The rate constants of the coupled chemical reactions are also assumed to satisfy the conditions of Eq. 29.

The surface concentration of $M^{m+}L_p$ is given by Eq. 27 and, similarly, that of M(Hg) by Eq. 51:

$$[M(Hg)]_s = \sqrt{7/3\pi} \, \frac{1}{\sqrt{D_a}} \int_0^t \frac{I/(zF)}{\sqrt{t^{7/3} - u^{7/3}}} u^{2/3} \, du \qquad (51)$$

where D_a denotes the diffusion coefficient of M(Hg). The equation of the polarographic reduction wave can be derived by solving the integral equation obtained by introducing Eqs. 27 and 51 into the equation representing the current density:

$$\frac{I}{zF} = \tilde{k}_p^c [M^{m+}L_p]_s - \tilde{k}_p^a [M(Hg)]_s [L]_s^p \qquad (52)$$

If the surface concentration of L is assumed to be constant, the instantaneous polarographic reduction current i can be represented by Eqs. 53–56 with reasonable accuracy:

$$\frac{i}{i_d} = \frac{\dfrac{1.61}{\sqrt{t}} \{\Lambda_p^{-1} \exp(\alpha_p \zeta) + \lambda_{ox}^{-1}\} + 1 + \exp(\zeta)}{\left(\dfrac{1.13}{\sqrt{t}} \{\Lambda_p^{-1} \exp(\alpha_p \zeta) + \lambda_{ox}^{-1}\} + 1 + \exp(\zeta)\right)^2} \qquad (53)$$

$$\Lambda_p = (\tilde{k}_p^{\ominus})_B \left(\sum_{i=0}^{\mu} \beta_i^{ox}\right)^{(1-\alpha_p)} (\sqrt{D_M})^{-(1-\alpha_p)} (\sqrt{D_a})^{-\alpha_p}$$

$$\times \exp\left(\frac{(1-\alpha_p)zF}{RT} \{E_{1/2}^r - (E_{1/2}^r)_M\}\right) \qquad (54)$$

$$\zeta = \frac{zF}{RT}(E - E_{1/2}^r) \qquad (55)$$

$$i_d = (7/3)^{1/2} \, zFA \, (\bar{D}_{ox})^{1/2} \, (\pi t)^{-1/2} \, c_{ox} \qquad (56)$$

where i_d is the instantaneous diffusion current given by the Ilkovič equation, and A the surface area of the DME. In these equations D_M is the diffusion coefficient of a simple metal ion M^{m+} and $E_{1/2}^r$ and $(E_{1/2}^r)_M$ are the polarographic reversible half-wave potentials of the complex species and the simple metal ion, respectively, as defined by the relations

$$E^r_{1/2} = (E^r_{1/2})_M + \frac{RT}{zF} \ln \frac{\sqrt{D_M}}{\sqrt{\bar{D}_{ox}}} - \frac{RT}{zF} \ln \left(\sum_{i=0}^{\mu} \beta_i^{ox}[L]^i \right) \quad (57)$$

$$(E^r_{1/2})_M = E_M^{\ominus} + \frac{RT}{zF} \ln \frac{\sqrt{D_a}}{\sqrt{D_M}} \quad (58)$$

where $E_M^{\ominus}$ refers to the standard potential of the $M^{m+}/M(Hg)$ system.

The average current $\bar{i}$ over a drop life t_d of the DME can be derived from Eq. 53:

$$\bar{i} = \frac{1}{t_d} \int_0^{t_d} i \, dt$$

$$\frac{\bar{i}}{\bar{i}_d} = \frac{1}{\frac{1.13}{\sqrt{t_d}} \Lambda_p^{-1} \exp(\alpha_p \zeta) + \frac{\bar{i}_d}{\bar{i}_l} + \exp(\zeta)} \quad (59)$$

$$\frac{\bar{i}_l}{\bar{i}_d} = \frac{\lambda_{ox} \sqrt{t_d}}{1.13 + \lambda_{ox} \sqrt{t_d}} \quad (60)$$

where $\bar{i}_d$ is the average diffusion current and $\bar{i}_l$ the average limiting current observed at sufficiently negative potentials at which the condition $\exp(\zeta) \to 0$ is maintained. Rearranging Eq. 59 we obtain Eq. 61, representing the so-called log plot of the reduction wave:

$$\ln \frac{\bar{i}}{\bar{i}_l - \bar{i}} = -\ln \left\{ \frac{1.13}{\sqrt{t_d}} \Lambda_p^{-1} \exp(\alpha_p \zeta) + \exp(\zeta) \right\} - \ln \frac{\bar{i}_l}{\bar{i}_d} \quad (61)$$

The parameter Λ_p involves the standard rate constant $(\tilde{k}_p^{\ominus})_B$ of the charge transfer process, and determines the reversibility of the electrode reaction. An electrode reaction with an extremely large Λ_p shows a totally reversible reduction wave which is represented by Eq. 62:

$$\zeta = \ln \frac{\bar{i}_l - \bar{i}}{\bar{i}} - \ln \frac{\bar{i}_l}{\bar{i}_d} \quad (62)$$

The half-wave potential $(E_{1/2})_{rev}$ of a totally reversible reduction wave is related to the reversible half-wave potential according to Eq. 63:

$$(E_{1/2})_{rev} = E^r_{1/2} - \frac{RT}{zF} \ln \frac{\bar{i}_l}{\bar{i}_d} \quad (63)$$

A totally irreversible reduction wave is observed when Λ_p is very small; its current–potential relationship and half-wave potential $(E_{1/2})_{irr}$ are given as

$$E = (E_{1/2})_{irr} + \frac{RT}{\alpha_p zF} \ln \frac{\bar{i}_l - \bar{i}}{\bar{i}} \qquad (64)$$

$$(E_{1/2})_{irr} = E^r_{1/2} - \frac{RT}{\alpha_p zF} \left(\ln \frac{1.13}{\Lambda_p \sqrt{t_d}} + \ln \frac{\bar{i}_l}{\bar{i}_d} \right) \qquad (65)$$

These equations for the current–potential relationship and half-wave potentials are of the same form as that for a simple electrode reaction without coupled chemical reactions except for the term $\bar{i}_l/\bar{i}_d$; $\bar{i}_l/\bar{i}_d$ is equal to unity for a system in which the kinetic effect of a coupled chemical reaction can be neglected.

2. Determination of Polarographic Parameters. When an electrode reaction is totally reversible, the reversible half-wave potential $E^r_{1/2}$ can be easily calculated from the half-wave potential $(E_{1/2})_{rev}$ according to Eq. 63, $(E_{1/2})_{rev}$ being determined by means of a log plot analysis.

The reversible half-wave potential of a so-called quasireversible system with a moderate value of Λ_p can be obtained by the following method (48, 49). Rearranging Eq. 61, we obtain the relation

$$E + \frac{RT}{zF} \ln \frac{\bar{i}}{\bar{i}_l - \bar{i}} + \frac{RT}{zF} \ln \frac{\bar{i}_l}{\bar{i}_d} = E^r_{1/2} + \frac{RT}{zF} \ln Z_p \qquad (66)$$

where the new parameter Z_p is defined by

$$Z_p = 1 + \frac{1.13}{\Lambda_p \sqrt{t_d}} \exp \{(\alpha_p - 1)\zeta\} \qquad (67)$$

At sufficiently positive potentials, the parameter Z_p is expected to be practically equal to unity, and the left-hand side of Eq. 66 approaches a constant value $E^r_{1/2}$. In practice, the plot of the left-hand side of Eq. 66 against the electrode potential can be used to determine the limiting constant value at positive potentials:

$$\lim_{\substack{E \to \text{pos} \\ (Z_p \to 1)}} (\text{Left-hand side of Eq. 66}) = E^r_{1/2} \qquad (68)$$

In a reversible system the plot is expected to give a straight line corresponding to $E^r_{1/2}$ and parallel to the E-axis (Fig. 1(*1*)). The plot of a quasireversible system reaches $E^r_{1/2}$ at positive potentials but deviates from the parallel line at more negative potentials (Fig. 1(*2*)). In an irreversible system with a very small Λ_p, the measurable reduction wave is observed only when

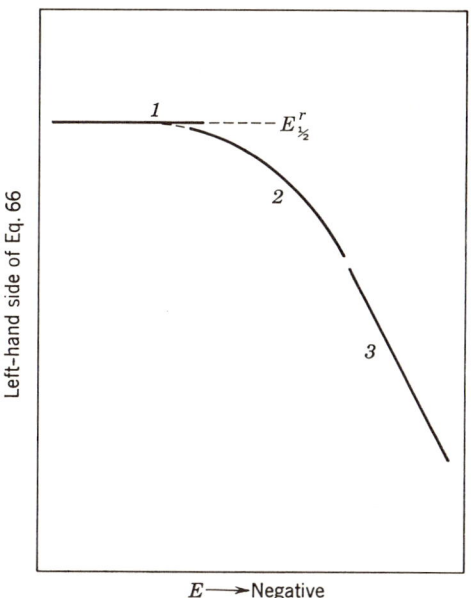

FIG. 1. Schematic presentation of the plot of the left-hand side of Eq. 66 against the electrode potential. (*1*) Reversible system; (*2*) quasireversible system; (*3*) irreversible system.

ζ has relatively large negative values; the plot mentioned above gives an inclined line as shown in Figure 1(*3*), and $E^r_{1/2}$ cannot be determined by this method. A typical example of the plot for a quasireversible system is reproduced in Figure 2. Essentially the same method was applied by Tamamushi and Tanaka (86) to the determination of $E^r_{1/2}$ of the zinc(II) system, in which the left-hand side of Eq. 66 was plotted against current $\bar{\imath}$, and $E^r_{1/2}$ was determined by the extrapolation of the plot to $\bar{\imath} = 0$.

The kinetic parameters Λ_p and α_p can also be determined by analysis of the polarographic reduction wave, provided that the reversible half-wave potential is available. Parameter Z_p, defined by Eq. 67, can be calculated from the experimental data according to Eq. 66; the value of $\log Z_p$ is considered to be a quantitative measure of the degree of reversibility. The plot of $\log (Z_p - 1)$ against ζ is expected to give a straight line as shown in Figure 3. Transfer coefficient α_p and parameter Λ_p are then determined from the slope and the intercept of the plot, respectively (86).

When an electrode reaction is totally reversible, the parameter Z_p becomes unity and the values of Λ_p and α_p cannot be obtained. In a totally irreversible

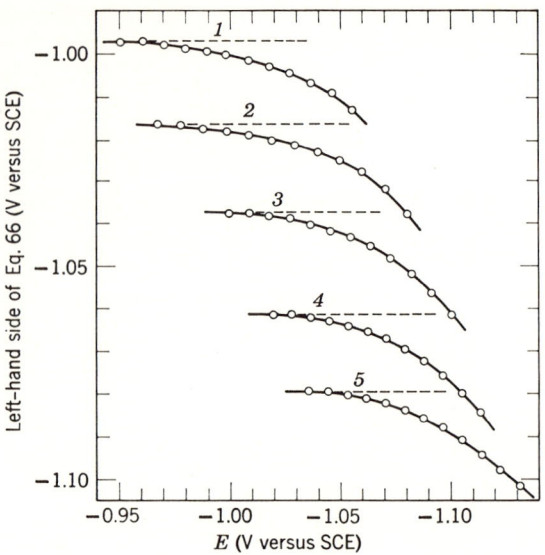

FIG. 2. Determination of reversible half-wave potentials of the DME–Zn(II)–acetate–NaNO$_3$ system at 25°C (ionic strength adjusted to 4 with NaNO$_3$). Concentration of acetate ion: (1) 0.4; (2) 0.8; (3) 1.4; (4) 2.4; (5) 3.4 M. [Matsuda, H., Y. Ayabe, and K. Adachi, *Ber. Bunsenges. Physik. Chem.*, **67**, 597 (1963).]

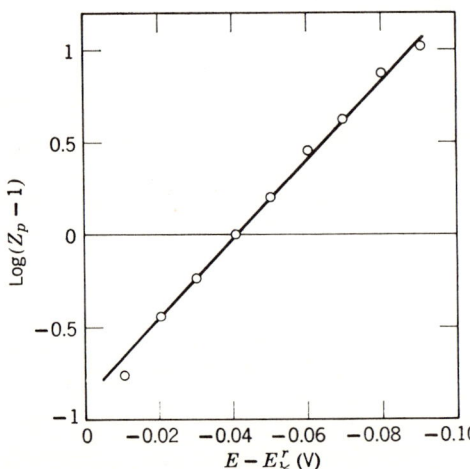

FIG. 3. Plot of log $(Z_p - 1)$ against $(E - E^r_{1/2})$ for the polarographic reduction of 1 mM Zn(NO$_3$)$_2$ in 1 M KBr solution containing 2 × 10^{-6} M polyoxyethylene lauryl ether at 25°C. [Tamamushi, R., and N. Tanaka, *Z. Physik. Chem.* (*Frankfurt*), **39**, 119 (1963).]

system, the transfer coefficient and the half-wave potential are easily determined by the conventional log plot analysis (cf. Eq. 64); the parameter Λ_p, however, cannot be obtained unless the reversible half-wave potential is available.

The cathodic limiting current and diffusion current can be used to determine the parameter λ_{ox} concerning the rates of the coupled chemical reactions. If the rates of these reactions are relatively large so that the condition $\lambda_{ox}\sqrt{t_d} \gg 1$ is satisfied, the instantaneous limiting current is practically identical with the diffusion current and its time dependence during the life of a mercury drop is represented by $t^{1/6}$ according to the Ilkovič equation. On the contrary, when the coupled chemical reactions are so slow that $\lambda_{ox}\sqrt{t_d} \ll 1$ is maintained, the limiting current is totally controlled by the rates of the coupled chemical reactions and is identical with the kinetic current i_k as given by Eq. 69:

$$\frac{\bar{i}_k}{\bar{i}_d} = \lambda_{ox}\sqrt{t_d} \qquad (69)$$

The instantaneous kinetic current is expected to be proportional to $t^{2/3}$ during a drop life.

Some characteristics of the polarographic limiting current are summarized in Table II. The highest value of λ that can be determined from the polarographic limiting current with reasonable accuracy is estimated to be about $5 \text{ sec}^{-1/2}$; the galvanostatic transition time is preferable to the polarographic limiting current when applied to the analysis of faster coupled chemical reactions.

TABLE II. Characteristics of Instantaneous Limiting Current i_l Observed at a DME with Drop Time $t_d \cong 4$ sec.

Parameter λ related to the rates of coupled chemical reactions	Limiting current controlled by	Time dependence of i_l
$\lambda > 5 \text{ sec}^{-1/2}$	Diffusion	Proportional to $t^{1/6}$
$5 > \lambda > 0.005 \text{ sec}^{-1/2}$	Diffusion and chemical reaction	Complicated
$0.005 \text{ sec}^{-1/2} > \lambda$	Chemical reaction	Proportional to $t^{2/3}$

G. Ac Polarography and Faradaic Impedance

The effect of coupled chemical reactions on the faradaic impedance of the electrode reaction at the equilibrium potential has been investigated by Gerischer (25), Matsuda (58), and others. The theory of ac polarographic waves of an electrode reaction involving coupled reactions, however, is a relatively recent development.

The faradaic admittance–potential relationship of a charge transfer process accompanied by an irreversible following reaction was analyzed by Aylward, Hayes, and Tamamushi (5) on the basis of the steady–state approach. Because of the assumptions introduced in the theoretical treatment, the application of their results is restricted to coupled chemical reactions that have first-order rate constants larger than 10 sec^{-1} but smaller than the angular frequency ω of the alternating signal.

A rigorous analysis of the ac polarographic current of an electrode reaction with a coupled chemical reaction was developed by Smith (78–80) on the basis of Matsuda's (47) treatment of ac polarography and the theory of kinetic current by Koutecký et al. (9, 10, 17, 42, 43, 119). The derivation of theoretical equations is mathematically complicated and is beyond the scope of this article. In the following discussion only the important assumptions and results are presented; the original papers and Smith's article should be consulted for the details.

Let us consider the following three types of electrode reaction in solutions containing no reduced species in the bulk: the charge transfer process

$$O + ze \rightleftharpoons R \tag{70}$$

coupled with (a) a preceding chemical reaction, (b) a following chemical reaction, and (c) a catalytic reaction. The coupled chemical reaction is always assumed to be first-order or pseudo-first-order with respect to the oxidized and reduced species.

In these electrode reactions the fundamental harmonic current $i(\omega t)$, which flows upon the superposition of sinusoidal alternating voltage $\Delta E \sin(\omega t)$ with a small amplitude of $\Delta E \lesssim 5 \text{ mV}$, is represented by the common equation

$$\begin{aligned} i(\omega t) &= zFA[O]\sqrt{D_O}\left(\frac{zF}{RT}\Delta E\right)\frac{H_0}{U^2 + V^2}\{U\sin(\omega t) + V\cos(\omega t)\} \\ &= zFA[O]\sqrt{D_O}\left(\frac{zF}{RT}\Delta E\right)\frac{H_0}{\sqrt{U^2 + V^2}}\sin\left(\omega t + \cot^{-1}\frac{U}{V}\right) \end{aligned} \tag{71}$$

where [O] is the concentration of O in the bulk of the solution and D_O the diffusion coefficient of O. Function H_0 concerns the dc component of the surface concentrations and has different forms according to the mechanism of the electrode reaction; its mathematical definition is very much involved and is not discussed here. In this treatment we also assume the stationary state with respect to the dc components of electrolysis; under this assumption function H_0 can be considered constant during the period of alternating signals. Parameters U and V involve the rate constants of the charge transfer and coupled chemical reactions and are given by the following equations for each type of electrode reaction. The common symbols used in the equations are:

$$K = \frac{\vec{\rho}}{\overleftarrow{\rho}}$$

$$g = (\vec{\rho} + \overleftarrow{\rho})/\omega$$

$$\zeta = zF(E - E^r_{1/2})/(RT) \tag{72}$$

$$E^r_{1/2} = E^\ominus - \frac{RT}{zF} \ln \frac{D_O}{D_R}$$

$$\Lambda = \tilde{k}^\ominus (\sqrt{D_O})^{-(1-\alpha)} (\sqrt{D_R})^{-\alpha} \{e^{-\alpha\zeta} + e^{(1-\alpha)\zeta}\}$$

where $E^r_{1/2}$ and $E^\ominus$ are the reversible half-wave potential and the standard potential, respectively, of the charge transfer process of Eq. 70, and $\tilde{k}^\ominus$ is the standard rate constant of the charge transfer process at $E^\ominus$.

Electrode reaction with a preceding reaction. For an electrode reaction in which the charge transfer process is preceded by a first-order chemical reaction

$$Y \underset{\overleftarrow{\rho}}{\overset{\vec{\rho}}{\rightleftarrows}} O \tag{73}$$

parameters U and V are given by Eqs. 74 and 75, respectively, provided that $D_Y = D_O$:

$$U = \frac{e^{-\alpha\zeta} + e^{(1-\alpha)\zeta}}{\Lambda} + \frac{1}{\sqrt{2\omega}} \frac{e^{-\alpha\zeta}}{1+K} \left\{ \frac{(1+g^2)^{1/2} + g}{1+g^2} \right\}^{1/2}$$

$$+ \frac{1}{\sqrt{2\omega}} \left\{ \frac{Ke^{-\alpha\zeta}}{1+K} + e^{(1-\alpha)\zeta} \right\} \tag{74}$$

$$V = \frac{1}{\sqrt{2\omega}} \frac{e^{-\alpha\zeta}}{1+K} \left\{ \frac{(1+g^2)^{1/2} - g}{1+g^2} \right\}^{1/2} + \frac{1}{\sqrt{2\omega}} \left\{ \frac{Ke^{-\alpha\zeta}}{1+K} + e^{(1-\alpha)\zeta} \right\} \tag{75}$$

Electrode reaction with a following reaction. Considering the electrode reaction that involves a first-order following reaction with respect to R:

$$Z \underset{\overleftarrow{p}}{\overset{\overrightarrow{p}}{\rightleftharpoons}} R \tag{76}$$

and assuming $D_R = D_Z$, we obtain

$$U = \frac{e^{-\alpha\zeta} + e^{(1-\alpha)\zeta}}{\Lambda} + \frac{1}{\sqrt{2\omega}} \frac{e^{(1-\alpha)\zeta}}{1+K} \left\{ \frac{(1+g^2)^{1/2} + g}{1+g^2} \right\}^{1/2}$$

$$+ \frac{1}{\sqrt{2\omega}} \left\{ \frac{Ke^{(1-\alpha)\zeta}}{1+K} + e^{-\alpha\zeta} \right\} \tag{77}$$

$$V = \frac{1}{\sqrt{2\omega}} \frac{e^{(1-\alpha)\zeta}}{1+K} \left\{ \frac{(1+g^2)^{1/2} - g}{1+g^2} \right\}^{1/2} + \frac{1}{\sqrt{2\omega}} \left\{ \frac{Ke^{(1-\alpha)\zeta}}{1+K} + e^{-\alpha\zeta} \right\} \tag{78}$$

Electrode reaction with a catalytic reaction. Consider the electrode reaction in which the oxidized species O is regenerated by the catalytic reaction

$$R + Z \underset{\overleftarrow{p}}{\overset{\overrightarrow{p}}{\rightleftharpoons}} O \tag{79}$$

When a large excess of Z is present in the solution and D_O is assumed to be equal to D_R, the parameters U and V are given by

$$U = \frac{e^{-\alpha\zeta} + e^{(1-\alpha)\zeta}}{\Lambda} + \frac{1}{\sqrt{2\omega}} \{e^{-\alpha\zeta} + e^{(1-\alpha)\zeta}\} \left\{ \frac{(1+g^2)^{1/2} + g}{1+g^2} \right\}^{1/2} \tag{80}$$

$$V = \frac{1}{\sqrt{2\omega}} \{e^{-\alpha\zeta} + e^{(1-\alpha)\zeta}\} \left\{ \frac{(1+g^2)^{1/2} - g}{1+g^2} \right\}^{1/2} \tag{81}$$

From the theoretical results mentioned above, the following conclusions are derived concerning the ac characteristics of the electrode reaction with a coupled chemical reaction.

(a) The formal presentation of fundamental harmonic current $i(\omega t)$, that is, Eq. 71, is exactly the same as that of a simple electrode reaction (47) in which the mass transfer is determined by diffusion only.

(b) The current $i(\omega t)$ has a phase angle ϕ with respect to the applied alternating signal. The cotangent of ϕ, being a function of the electrode potential and the frequency, is given by

$$\cot \phi = U/V \qquad (82)$$

and is related to the mechanism of the electrode reaction.

(c) The ac characteristics of the electrode reaction may be represented by assuming an equivalent faradaic impedance which is a complicated vector quantity with frequency dependence. If a series combination of a resistance and a capacitance is assumed for the faradaic impedance, the resistive component is proportional to parameter U and the capacitive component to parameter V.

(d) The series resistance R_s, called the polarization resistance, is given by the sum of the charge transfer resistance R_{ct} and the mass transfer resistance R_m:

$$R_s = R_{ct} + R_m$$

R_{ct} being determined by the rate constant of the charge transfer process and independent of frequency.

(e) The series capacitance C_s, called electrolytic or pseudocapacity, is determined by the nature of the mass transfer and is independent of the rate of the charge transfer process.

(f) By combining the components R_m and C_s attributable to the mass transfer process, the faradaic impedance may be represented as

where Z_m is called the mass transfer impedance. The mass transfer impedance is reduced to the Warburg impedance when the mass transfer is totally controlled by diffusion.

The fundamental harmonic current and its vector components, the polarization resistance and the electrolytic capacity, can be determined by the modified ac polarographic method or the bridge technique. These quantities, however, are not easily accessible to quantitative analysis of the electrode reaction unless sufficient information about function H_0 is available. However, function H_0 is eliminated in the expression of phase angle ϕ,* and the cotangent of ϕ is expected to play an important role in elucidating

* This is probably true for the electrode reaction with a single charge transfer process coupled with first- or pseudo first-order homogeneous reactions but may not be so for the processes involving higher order chemical reactions and/or some heterogeneous steps such as adsorption (78).

the mechanism and determining the kinetic parameters of the electrode reaction.

Figures 4 and 5 present an example of the effect of a coupled chemical reaction on the frequency dependence and the potential dependence of cot ϕ (78). It can be understood that when the kinetic contribution of the coupled reaction is neglected ($K \to \infty$ or $g \to 0$) the phase angle satisfies the relation

$$\cot \phi = 1 + \frac{\sqrt{2\omega}}{\Lambda} \quad (83)$$

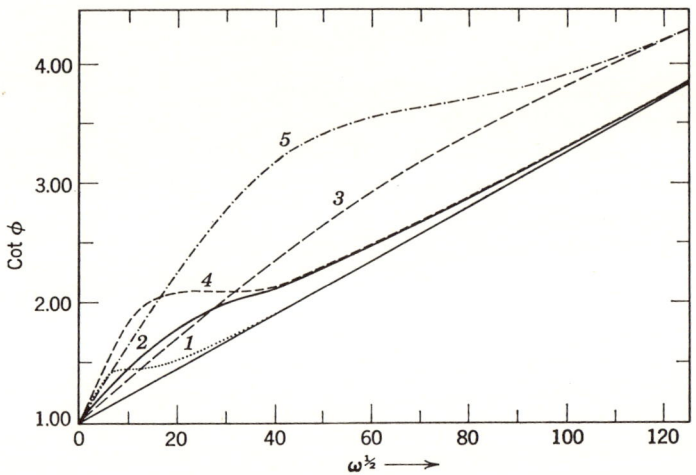

FIG. 4. Effect of preceding or following reactions with varying $\vec{\rho}$ and K on the frequency dependence of cot ϕ at $E = E^r_{1/2}$ ($\tilde{k}^\ominus = 0.100$ cm sec^{-1}, $\alpha = 0.500$, $z = 1$, $D = 1.00 \times 10^{-5}$ cm^2 sec^{-1}, $T = 298°$K). (1) $K = 1.00$, $\vec{\rho} = 50.0$ sec^{-1}; (2) $K = 1.00$, $\vec{\rho} = 500$ sec^{-1}; (3) $K = 1.00$, $\vec{\rho} = 5000$ sec^{-1}; (4) $K = 0.100$, $\vec{\rho} = 50$ sec^{-1}; (5) $K = 0.0100$, $\vec{\rho} = 50$ sec^{-1}. [Smith, D. E., *Anal. Chem.*, **35**, 606 (1963).]

and the cot ϕ versus $\sqrt{\omega}$ plot shows a straight line with a slope of $\sqrt{2}/\Lambda$, from which the kinetic parameter Λ for the charge transfer process can be determined. However, if the rate constants of the coupled chemical reaction are relatively large so that the value of g cannot be neglected against unity, the cot ϕ versus $\sqrt{\omega}$ plot will deviate from the linearity as shown in Figure 4. Extending the measurement to higher frequencies, we will be able to analyze the coupled chemical reaction with larger reaction rates.

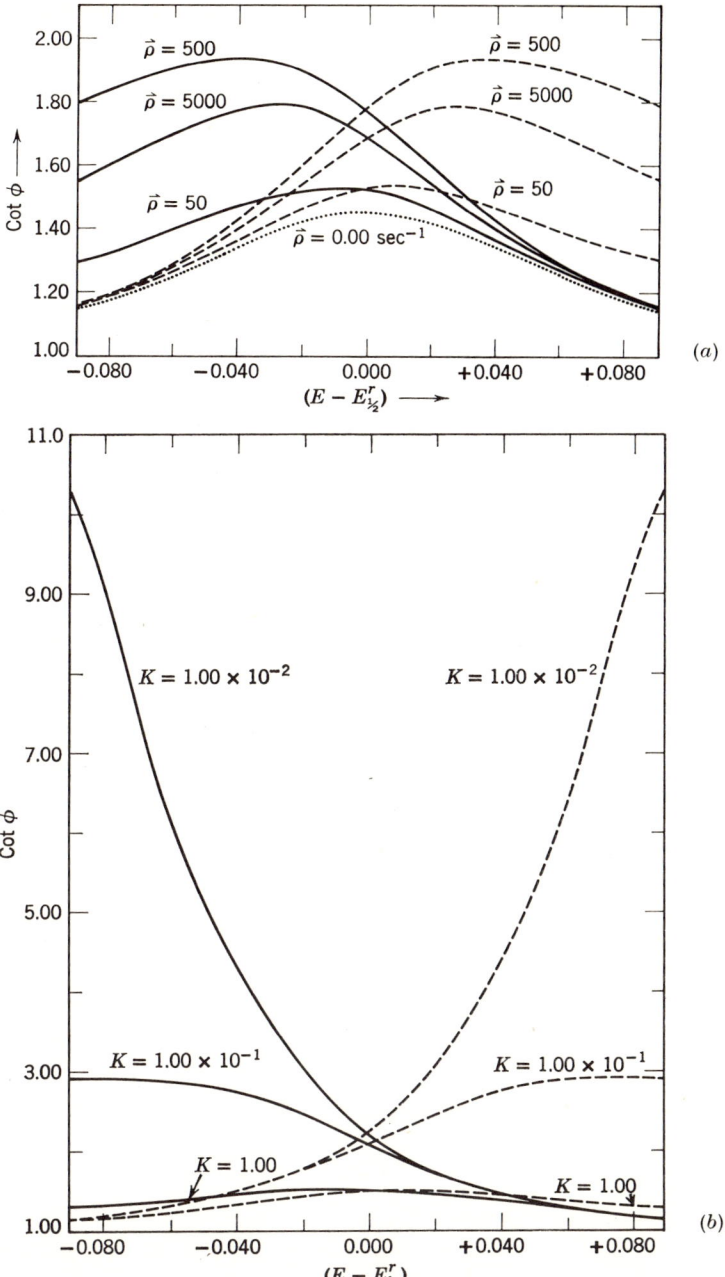

FIG. 5. Potential dependence of cot ϕ with preceding or following reactions with (a) varying $\vec{\rho}$ ($K = 1.00$) and (b) varying K ($\vec{\rho} = 50 \text{ sec}^{-1}$). ($\sqrt{\omega} = 20.0 \text{ sec}^{-1/2}$, $\tilde{k}^{\ominus} = 0.100 \text{ cm sec}^{-1}$, $\alpha = 0.500$, $z = 1$, $D = 1.00 \times 10^{-5} \text{ cm}^2 \text{ sec}^{-1}$, $T = 298°\text{K}$). ———, Preceding reaction; – –, following reaction; ...., no chemical reaction. [Smith, D. E., *Anal. Chem.*, **35**, 606, 607 (1963).]

III. Equilibrium Studies of Labile Complexes

A. Determination of Stability Constants—Use of the Reversible Half-Wave Potential

There are two approaches to the polarographic determination of equilibrium constants of labile complexes: the potentiometric approach or the use of the reversible half-wave potential, and the amperometric approach or the use of the diffusion current. The latter is discussed in Section III-B and C.

In the cases of reversible and quasireversible electrode reactions, the reversible half-wave potential $E^r_{1/2}$ can be determined experimentally. Methods for the necessary extrapolation in the quasireversible cases are described in Section II-F-2.

The method of determining stability constants from the reversible half-wave potential is essentially the same as the classic one developed by DeFord and Hume (16, 30) for reversible cases.

Equation 57 is written:

$$\sqrt{D_M/\bar{D}_{ox}} \exp\left(\frac{zF}{RT}\{(E^r_{1/2})_M - E^r_{1/2}\}\right) = \sum_{i=0}^{\mu} \beta_i^{ox}[L]^i \qquad (84)$$

The left-hand side of Eq. 84 can be determined if $(E^r_{1/2})_M$ is known, the factor $D_M/\bar{D}_{ox}$ being obtained from the ratio of the diffusion current of the simple ion to that of the complex ions. The right-hand side is the function $F_0^{ox}([L])$ defined by Eq. 6a. Hereafter the superscript "ox" of β and $F([L])$ is omitted for the sake of simplicity.

We define a new function:

$$F_1([L]) \equiv \{F_0([L]) - 1\}/[L] \qquad (85)$$

When $F_1([L])$ is plotted against $[L]$ and extrapolated to $[L] = 0$,* the intercept gives β_1,

$$\lim_{[L] \to 0} F_1([L]) = \beta_1 \qquad (86)$$

By using β_1 thus obtained, the function $F_2([L])$

$$F_2([L]) \equiv \{F_1([L]) - \beta_1\}/[L] \qquad (87)$$

is calculated and plotted against $[L]$; β_2 is then obtained by extrapolating the curve to $[L] = 0$. The successive function $F_i([L])$, which gives β_i, is

---

* Since the solution contains a large excess of L, [L] can be replaced by the total ligand concentration as the first approximation; the method of successive approximation is applied if necessary.

calculated from the preceding function $F_{i-1}([L])$ and the value of β_{i-1} by Eq. 88:

$$F_i([L]) \equiv \{F_{i-1}([L]) - \beta_{i-1}\}/[L] \tag{88}$$

The procedure is repeated until $F_\mu([L])$ becomes a constant $(=\beta_\mu)$. An example of analysis of the $F_i([L])$ function is reproduced in Figure 6.

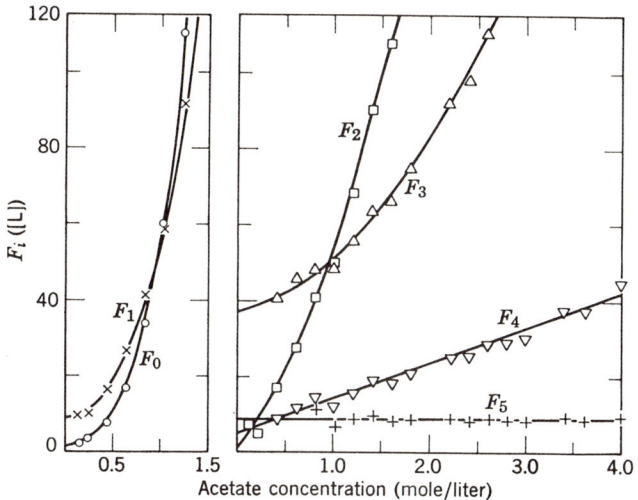

FIG. 6. Analysis of $F_i([L])$ $(i = 0, 1, \ldots, 5)$ functions for zinc(II)–acetate system. Ionic strength = 4.0 (NaNO$_3$); 25°C. [Matsuda, H., and Y. Ayabe, *Ber. Bunsenges. Physik. Chem.*, **67**, 598 (1963).]

The analysis of the reversible half-wave potential was applied by Matsuda and Ayabe to the determination of the formation constants of hydroxo- (55), tartrato- (56), and acetato- (57) complexes of zinc(II).

Matsuda (48) extended the method to cases in which mixed complex ions $ML_pL'_{p'}$ $(p = 0, 1, 2, \ldots, \mu; p' = 0, 1, 2, \ldots, \mu')$ are reduced to the metal amalgam in the presence of a large excess of the ligands L and L', the electrode reaction always being assumed to be reversible. The relation corresponding to Eq. 84 then becomes

$$\sqrt{D_M/\bar{D}_{ox}} \exp\left(\frac{zF}{RT}\{(E^r_{1/2})_M - E^r_{1/2}\}\right) = \sum_{i=0}^{\mu} \sum_{i'=0}^{\mu'} \beta_{ii'}[L]^i[L']^{i'} \equiv F_{00}([L], [L']) \tag{89}$$

where $\beta_{ii'}$ is the overall stability constant of $ML_iL'_{i'}$ and $\bar{D}_{ox}$ is the average diffusion coefficient of the oxidized form:

$$\beta_{ii'} = \frac{[ML_iL'_{i'}]}{[M][L]^i[L']^{i'}} \qquad (90)$$

$$\bar{D}_{ox} = \frac{\sum_{i=0}^{\mu}\sum_{i'=0}^{\mu'} \beta_{ii'}[L]^i[L']^{i'} D_{ML_iL'_{i'}}}{\sum_{i=0}^{\mu}\sum_{i'=0}^{\mu'} \beta_{ii'}[L][L']^{i'}} \qquad (91)$$

The function $F_{00}([L], [L'])$ is determined by measuring the reversible half-wave potentials at varied concentrations of one of the ligands, for example, L', while the concentration of the other is kept at a constant value, which is denoted here by $[L]_1$. Extrapolating $F_{00}([L]_1, [L'])$ to $[L'] = 0$, we obtain $F_{00}([L]_1, 0)$:

$$F_{00}([L]_1, 0) \equiv \lim_{[L'] \to 0} F_{00}([L]_1, [L']) = \sum_{i=0}^{\mu} \beta_{i0}[L]_1^i \qquad (92)$$

The successive functions $F_{01}([L]_1, [L'])$, $F_{02}([L]_1, [L'])$, ...

$$F_{0i'}([L]_1, [L']) \equiv \{F_{0(i'-1)}([L]_1, [L']) - F_{0(i'-1)}([L]_1, 0)\}/[L'] \qquad (93)$$

are calculated by extrapolating the preceding function to $[L'] = 0$. Thus we obtain a set of the limiting values:

$$\left.\begin{aligned}
F_{00}([L]_1, 0) &= \sum_{i=0}^{\mu} \beta_{i0}[L]_1^i \\
F_{01}([L]_1, 0) &= \sum_{i=0}^{\mu} \beta_{i1}[L]_1^i \\
&\cdots \qquad \cdots \\
F_{0i'}([L]_1, 0) &= \sum_{i=0}^{\mu} \beta_{ii'}[L]_1^i \\
&\cdots \qquad \cdots \\
F_{0\mu'}([L]_1, 0) &= \sum_{i=0}^{\mu} \beta_{i\mu'}[L]_1^i
\end{aligned}\right\} \qquad (94)$$

Now the concentration of L is changed to another constant value $[L]_2$ and the reversible half-wave potentials are determined at varied concentrations of L'. Applying the same procedure we obtain another set of $F_{0i'}([L]_2, 0)$ ($i' = 0, 1, 2, \ldots, \mu'$).

In this way we have the series of function $F_{0i'}([L]_k, 0)$:

$$\left.\begin{array}{l} F_{00}([L]_1, 0), F_{00}([L]_2, 0), \ldots, F_{00}([L]_k, 0), \ldots \\ F_{01}([L]_1, 0), F_{01}([L]_2, 0), \ldots, F_{01}([L]_k, 0), \ldots \\ \quad \ldots \qquad \quad \ldots \qquad \ldots \qquad \quad \ldots \qquad \ldots \\ F_{0i'}([L]_1, 0), F_{0i'}([L]_2, 0), \ldots, F_{0i'}([L]_k, 0), \ldots \\ \quad \ldots \qquad \quad \ldots \qquad \ldots \qquad \quad \ldots \qquad \ldots \\ F_{0\mu'}([L]_1, 0), F_{0\mu'}([L]_2, 0), \ldots, F_{0\mu'}([L]_k, 0), \ldots \end{array}\right\} \quad (95)$$

As seen from Eq. 94, each $F_{0i'}([L], 0)$ has the same form as the $F_0([L])$ function (Eq. 6a) and can be treated in the same way. For example, $F_{10}([L], 0)$ is calculated by

$$F_{10}([L], 0) \equiv \{F_{00}([L], 0) - 1\}/[L] \qquad (96)$$

and plotted against [L]. The extrapolation to [L] = 0 gives β_{10}, which is used to calculate the next function $F_{20}([L], 0)$; by extrapolating the latter function, we obtain β_{20}, and so on. Starting from $F_{0i'}([L], 0)$, we can determine $\beta_{0i'}, \beta_{1i'}, \ldots, \beta_{ii'}, \ldots, \beta_{\mu i'}$.

To obviate the uncertainty involved in the graphical method, Momoki et al. (62) utilized a high-speed digital computer in analyzing the $F_0([L])$ function; by a calculation based on the least squares method they determined the most probable values of the stability constants for the cadmium(II)–thiocyanate system with 95% confidence.

The determination of stability constants from the reversible half-wave potential presupposes knowledge of the reversible half-wave potential of the uncomplexed metal ion. However, the reduction of a metal ion may become irreversible in the absence of ligand, whereas it proceeds reversibly or quasi-reversibly in the presence of ligand. In such a case the experimental determination of $(E^r_{1/2})_M$ is impossible and it should be estimated by other means.

Momoki and Ogawa (61) proposed a method for solving this problem by a computerized calculation. From Eq. 84 we have

$$\frac{\sum_{i=0}^{\mu} \beta_i [L]^i}{\sum_{i=0}^{\mu} \beta_i [L]_1^i} = \ln\left\{\frac{\bar{D}_{ox}}{(\bar{D}_{ox})_{[L]=[L]_1}}\right\} + \frac{zF}{RT}\{(E^r_{1/2})_{[L]=[L]_1} - E^r_{1/2}\} \qquad (97)$$

where $[L]_1$ is an arbitrary, fixed concentration of the ligand. The right-hand side of Eq. 97 can be determined experimentally. The stability constants on the left-hand side are so adjusted by the least squares method that the experimental values of the right-hand side for varied concentrations of the ligand are best reproduced.

Completely irreversible systems cannot be treated by the potentiometric method, but the use of diffusion currents may be successful in such cases.

B. Determination of Stability Constants—Use of the Diffusion Current

In this section we discuss the use of the polarographic diffusion current as an analytical measure for determining stability constants of labile complexes. The diffusion current is also a measure of the average diffusion coefficient; this subject is discussed in Section III-C.

When a metal ion M' is added to a buffered solution containing a complex MX, an exchange reaction takes place, and its equilibrium may be represented by

$$MX + M' \rightleftharpoons M + M'X \tag{98}$$

If either of the simple ions produces a diffusion current measurable in the presence of other species, its concentration can be determined polarographically. The equilibrium constant of reaction 98 is calculated from this value and the known total concentrations of M, M', and X. In order to obtain the stability constant of MX, it is necessary that the stability constant of the other complex, M'X, be known as well as the constants required to make corrections for the formation of protonated species and of complexes with the anion of the buffer.* Schwarzenbach and his co-workers (75–77) determined the stability constants of aminopolycarboxylato complexes of a wide variety of metal ions by this principle (correction for the existence of complexes with the anion of the buffer being neglected).

Exchange reactions are also used for the determination of the stability constants of relatively weak complexes, such as acetato complexes.

1. Ligand Exchange Reaction. When we add a ligand L to a solution containing a complex MX, the ligand exchange reaction

$$MX + xL \rightleftharpoons X + ML + ML_2 + \cdots \tag{99}$$

takes place. By analyzing the shift in the equilibrium of the reaction, it is possible to determine the stability constants of L-complexes. The case in which X forms only 1:1 complexes is of particular interest.

Consider a system of constant ionic strength buffered at a given pH in which the following equilibria exist (the ionic charge for each ion being

---

* For a discussion of these corrections, see Section III-B-2.

omitted except for hydrogen ion):

$$M + L \rightleftharpoons ML, ML + L \rightleftharpoons ML_2, \ldots, ML_{\mu-1} + L \rightleftharpoons ML_\mu \quad (100)$$

$$M + X \rightleftharpoons MX \quad (101)$$

$$MX + H^+ \rightleftharpoons MHX \quad (102)$$

$$H_nX \rightleftharpoons H_{n-1}X + H^+, \ldots, H_2X \rightleftharpoons HX + H^+, HX \rightleftharpoons X + H^+ \quad (103)$$

where X is a ligand that forms 1:1 complexes with M (e.g., X = nitrilotriacetate) and L is another ligand such as acetate. Equilibria 102 and 103 are included in order to take into account the formation of the protonated X-complex and the dissociation of the acid H_nX. The corresponding equilibrium constants are given by

$$\beta_{ML_i} = [ML_i]/[M][L]^i \quad (i = 1, 2, \ldots, \mu) \quad (104)$$

$$\beta_{MX} = [MX]/[M][X] \quad (105)$$

$$K_{MX}^H = [MHX]/[MX][H^+] \quad (106)$$

$$K_{H_kX} = [H_{k-1}X][H^+]/[H_kX] \quad (k = 1, 2, \ldots, n) \quad (107)$$

If the complexes ML_i are relatively weak, they will be reduced at almost the same potential as the reduction potential of the simple ion and will give a single polarographic wave, which precedes the reduction of the species MX and MHX. The limiting current of this wave is controlled by the diffusion of M and ML_i if the rates of dissociation of MX and MHX into M and (X + HX) are sufficiently small. The diffusion current is then proportional to the sum of the concentrations of the metal species not bound to the ligand X:

$$\bar{i}_d = \text{constant} \times \sqrt{\bar{D}_{ox}}[M]_{app} \quad (108)$$

where

$$\bar{D}_{ox} = \frac{D_M + D_{ML}\beta_1[L] + \cdots + D_{ML_\mu}\beta_\mu[L]^\mu}{1 + \beta_1[L] + \cdots + \beta_\mu[L]^\mu} \quad (109)$$

and

$$[M]_{app} = [M] + [ML] + \cdots + [ML_\mu] \quad (110)$$

Thus $[M]_{app}$ can be determined by comparing the diffusion current at a given concentration of L in the presence of X with the diffusion current of M of a known concentration at the same concentration of L but in the absence of X.

From Eqs. 104–107 and Eq. 110, we obtain Eq. 111:

$$\frac{([M]_{app} - c_M + c_X)[M]_{app}}{c_M - [M]_{app}} = \frac{\alpha([H^+])}{\gamma([H^+])\beta_{MX}} \{1 + \beta_{ML}[L] + \cdots + \beta_{ML_\mu}[L]^\mu\} \quad (111)$$

where c_M and c_X are the total concentrations of the metal and of the ligand X, respectively,

$$c_M = [M] + [MX] + [MHX] + \sum_{i=1}^{\mu} [ML_i]$$

$$c_X = [MX] + [MHX] + \sum_{k=1}^{n} [H_k X]$$

and $\alpha([H^+])$ and $\gamma([H^+])$ are the functions of pH as defined by Eqs. 112 and 113:

$$\alpha([H^+]) = 1 + \frac{[H^+]}{K_{HX}} + \frac{[H^+]^2}{K_{HX} K_{H_2X}} + \cdots + \frac{[H^+]^n}{K_{HX} K_{H_2X} \cdots K_{H_nX}} \quad (112)$$

$$\gamma([H^+]) = 1 + K_{MX}^H [H^+] \quad (113)$$

The left-hand side of Eq. 111 contains only experimentally accessible quantities, and the quantity between the braces of the right-hand side is the $F_0([L])$ function, which is then subject to the determination of the stability constants β_{ML_i}.

Tanaka and Kato (92) noted that the factor $\alpha([H^+])/\gamma([H^+])$ is a constant at a given pH and ionic strength, and that it can be determined by extrapolating the left-hand side of Eq. 111 to $[L] = 0$. An example of the plots of the left-hand side of Eq. 111 is reproduced in Figure 7, in which the plots are straight lines indicating that no higher complexes ($i \geq 2$) are formed under the experimental conditions. When $F_0([L])$ is obtained, the procedure described in Section III-A is then applied to evaluating the successive functions, $F_1([L])$, $F_2([L])$, and so on, each of which gives $\beta_1, \beta_2, \ldots$.

One of the advantages of this method is that no information about the equilibrium constants concerning the ligand X is required in the determination of the formation constants of ML_i. This method can also be applied to cases in which the electrode reactions are irreversible. For successful application of the method, however, a system should give a diffusion current without kinetic contribution. The above-mentioned authors applied this method to the system nickel(II) (M)/nitrilotriacetate (X)/acetate (L), and determined the formation constants of acetatonickel ion.

A ligand exchange reaction between two chelating ligands

$$MX + \sum_j H_j Y \rightleftharpoons MY + \sum_k H_k X$$

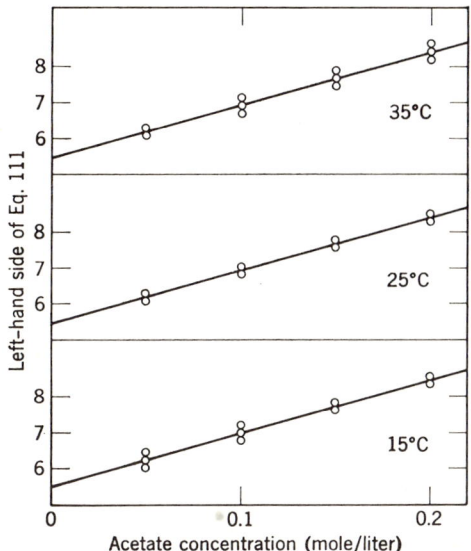

FIG. 7. Left-hand side of Eq. 111 as a function of [L]. $M = Ni^{2+}$, X = nitrilotriacetate, L = acetate; 25°C; ionic strength = 0.2 (KNO_3). [Tanaka, N., and K. Kato, *Bull. Chem. Soc. Japan*, **32**, 519 (1959).]

can be used for determining the relative stability of MY to MX if either of the unbound ligands gives an anodic wave which permits its polarographic determination (67). The procedure is quite similar to that for metal exchange reactions; for details the original paper should be consulted.

2. Metal Exchange Reaction. When a second metal ion M' is added to the system considered in Section III-B-1, the equilibria may be represented as

$$\left\{ \begin{array}{c} M \\ \updownarrow \\ \sum_i ML_i \end{array} \right\} + \left\{ \begin{array}{c} M'X \\ \updownarrow \\ M'HX \end{array} \right\} \rightleftharpoons \left\{ \begin{array}{c} MX \\ \updownarrow \\ MHX \end{array} \right\} + \left\{ \begin{array}{c} M' \\ \updownarrow \\ \sum_j M'L_j \end{array} \right\} \quad (114)$$

The concentrations of the metal species not bound to the ligand X are given by

$$\begin{array}{l} [M]_{app} = [M] + [ML] + [ML_2] + \cdots + [ML_\mu] \\ [M']_{app} = [M'] + [M'L] + [M'L_2] + \cdots + [M'L_{\mu'}] \end{array} \right\} \quad (115)$$

When either of them, for example, $[M]_{app}$, is determined by polarographic measurement, the other can be calculated from the known values of the total concentrations c_M, $c_{M'}$, and c_X provided $c_M + c_{M'} > c_X$, because under these conditions the concentration of free ligand X unbound to the metal ions is negligible; accordingly, we have as a good approximation:

$$[M']_{app} = c_M + c_{M'} - c_X - [M]_{app} \tag{116}$$

We can thus determine the equilibrium constant of the overall reaction 114:

$$(K_{M'}^M)_{app} = \frac{[M']_{app}([MX] + [MHX])}{[M]_{app}([M'X] + [M'HX])}$$

$$= \frac{[M']_{app}(c_M - [M]_{app})}{[M]_{app}(c_{M'} - [M']_{app})} \tag{117}$$

However, the equilibrium constant of reaction 118

$$M'X + M \rightleftharpoons M' + MX \tag{118}$$

is given by

$$K_{M'}^M = [M'][MX]/[M][M'X] \tag{119}$$

and related to $(K_{M'}^M)_{app}$ by the equation

$$(K_{M'}^M)_{app} = K_{M'}^M \frac{\gamma}{\gamma'} \frac{\left(1 + \sum_{j=1}^{\mu'} \beta'_j[L]^j\right)}{\left(1 + \sum_{i=1}^{\mu} \beta_i[L]^i\right)} \tag{120}$$

where

$$\left.\begin{array}{ll} \beta_i = [ML_i]/[M][L]^i & (i = 1, 2, \ldots, \mu) \\ \beta'_j = [M'L_j]/[M'][L]^j & (j = 1, 2, \ldots, \mu') \end{array}\right\} \tag{121}$$

and

$$\left.\begin{array}{l} \gamma = 1 + K_{MX}^H[H^+] \\ \gamma' = 1 + K_{M'X}^H[H^+] \end{array}\right\} \tag{122}$$

If the formation constants of the L-complexes of either metal are known, those of the other can be determined according to the following procedure. Suppose the stability constants for ML_i are known; Eq. 120 is conveniently rearranged to Eq. 123:

$$(K_{M'}^M)_{app}\left(1 + \sum_{i=1}^{\mu} \beta_i[L]^i\right) = K_{M'}^M \frac{\gamma}{\gamma'}\left(1 + \sum_{j=1}^{\mu'} \beta'_j[L]^j\right) \tag{123}$$

the left-hand side of which is known. We see that the factor $K_{M'}^M \gamma/\gamma'$, being independent of [L], is determined by extrapolating the plot of the left-hand side against [L] to [L] = 0 and that the $F_0([L])$ function for M′ is obtained*; an example of the plot is illustrated in Figure 8.

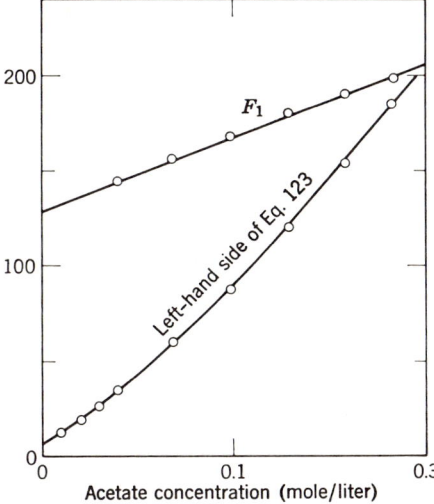

FIG. 8. Left-hand side of Eq. 123 and $F_1([L])$ as functions of [L]. M = Cu^{2+}, M′ = Pb^{2+}, X = ethylenediaminetetraacetate, L = acetate; 25°C; ionic strength = 0.2 (KNO_3). [Tanaka, N., and K. Kato, *Bull. Chem. Soc. Japan*, **33**, 420 (1960).]

When the formation constants of $M'L_j$ are known, the same procedure is applied to treatment of the inverse form of Eq. 123, and the $F_0([L])$ function for M is obtained.

Starting from the stability constants of acetatonickel(II) obtained by applying the method described in Section III-B-1), Tanaka and his coworkers (91, 94) determined the stability constants of acetato complexes of several metal ions. They also extended both methods described above to the systems in which another ligand L′ is involved to form $ML'_{i'}$ and/or $ML_iL'_{i'}$ and determined the stability constants of sulfato and sulfatoacetato complexes of copper(II) and nickel(II) ions. The principle of the extended methods is similar to that discussed in Section III-A, and the reader may refer to the original papers (99, 106, 107) for details.

The stability constants of acetato, sulfato, and sulfatoacetato complexes of several metal ions are summarized in Table III.

* Note that by knowing γ/γ' we obtain $K_{M'}^{M}$, the equilibrium constant of the exchange reaction of Eq. 118, corrected for the formation of ML_i and $M'L_j$ and of the protonated species as well (67, 101).

TABLE III. Overall Stability Constants of Acetato, Sulfato, and Sulfatoacetato Complexes Determined by the Amperometric Method

Complex[a]	Ionic strength	Temperature, °C			Reference
		15	25	35	
CdOAc$^+$	0.2	27	16	20	91
CoOAc$^+$	0.2	2.1	0.6	0.6	91
CuOAc$^+$	0.2	41	52	69	94
	1.0	—	20	—	107
Cu(OAc)$_2$	0.2	190	93	82	94
	1.0	—	110	—	107
CuSO$_4$	1.0	—	<3	—	107
Cu(SO$_4$)$_2^{2-}$	1.0	—	~30	—	107
Cu(OAc)(SO$_4$)$^-$	1.0	—	40	—	107
Cu(OAc)(SO$_4$)$_2^{3-}$	1.0	—	~70	—	107
NiOAc$^+$	0.2	2.6	2.6	2.4	92
	1.0	—	1.9	—	106
NiSO$_4$	0.2	14.3 ± 0.4	11.6 ± 0.9	14.6 ± 0.7	99
	1.0	—	3.7	—	706
Ni(SO$_4$)$_2^{2-}$	1.0	—	26	—	106
Ni(OAc)(SO$_4$)$^-$	1.0	—	1	—	106
Ni(OAc)(SO$_4$)$_2^{3-}$	1.0	—	~3	—	106
PbOAc$^+$	0.2	130	130	120	94
Pb(OAc)$_2$	0.2	750	390	380	94
ZnOAc$^+$	0.2	5.7	4.6	3.7	94

[a] OAc$^-$ denotes acetate ion.

C. Study of Ion Association by Analysis of the Average Diffusion Coefficient

In principle, ion association can be analyzed by the same methods as those used in the study of labile complex formation, although equilibrium constants of ion association are usually smaller and the formation of higher associated species (triple ions, and so on) is rare in aqueous solutions of moderate concentration.

The shift in the half-wave potential of reversible waves was used for the determination of association constants (35, 45). When the electrode reaction is irreversible, the presence of pair-forming counterions affects the half-wave potential not only through ion association but also through change in the structure of the double layer; it is difficult to evaluate the effect of ion association separately. Although the shift in the half-wave potential of an irreversible wave may serve as a qualitative indication of ion association, its quantitative interpretation for determining the association constants is still open to question. At present, the change in the diffusion current attributable to ion association appears to be a more useful quantity for the determination of association constants. The change is related to the average diffusion coefficient defined by Eq. 109. The average diffusion coefficient is a function of the concentration of a pair-forming counterion, and its analysis gives the association constants (103).

In such experiments the ionic strength of solution is kept constant by adding an appropriate amount of an indifferent electrolyte. Its effect is worth attention particularly in polarography, which requires a relatively high ionic strength. The ion of the indifferent electrolyte may be involved in ion association equilibria and may disturb the determination of the association constant of the ion pair under study. Let us examine this effect in a simple case (74).

Consider a solution containing three electrolytes: MA_m, N_nB, and the indifferent electrolyte NA. Here M is an m-valent cation (anion), N and A are univalent cations (anions) and anions (cations), respectively, and B is an n-valent anion (cation). The association between M and B is to be studied by measuring the average diffusion coefficient of M at varied concentrations of N_nB.

We assume that the concentration of MA_m is sufficiently small and its contribution to the ionic strength of the solution is negligible and that the ionic strength is kept constant by adding NA when the concentration of N_nB is changed. Then we have

$$\text{Ionic strength} = \bar{c} = c_A + kc_B \qquad (124)$$

where c_A and c_B are the concentrations of NA and N_nB, respectively, and k is a constant determined by the ionic charge of B,

$$k = (n + n^2)/2 \qquad (125)$$

We further assume that the only possible ion pairs are MA and MB and that higher-order association can be ignored. The stability constants of the ion pairs are defined as:

$$\beta_{MA} = [MA]/[M][A] \qquad (126)$$

and

$$\beta_{MB} = [MB]/[M][B] \qquad (127)$$

The average diffusion coefficient of this system is represented by

$$\bar{D}_{ox} = \frac{D_M + D_{MA}\beta_{MA}c_A + D_{MB}\beta_{MB}c_B}{1 + \beta_{MA}c_A + \beta_{MB}c_B} \quad (128)$$

where D_M, D_{MA}, and D_{MB} are the individual diffusion coefficients of the respective species. In Eq. 128 the concentrations of the free counterions [A] and [B] are assumed to be identical with their total concentrations c_A and c_B, respectively; this is justified by the assumption that the concentration of MA_m is very small.

When the diffusion coefficient is measured in the absence of N_nB, that is, in a solution containing only $\bar{c}$ mole/liter of NA and a small amount of MA_m, the observed diffusion coefficient should be:

$$\bar{D}_{ox}^0 = \frac{D_M + \beta_{MA}D_{MA}\bar{c}}{1 + \beta_{MA}\bar{c}} \quad (129)$$

Combining Eqs. 124, 128, and 129, we obtain:

$$\bar{D}_{ox} = \frac{\bar{D}_{ox}^0(1 + \beta_{MA}\bar{c}) + (D_{MB}\beta_{MB} - kD_{MA}\beta_{MA})c_B}{1 + \beta_{MA}\bar{c} + (\beta_{MB} - k\beta_{MA})c_B} \quad (130)$$

which can be rearranged into a form convenient for graphical treatment:

$$\frac{c_B}{\bar{D}_{ox}^0 - \bar{D}_{ox}} = \frac{(1 + \beta_{MA}\bar{c}) + \beta_{MB}\{1 - k\beta_{MA}/\beta_{MB}\}c_B}{\beta_{MB}(\bar{D}_{ox}^0\{1 - k\beta_{MA}/\beta_{MB}\} - D_{MB}\{1 - kD_{MA}\beta_{MA}/D_{MB}\beta_{MB}\})} \quad (131)$$

Plotting the left-hand side of Eq. 131 against c_B results in a straight line as shown in Figure 9. The ratio of the slope to the intercept of the straight line gives the apparent association constant of the ion pair MB:

$$\frac{\text{Slope}}{\text{Intercept}} = \frac{\beta_{MB} - k\beta_{MA}}{1 + \beta_{MA}\bar{c}} \equiv (\beta_{MB})_{app} \quad (132)$$

If the ion association between M and A is negligible, the slope/intercept ratio of the experimental plot should give the true association constant β_{MB}.

The extrapolation of $(\beta_{MB})_{app}$ to $\bar{c} = 0$ does not give the true association constant but the limiting value of $(\beta_{MB} - k\beta_{MA})$. In order to obtain β_{MB}, the value of β_{MA} should be estimated experimentally or theoretically. For example, Tanaka and Koseki (98) determined the association constant of $NH_4^+[Co(oxalato)_3]^{2-}$ by the conductivity measurement, and using this value they corrected the apparent association constants for various cation/$[Co(oxalato)_3]^{2-}$ systems determined by the measurement of the diffusion current in ammonium salt solutions.

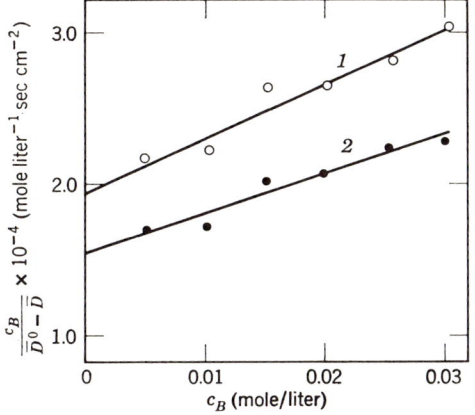

FIG. 9. Left-hand side of Eq. 131 as a function of c_B. M = trioxalatochromate(III), N = NH_4^+, A = NO_3^-; (1) B = Mg^{2+}; (2) B = Ba^{2+}; 25°C; ionic strength = 0.1. [Tanaka, N., and K. Koseki, *Bull. Chem. Soc. Japan*, **41**, 2069 (1968).]

The determination of the average diffusion coefficient from the polarographic diffusion current involves complications inherent to the DME. The well-known two effects that make the observed diffusion current deviate from the theoretical value are the dilution effect and the shielding effect (44). Using a thin-walled capillary (22, 82) and measuring the current at the first drop, we can minimize both effects. However, the convective flow of solution resulting from the dragging by falling drops (82) is another source of unpredictable error in the polarographic determination of diffusion coefficients.

It is interesting to examine the values of diffusion coefficients determined by the polarographic method. In Figure 10 the diffusion coefficients of thallium(I) and cadmium(II) ions in potassium nitrate solutions at 25°C measured by the diaphragm cell method (27) are compared with two sets of corresponding values obtained by the polarographic method. In each set of polarographic measurements, diffusion currents were measured at the first drops. However, in one set (82) thin-walled capillaries were used and correction was made for the shielding effect; and in the other (31) an ordinary thick-walled capillary in vertical position was used and no correction was made for the shielding effect. In the case of thallium(I) ion, the three sets of data are in fair agreement; in the case of cadmium(II), however, disagreement is observed between the data of the diaphragm cell method and other two sets of polarographic data. More investigation is required in order to explain such a discrepancy.

In this respect the use of stationary (or nearly stationary) electrodes is of interest, since the hydrodynamics may be less complicated with this type of electrode than with dropping electrodes.

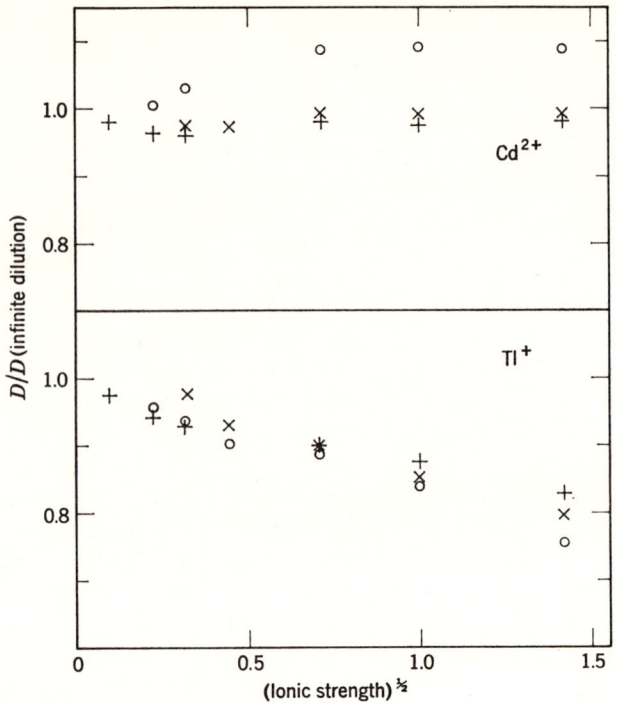

FIG. 10. Diffusion coefficients (relative to the values of infinite dilution) of cadmium(II) and thallium(I) in potassium nitrate solutions at 25°C. (○) Diaphragm-cell method; (+) polarographic method (thin-walled capillary)[82]; (×) polarographic method (thick-walled capillary)[31].

As seen in Section II-E, the galvanostatic transition time is a measure of the average diffusion coefficient. When the rates of association–dissociation of ion pairs are sufficiently large, the second term of the right-hand side of Eq. 45 can be neglected, and we have

$$\sqrt{\bar{D}_{ox}} = \frac{2I_c}{zF\sqrt{\pi}} \frac{\sqrt{\tau}}{c_{ox}} \qquad (133)$$

The galvanostatic method was applied by Tanaka and Yamada (110) to the determination of the stability constants of the ion pairs of sulfate ion with several cobalt(III) and chromium(III) complex cations. They used a DME with a long drop time (61 sec) and measured the transition times (shorter than 1.5 sec) at a given moment in the life of a drop. The effect of the supporting electrolyte (0.1 mole/liter $NaClO_4$) was neglected, and the following equation was used for evaluating β_{MB} and D_{MB}/D_M by the least squares method:

$$\frac{\bar{D}}{D_M} = \frac{1 + \beta_{MB}\left(\dfrac{D_{MB}}{D_M}\right)c_B}{1 + \beta_{MB}c_B} \tag{134}$$

Here D_M was obtained from the transition time observed in the absence of the pair-forming counterion.

The association constants of some ion pairs as determined by polarographic and chronopotentiometric methods are summarized in Table IV.

TABLE IV. Association Constants β of Some Ion Pairs at 25°C Determined by Polarographic and Chronopotentiometric Methods

Ion pair[a]	β [b]	$\log \beta$ [c]	$D_{(ion\ pair)}$[b] / $D_{(complex\ ion)}$	Method	Reference
$[Co(NH_3)_6]^{3+}SO_4^{2-}$	98 ± 8	3.30	0.6$_8$	Chrono.	110
	—	3.21	—	Polarog.	103
	—	3.46	—	Polarog.	116
$[Cr(NH_3)_6]^{3+}SO_4^{2-}$	61 ± 7	3.15	0.6$_7$	Chrono.	110
	—	3.13	—	Polarog.	103
$[Co(En)_3]^{3+}SO_4^{2-}$	99 ± 15	3.22	0.8$_2$	Chrono.	110
	—	2.72	—	Polarog.	45
$[Cr(En)_3]^{3+}SO_4^{2-}$	58 ± 8	2.99	0.8$_7$	Chrono.	110
$Na^+[Co(Ox)_3]^{3-}$	22 ± 1	1.9$_4$	0.72	Polarog.	98
$Mg^{2+}[Co(Ox)_3]^{3-}$	56 ± 6	2.8$_9$	0.59	Polarog.	98
$Ca^{2+}[Co(Ox)_3]^{3-}$	55 ± 4	2.9$_2$	0.54	Polarog.	98
$Sr^{2+}[Co(Ox)_3]^{3-}$	53 ± 3	2.9$_4$	0.52	Polarog.	98
$Ba^{2+}[Co(Ox)_3]^{3-}$	55 ± 7	2.9$_5$	0.52	Polarog.	98
$Al^{3+}[Co(Ox)_3]^{3-}$	118 ± 17	3.7$_0$	0.47	Polarog.	98
$(CH_3)_4N^+[Co(Ox)_3]^{3-}$	7.6	1.5$_2$	0.53	Polarog.	98
$(C_2H_5)_4N^+[Co(Ox)_3]^{3-}$	3.7 ± 0.4	1.1$_8$	0.30	Polarog.	98

[a] En, ethylenediamine; Ox, oxalate ion.
[b] At ionic strength of 0.1.
[c] At infinite dilution.

IV. Kinetic Studies of Labile Complexes

A. Mechanistic Study of Electrode Reactions

The analysis of various parameters introduced in Section II offers useful information about the mechanism of electrode reactions; the fundamental methods of analysis are summarized in Table V. A description of mechanistic study of the charge transfer process is not our main purpose, and the following discussion is limited to that related to the coupled chemical reaction.

The exchange current density refers to the kinetic properties of electrode reaction at the equilibrium state, from which the transfer coefficient in the neighborhood of the equilibrium potential can be determined according to the relations given in Table V (Eqs. 135–137). When the metal complexes are reduced to the metal amalgams according to reaction scheme 50, in which $F_0^{red}([L]) = 1$, the number of ligands p of the complex species participating in the charge transfer process can be estimated from the dependences of $\ln I_p^0$ and E_e on the logarithm of ligand concentration (see Table V, Eqs. 138, 139) if the transfer coefficient is available. Particularly when only one complex species $M^{m+}L_s$ is assumed to exist in the solution ($\beta_s^{ox} \gg \beta_{i \neq s}^{ox}$), Eq. 139 in Table V is simplified as

$$s = -\frac{zF}{RT}\frac{\partial E_e}{\partial \ln [L]} \tag{150}$$

from which the value of s can be easily obtained.

This method of analyzing the exchange current was applied by Gerischer (26) to the electrode reactions of some zinc(II) and cadmium(II) complexes at the corresponding metal amalgam electrodes, the results of which are reproduced in Table VI.

The accurate determination of exchange currents is often difficult or even impossible in systems in which either one of the oxidized and reduced forms is unstable in solution or difficult to prepare. In such systems polarographic and related methods play an important role, in which either one of the oxidized and reduced forms can be generated in situ at the electrode surface.

Consider the electrode reaction represented by scheme 50 as an example, in which the metal complex is reduced to the metal amalgam. Analyzing the dependence of parameter Λ and polarographic reversible half-wave potential $E_{1/2}^r$, defined by Eqs. 54 and 57 respectively, on the ligand concentration, we can estimate the species prevailing in the solution and that participating in the charge transfer process according to the relations given by Eqs. 141–144 in Table V. Some of these equations involve the terms representing the concentration dependence of diffusion coefficients whose contribution, however, may be ignored in many instances without introducing serious errors.

The theoretical treatment of the polarographic current–potential relationship given in Section II-F has been extended by Matsuda et al. (48, 56, 57) to the following electrode reactions under the assumption that there is no kinetic contribution of the coupled chemical reactions (i.e., $\bar{\imath}_l = \bar{\imath}_d$).

(a) Each complex species of a series of $M^{m+}L_i$ ($i = 0, 1, 2, \ldots, \mu$) is subject to the charge transfer process

$$M^{m+}L_i + ze + Hg \rightleftharpoons M(Hg) + iL \qquad (i = 0, 1, 2, \ldots, \mu) \qquad (151)$$

with different kinetic parameters Λ_i and α_i.

(b) The simple metal ion M^{m+} can be coordinated with more than two kinds of ligands; for example, each one of a series of $M^{m+}L_iL'_{i'}$ ($i = 0, 1, 2, \ldots, \mu; i' = 0, 1, 2, \ldots, \mu'$) is reduced to metal amalgam $M(Hg)$:

$$M^{m+}L_iL'_{i'} \rightleftharpoons M^{m+} + iL + i'L' \qquad (152a)$$

$$M^{m+}L_iL'_{i'} + ze + Hg \rightleftharpoons M(Hg) + iL + i'L' \qquad (152b)$$

The polarographic average current of these electrode reactions is represented by a common equation:

$$\frac{\bar{\imath}}{\bar{\imath}_d} = \frac{1}{\dfrac{1.13}{\sqrt{t_d}}\Phi(\Lambda) + 1 + e^\zeta} \qquad (153)$$

with parameter $\Phi(\Lambda)$ as given in Table VII.

The application of the theoretical equations to the study of the electrode reaction is demonstrated with the polarographic reduction of zinc(II)–acetate complexes (57). From Eq. 153 of the polarographic reduction current, we can derive the relation

$$\frac{1.13 t_d^{-1/2}}{\left(\dfrac{\bar{\imath}_d - \bar{\imath}}{\bar{\imath}}\right) e^{-\zeta} - 1} = \sum_{i=0}^{\mu} M_i(E)[L]^i \qquad (154)$$

with

$$M_i(E) = (\tilde{k}_i^{\ominus})_B \left(\sum_{i=0}^{\mu} \beta_i^{ox}\right)^{(1-\alpha_i)} (\sqrt{D_M})^{(\alpha_i-1)} (\sqrt{D_a})^{-\alpha_i}$$

$$\times \exp\left(\frac{(1-\alpha_i)zF}{RT}\{E - (E^r_{1/2})_M\}\right) \qquad (155)$$

TABLE V. Mechanistic Information Obtained from the Analysis of Kinetic Parameters and Related Quantities

Parameter ξ to be examined	Variable x	Conditions	Information obtained	
$\xi = \ln I_p^0$	$x = \ln c_{ox}$	$c_{red}, [L] = $ constant	$\partial \xi / \partial x = (1 - \alpha_p) \dfrac{zF}{RT}$	(135)
$\xi = \ln I_p^0$	$x = \ln c_{red}$	$c_{ox}, [L] = $ constant	$\partial \xi / \partial x = -\alpha_p \dfrac{zF}{RT}$	(136)
$\xi = \ln (I_p^0/c_{ox})$	$x = \ln (c_{red}/c_{ox})$	$[L] = $ constant	$\partial \xi / \partial x = \alpha_p$	(137)
$\xi = \ln I_p^0$	$x = \ln [L]$	$c_{ox}, c_{red} = $ constant	$\partial \xi / \partial x = p - \dfrac{\alpha_p zF}{RT} \left(\dfrac{\partial E_e}{\partial x}\right)_{c_{ox}, c_{red}} - \dfrac{\partial \ln F_0^{ox}([L])}{\partial x}$ $= p - (1-\alpha_p) \dfrac{\partial \ln F_0^{ox}([L])}{\partial x} - \alpha_p \dfrac{\partial \ln F_0^{red}([L])}{\partial x}$	(138)
$\xi = E_e$	$x = \ln [L]$	$c_{ox}, c_{red} = $ constant	$\partial \xi / \partial x = \dfrac{RT}{zF} \left(\dfrac{\partial \ln F_0^{red}([L])}{\partial x} - \dfrac{\partial \ln F_0^{ox}([L])}{\partial x} \right)$	(139)
$\xi = \ln \dfrac{(\tilde{k}_p^c)_B}{(\tilde{k}_p^a)_B} + \dfrac{zF}{RT} E$	$x = \ln [L]$	$c_{ox}, c_{red} = $ constant	$\partial \xi / \partial x = \dfrac{\partial \ln F_0^{red}([L])}{\partial x} - \dfrac{\partial \ln F_0^{ox}([L])}{\partial x}$	(140)
$\xi = E_{1/2}^r$	$x = \ln [L]$	$F_0^{red}([L]) = 1$	$\partial \xi / \partial x = -\dfrac{RT}{zF} \dfrac{\partial \ln \sqrt{D_{ox}}}{\partial x} - \dfrac{RT}{zF} \dfrac{\partial \ln F_0^{ox}([L])}{\partial x}$	(141)
$\xi = \ln \sqrt{\dfrac{D_M}{D_{ox}}} + \dfrac{zF}{RT} \{(E_{1/2}^r)_M - E_{1/2}^r\}$	$x = \ln [L]$	$F_0^{red}([L]) = 1$	$\partial \xi / \partial x = \dfrac{\partial \ln F_0^{ox}([L])}{\partial x}$	(142)

$\xi = \ln \Lambda_p$	$x = \ln [\mathrm{L}]$	$F_0^{\mathrm{red}}([\mathrm{L}]) = 1$	$\partial \xi/\partial x = p - (1 - \alpha_p) \dfrac{\partial \ln F_0^{\mathrm{ox}}([\mathrm{L}])}{\partial x}$	(143)
$\xi = \ln \Lambda_p - \dfrac{(1 - \alpha_p)zF}{RT}$ $\times \{E_{1/2}^{\mathrm{r}} - (E_{1/2}^{\mathrm{r}})_{\mathrm{M}}\}$	$x = \ln [\mathrm{L}]$	$F_0^{\mathrm{red}}([\mathrm{L}]) = 1$	$\partial \xi/\partial x = p - (1 - \alpha_p)\dfrac{\partial \ln \sqrt{\bar{D}_{\mathrm{ox}}}}{\partial x}$	(144)
$\xi = \ln (Z_p - 1)$	$x = \zeta$	$F_0^{\mathrm{red}}([\mathrm{L}]) = 1$; $[\mathrm{L}] = \mathrm{constant}$	$\partial \xi/\partial x = \alpha_p - 1$	(145)
$\xi = \ln (Z_p - 1)$	$x = \ln [\mathrm{L}]$	$F_0^{\mathrm{red}}([\mathrm{L}]) = 1$; $\zeta = \mathrm{constant}$	$\partial \xi/\partial x = -\partial \ln \Lambda_p/\partial x$	(146)
$\xi = (E_{1/2})_{\mathrm{irr}}$	$x = \ln [\mathrm{L}]$	$F_0^{\mathrm{red}}([\mathrm{L}]) = 1$	$\partial \xi/\partial x = \dfrac{\alpha_p zF}{RT}\left\{p - \dfrac{\partial \ln F_0^{\mathrm{ox}}([\mathrm{L}])}{\partial x} - \dfrac{\partial \ln \sqrt{\bar{D}_{\mathrm{ox}}}}{\partial x}\right.$ $\left. - \dfrac{\partial \ln (\bar{i}_l/\bar{i}_d)}{\partial x}\right\}$	
			$\cong \dfrac{\alpha_p zF}{RT}\left\{p - \dfrac{\partial \ln F_0^{\mathrm{ox}}([\mathrm{L}])}{\partial x} - \dfrac{\partial \ln \bar{i}_l}{\partial x}\right\}$	(147)
$\xi = (E_{1/2})_{\mathrm{irr}} + \dfrac{RT}{\alpha_p zF}\ln \dfrac{\bar{i}_l}{\bar{i}_d}$	$x = \ln [\mathrm{L}]$	$F_0^{\mathrm{red}}([\mathrm{L}]) = 1$	$\partial \xi/\partial x = \dfrac{\alpha_p zF}{RT}\left\{p - \dfrac{\partial \ln F_0^{\mathrm{ox}}([\mathrm{L}])}{\partial x} - \dfrac{\partial \ln \sqrt{\bar{D}_{\mathrm{ox}}}}{\partial x}\right.$ $\left. - \dfrac{\partial \ln \bar{i}_d}{\partial x}\right\}$	
			$\cong \dfrac{\alpha_p zF}{RT}\left\{p - \dfrac{\partial \ln F_0^{\mathrm{ox}}([\mathrm{L}])}{\partial x} - \dfrac{\partial \ln \bar{i}_d}{\partial x}\right\}$	(148)
$\xi = (E_{1/2})_{\mathrm{irr}} + \dfrac{RT}{\alpha_p zF}\ln \bar{i}_l$	$x = \ln [\mathrm{L}]$	$F_0^{\mathrm{red}}([\mathrm{L}]) = 1$	$\partial \xi/\partial x \cong p - \dfrac{\partial \ln F_0^{\mathrm{ox}}([\mathrm{L}])}{\partial x}$	(149)

TABLE VI. Mechanism of Electrode Reactions of Some Zinc(II) and Cadmium(II) Complexes (26)

Electrode system[a]	Main species in solution	Species participating in charge transfer	$1 - \alpha$
$Zn(Hg)/Zn^{2+}$, OH^-	$Zn(OH)_4^{2-}$	$Zn(OH)_2$	0.52
$Zn(Hg)/Zn^{2+}$, Ox^{2-}	$Zn(Ox)_3^{4-}$	$Zn(Ox)$ [and Zn_{aq}^{2+}]	0.75
$Zn(Hg)/Zn^{2+}$, CN^-, (OH^-)	$Zn(CN)_4^{2-}$	$Zn(OH)_2$	0.54
$Zn(Hg)/Zn^{2+}$, NH_3, OH^-	$Zn(NH_3)_3(OH)_2$	$Zn(NH_3)_2^{2+}$ [and $Zn(OH)_2$]	0.42
$Cd(Hg)/Cd^{2+}$, CN^-	$Cd(CN)_4^{2-}$	In a large excess of CN^-: $Cd(CN)_3^-$	0.70
	$Cd(CN)_4^{2-}$	In a small excess of CN^-: $Cd(CN)_2$	0.75

[a] Ox^{2-} denotes oxalate ion.

the function $M_i(E)$ ($i = 0, 1, 2, \ldots, \mu$) being dependent on the electrode potential but independent of the ligand concentration.

If the reversible half-wave potential is known, the left-hand side of Eq. 154, being denoted by $L_0(E, [L])$, can be determined experimentally. From a series of $L_0(E, [L])$ versus E curves corresponding to various ligand concentrations (Fig. 11), we can plot $L_0(E, [L])$ against [L] at a given electrode potential as shown in Figure 12. The extrapolation of the $L_0(E, [L])$ versus [L] plot to [L] = 0 then gives the value of $M_0(E)$:

$$\lim_{[L] \to 0} L_0(E, [L]) = M_0(E) \qquad (156)$$

The values of $M_i(E)$ ($i = 1, 2, \ldots, \mu$) can be obtained successively by applying a similar method to that of DeFord and Hume (cf. Section III-A) to the function $L_i(E, [L])$ as defined by

$$L_i(E, [L]) = \frac{\{L_{i-1}(E, [L]) - \lim L_{i-1}(E, [L])\}}{[L]} \qquad (157)$$

Extrapolating these functions to [L] = 0, we can determine $M_i(E)$ according to the relation

$$\lim_{[L] \to 0} L_i(E, [L]) = M_i(E) \qquad (158)$$

TABLE VII. Parameters in Eq. 153 Representing the Average Current Attributable to the Polarographic Reduction of Labile Complexes

Parameter	Electrode reaction
	$M^{m+}L_i \rightleftharpoons M^{m+} + iL$ (equilibrium)
	$M^{m+}L_i + ze + Hg \rightleftharpoons M(Hg) + iL$
	$i = 0, 1, 2, \ldots, \mu$
$\Phi(\Lambda)$	$\left(\sum_{i=0}^{\mu} \Lambda_i \exp\left\{ -\frac{\alpha_i z F}{RT}(E - E_{1/2}^r) \right\} \right)^{-1}$
Λ	$\Lambda_i = (\tilde{k}_i^{\ominus})_B \left(\sum_{i=0}^{\mu} \beta_i^{ox} \right)^{(1-\alpha_i)} (\sqrt{D_M})^{(\alpha_i - 1)} (\sqrt{D_a})^{-\alpha_i}$
	$\times [L]^i \exp\left(\frac{(1-\alpha_i)zF}{RT} \{ E_{1/2}^r - (E_{1/2}^r)_M \} \right)$
$E_{1/2}^r$	$(E_{1/2}^r)_M + \frac{RT}{zF} \ln \sqrt{D_M/\bar{D}} - \frac{RT}{zF} \ln F_0^{ox}([L])$
$\bar{D}$	$\sum_{i=0}^{\mu} \beta_i^{ox} [L]^i D_i / F_0^{ox}([L])$

Parameter	Electrode reaction
	$M^{m+}L_iL'_{i'} \rightleftharpoons M^{m+} + iL + i'L'$ (equilibrium)
	$M^{m+}L_iL'_{i'} + ze + Hg \rightleftharpoons M(Hg) + iL + i'L'$
	$i = 0, 1, 2, \ldots, \mu; i' = 0, 1, 2, \ldots, \mu'$
$\Phi(\Lambda)$	$\left(\sum_{i=0}^{\mu} \sum_{i'=0}^{\mu'} \Lambda_{ii'} \exp\left\{ -\frac{\alpha_{ii'} zF}{RT}(E - E_{1/2}^r) \right\} \right)^{-1}$
Λ	$\Lambda_{ii'} = (\tilde{k}_{ii'}^{\ominus})_B \left(\sum_{i=0}^{\mu} \sum_{i'=0}^{\mu'} \beta_i^{ox} \beta_{i'}^{ox} \right)^{(1-\alpha_{ii'})} (\sqrt{D_M})^{(\alpha_{ii'} - 1)}$
	$\times (\sqrt{D_a})^{-\alpha_{ii'}} [L]^i [L']^{i'}$
	$\times \exp\left(\frac{(1-\alpha_{ii'})zF}{RT} \{ E_{1/2}^r - (E_{1/2}^r)_M \} \right)$
$E_{1/2}^r$	$(E_{1/2}^r)_M + \frac{RT}{zF} \ln \sqrt{D_M/\bar{D}}$
	$- \frac{RT}{zF} \ln \left(\sum_{i=0}^{\mu} \sum_{i'=0}^{\mu'} \beta_{ii'}^{ox} [L]^i [L']^{i'} \right)$
$\bar{D}$	$\sum_{i=0}^{\mu} \sum_{i'=0}^{\mu'} \beta_{ii'}^{ox} [L]^i [L']^{i'} D_{ii'} \Big/ \sum_{i=0}^{\mu} \sum_{i'=0}^{\mu'} \beta_{ii'}^{ox} [L]^i [L']^{i'}$

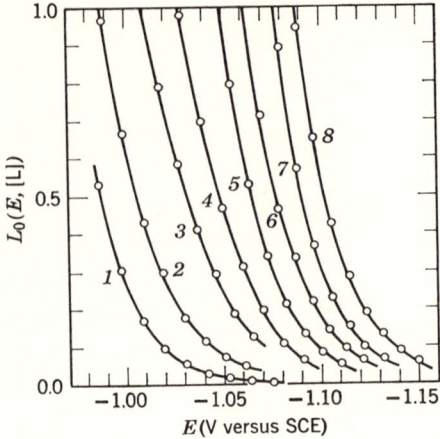

FIG. 11. Plots of $L_0(E, [L])$ against the electrode potential for the polarographic reduction of zinc(II) in acetate solutions at 25°C (ionic strength adjusted to 4 with $NaNO_3$). Concentration of acetate ion: (1) 0; (2) 0.1; (3) 0.4; (4) 0.8; (5) 1.4; (6) 2.2; (7) 3.0; (8) 4.0 M. [Matsuda, H., Y. Ayabe, and K. Adachi, Ber. Bunsenges. Physik. Chem., **67**, 598 (1963).]

FIG. 12. Plots of $L_0(E, [L])$ against the concentration of ligand (acetate ion) for the polarographic reduction of zinc(II) in acetate solutions at 25°C (ionic strength adjusted to 4 with $NaNO_3$). $E =$ (1) -1.000; (2) -1.020; (3) -1.040; (4) -1.050; (5) -1.060; (6) -1.070; (7) -1.080; (8) -1.090; (9) -1.100; (10) -1.110; (11) -1.120 V versus SCE. [Matsuda, H., Y. Ayabe, and K. Adachi, Ber. Bunsenges. Physik. Chem., **67**, 599 (1963).]

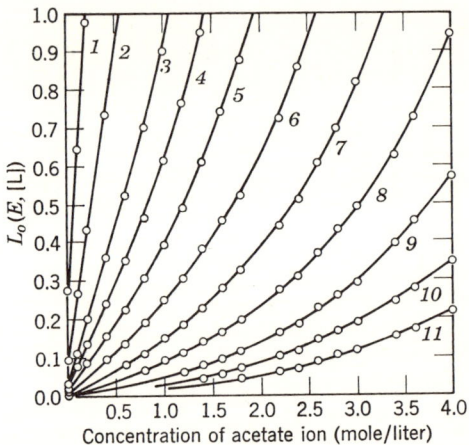

The transfer coefficient α_i ($i = 0, 1, 2, \ldots, \mu$) can be easily determined from the potential dependence of $M_i(E)$ (see Fig. 13) according to the relation

$$\alpha_i = 1 - \frac{RT}{zF} \frac{d \ln M_i(E)}{dE} \tag{159}$$

and the standard rate constants $(\tilde{k}_i^{\ominus})_B$ are also calculated from $M_i(E)$ by means of Eq. 155 if the stability constants of the complex species and the reversible half-wave potential $(E_{1/2}^r)_M$ of the charge transfer process of simple metal ion M^{m+} are available.

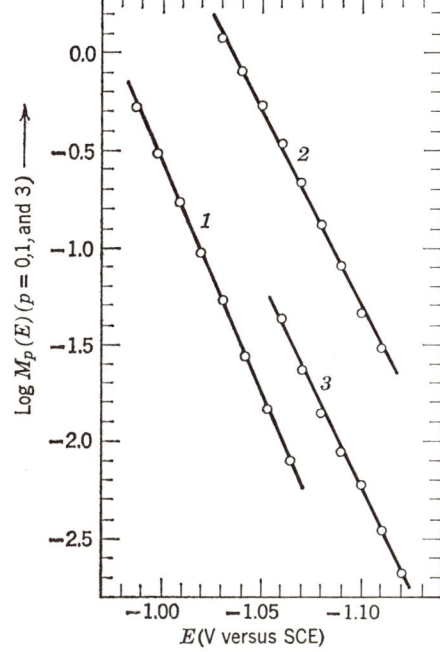

FIG. 13. Plots of log $M_p(E)$ ($p = 0$, 1, and 3) against the electrode potential for the polarographic reduction of zinc(II) in acetate solutions at 25°C (ionic strength adjusted to 4 with $NaNO_3$). (1) log $M_0(E)$, (2) log $M_1(E)$, (3) log $M_3(E)$. [Matsuda, H., Y. Ayabe, and K. Adachi, Ber. Bunsenges. Physik. Chem., **67**, 600 (1963).]

The analysis of zinc(II)–acetate complexes leads to the conclusion that these complexes are reduced at a DME to zinc amalgam through the following three simultaneous charge transfer processes with the kinetic parameters given in Table VIII:

$$Zn^{2+} + 2e + Hg \rightleftharpoons Zn(Hg)$$

$$ZnOAc^+ + 2e + Hg \rightleftharpoons Zn(Hg) + OAc^-$$

$$Zn(OAc)_3^- + 2e + Hg \rightleftharpoons Zn(Hg) + 3OAc^-$$

Similar but slightly simpler treatments were applied by Matsuda and Ayabe (55, 56) to the polarographic reduction processes of the hydroxo and tartrate complexes of zinc(II), the results of which are shown in Table VIII together with the successive dissociation constants of each complex determined by applying the DeFord–Hume method to the reversible half-wave potential.

In most of the irreversible electrode reactions, only the log plot and the half-wave potential $(E_{1/2})_{irr}$ are available for the mechanistic study of the electrode reaction. The change in the half-wave potential with the ligand

TABLE VIII. Mechanistic Information on the Polarographic Reduction of Some Zinc(II) Complexes to Zinc Amalgam at 25°C (55–57)

Cell solution	Complex species present in solution and their dissociation constants	Species participating in the charge transfer process and the corresponding kinetic parameters
Zn(II), 0.1–2 M OH$^-$, Cl$^-$; ionic strength = 2	Zn(OH)$_4^{2-}$; log $\beta_4 = -15.3 \pm 0.1$	Zn(OH)$_2$; $\alpha = 0.42$; log $(\tilde{k}^\ominus)_B = -3.3 \pm 0.05$
Zn(II), 0.1–1.5 M tartrate, Cl$^-$; ionic strength = 1.5; pH = 6.9 ± 0.1	Zn(tart)$_i$; $i = 1, 2, 3, 4$; $K_1 = 2.94 \times 10^{-3}$; $K_2 = 0.213$; $K_3 = 0.356$; $K_4 = 1.73$	Zn$_{aq}^{2+}$; $\alpha_0 = 0.22$; $(\tilde{k}_0^\ominus)_B = 3.4 \times 10^{-5}$ cm sec^{-1} Zn(tart)$_2$; $\alpha_2 = 0.22$; $(\tilde{k}_2^\ominus)_B = 3.4 \times 10^{-4}$ cm liter^2 mole^{-2} sec^{-1}
Zn(II), 0.1–4 M acetate, NO$_3^-$; ionic strength = 4; pH = 6.0 ± 0.1	Zn(OAc)$_i^{(2-i)+}$; $i = 0, 1, 2, 3, 4, 5$; $K_1 = 0.11$; $K_2 = 9(?)$; $K_3 = 0.026$; $K_4 = 7.4$; $K_5 = 0.56$	Zn$_{aq}^{2+}$; $\alpha_0 = 0.30$; $(\tilde{k}_0^\ominus)_B = 1.8 \times 10^{-4}$ cm sec^{-1} Zn(OAc)$^+$; $\alpha_1 = 0.41$; $(\tilde{k}_1^\ominus)_B = 4.4 \times 10^{-3}$ cm liter^2 mole^{-2} sec^{-1} Zn(OAc)$_3^-$; $\alpha_3 = 0.37$; $(\tilde{k}_3^\ominus)_B = 6.2 \times 10^{-4}$ cm liter^3 mole^{-3} sec^{-1}
Zn(II), 0.25–2.8 M ammonia, Cl$^-$; ionic strength = 4; pH < 9.2	Zn(NH$_3$)$_4^{2+}$ (estimated from K-values published in the literature)	Zn(NH$_3$)$_4^{2+}$; $\alpha = 0.65 \pm 0.02$; log $(\tilde{k}^\ominus)_B = -4.9 \pm 0.1$

concentration was examined by Matsuda and Ayabe (55) with the polarographic irreversible reduction of zinc(II) in solutions containing 0.25–2.8 M ammonia, the result of which is expressed by the relation

$$(E_{1/2})_{\text{irr}} = -1.368 - 0.101 \log [\text{NH}_3] \qquad \text{(V versus SCE)}$$

at 25°C and at the ionic strength of 4 adjusted by ammonium chloride. Combining this result with the value of α ($= 0.65 \pm 0.02$) determined by log plot analysis and with the dissociation constants of zinc(II)–ammine complexes reported in the literature, they concluded that the charge transfer process proceeds from the complex species $\text{Zn}(\text{NH}_3)_2^{2+}$, which is in agreement with the conclusion obtained by Gerischer (26) from the analysis of the exchange current (see Table VI).

B. Kinetics of Complex Formation and Substitution Reactions

1. Application of Polarographic Limiting Current and Galvanostatic Transition Time. The analysis of limiting currents and transition times provides the most useful method of determining the rate constants of coupled chemical reactions; the fundamental method of analysis and the information obtained are given in Table IX. In principle, these methods are based upon the fact that the reaction is forced to deviate from its equilibrium state by the continuous and stationary removal of the reacting species through the charge transfer process, and can be applied to the study of relatively rapid chemical reactions with first-order rate constants up to about 500 sec^{-1} (cf. Section II-D, E, and F).

Although the general expression of parameter λ, which is related to the kinetics of coupled chemical reactions, is very much involved and not easily accessible to the analysis, it can be reduced to a much simpler form in some particular cases. Let us assume that the rate-determining step of preceding reaction 1a is the dissociation–association process

$$M^{m+}L_{k-1} + L \underset{\overleftarrow{p}_k}{\overset{\overrightarrow{p}_k}{\rightleftharpoons}} M^{m+}L_k \qquad (160)$$

and that all the other steps in reaction 1a proceed so rapidly that they are practically kept in the equilibrium state even under the flow of current. In such a case parameter λ_{ox} is given by Eqs. 161 and 162, respectively, according to the relative magnitude of the number of ligands p of the complex species $M^{m+}L_p$ involved in the charge transfer process with respect to k of the species $M^{m+}L_k$ (49).

TABLE IX. Kinetic Information Obtained from the Analysis of Limiting Currents and Transition Times

Parameter ξ to be examined	Variable x	Conditions	Information obtained
$\xi = i_l/i_d$	$x = t$	Potentiostatic measurement at a constant potential; no convective flow	$\sqrt{\pi}\lambda\sqrt{t}\exp(\lambda^2 t)\operatorname{erfc}(\lambda\sqrt{t}) = \xi$ $\sqrt{\pi}\lambda = d\xi/d\sqrt{x}$ (when $\lambda\sqrt{t} < 0.03\ \text{sec}^{-1/2}$)
$\xi = i_l/i_d$	—	Polarographic instantaneous current measurement at a given t ($0 < t < t_{dl}$)	$\lambda\sqrt{t} = \dfrac{1}{2(1-\xi)} - \{(2.26\xi - 1.61)$ $\qquad\qquad + \{(1.61 - 2.26\xi) + 5.11\xi(1-\xi)\}^{1/2}\}$ $\cong \dfrac{1}{2(1-\xi)} - \{(2.26\xi - 1.61)$ $\qquad\qquad + (1.35 - 0.13\xi - 0.02\xi^2)^{1/2}\}$
$\xi = \bar{i}_l/(\bar{i}_d - \bar{i}_l)$	—	Polarographic average current measurement	$\lambda\sqrt{t_d} = 1.13\xi$
$\xi = I\sqrt{\tau}$	$x = I$	Galvanostatic measurement at a constant current density; no convective flow	$\sqrt{\pi}/(2\lambda) = -d\xi/dx$

When $k > p$:

$$\lambda_{ox} = \frac{\bar{D}_{ox}}{\bar{D}_{II}} \left(\frac{\bar{D}_{I}}{\bar{D}_{II}}\right)^{1/2} \frac{\sum\limits_{i=0}^{k-1} \beta_i^{ox}[L]^i}{\sum\limits_{i=k}^{\mu} \beta_i^{ox}[L]^i} \left\{\beta_k^{ox}[L]^k \frac{F_0^{ox}([L])}{\left(\sum\limits_{i=0}^{k-1} \beta_i^{ox}[L]^i\right)\left(\sum\limits_{i=k}^{\mu} \beta_i^{ox}[L]^i\right)}\right\}^{1/2} \sqrt{\bar{\rho}_k}$$

(161)

and when $k \leq p$:

$$\lambda_{ox} = \frac{\bar{D}_{ox}}{\bar{D}_{I}} \left(\frac{\bar{D}_{II}}{\bar{D}_{I}}\right)^{1/2} \frac{\sum\limits_{i=k}^{\mu} \beta_i[L]^i}{\sum\limits_{i=0}^{k-1} \beta_i[L]^i} \left\{\beta_k^{ox}[L]^k \frac{F_0^{ox}([L])}{\left(\sum\limits_{i=0}^{k-1} \beta_i^{ox}[L]^i\right)\left(\sum\limits_{i=k}^{\mu} \beta_i^{ox}[L]^i\right)}\right\}^{1/2} \sqrt{\bar{\rho}_k}$$

(162)

where

$$\bar{D}_{I} = \frac{\sum\limits_{i=0}^{k-1} (\beta_i^{ox} D_i)}{\sum\limits_{i=0}^{k-1} \beta_i^{ox}}$$

$$\bar{D}_{II} = \frac{\sum\limits_{i=k}^{\mu} (\beta_i^{ox} D_i)}{\sum\limits_{i=k}^{\mu} \beta_i^{ox}}$$

Equation 161 is reduced to the relation 163 derived by Koryta (37, 39) under the condition that

$$\sum\limits_{i=k}^{\mu} \beta_i^{ox}[L]^i \gg \sum\limits_{i=0}^{k-1} \beta_i^{ox}[L]^i$$

and that the diffusion coefficients of all the species with a number of ligands smaller than $(k - 1)$ are the same:

$$\lambda_{ox} = \left(\frac{D_{k-1}}{\bar{D}_{ox}}\right)^{1/2} \frac{\left(\beta_k[L]^k \sum\limits_{i=0}^{k-1} \beta_i^{ox}[L]^i\right)^{1/2}}{\sum\limits_{i=k}^{\mu} \beta_i^{ox}[L]^i} \sqrt{\bar{\rho}_k} \quad (163)$$

These simplified forms of λ have been applied to the kinetic studies of substitution and complex formation reactions in solution. A typical example may be found in the dissociation reaction of nitrilotriacetatocadmate(II) complexes in aqueous solutions (88). In acetate buffer solutions containing an excess of nitrilotriacetic acid (NTA), cadmium(II) is reduced at a DME to cadmium amalgam in two steps, the first step with a kinetic nature being considered to be attributable to the reduction of cadmium(II) coupled with the dissociation of Cd(II)–NTA complexes. The electrode reaction is represented by the scheme

$$Cd^{2+} + H_jX^{(3-j)-} \rightleftharpoons CdX^- + jH^+ \quad (j = 0, 1, \ldots) \quad (164a)$$

$$Cd^{2+} + OAc^- \rightleftharpoons CdOAc^+ \quad (164b)$$

$$\left\{ \begin{array}{c} Cd^{2+} \\ \updownarrow \\ CdOAc^+ \end{array} \right\} + 2e + Hg \rightleftharpoons \left\{ \begin{array}{c} Cd(Hg) \\ \\ Cd(Hg) + OAc^- \end{array} \right\} \quad (164c)$$

where X^{3-} denotes a tervalent NTA anion, OAc^- an acetate ion, and $\bar{k}_j$ the rate constant of the dissociation reaction of Cd(II)–NTA complexes.

Under the assumption that (a) reaction 164b is rapid enough to maintain the equilibrium state, (b) Cd^{2+} and $CdOAc^+$ are reduced while Cd(II)–NTA complexes are not, at the potentials under consideration, and (c) all the cadmium(II) species have the same diffusion coefficient, the polarographic average limiting current attributable to reaction 164 is represented by Eq. 165 according to Koryta's (37, 39) treatment:

$$\frac{\bar{i}_l}{\bar{i}_d - \bar{i}_l} = 0.886 \left\{ \frac{\bar{\rho} t_d (1 + K_{CdOAc}[OAc^-])}{K_{CdX}[X^{3-}]} \right\}^{1/2} \quad (165a)$$

$$\bar{\rho} = \sum_{j=0}^{\mu} \bar{k}_j[H^+]^j \quad (165b)$$

where $\bar{i}_d$ is the hypothetical average diffusion current corresponding to the total concentration of cadmium(II) species and K_{CdX} and $K_{Cd(OAc)}$ are the formation constants of CdX^- and $CdOAc^+$, respectively.

The overall rate constant $\bar{\rho}$ determined from the limiting current measurement by means of Eq. 165a gives different values with different pH values of the solution; the plot of $\bar{\rho}$ against the hydrogen ion concentration shows a straight line as illustrated in Figure 14, from which the overall rate constant of dissociation reaction 164a at pH 5.3–5.9 and at 25°C can be expressed as

$$\bar{\rho} = 3.4 + 2.5 \times 10^5 [H^+] \quad (\text{sec}^{-1})$$

A similar analysis has been applied to the chronopotentiometric determination of the dissociation rate constant $\bar{\rho}$ of Cd(II)–NTA complexes in solutions of pH 2.8–3.4 containing an excess of NTA (88). When NTA ions are only one complex-forming species present in the solution, the following relation is

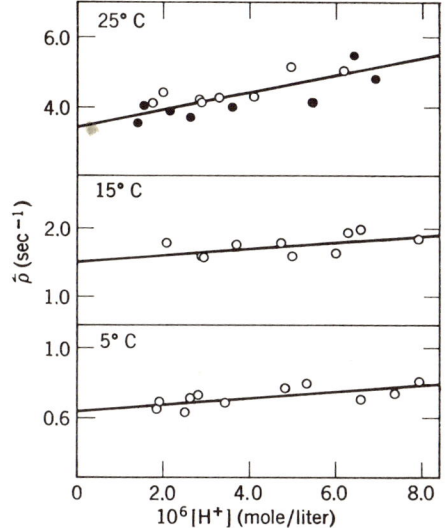

FIG. 14. Overall dissociation rate constants $\bar{\rho}$ of Cd(II)–NTA complexes as a function of hydrogen ion concentration at 5, 15, and 25°C. Total concentration of cadmium(II) = 0.4 mM; total concentration of NTA = 0.01 M; concentration of acetate ion = 0.05 M; ionic strength = 0.2 (adjusted with KNO_3). (○)First experimental run; (●) second experimental run. [Tanaka, N., K. Ebata, T. Takahari, and T. Kumagai, *Bull. Chem. Soc. Japan*, **35**, 1838 (1962).]

obtained for the transition time for the reduction of cadmium(II):

$$I_c\sqrt{\tau} = \sqrt{\pi}F\sqrt{D}c_{ox} - \frac{\sqrt{\pi}\sqrt{K_{CdX}[X^{3-}]}}{2\sqrt{\bar{\rho}}}I_c \quad (166)$$

where I_c is the cathodic current density, D the common value of the diffusion coefficients of various cadmium(II) species, and c_{ox} the total concentration of cadmium(II). Figure 15 presents the plot of $I_c\sqrt{\tau}$ against I_c, from which the rate constant can be calculated. The following conclusion has been derived for $\bar{\rho}$ at pH 2.8–3.4 from the analysis of its pH dependence:

$$\bar{\rho} = 3 \times 10^5 [H^+] + 2 \times 10^8 [H^+]^2 \quad (\text{sec}^{-1})$$

By combining the results obtained from the polarographic and chronopotentiometric methods, it can be concluded that the overall rate constant of the dissociation of Cd(II)–NTA complex in solutions of pH 2–6 containing an excess of NTA is given by

$$\bar{\rho} = 3.4 + 2.5 \times 10^5 [H^+] + 2 \times 10^8 [H^+]^2 \quad (\text{sec}^{-1})$$

This suggests that the dissociation reaction proceeds through the following three simultaneous paths:

$CdX^- \rightarrow Cd^{2+} + X^{3-}$ $k_0 = 3.4 \text{ sec}^{-1}$

$CdX^- + H^+ \rightarrow Cd^{2+} + HX^{2-}$ $k_1 = 2.5 \times 10^5 \text{ liter mole}^{-1} \text{ sec}^{-1}$

$CdX^- + 2H^+ \rightarrow Cd^{2+} + H_2X^-$ $k_2 = 2 \times 10^8 \text{ liter}^2 \text{ mole}^{-2} \text{ sec}^{-1}$

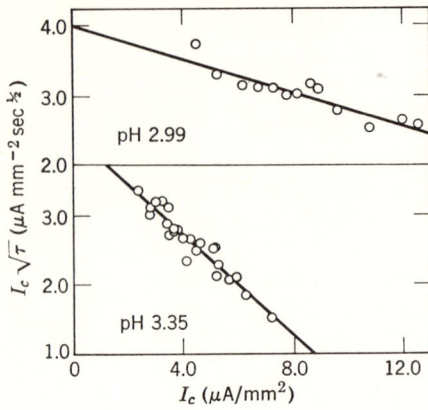

Fig. 15. Plots of $I_c/\sqrt{\tau}$ against I_c for the galvanostatic reduction of Cd-(II)–NTA complexes in acetate buffer solutions containing excess NTA at 25°C. [Tanaka, N., K. Ebata, T. Takahari, and T. Kumagai, *Bull. Chem. Soc. Japan*, **35**, 1839 (1962).]

The analysis of limiting currents and transition times attributable to the electrode reactions with the ECE mechanism can be applied to the kinetic study of some substitution reactions (2, 21, 28, 64, 65, 66, 83, 87, 113, 114); the method is considered to be particularly suitable for the reactions of unstable species which cannot exist in solution under normal conditions but can be formed only at the electrode surface as the result of a charge transfer process. Consider the electrode reaction

$$O + z_1 e \rightarrow R \tag{167a}$$

$$R + X \underset{\overleftarrow{\rho}}{\overset{\overrightarrow{\rho}}{\rightleftharpoons}} RX \tag{167b}$$

$$RX \rightarrow OX + z_2 e \tag{167c}$$

in which reaction 167b is the following reaction of the reduction process of O. If the anodic rate constant of the charge transfer process of Eq. 167c is sufficiently large at the potentials at which the reduction of O gives the limiting current, the effect of the following reaction results in a decrease in the limiting current $\bar{i}_l$ attributable to the reduction of O; $\bar{i}_l$ becomes smaller than $\bar{i}_d$ as shown in Figure 16, $\bar{i}_d$ being the diffusion current that should be observed either in the absence of reactions 167b and 167c or when rate constant $\overrightarrow{\rho}$ of reaction 167b is extremely small.

The theoretical equations for the limiting current of electrode reaction 167 were derived by Alberts and Shain (2) and discussed in detail by Ebata (21). These equations were applied by Ogino and Tanaka to the one-electron reduction wave of hexaamminochromium(III) ions at a DME in solutions

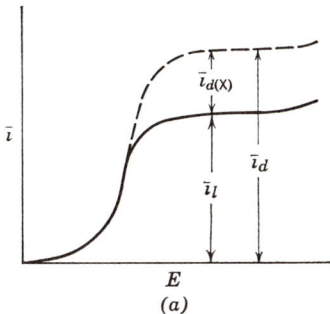 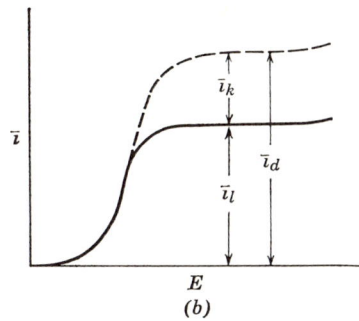

FIG. 16. Polarographic limiting current, $\bar{i}_l$, of O from the reaction scheme of Eq. 167: (a) when [X] < [O] and $\vec{\rho}$ is sufficiently large; (b) when [X] ≫ [O] and $\vec{\rho}$ is relatively small. $\bar{i}_{d(X)}$, Current controlled by the diffusion of X; $\bar{i}_k$, current controlled by the rate of the following reaction of Eq. 167b.

containing manganese(II) and ethylenediaminetetraacetic acid (EDTA) (70); the electrode reaction can be represented by

$$Cr(NH_3)_6^{3+} + e \rightarrow Cr^{2+} + 6NH_3 \tag{168a}$$

$$Cr^{2+} + MnY^{2-} \rightleftharpoons CrY^{2-} + Mn^{2+} \tag{168b}$$

$$CrY^{2-} \rightleftharpoons CrY^{-} + e \tag{168c}$$

where Y^{4-} is an ethylenediaminetetraacetate anion. The primary product, $Cr(NH_3)_6^{2+}$, of the reduction of $Cr(NH_3)_6^{3+}$ at a DME may be subject to a rapid aquation; the aquation is known to be faster than the direct substitution reaction between $Cr(NH_3)_6^{2+}$ and MnY^{2-} (87). Therefore the reaction of chromium(II) with MnY^{2-} is considered to proceed via an aquo complex. The rate constants of the substitution reaction of Eq. 168b were determined from analysis of the polarographic limiting current, and then those of the dissociation–association reaction

$$Cr^{2+} + HY^{3-} \underset{\overleftarrow{k}_1}{\overset{\vec{k}_1}{\rightleftharpoons}} CrY^{2-} + H^+$$

were calculated to be $\vec{k}_1 = 8 \times 10^7$ liter mole^{-1} sec^{-1} and $\overleftarrow{k}_1 = 4 \times 10^4$ liter mole^{-1} sec^{-1} at an ionic strength of 1.0 and at 25°C.

Galvanostatic (chronopotentiometric) methods have proved to be useful when applied to the study of electrode reactions with the ECE mechanism (28, 111, 113, 114). The theoretical equations representing the potential–time relationship and the transition time attributable to the ECE mechanism are relatively complicated, and the original papers should be consulted for details. Tanaka and Yamada (111) determined the rate constants of complex formation reactions between chromium(II) and EDTA at 25°C,

$$Cr^{2+} + HY^{3-} \xrightarrow{k_{HY}} CrY^{2-} + H^+$$

$$k_{HY} = 2.2 \times 10^8 \text{ liter mole}^{-1} \text{ sec}^{-1}$$

$$Cr^{2+} + H_2Y^{2-} \xrightarrow{k_{H_2Y}} CrY^{2-} + 2H^+$$

$$k_{H_2Y} = 2.5 \times 10^6 \text{ liter mole}^{-1} \text{ sec}^{-1}$$

from the chronopotentiometric study of the cathodic reaction of hexaamminochromium(III) ions at a DME in acetate buffer solutions (pH = 3–5) containing calcium and EDTA ions. The overall electrode reaction of this system is considered to be represented by similar scheme to reaction 168:

$$Cr(NH_3)_6^{3+} + e \rightarrow Cr(NH_3)_6^{2+} \tag{169a}$$

$$Cr(NH_3)_6^{2+} + EDTA + 6H^+ \underset{\overleftarrow{\rho}}{\overset{\overrightarrow{\rho}}{\rightleftharpoons}} Cr(II)\text{--}EDTA + 6NH_4^+ \tag{169b}$$

$$Cr(II)\text{--}EDTA \rightarrow Cr(III)\text{--}EDTA + e \tag{169c}$$

If the conditions $K = \overrightarrow{\rho}/\overleftarrow{\rho} \gg 1$ and $(\overrightarrow{\rho} + \overleftarrow{\rho})t \gg 1$ are satisfied, the transition time is represented by

$$\tau = \frac{zF[Cr(NH_3)_6^{3+}]\sqrt{D_{Cr(NH_3)_6^{3+}}}}{\sqrt{\overrightarrow{\rho}}} \frac{1}{I_c} - \frac{1}{2\overrightarrow{\rho}} \tag{170}$$

and the plot of τ against $1/I_c$ is expected to give a straight line, which was verified experimentally as shown in Figure 17. The overall rate constant $\overrightarrow{\rho}$ can then be obtained from the slope of the plot.

By considering the ionic species of EDTA and Ca(II)–EDTA complexes present in the solution, reaction 169b can be assumed to proceed according to the mechanism

$$Cr(NH_3)_6^{2+} + \begin{Bmatrix} H_4Y \\ \updownarrow \\ H_3Y^- \\ \updownarrow \\ H_2Y^{2-} \\ \updownarrow \\ HY^{3-} \\ \updownarrow \\ Y^{4-} \end{Bmatrix} \rightleftharpoons \begin{Bmatrix} CaHY^- \\ \updownarrow \\ CaY^{2-} \end{Bmatrix} + qH^+ \overset{\overrightarrow{\rho}}{\rightleftharpoons} \begin{Bmatrix} CrHY^- \\ \updownarrow \\ CrY^{2-} \end{Bmatrix} + 6NH_4^+ \tag{171}$$

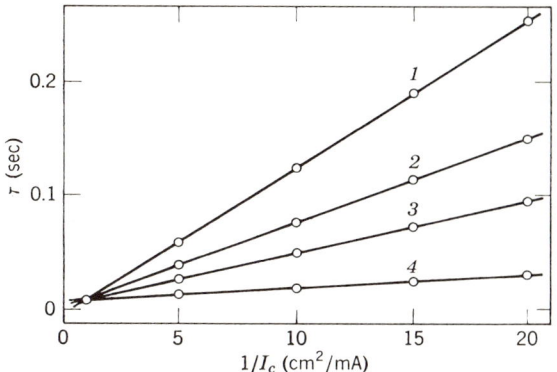

FIG. 17. Transition time τ versus $1/I_c$ relationship for the cathodic reaction of 0.500 mM [Cr(NH$_3$)$_6$]Cl$_3$ in 0.05 M acetate buffer solutions containing 5.50 mM EDTA and 0.14 M Ca(ClO$_4$)$_2$ at 25°C (ionic strength adjusted to 0.5 with NaClO$_4$). (1) pH 5.08; (2) pH 4.63; (3) pH 4.35; (4) pH 3.43. [Tanaka, N., and A. Yamada, *Rev. Polarog. (Kyoto)*, **14**, 242 (1967).]

and the overall rate constant $\vec{\rho}$ may be described by

$$\vec{\rho} = k_{\text{CaY}}[\text{CaY}^{2-}] + k_{\text{CaHY}}[\text{CaHY}^-] + \sum_{j=0}^{4} k_{\text{H}_j\text{Y}}[\text{H}_j\text{Y}^{(4-j)-}] \quad (172)$$

where k is the second-order rate constant of the reaction involving the species specified by the subscript. Introducing the successive and overall dissociation constants of EDTA, $K_{\text{H}_j\text{Y}}$ and $\beta_{\text{H}_j\text{Y}}$, respectively, and the stability constants of CaY^{2-} and CaHY$^-$, we can define the function $G_0([\text{H}^+])$ from Eq. 172:

$$G_0([\text{H}^+]) \equiv \frac{\vec{\rho}}{[\text{CaY}^{2-}]} = \left(k_{\text{CaY}} + \frac{k_\text{Y}}{K_{\text{CaY}}[\text{Ca}^{2+}]} \right)$$

$$+ \left(k_{\text{CaHY}} K_{\text{CaHY}}^\text{H} + \frac{k_{\text{HY}}}{K_{\text{HY}} K_{\text{CaY}}[\text{Ca}^{2+}]} \right)[\text{H}^+]$$

$$+ \sum_{j=2}^{4} \frac{k_{\text{H}_j\text{Y}}}{\beta_{\text{H}_j\text{Y}} K_{\text{CaY}}[\text{Ca}^{2+}]}[\text{H}^+]^j \quad (173)$$

Each rate constant k can be determined successively by analyzing the $G_0([\text{H}^+])$ function according to a method similar to that of DeFord and Hume.

The analysis of polarographic catalytic current, which is considered to be a special type of kinetically controlled limiting current, can be also applied to the study of reactions involving unstable complex species. We consider, for example, the reduction process of O to R coupled with an irreversible catalytic regeneration of O at the electrode surface:

$$O + ze \rightarrow R \tag{174a}$$

$$R + Z \xrightarrow{\vec{\rho}} O \tag{174b}$$

The theory of the catalytic current was developed by Delahay and Stiehl (20) and Koutecký (41), and the results were discussed and summarized in articles by Koutecký and Koryta (43) and of Brdička et al. (9). Under some simplified conditions the average limiting current $\bar{i}_l$ attributable to the electrode reaction of Eq. 174 is given by

$$\frac{\bar{i}_l}{\bar{i}_d} = 0.81 \sqrt{\vec{\rho} t_d} \tag{175}$$

where $\bar{i}_d$ is the hypothetical diffusion current that should be observed in the absence of the coupled catalytic reaction, and $\vec{\rho}$ the pseudo-first-order rate constant of reaction 174b.

The rate constant of the oxidation of Cr(II)–EDTA by nitrate ions was determined by Tanaka and Ito to be 30 liter mole^{-1} sec^{-1} at 0°C and at the ionic strength of 0.1 from the analysis of the polarographic catalytic current attributable to the reduction of ethylenediaminetetraacetato-aquochromium(III) ions in acetate buffer solutions containing nitrate ions (89).

2. Ac Polarography and Related Methods. The application of ac polarography and related methods to the kinetic study of labile complexes is a relatively recent development, and only a few examples have been reported in the literature.

The theoretical treatment of Aylward, Hayes, and Tamamushi (5) of the peak admittance of an ac polarographic wave of a charge transfer process followed by an irreversible first-order (or pseudo-first-order) chemical reaction predicts the relation

$$\vec{\rho} = \frac{(2\sqrt{Y_d/Y_f} - 1)^4}{1.34 \, t_d} \tag{176}$$

where $\vec{\rho}$ is the first-order rate constant of the following reaction, Y_d the admittance at the summit potential in the absence of the following reaction, and Y_f the admittance at the summit potential in the presence of the following reaction. This equation holds under the assumption that the charge transfer process is reversible in the dc polarographic sense, and that the rate constant $\vec{\rho}$ of the following reaction is greater than 10 sec^{-1} but less than the angular frequency ω of the superimposed alternating voltage.

The ac polarographic behavior of the oxidation of cadmium amalgam was examined by Aylward and Hayes (3) in acetate buffer solutions containing calcium and EDTA ions, and the rate constant of the reaction

$$\text{Cd}^{2+} + \text{HY}^{3-} \xrightarrow{\vec{k}} \text{CdHY}^{-}$$

was determined to be 6.1×10^8 liter mole^{-1} sec^{-1} at 25°C and at the ionic strength of 0.5 by means of Eq. 176. The rate constant thus obtained was proved to be practically independent of ω if $\vec{k}[\text{HY}^{3-}] < \omega$, and is in relatively good agreement with the values determined by other methods (see Table 10).

A similar treatment was applied by the same authors to the electrode reaction

$$\text{Eu(III)-Y}^{-} + e \rightleftharpoons \text{Eu(II)-Y}^{2-} \qquad (177a)$$

$$\text{Eu(II)-Y}^{2-} + \text{Ca}^{2+} \xrightarrow{\vec{k}} \text{Eu}^{2+} + \text{CaY}^{2-} \qquad (177b)$$

in an ammonia–ammonium chloride buffer solution of pH 9.3 containing Ca–EDTA complex (4). The rate constant of the substitution reaction (Eq. 177b) calculated from the ac measurement was reported to be approximately 1/10 of the value ($\vec{k} = 1580 \pm 180$ liter mole^{-1} sec^{-1}) determined from the dc polarographic half-wave potential according to the equation derived by Koryta and Zábranský (40). The ac method also gave different values of $\vec{k}$ with different ionic strengths, calcium ion concentrations, and frequencies. A tentative explanation of these discrepancies has been offered in terms of the effect of double-layer structure on the magnitude of the faradaic admittance and its phase angle.

The applicability of Smith's (78–80) rigorous treatment to the kinetic study of preceding chemical reactions was examined by Matsuda and Tamamushi (59) with the Cd(II)–EDTA system. From the frequency dependence of the faradaic impedance and phase angle ϕ of the ac polarographic reduction of cadmium(II) in acetate buffer solutions containing calcium and EDTA ions (Fig. 18), the rate constant of the reaction

$$\text{Cd}^{2+} + \text{HY}^{3-} \rightarrow \text{CdY}^{2-} + \text{H}^{+}$$

was calculated to be $2._3 \times 10^9$ liter mole^{-1} sec^{-1} at 25°C and at the ionic strength of 0.5, which is in good agreement with the value obtained by

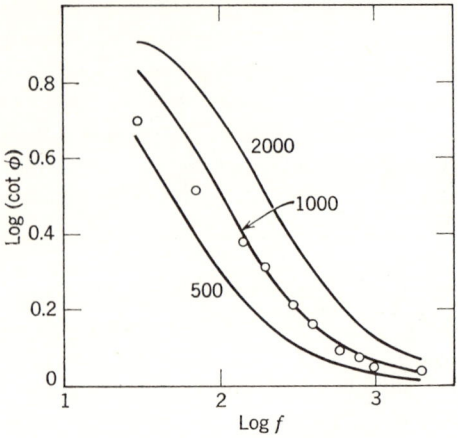

Fig. 18. Relation between log (cot ϕ) and log f (f, frequency) at the summit potential of the Cd(II)–EDTA system. (○) Experimental values; (—) theoretical curves calculated with the values of ($\vec{\rho} + \overleftarrow{\rho}$) given on each curve. [Matsuda, K., and R. Tamamushi, Bull. Chem. Soc. Japan, **41**, 1567 (1968).]

Fujisawa and Tanaka (23) from the polarographic limiting current under similar conditions (see Table 10).

In principle, the analysis of the ac behavior is expected to provide a useful method of studying the mechanism of electrode reaction and the kinetics of coupled chemical reactions involving labile complexes. In practice, however, ac polarography and related methods are expected to have little advantage over the dc methods when applied to the determination of rates of coupled chemical reactions unless further knowledge is available concerning the effect of the double-layer structure, this effect being more and more pronounced at higher frequencies. In this respect, the comment by Aylward and Hayes (3) that "studies of the rates of coupled chemical reactions by ac polarography, as well as throwing further light on electrochemical mechanisms, may help in the elucidation of double-layer structure" is likely true.

3. Application of Dc Polarography to the Kinetic Study of Labile Complexes. If a chemical reaction involves one or more substances that give polarographic waves, it can be followed by measuring the limiting current of the wave.

In such experiments the reaction is usually started by injecting a small volume of a deoxygenated solution containing one of the reactants into a solution of other reactants in a polarographic cell. Mixing is accomplished by bubbling deoxygenating gas for a short time. The cell should be designed so that the agitated solution calms down quickly. The limiting current at a suitable potential is then recorded in order to follow production or consumption of the polarographically active species. This technique was first applied by Ackermann and Schwarzenbach (1) to the study of the exchange reactions of ethylenediaminetatraacetato complexes and has been extensively used by Tanaka and others (11, 32–34, 68, 69, 90, 93, 95–97, 100, 104, 105).

Diffusion currents are mostly used for this purpose, but limiting currents of other types (e.g., kinetic or catalytic currents) may possibly be utilized provided the relationship between the current and the concentration is fully established under the experimental condition.

By this technique reactions having half-lives longer than about 15 sec can be followed. Faster reactions can be followed by analyzing instantaneous currents of a DME. During a reaction the diffusion current of a reactant depends not only on the time that has elapsed after a drop has begun to grow but also on the time elapsed after the reaction started. According to Reinert (73), the reaction-controlled diffusion current i_{rd} at the time t_r after the reaction has started is represented as the product of two terms, $i_d(\Delta t)$ and $f(t_r)$:

$$i_{rd}(t_r) = i_d(\Delta t) f(t_r) \tag{178}$$

with

$$t_r = Nt_d + \Delta t - t_0 \tag{179}$$

where t_0 is the time interval between the start of the reaction and the beginning of growth of the zeroth drop and Δt is the time of electrolysis at the Nth drop (see Fig. 19). The term $i_d(\Delta t)$ represents the value of the diffusion current observed at time Δt of a drop when no reaction has occurred. The form of the function $f(t_r)$ depends on the type of reaction and has been determined for several reaction schemes; for example, it is an exponential

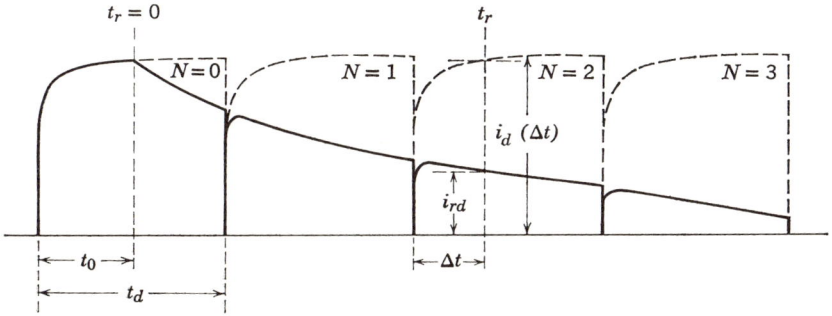

FIG. 19. Instantaneous diffusion currents at a DME during a reaction (schematic).

TABLE X. Rate Constants for the Dissociation–Association Reaction $M^{m+}L + jH^+ \underset{\overleftarrow{k_j}}{\overset{\overrightarrow{k_j}}{\rightleftharpoons}} M^{m+} + H_jL$ ($j = 0, 1, 2, \ldots$) Determined by Polarography and Related Electrochemical Methods[a]

Metal ion (M^{m+})	Ligand[b] (L)	Temp., °C	Ionic strength	$\overrightarrow{k_j}$		$\overleftarrow{k_j}$		Method[c]	Reference
Cd^{2+}	EDTA	25	0.5	1.1×10^3	($j = 1$)	2.3×10^9	($j = 1$)	$\bar{v}_l$	23
	EDTA	25	0.5	—		4.0×10^9	($j = 1$)	$E_{1/2}$	3
	EDTA	25	0.5	—		5.3×10^8	($j = 1$)	$E_{1/2}$	3
	EDTA	25	0.5 (pH 4.5)	—		6.1×10^8	($j = 1$)	Y_f	3
	EDTA	25	0.5 (pH 3.7)	—		2.3×10^9	($j = 1$)	ϕ	59
	EDTA	25	0.5 (pH 2.9–4.4)	—		3.7×10^9	($j = 1$)	$\bar{v}_l$	109
	EDTA	25	0.1	4.0×10^2	($j = 1$)	8.5×10^8	($j = 1$)	$E_{1/2}$	40
	EDTA	25	0.1	—		6.4×10^9	($j = 1$)	$E_{1/2}$	3
	NTA	25	0.2 (pH 5.3–5.9)	3.4	($j = 0$)	4.3×10^{10}	($j = 0$)	$\bar{v}_l$	29, 88
	NTA	25	0.1 (pH 4.5–6.0)	1.6	($j = 0$)	2.0×10^{10}	($j = 0$)	$\bar{v}_l$	29, 36, 38
	NTA	25	3.0 (pH 4.6–5.7)	$(1.5 \pm 0.3) \times 10^6$	($j = 1$)	—		$\bar{v}_l$	71
	NTA	25	0.3	$(1.3 \pm 0.3) \times 10^6$	($j = 1$)	—		$\bar{v}_l$	71
	NTA	25	0.2 (pH 5.3–5.9)	2.5×10^5	($j = 1$)	5.9×10^5	($j = 1$)	$\bar{v}_l$	29, 88
	NTA	25	0.1 (pH 4.5–6.0)	3.0×10^5	($j = 1$)	7.0×10^5	($j = 1$)	$\bar{v}_l$	29, 36, 38
	NTA	25	0.1 (pH 4.2–4.9)	$(6.5 \pm 1.7) \times 10^4$	($j = 1$)	—		$\bar{v}_l$	63
	NTA	25	0.2 (pH 2.8–3.4)	2×10^8	($j = 2$)	—		τ	88
	HEDTA	25	0.5	4×10^5	($j = 1$)	3×10^8	($j = 1$)	$\bar{v}_l$	24

Ion	Ligand	T (°C)	μ	$\overleftarrow{k}_j$	$\overrightarrow{k}_j$	Method	Ref.
Co^{2+}	EDTA	25	0.2		3×10 ($j = 1$)	8×10^6 ($j = 1$)	$\bar{i}_d$	102
Cr^{2+}	EDTA	25	1.0		4×10^4 ($j = 1$)	8×10^7 ($j = 1$)	$\bar{i}_l$	70, 102
	EDTA	25	0.5 (pH 3–5)	—	2.2×10^8 ($j = 1$)	τ	111
	EDTA	25	0.5 (pH 3–5)	—	2.5×10^6 ($j = 2$)	τ	111
	HEDTA	25	1.0 (pH 4.1–5.1)	—	5.7×10^{10} ($j = 0$)	$\bar{i}_l$	29
	HEDTA	25	1.0 (pH 4.1–5.1)	—	1.2×10^5 ($j = 1$)	$\bar{i}_l$	29
Cu^{2+}	EDTA	25	0.2	9.5 ($j = 1$)	3×10^9 ($j = 1$)	$\bar{i}_d$	93
Mn^{2+}	NTA			5.3 ($j = 0$)	—	$\bar{i}_l$	8
	NTA			$10^{6.96}$ ($j = 1$)	—	$\bar{i}_l$	8
Ni^{2+}	EDTA	25	0.2	5×10^{-5} ($j = 1$)	1.1×10^4 ($j = 1$)	$\bar{i}_d$	108
	NTA	25	1.25	3.5×10^{-6} ($j = 0$)	4.8×10^5 ($j = 0$)	—	12, 29, 97
	NTA	25	1.25	4.34×10^{-1} ($j = 1$)	7.5 ($j = 1$)	—	12, 29, 97
	EDTA	0	0.2	8.0×10^{-7} ($j = 0$)	5.6×10^4 ($j = 0$)	$\bar{i}_d$	97
	EDTA	0	0.2	1×10^{-6} ($j = 0$)	8.7×10^4 ($j = 0$)	$\bar{i}_l$	97
	EDTA	0	0.2	—	4 ($j = 1$)	$\bar{i}_d$	97
Pb^{2+}	EDTA	25	0.5	0.7×10^3 ($j = 1$)	2.4×10^{10} ($j = 1$)	$\bar{i}_l$	23
	EDTA	25	0.2	2.2×10^2 ($j = 1$)	1.1×10^{10} ($j = 1$)	$\bar{i}_d$	100
	HEDTA	25	0.5	0.75×10^5 ($j = 1$)	1.6×10^{10} ($j = 1$)	$\bar{i}_l$	24
Zn^{2+}	EDTA	25	0.5	7×10^2 ($j = 1$)	1.6×10^9 ($j = 1$)	$\bar{i}_l$	23

[a] Rate constants $\overleftarrow{k}_j$ for $j = 0$ in sec^{-1}, $\overleftarrow{k}_j$ for $j = 0$ and 1 in liter mole^{-1} sec^{-1}, and $\overrightarrow{k}_j$ for $j = 2$ in liter^2 mole^{-2} sec^{-1}.

[b] EDTA, ethylenediaminetetraacetate ion; NTA, nitrilotriacetate ion; HEDTA, hydroxyethylethylenediaminetriacetate ion.

[c] $\bar{i}_d$, Polarographic diffusion current; $\bar{i}_l$, polarographic limiting current; $E_{1/2}$, half-wave potential; τ, chronopotentiometric transition time; Y_f, faradaic admittance; ϕ, phase angle.

function when the depolarizer is consumed by a first-order reaction (7, 73):

$$i_{rd} = i_d(\Delta t) \exp(-k_1 t_r) \tag{180}$$

where k_1 is the rate constant.

The rate constant is determined by comparing the current–time curves of single drops recorded during a reaction with those recorded in the absence of the reaction. This technique was used in the study of photoreactions (6). The experimental difficulty with fast chemical reactions lies in the starting of reaction.

With reactions of second or higher order, the rate of the change in concentration of reactants is decreased by reducing the initial concentrations. Thus the use of an electrode more sensitive than a DME can extend the applicability of the polarographic method to reactions of larger rate constants. A rotated DME has been used to investigate the reaction between nickel(II) and ethylenediaminetetraacetate ions at 0°C (108).

Stationary electrodes (disk, wedge, cone, and tube electrodes) placed in laminar flow were investigated by Matsuda (50–53). Such electrodes are useful in the study of fast reactions; they can be used as detectors in the flow method.

Table 10 summarizes the rate constants of the dissociation–association reactions

$$M^{m+}L + jH^+ \underset{\overleftarrow{k_j}}{\overset{\overrightarrow{k_j}}{\rightleftharpoons}} M^{m+} + H_jL \qquad (j = 0, 1, 2, \ldots)$$

determined by polarography and related electrochemical methods.

List of Symbols

A:	Surface area of electrode
C_s:	Electrolytic (or pseudo-) capacity
c:	Total concentration of the species specified by the subscript
D:	Diffusion coefficient of the single species specified by the subscript [subscript "M" refers to a simple metal ion M^{m+}, and subscript "i" to the species $M^{m+}L_i$ ($i = 1, 2, \ldots, p, \ldots$)]
$\bar{D}$:	Average diffusion coefficient weighted with respect to the stability constant of each species involved
E:	Electrode potential referred to a given reference electrode
E_e:	Equilibrium potential
$E^{\ominus}$:	Standard potential [subscript "M" refers to the charge transfer process involving a simple metal ion M^{m+}, and

	subscript "i" to the charge transfer process involving $M^{m+}L_i$ ($i = 1, 2, \ldots, p, \ldots$)]
$(E^{\ominus})_B$:	Standard potential of the overall electrode reaction
$E_{1/2}$:	Polarographic half-wave potential
$E^r_{1/2}$:	Polarographic reversible half-wave potential
$(E^r_{1/2})_M$:	Polarographic reversible half-wave potential of the charge transfer process $M^{m+} + ze + Hg = M(Hg)$
F:	Faraday constant
I:	Current density
I_d:	Diffusion current density
I_k:	Kinetic current density
I_l:	Limiting current density
I^0:	Exchange current density [subscript "i" refers to a charge transfer process involving $M^{m+}L_i$ ($i = 0, 1, 2, \ldots, p, \ldots$)]
i:	Instantaneous current
i_d:	Instantaneous diffusion current
i_k:	Instantaneous kinetic current
i_l:	Instantaneous limiting current
$\bar{i}$:	Polarographic average current
$\bar{i}_d$:	Polarographic average diffusion current
$\bar{i}_k$:	Polarographic average kinetic current
$\bar{i}_l$:	Polarographic average limiting current
$i(\omega t)$:	Sinusoidal alternating current of angular frequency ω
K:	Equilibrium constant
$\vec{k}, \vec{k}$:	Rate constants of chemical reactions
$\tilde{k}^a, \tilde{k}^c$:	Anodic and cathodic electrochemical rate constants, respectively [subscript "i" refers to the charge transfer process involving $M^{m+}L_i$ ($i = 0, 1, 2, \ldots, p, \ldots$)]
k^a, k^c:	Values of $\tilde{k}^a$ and $\tilde{k}^c$ at $E = 0$, respectively
$(\tilde{k}^a)_B, (\tilde{k}^c)_B$:	Overall electrochemical rate constants [subscript "i" refers to the charge transfer process involving $M^{m+}L_i$ ($i = 0, 1, 2, \ldots, p, \ldots$)]
$\tilde{k}^{\ominus}$:	Electrochemical standard rate constant [subscript "i" refers to the charge transfer process involving $M^{m+}L_i$ at E_i^{θ} ($i = 0, 1, 2, \ldots, p, \ldots$)]
$(\tilde{k}^{\ominus})_B$:	Electrochemical standard rate constant at $(E^{\ominus})_B$ [subscript "i" refers to the charge transfer process involving $M^{m+}L_i$ ($i = 0, 1, 2, \ldots, p, \ldots$)]
Q:	Parameter concerning the kinetics of electrode reaction
R:	Gas constant
R_{ct}:	Charge transfer resistance

R_m: Mass transfer resistance
R_s: Polarization resistance
T: Thermodynamic temperature
t: Time (of electrolysis or reaction)
t_d: Drop time of a DME
u: Arbitrary variable
x: Distance from the electrode surface to the bulk of solution
Y: Admittance
Z_m: Mass transfer impedance
z: Number of electrons participating in the electrode reaction
α: Cathodic transfer coefficient [subscript "i" refers to the charge transfer process involving $M^{m+}L_i$ ($i = 0, 1, 2, \ldots, \mu$)]
β: Overall stability constant, unless otherwise stated [subscripts "i" and "j" refer to the complex species $M^{m+}L_i$ ($i = 0, 1, 2, \ldots, \mu$) and $M^{(m-z)+}L_j$ ($j = 0, 1, 2, \ldots, \nu$), respectively]
η: Overvoltage
Λ: Parameter related to the kinetics of electrode reaction [subscript "i" refers to the charge transfer process involving $M^{m+}L_i$ ($i = 0, 1, 2, \ldots, p, \ldots$)]
λ: Parameter related to the kinetics of coupled chemical reactions (subscript "ox" refers to reactions involving the oxidized species, and subscript "red" to reactions involving the reduced species)
$\vec{\rho}, \overleftarrow{\rho}$:   First-order (or pseudo-first-order) rate constants of coupled chemical reactions
τ: Chronopotentiometric transition time
ϕ: Phase angle
ω: Angular frequency
[ ]:   Molar concentration of a single species in the bulk of the solution
[ ]$_s$:   Molar concentration of a single species at the electrode surface

References

1. Ackermann, H., and G. Schwarzenbach, *Helv. Chim. Acta,* **35**, 485 (1952).
2. Alberts, G. S., and I. Shain, *Anal. Chem.,* **35**, 1859 (1963).
3. Aylward, G. H., and J. W. Hayes, *Anal. Chem.,* **37**, 195 (1965).
4. Aylward, G. H., and J. W. Hayes, *Anal. Chem.,* **37**, 197 (1965).
5. Aylward, G. H., J. W. Hayes, and R. Tamamushi, in *Proceedings of the 1st*

Australian Conference on Electrochemistry (J. A. Friend, F. Gutmann, and J. W. Hayes, Eds.), Pergamon Press, Oxford, 1965, p. 323.
6. Berg, H., Z. Chem., **2**, 237 (1962).
7. Berg, H., and H. Kapulla, Z. Elektrochem., **64**, 44 (1960).
8. Biernat, J., and J. Koryta, Collection Czech. Chem. Commun., **25**, 38 (1960).
9. Brdička, R., V. Hanuš, and J. Koutecký, in Progress in Polarography, Vol. 1 (P. Zuman and I. M. Kolthoff, Eds.), Interscience, New York, 1962, p. 145.
10. Brdička, R., and K. Wiesner, Collection Czech. Chem. Commun., **12**, 138 (1947).
11. Bril, K., S. Bril, and P. Krumholz, J. Phys. Chem., **59**, 596 (1955).
12. Bydalek, T. J., and M. L. Blomster, Inorg. Chem., **3**, 667 (1964).
13. Crow, D. R., Polarography of Metal Complexes, Academic Press, London, 1969.
14. Crow, D. R., and J. V. Westwood, Quart. Rev. (London), **19**, No. 1, 57 (1965).
15. Crow, D. R., and J. V. Westwood, Polarography, Methuen, London, 1968, Chap. 5.
16. DeFord, D. D., and D. N. Hume, J. Am. Chem. Soc., **73**, 5321 (1951).
17. Delahay, P., New Instrumental Methods in Electrochemistry, Interscience, New York, 1954.
18. Delahay, P., in Advances in Electrochemistry and Electrochemical Engineering, Vol. 1 (P. Delahay, Ed.), Interscience, New York, 1961, p. 233.
19. Delahay, P., and T. Berzins, J. Am. Chem. Soc., **75**, 2486 (1953).
20. Delahay, P., and G. L. Stiehl, J. Am. Chem. Soc., **74**, 3500 (1952).
21. Ebata, K., Sci. Rept. Tohoku Univ., Ser. I, **47**, 191 (1964).
22. Flemming, J., and H. Berg, J. Electroanal. Chem., **8**, 291 (1964).
23. Fujisawa, T., and N. Tanaka, Nippon Kagaku Zasshi (J. Chem. Soc. Japan, Pure Chem. Sect.), **87**, 965 (1966).
24. Fujisawa, T., and N. Tanaka, Nippon Kagaku Zasshi (J. Chem. Soc. Japan, Pure Chem. Sect.), **88**, 734 (1967).
25. Gerischer, H., Z. Physik. Chem. (Leipzig), **198**, 286 (1951).
26. Gerischer, H., Z. Elektrochem., **57**, 604 (1953).
27. Hashitani, T., unpublished data.
28. Herman, H. B., and A. J. Bard, J. Phys. Chem., **70**, 396 (1966).
29. Hikichi, H., and N. Tanaka, Nippon Kagaku Zasshi (J. Chem. Soc. Japan, Pure Chem. Sect.), **88**, 1154 (1967).
30. Hume, D. N., D. D. DeFord, and G. C. Cave, J. Am. Chem. Soc., **73**, 5323 (1951).
31. Ikeuchi, H., and G. Satô, 12th Symposium on Polarography, Kyoto, 1965.
32. Kato, K., Bull. Chem. Soc. Japan, **33**, 600 (1960).
33. Kodama, M., Bull. Chem. Soc. Japan, **42**, 2532 (1969).

34. Kodama, M., *Bull. Chem. Soc. Japan,* **42,** 3330 (1969).
35. Konràd, D., and A. A. Vlček, *Collection Czech. Chem. Commun.,* **28,** 595 (1963).
36. Koryta, J., *Z. Physik. Chem. (Leipzig),* Sonderheft, 157 (1958).
37. Koryta, J., *Collection Czech. Chem. Commun.,* **23,** 1408 (1958).
38. Koryta, J., *Collection Czech. Chem. Commun.,* **24,** 3057 (1959).
39. Koryta, J., in *Advances in Electrochemistry and Electrochemical Engineering,* Vol. 6, (P. Delahay, Ed.), Interscience, New York, 1967, p. 289.
40. Koryta, J., and Z. Zábranský, *Collection Czech. Chem. Commun.,* **25,** 3153 (1960).
41. Koutecký, J., *Collection Czech. Chem. Commun.,* **18,** 311 (1953).
42. Koutecký, J., and R. Brdička, *Collection Czech. Chem. Commun.,* **12,** 337 (1947).
43. Koutecký, J., and J. Koryta, *Electrochim. Acta,* **3,** 318 (1961).
44. Koutecký, J., and M. von Stackelberg, in *Progress in Polarography,* Vol. 1 (P. Zuman and I. M. Kolthoff, Eds.), Interscience, New York, 1962, p. 21.
45. Laitinen, H. A., and N. W. Grieb, *J. Am. Chem. Soc.,* **77,** 5201 (1955).
46. Lingane, J. J., *Chem. Rev.,* **29,** 1 (1941).
47. Matsuda, H., *Z. Elektrochem.,* **62,** 977 (1958).
48. Matsuda, H., *Tokyo Kogyo Shikensho Hokoku* (Rept. Govt. Chem. Ind. Res. Inst., Tokyo), **61,** 315 (1966).
49. Matsuda, H., *Tokyo Kogyo Shikensho Hokoku* (Rept. Govt. Chem. Ind. Res. Inst., Tokyo), **62,** 107 (1967).
50. Matsuda, H., *J. Electroanal. Chem.,* **15,** 109 (1967).
51. Matsuda, H., *J. Electroanal. Chem.,* **15,** 325 (1967).
52. Matsuda, H., *J. Electroanal. Chem.,* **21,** 433 (1969).
53. Matsuda, H., *J. Electroanal. Chem.,* **22,** 413 (1969).
54. Matsuda, H., and Y. Ayabe, *Bull. Chem. Soc. Japan,* **29,** 134 (1956).
55. Matsuda, H., and Y. Ayabe, *Z. Elektrochem.,* **63,** 1164 (1959).
56. Matsuda, H., and Y. Ayabe, *Z. Elektrochem.,* **66,** 469 (1962).
57. Matsuda, H., Y. Ayabe, and K. Adachi, *Ber. Bunsenges. Physik. Chem.,* **67,** 593 (1963).
58. Matsuda, H., P. Delahay, and M. Kleinerman, *J. Am. Chem. Soc.,* **81,** 6379 (1959).
59. Matsuda, K., and R. Tamamushi, *Bull. Chem. Soc. Japan,* **41,** 1563 (1968).
60. Milner, G. W. C., *The Principles and Applications of Polarography,* Longmans Green, London, 1957.
61. Momoki, K., and H. Ogawa, 22nd Annual Meeting of the Chemical Society of Japan, Tokyo, 1969.
62. Momoki, K., H. Sato and H. Ogawa, *Anal. Chem.,* **39,** 1072 (1967).

63. Morinaga, K., and T. Nomura, *Nippon Kagaku Zasshi* (J. Chem. Soc. Japan, Pure Chem. Sect.), **79**, 200 (1958).
64. Nicholson, R. S., and I. Shain, *Anal. Chem.*, **37**, 178 (1965).
65. Nicholson, R. S., and I. Shain, *Anal. Chem.*, **37**, 190 (1965).
66. Nicholson, R. S., J. W. Wilson, and M. L. Olmstead, *Anal. Chem.*, **38**, 542 (1966).
67. Ogino, H., *Bull. Chem. Soc. Japan*, **38**, 771 (1965).
68. Ogino, H., and N. Tanaka, *Bull. Chem. Soc. Japan*, **40**, 852 (1967).
69. Ogino, H., and N. Tanaka, *Bull. Chem. Soc. Japan*, **40**, 857 (1967).
70. Ogino, K., and N. Tanaka, *Bull. Chem. Soc. Japan*, **39**, 2672 (1966).
71. Papoff, P., *J. Am. Chem. Soc.*, **81**, 3254 (1958).
72. Randles, J. E. B., *Can. J. Chem.*, **37**, 238 (1959).
73. Reinert, K. E., *Z. Elektrochem.*, **66**, 379 (1962).
74. Satô, G., *Electrochim. Acta*, **15**, 179 (1970).
75. Schwarzenbach, G., and R. Gut, *Helv. Chim. Acta*, **39**, 1589 (1956).
76. Schwarzenbach, G., R. Gut, and G. Anderegg, *Helv. Chim. Acta*, **37**, 937 (1954).
77. Schwarzenbach, G., and J. Sandera, *Helv. Chim. Acta*, **36**, 1089 (1953).
78. Smith, D. E., *Anal. Chem.*, **35**, 602 (1963).
79. Smith, D. E., *Anal. Chem.*, **35**, 610 (1963).
80. Smith, D. E., in *Electroanalytical Chemistry*, Vol. 1, (A. J. Bard, Ed.), Arnold, London, 1966, p. 1.
81. Stackelberg, M. von, and H. von Freyhold, *Z. Elektrochem.*, **46**, 120 (1940).
82. Štráfelda, F., and M. Šťastný, *Collection Czech. Chem. Commun.*, **32**, 1836 (1967).
83. Tachi, I., and M. Senda, in *Advances in Polarography*, (I. S. Langmuir, Ed.), Vol. 2, Pergamon Press, Oxford, 1960, p. 454.
84. Tamamushi, R., *Ochanomizu Joshidaigaku Shizenkagaku Hokoku* (Natural Sci. Rept. Ochanomizu Univ.), **5**, 239 (1955).
85. Tamamushi, R., and N. Tanaka, *Bull. Chem. Soc. Japan*, **23**, 110 (1950).
86. Tamamushi, R., and N. Tanaka, *Z. Physik. Chem. (Frankfurt)*, **39**, 117 (1963).
87. Tanaka, N., and K. Ebata, *J. Electroanal. Chem.*, **8**, 120 (1964).
88. Tanaka, N., K. Ebata, T. Takahari, and T. Kumagai, *Bull. Chem. Soc. Japan*, **35**, 1836 (1962).
89. Tanaka, N., and T. Ito, *Bull. Chem. Soc. Japan*, **39**, 1043 (1966).
90. Tanaka, N., and M. Kamada, *Bull. Chem. Soc. Japan*, **35**, 1596 (1962).
91. Tanaka, N., M. Kamada, H. Osawa, and G. Satô, *Bull. Chem. Soc. Japan*, **33**, 1412 (1960).
92. Tanaka, N., and K. Kato, *Bull. Chem. Soc. Japan*, **32**, 516 (1959).

93. Tanaka, N., and K. Kato, *Bull. Chem. Soc. Japan,* **32,** 1376 (1959).
94. Tanaka, N., and K. Kato, *Bull. Chem. Soc. Japan,* **33,** 417 (1960).
95. Tanaka, N., and K. Kato, *Bull. Chem. Soc. Japan,* **33,** 1236 (1960).
96. Tanaka, N., K. Kato, and R. Tamamushi, *Bull. Chem. Soc. Japan,* **31,** 283 (1958).
97. Tanaka, N., and M. Kimura, *Bull. Chem. Soc. Japan,* **40,** 2100 (1967).
98. Tanaka, N., and K. Koseki, *Bull. Chem. Soc. Japan,* **41,** 2067 (1968).
99. Tanaka, N., and H. Ogino, *Bull. Chem. Soc. Japan,* **34,** 1040 (1961).
100. Tanaka, N., and H. Ogino, *Bull. Chem. Soc. Japan,* **36,** 175 (1963).
101. Tanaka, N., and H. Ogino, *Bull. Chem. Soc. Japan,* **38,** 439 (1965).
102. Tanaka, N., and H. Ogino, 15th Annual Symposium on Coordination Chemistry, Kanazawa, Japan, 1965.
103. Tanaka, N., K. Ogino, and G. Satô, *Bull. Chem. Soc. Japan,* **39,** 366 (1966).
104. Tanaka, N., H. Osawa, and M. Kamada, *Bull. Chem. Soc. Japan,* **36,** 67 (1963).
105. Tanaka, N., H. Osawa, and M. Kamada, *Bull. Chem. Soc. Japan,* **36,** 530 (1963).
106. Tanaka, N., Y. Saito, and H. Ogino, *Bull. Chem. Soc. Japan,* **36,** 794 (1963).
107. Tanaka, N., Y. Saito, and H. Ogino, *Bull. Chem. Soc. Japan,* **38,** 984 (1965).
108. Tanaka, N., and Y. Sakuma, *Bull. Chem. Soc. Japan,* **32,** 578 (1959).
109. Tanaka, N., R. Tamamushi, and M. Kodama, *Z. Physik. Chem. (Frankfurt),* **14,** 141 (1958).
110. Tanaka, N., and A. Yamada, *Z. Anal. Chem.,* **224,** 117 (1967).
111. Tanaka, N., and A. Yamada, *Rev. Polarog. (Kyoto),* **14,** 234 (1967).
112. Tanaka, N., and A. Yamada, *Electrochim. Acta,* **14,** 491 (1969).
113. Testa, A. C., and W. H. Reinmuth, *J. Am. Chem. Soc.,* **83,** 784 (1961).
114. Testa, A. C., and W. H. Reinmuth, *Anal. Chem.,* **33,** 1320 (1961).
115. Vetter, K. J., *Elektrochemische Kinetik,* Springer, Berlin, 1961.
116. Vlček, A. A., in *Advances in the Chemistry of the Coordination Compounds,* (S. Kirschner, Ed.), Macmillan, New York, 1961, p. 590.
117. Vlček, A. A., in *Progress in Polarography,* Vol. 1 (P. Zuman and I. M. Kolthoff, Eds.), Interscience, New York, 1962, p. 269.
118. Vlček, A. A., in *Progress in Inorganic Chemistry,* Vol. 5 (F. A. Cotton, Ed.), Interscience, New York, 1963, p. 211.
119. Wiesner, K., *Z. Elektrochem.,* **49,** 164 (1943).
120. Yamada, A., and N. Tanaka, *Bull. Chem. Soc. Japan,* **42,** 1600 (1969).

CHAPTER II

EFFECTS OF ACIDITY IN THE ELUCIDATION OF ORGANIC ELECTRODE PROCESSES

P. Zuman

Department of Chemistry, Clarkson College of Technology Potsdam, New York

Contents

I. Introduction	73
II. Experimental Conditions	78
III. Dependence of Half-Wave Potentials and Limiting Currents on pH	83
IV. Systems Showing One Reduction Wave	93
V. Systems Showing Two Reduction Waves	116
VI. Systems Showing Three or More Reduction Waves	136
A. Dibasic acids	136
B. Tribasic acids	141
C. Aryl ketones	145
D. α, β-Unsaturated aldehydes and ketones	148
VII. Conclusions	152
References	153

I. Introduction

The polarographic reduction of organic compounds can be studied either in the absence of proton donors in strictly non-aqueous media, or in the presence of water or other proton donors. Studies of organic compounds in non-aqueous media offer some advantages both from the practical point of view (because of increased solubility) and from certain theoretical aspects. In particular, the uptake of the first electron is in some cases not accompanied by a protonation antecedent or subsequent to the electron transfer. The electrode process under these conditions may be reversible, allowing such systems to be treated thermodynamically. The half-wave potential of the first one-electron wave measured under such conditions is a direct measure of the difference between the standard free energies of the organic molecule and the radical anion formed. Nevertheless, when we are interested not only in the thermodynamics of the first electron uptake, but in the mechanism of the whole electrode process, including chemical and electrochemical deactivation of the primary electrolysis product, the possibilities offered by non-aqueous systems are more restricted. The radical anions or primary products of the first electron uptake are strong bases that are able to extract protons from solvents usually not considered acidic, or from tetraalkylammonium salts used as supporting electrolytes, or from other organic molecules, including other molecules of the same organic compound that undergoes electroreduction. As a result, the overall processes involved in the second and subsequent electron transfers in non-aqueous media are frequently irreversible. These processes often involve chemical reactions in addition to the electrode process proper. Because ways of keeping the concentrations of proton donors constant during the course of electrolysis by buffering are usually not available in non-aqueous media, the resulting systems involve reactions of second and higher order. The exact treatment of such systems is complicated and in most cases not accessible. For these reasons the electroreductions of organic systems in aprotic non-aqueous systems are often difficult to interpret because the diagnostic tools are not yet available, if our interest is in the mechanism of the electrode process more than in the thermodynamic treatment. In the future, non-aqueous electrochemistry will undoubtedly offer an important and interesting extension of our present knowledge, but before these systems can be fully exploited, a deeper understanding of the chemistry in such solvents is essential. There is special need for a better understanding of acid–base equilibria and of ion pair formation. Because of this, a study of the polarographic behavior of organic compounds in aqueous and water-containing systems seems to be still a reasonable target at present.

In addition to the problem of buffering, which deprives us of diagnostic tools for the investigation of organic electrode processes in non-aqueous aprotic solvents, there are practical problems involved. The limited solubility of inorganic salts in non-aqueous solvents restricts the number of salts that can be used for supporting electrolytes. The systems used often have considerably higher resistances than water-containing solutions, but there are modifications of apparatus that enable us to deal with the resulting problems. Furthermore, to keep the concentration of water at least one order of magnitude below the concentration of the electroactive species, special experimental procedures are necessary. All these practical problems are surmountable but make the investigations less attractive than those carried out in the presence of water.

However, apart from such practical problems, there is still one aspect that makes less reasonable the widespread suggestions heard several years ago that studies of aqueous systems be abolished or restricted in favor of concentration on non-aqueous systems. This is the utilitarian aspect; organic chemists wish to obtain information about reactivity and other properties of organic compounds from polarographic investigations or to use polarography as a means of determining the optimum conditions for electrosynthetic work. The reactivities at a dropping mercury electrode and in homogeneous reactions can be compared only if comparisons are carried out under identical conditions. Because organic chemists rarely exclude the last traces of water from their reaction mixtures, it would be illogical to try to compare the reactivity in homogeneous reactions with electrochemical data obtained under a careful exclusion of the last traces of water. Furthermore, the costs of organic solvents for large-scale electrosyntheses are rather prohibitive.

For all the above reasons, we may reasonably hope for further development of the electrochemistry of organic compounds in water-containing solutions, and the following discussion of techniques for elucidating organic electrode processes is restricted to aqueous and water-containing solutions.

For the elucidation of an organic electrode process, it is first necessary to identify the individual steps. Proofs of composition and structure of intermediates and products can be given by electrochemical and preparative techniques. Information about composition and spatial orientation of the transition state can be obtained from studies of structural effects. These aspects have been dealt with in other reviews (1, 2) and monographs (3, 4). This chapter is restricted to the investigation of the chemical reactions accompanying the electrode process proper. Because the chemical reactions involved are often either acid–base equilibria or reactions in which rates depend on acidity, their extents or rates are pH-dependent. Hence the investigation of such systems aims at controlling the pH value (usually by buffering) and following the change in polarographic curves with pH.

To distinguish the individual chemical steps accompanying an electrode process proper and to identify the electroactive species in the electroreduction of organic compounds in aqueous or water-containing media, measurements of wave heights, half-wave potentials, and of wave shapes as a function of pH proved to be useful.

In the past the pH-dependence of either the half-wave potential or the limiting current was generally applied in the study of organic electrode processes. The use of each of these relationships separately for the elucidation of the electrode process often proved useful for distinguishing the type of chemical reaction involved. However, in many instances the use of only one of these relationships left several mechanisms as equally tenable explanations of the experimental findings; it was impossible to decide among the various explanations by measuring only wave heights or the half-wave potentials.

Only recently has it been recognized that a combination and comparison of the pH-dependences of both half-wave potentials and wave heights provide a most powerful tool in interpreting acid–base equilibria and chemical reactions accompanying the electrode process proper. In this way, and in combination with the identification of intermediates and products and with the study of structural effects mentioned earlier, the number of possible mechanisms explaining all the experimental results is usually reduced to one or a few. This combination of pH-dependences of wave heights and half-wave potentials is therefore the main content of this discussion. Before approaching this discussion in a systematic way, however, it is necessary to pay attention to two questions—the use of wave shape and the role of adsorption.

The shapes of polarographic waves are usually characterized by plotting potential as a function of $\log[(i_d - i)/i]$ (or an analogous expression for more complex systems) in a so-called logarithmic analysis. The use of these plots in the identification of individual steps of the electrode process is more restricted than that of i–pH and $E_{1/2}$–pH plots, even though they offer information about the electrode process proper. It is nevertheless often forgotten that logarithmic analysis can be applied to the determination of rate parameters of the electrode process proper only when the nature of the electroactive species is known, when the role of the chemical reactions accompanying the electrode process proper is understood, and when it can be assumed that these reactions do not markedly affect the shape of the polarographic wave.

The measurement of limiting currents and the concept of half-wave potentials measured under potentiostatic conditions with current flowing can be considered important contributions of polarography to electrochemical techniques. It has been proved conclusively that in particular the combination of measurements of both limiting currents and half-wave potentials and their pH plots offers more and clearer information about the individual steps in complex electrode processes and the nature of participating species than logarithmic analyses or Tafel plots. To restrict oneself to measurements of Tafel plots and to omit measurements of limiting currents can be considered nonprogressive at the present stage of development of theory and experimental techniques. Furthermore, the differences between the logarithmic analyses for different types of electrode processes are usually smaller than those between limiting currents. Decisions as to which reaction scheme and which corresponding logarithmic plot best represent the experimental points frequently involve a high degree of arbitrariness. Furthermore, experimental verification of logarithmic analyses is subject to greater experimental error because of the effects of adsorption, and so on. The measurement of wave heights and, with limited precision (when good-quality instruments are used), even of half-wave potentials can be carried out from current–voltage curves recorded automatically. Wave shapes, however, are strongly affected by the hysteresis of recording instruments, as can be proved by recording the current–voltage curve for the same solution first from positive to negative and then from negative to positive potentials. For logarithmic analyses it is hence necessary to measure the current–voltage curves manually point by point, or to use a very slow rate of voltage scanning (of the order of 10–50 mV/min). Hence logarithmic analysis can be considered a tool for the confirmation of a possible mechanism rather than for primary diagnosis.

The role of adsorption in organic electrode processes has probably been underestimated in the past and currently is sometimes overstressed. One of the inherent difficulties is that the term adsorption encompasses a wide spectrum of interactions between the electrode and the species participating in an electrolysis. If the orientation of the organic molecule at the electrode surface is also termed adsorption, this is naturally one aspect of adsorption which is of great interest in the study of mechanisms of electrode processes. The orientation is considered to be equivalent to the stereochemistry of the transition state. Information about this type of adsorption of the electroactive species is best obtained from the study of effects of structure of the organic molecule on polarographic behavior, in particular of the effects of conformation, of bulky alkyl groups, and other steric factors. Interactions of the π-electrons of benzene or pyridine rings with the electrode can also contribute to the stereochemistry of the transition state. The effect of pH on surface orientation of organic molecules is only indirect in that it changes the degree of ionization of the electroactive species. This is important because (as a result of Coulombic interactions) species bearing unit charges are oriented at the electrode surface in a different way than are corresponding uncharged molecules.

Most of the stronger types of adsorption interaction between organic molecules and the electrode surface can affect wave heights, cause a separation of new waves, or change the shape of the plateau of the wave. The role of adsorption of the electroactive species, of the electrolysis product, of the solvent, of the buffer component, or of other component of the supporting electrolyte, can usually be determined experimentally in these cases.

This type of adsorption can affect the quantitative treatment for determination of rate or equilibrium constants of the accompanying chemical reactions which in such cases take place as "surface reactions" (5). However, for the qualitative determination of the nature of the chemical reactions involved, these adsorption phenomena usually do not play an important role.

In two main cases adsorption can play a direct role in the elucidation of the mechanism of the electrode processes: when radicals or other species formed in the electrode reaction interact with the material of the electrode to give organometallic compounds, and when proton donors present in the buffer used are strongly adsorbed at the electrode surface (as was observed for imidazole and phenylacetic acid).

Even though adsorption does not usually interfere with the elucidation of the course and mechanism of the electrode process, it is useful to record polarographic curves for such studies under conditions in which the adsorption phenomena are less marked. The current–voltage curves may then be simpler and easier to measure and interpret. This is of particular importance for groups of compounds known to be involved in strong adsorption phenomena, such as pyridine derivatives (because of the above mentioned π-electron interaction) and polynuclear and alicyclic compounds. It has been proved that adsorption phenomena are less marked when the concentration of the electroactive species is sufficiently low (about $1–2 \times 10^{-4}$ M) and the concentration of the organic solvent in the mixture with water higher. Because various other reasons (definition of acidity scale, resistance, and so on) make it advantageous to have in the water mixture as little of the organic solvent as possible, a compromise solution must be found for each type of organic molecule, corresponding to the lowest content of the organic solvent that practically eliminates adsorption phenomena. For numerous organic compounds mixtures containing 5–70% ethanol with water have proved useful.

In the preparation of the supporting electrolyte, strongly adsorbable components (such as imidazole, phenylacetic acid, and so on) are omitted. When tetraalkylammonium salts are used, it should be considered that tetrabutylammonium ion shows considerably more adsorption properties of the type exerted by hydrocarbons and large alkyl groups than tetramethylammonium ion. It is recommended that the behavior of an organic compound studied in a given solvent in the presence of tetraalkylammonium ions be compared with that in the presence of alkali metal ions (lithium salts are usually most soluble) to ensure that the adsorption of tetraalkylammonium ions (which can exert a number of other effects on the electrode process) does not fundamentally alter the course of the electrolysis.

II. Experimental Conditions

To follow the changes in wave heights and half-wave potentials with pH, the polarographic curves of an investigated compound are recorded over as broad a range of pH values as possible. The limits toward low and high pH values may be governed by the solubility of the compound studied and by the possibility that chemical cleavage occurs in very strongly acidic or very strongly alkaline media. The effects of homogeneous chemical reactions of the compound being studied can sometimes be diminished or eliminated by working at lower temperatures.

The pH range in which a given compound can be studied can also be limited by the overlapping of the waves by the current attributable to the electrolysis of the supporting electrolyte. In acidic media the final rise in current corresponds to hydrogen evolution. The potential at which the final rise occurs shifts with increasing acidity to a more positive potential (at a given current density by 59 mV/pH unit). If care is taken to eliminate the presence of substances that catalytically lower the hydrogen overvoltage (e.g., nitrogen-containing heterocyclics, amino acids, sulfur derivatives), the potential of the final rise is hence only a function of acidity and usually cannot be shifted if the current of the supporting electrolyte is superimposed on the studied wave.

Sometimes the final rise in current at lower pH values occurs in solutions containing the studied substance at a more positive potential than it does in the blank (pure supporting electrolyte of the same composition). When the possibility of another, subsequent, more negative reduction wave of the studied substance is excluded, this shift in the final rise in current may be attributable to a catalytic hydrogen evolution caused by the studied compound. This current of catalytic hydrogen evolution can in some cases overlap the reduction wave and prevent measurement of the height of the latter. If this happens in buffered solutions, a dilution of the buffer sometimes results in a better separation of the studied wave. If the catalytic wave is superimposed on the reduction wave in more strongly acidic media (at pH < 1, in solutions of acids), no remedy is known that permits resolution of the overlap.

At pH > 8–10 the final rise in current is no longer attributable to hydrogen evolution; it is instead governed by the reduction of the alkali metal or other cation of the supporting electrolyte (Fig. 1). If the studied wave is super-

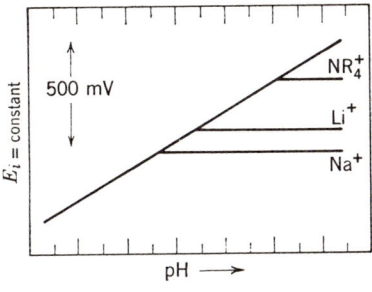

FIG. 1. Dependence of the potential of final rise in current (measured at constant current density) on pH for buffers containing sodium, lithium, and tetraalkylammonium cations.

imposed on the current of the supporting electrolyte in the alkaline region in buffers containing sodium or potassium ions, some improvement in wave measurability can be achieved when analogous solutions containing lithium ions are used. The use of lithium salts is limited with some buffers (e.g., phosphate) the lithium salts of which are only slightly soluble. Even more negative potentials of the final rise in current can be reached by using tetraalkylammonium salts, although their adsorption and other specific effects should be kept in mind.

Finally, instead of shifting the final rise in current to more negative potentials, it is sometimes possible to improve the measurability of the studied waves by the addition of components to the supporting electrolyte, which causes a shift in those waves to more positive potentials. The most common example of this type of choice of composition of the supporting electrolyte is the addition of polyvalent ions (e.g., Ca^{2+} or Sr^{2+}) to solutions in which the studied substance is present as an anion (Fig. 2). Limitations resulting from slight solubility of calcium or strontium salts of buffer components should again be kept in mind. This type of experimental approach may be particularly successful when with increasing pH the wave becomes less measurable not because its half-wave potential is shifted to more negative values than the final current rise for the supporting electrolyte, but because

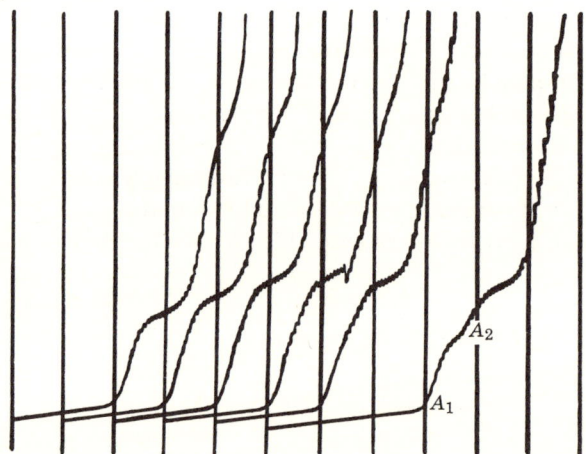

FIG. 2. Effect of calcium ions. $4 \times 10^{-4}\,M$ picolinic acid in barbital-acetate buffer, $CaCl_2$ concentration (from right): 0; $0.005\,M$; $0.01\,M$; $0.02\,M$; $0.04\,M$, and $0.08\,M$. First curve starting at -0.8 V, all others at -1.0 V, SCE, 200 mV/ absc., $h = 50$ cm: sens. 1 μA. According to J. Volke [*Collect. Czechoslov. Chem. Comm.*, **20**, 1334 (1955)].

a wave at more positive potentials decreases with a simultaneous increase in a more negative wave. In this rather common pattern, a more negative wave, which can be superimposed on the current attributable to the electrolysis of the supporting electrolyte, corresponds to a conjugate base of the species that gives a more positive wave at lower pH values. As these basic forms often bear a negative unit charge, their waves are susceptible to the effects of polyvalent cations mentioned above.

The study of the pH-dependences of wave heights and half-wave potentials usually consists of preliminary tests, an overall study, and a detailed investigation. In all cases a stock solution of the investigated substance is added to the appropriate supporting electrolyte so as to achieve a final concentration of about 2×10^{-4} M. This concentration has been proved in many cases to be optimum for the investigation of pH-dependence, as it is high enough to avoid marked effects of capacity currents but low enough to avoid complications resulting from adsorption and other complicating phenomena sometimes observed in millimolar and more concentrated solutions. An investigation at the 2×10^{-4} M concentration level also limits the amount of organic substance needed for an extensive examination to approximately 50–100 mg—considerably less than that required for work with millimolar solutions. Moreover, lower concentrations make it possible to use a lower proportion of organic solvent in the examined solution and, because the ratio electroactive substance/buffer is more favorable, ensure better buffering capacity.

The stock solution then added to various supporting electrolytes is usually prepared 0.01 M with enough organic solvent (most frequently ethanol*) to ensure complete solution.

Whenever possible, the stock solution is prepared without heating, but if application of heat is necessary to accelerate the dissolution, it is essential to make sure that the behaviors of the "saturated" solution before heating and the warmed solution after heating are identical. Furthermore, it is important to control the stability of the stock solution toward prolonged periods of standing and toward the effects of diffuse daylight. These factors are sometimes overlooked and cause confusion. Less stable stock solutions should be prepared fresh daily and kept in the dark at low temperatures.

The stock solution prepared as described is then added to a given volume of a deaerated supporting electrolyte so that the final concentration of the electroactive species is about 2×10^{-4} M. For the preliminary test that should offer information about the pH range in which the compound is

---

* Ethanol is frequently used because it mixes with water, is a hydroxylic solvent not too different in its properties from water, and is inexpensively available in a high grade of purity.

electroactive and gives polarographic reduction waves, the following typical supporting electrolytes, covering a wide pH range and containing a sufficient portion of the organic solvent, proved to be useful: 0.1 M sulfuric acid, acetate buffer of pH 4.7 (0.2 M sodium acetate, 0.2 M acetic acid), phosphate buffer of pH 6.8 (0.1 M monosodium or potassium phosphate and 0.1 M disodium or -potassium phosphate), borate buffer of pH 9.3 (0.05 M borax), and 0.1 M sodium hydroxide. According to the character of the substance under study, some other buffers may be added to these, for example, for α-hydroxy carbonyl compounds that form borate complexes, the inclusion of a barbital buffer of pH 8.5 and of an ammonia buffer of pH 9.3 (0.1 M ammonia, 0.1 M ammonium chloride) is useful, whereas for some saturated aldehydes sodium hydroxide can be replaced by lithium hydroxide.

The resulting curves in these preliminary tests are inspected to determine whether they show polarographic waves and in which pH region. When a reduction wave is obtained, it is useful to check that it does not change with time. If two or more waves are observed, or if the recorded wave is abnormally low or high, the dependence on concentration of the electroactive substance, on mercury pressure and other discriminative experiments are carried out to determine the processes governing the height of the wave.

The aim of the more general investigation of pH-dependence described is to determine the character of this dependence over the pH range in which waves were observed in preliminary experiments. Over the range of pH 2–12, Britton–Robinson buffers (consisting of 0.04 M acetic, phosphoric, and boric acids to which varying volumes of 0.2 M sodium hydroxide are added) proved useful. For more acidic media 0.01 M and more concentrated solutions of strong acids (e.g., sulfuric or perchloric) are used, up to concentrated acid solutions; for the alkaline region 0.01 M and more concentrated solutions of sodium or lithium hydroxides are used, up to saturated solutions. Instead of pH, acidity functions such as H_0 or H_- are used in this region when plotting the changes in limiting currents or half-wave potentials as a function of acidity.

To distinguish the shape of these plots and to attribute correctly the corresponding waves at different pH values, the gaps in the pH values in the sequence of solutions compared should not be too wide. For the pH region in which no significant changes in the wave height and no breaks on the $E_{1/2}$–pH plot are observed, it is possible to investigate solutions differing by one or two pH units. However, in pH regions in which a change in the limiting current with pH is observed or in which the pH dependence of half-wave potentials shows a change in slope, the differences between consecutively recorded solutions should not be greater than 0.3–0.5 pH units (or corresponding acidity function units in strongly acidic or strongly alkaline media.)

Even though the result of such an investigation usually provides a correct qualitative picture of the shape of the pH-dependence of limiting currents and half-wave potentials, it should be used as a final result and for quantitative evaluation only with care and under special conditions. This is because polarographic curves are frequently affected not only by pH but also by other factors which are not kept under control in universal buffers of the Britton–Robinson type. In some cases the buffer components (e.g., boric acid) affect polarographic behavior even in pH ranges in which other buffer components (acetate, phosphate) exert their buffering properties. Some polarographic currents depend on ionic strength which is usually not kept constant in universal buffers or in solutions of strong acids or bases (even though this can be rather easily achieved). In some cases it is not sufficient just to keep the ionic strength constant but it is important to have in the solution only one type of cation in the supporting electrolyte and to keep its concentration constant in all compared solutions. For some acid–base equilibria, only the hydrogen ion concentration is essential, whereas for the study of others it is necessary to keep the concentration of the base component of the buffer constant and to change only the pH by varying the concentration of the acidic buffer component.

For all these reasons it is suggested that the final quantitative and precise measurements be carried out in simple buffers containing only one acidic and one basic component. The choice of buffer (or strong acid or strong base) depends on the pH region in which the measurements are to be made. Buffers are selected in such a way that the pK_a value of the acidic form of the buffer lies within 1 pH unit of the center of the pH range in which the change in the limiting current with pH or the change in the shape of the $E_{1/2}$–pH plot has been observed. The most frequently used buffers are (in order of increasing pH): phosphate, formate, acetate, citrate, barbital, triethanolamine, tris, glycine, borate, ammonia, carbonate, and trimethylamine. After selection of the buffering system based on preliminary tests, the buffers are prepared with constant ionic strength and so as to correspond to the requirements of the particular investigation, that is, with the concentration of a given cation or basic buffer component constant. In such buffers prepared in intervals of 0.2–0.4 pH units, the final quantitative and precise measurements of wave heights and half-wave potentials are carried out.

III. Dependence of Half-Wave Potentials and Limiting Currents on pH

The final, accurate values of half-wave potentials and wave heights are next compared with pH and the type and shape of these changes observed. It is useful to depict these relationships graphically by plotting i against pH and $E_{1/2}$ against pH (or the corresponding acidity function). From these plots deductions as to the type of scheme involved in the electrode process are drawn.

Half-wave potentials are found to be pH-independent (Fig. 3a) only in pH regions in which the species that predominates in the bulk of the solution is also the one that is electroactive and takes part in the electrode process proper. This applies to organic compounds that do not have groups prone to ionization (e.g., hydrocarbons, alkyl halides, and so on). For compounds with acidic functions, this condition is fulfilled at pH $< (pK_a - 1)$. However, for the basic component of an electroactive acid–base couple, the experimental conditions under which this applies are found only at very high pH values at which the rate of the recombination reaction converting this base into

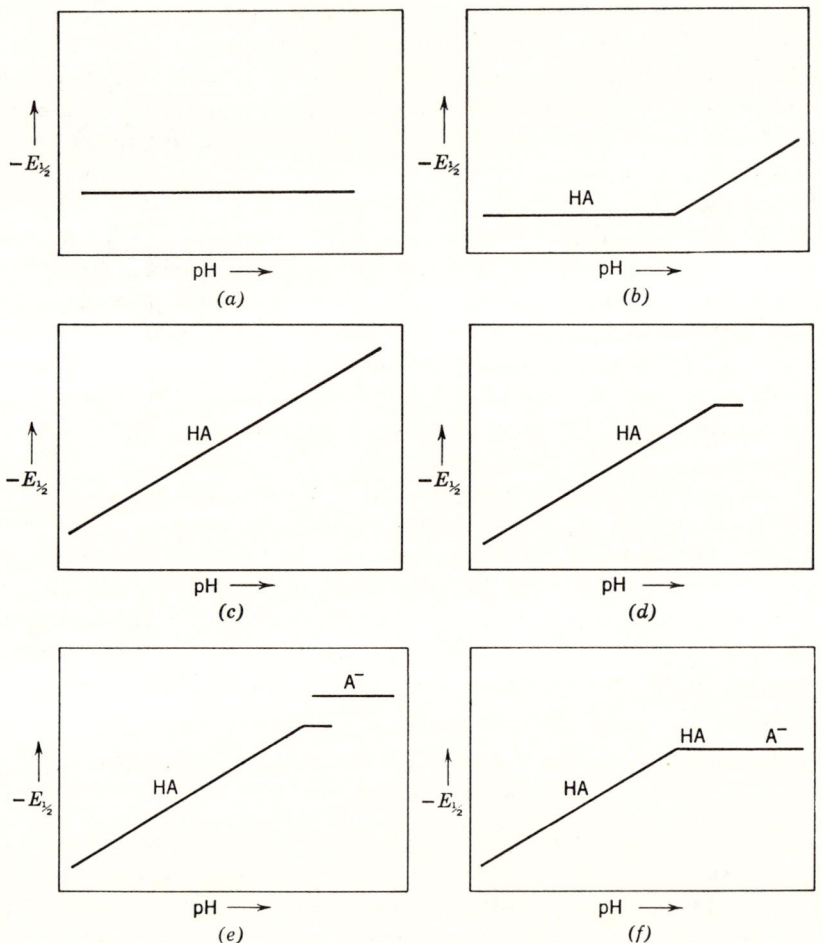

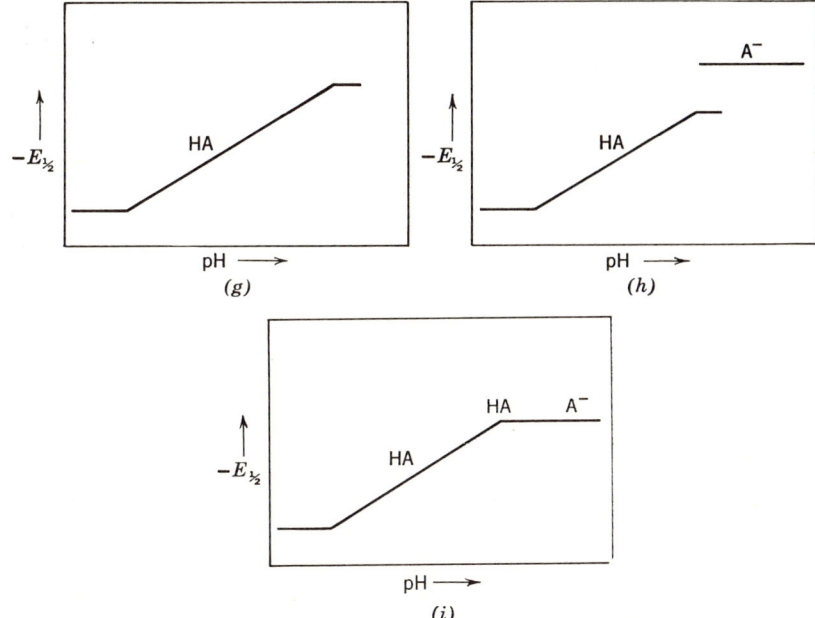

FIG. 3. The most common types of pH dependence of half-wave potentials for irreversible reduction accompanied by simple acid–base equilibria in which the uptake of the first electron is potential-determining. See discussion in the text.

the conjugate acid is very low. In all these cases the electroactive species is transported to the electrode surface and does not undergo an acid–base or another chemical reaction before electroreduction takes place.

Every observed shift in the half-wave potential toward more negative potentials with increasing pH indicates the occurrence of a proton transfer preceding the potential-determining step. For the sake of simplicity in this discussion, it is considered that the uptake of the first electron is potential determining. Nevertheless, systems involving a proton transfer reaction interposed between two electron transfers, the second of which is potential determining, can be discussed in a completely analogous way.

If—with increasing pH—the half-wave potential of the reduction of an organic compound, which was found to be independent of pH at low pH values, is observed to shift to more negative values in a manner indicated in Figure 3b, it is possible to conclude that the conjugate acid form is reduced and that the establishment of the equilibrium between the acid and corresponding base forms can be considered fast when compared with the transport of both species by diffusion toward the electrode surface and the removal of the acid form by reduction. The conclusion concerning the rate of the establishment of the equilibrium remains valid for pH values smaller at least by 1 pH unit than the pH at which another change in the shape of the $E_{1/2}$–pH plot is observed. For this pH range the $E_{1/2}$–pH plot corresponds to Figure 3b and the half-wave potential shifts are given by Eq. 1 (6):

$$E_{1/2} = C + \frac{RT}{\alpha nF} \ln \frac{[\text{H}^+]^m}{K_1 + [\text{H}^+]^m} \qquad (1)$$

In this equation C is a constant, α the transfer coefficient, n the number of electrons in the potential-determining step of an irreversible reaction, m the number of protons transferred before the potential-determining step, K_1 the acid dissociation constant at a given ionic strength, and RT/F has the usual meaning. For reversible reactions the equation has an analogous form, only $\alpha = 1$, n represents the total number of electrons consumed, and the constant C is related to the standard oxidation–reduction potential of the system. The simple equation (Eq. 1) remains valid for reversible systems provided that in the investigated pH range there is no appreciable change in the degree of ionization of the reduced form (i.e., that the numerical value of pK is smaller than the smallest pH or greater than the greatest pH investigated) and that the form of the reduced compound that undergoes oxidation is the same as the form that predominates in the bulk of the solution. Usually $m = 1$ and this is assumed henceforth.

At pH $<$ (p$K_1 - 1$) the value of K_1 can be neglected when compared with [H$^+$], and the half-wave potential becomes pH-independent ($E_{1/2} = C$). At pH $>$ (p$K_1 + 1$), but still in the pH region in which the rate of protonation of the conjugate base remains sufficiently fast, hence in which the wave height remains practically constant, it is possible to neglect [H$^+$] in the denominator as compared with K_1, and the half-wave potential is shifted to more negative values according to Eq. 2a:

$$E_{1/2} = C - \frac{RT}{\alpha nF} \ln K_1 + \frac{RT}{\alpha nF} \ln [\text{H}^+] \qquad (2a)$$

This predicts a linear dependence of the half-wave potential on pH (at 20°C):

$$E_{1/2} = \text{constant} - \frac{0.058}{\alpha n} \text{pH} \qquad (2b)$$

Equation 2b indicates the negative direction of the potential shift with increasing pH and the possibility of determining from $E_{1/2}$–pH plots either the number of protons consumed per molecule of electroactive species (m) provided the value of αn is known, or vice versa.

The intersection of the two linear sections, one pH-independent and the other pH-dependent (Fig. 3b), occurs at a pH value equal to pK_1. Hence from the pH-dependence of the type depicted in Figure 3b it is possible to conclude: (a) that an acid form of the electroactive species is reduced; (b) that the equilibrium between this acid and the conjugate base is rapidly established; (c) that it is possible to make an estimate of the pK_1 value corresponding to the acid dissociation constant of the investigated compound (for a more accurate value of pK_1, potentiometric or spectrophotometric measurements are recommended).

In some instances it is experimentally impossible to reach the acidity region corresponding to pK_1. This happens when solutions of sufficiently high acidity cannot be prepared, when the waves at higher hydrogen ion concentration are overlapped by the final rise in the current of the supporting electrolyte, or when the investigated compound undergoes cleavage in strongly acidic media. In such cases the observed $E_{1/2}$–pH plot has the shape shown in Figure 3c. Such a plot can also be obtained when the investigation is not extended to sufficiently acidic media.

A plot of the type shown in Figure 3c allows us to conclude that: (a) the form reduced is an acid when compared with the form predominating in the bulk of the solution, that is, a form that bears more hydrogen ions than the form predominating in the bulk; (b) the equilibrium by which the form predominating in the bulk and transported to the electrode surface by diffusion is transformed into the electroactive form is rapidly established; (c) the pK_1 value corresponding to this reaction occurring in the vicinity of the electrode surface is smaller than the smallest pH value investigated (plots of the type depicted in Figure 3c do not permit any conclusions concerning the value of K_1); (d) from plots of $E_{1/2}$ against pH of this type it is still possible to determine the ratio (in general) $m/\alpha n$.

With systems having half-wave potentials that at lower pH values correspond to plots given in Figure 3b or c, it may be possible to reach pH values so high that the equilibrium (Eq. 3) between the acid (HA) and base (A$^-$) forms* can no longer be established very rapidly. The behavior then conforms to the equations

$$\text{HA} + \text{S} \underset{k_{-1}}{\overset{k_1}{\rightleftharpoons}} \text{A}^- + \text{SH}^+ \qquad pK_1 \qquad (3a)$$

$$\text{HA} + n_1 e \xrightarrow{E_1} \text{P}_1 \qquad i_{\text{HA}} \qquad (3b)$$

$$\text{A}^- + n_2 e \xrightarrow{E_2} \text{P}_2 \qquad i_{\text{A}} \qquad (3c)$$

* The symbols HA and A$^-$ are intended to represent the acid and the conjugate base, respectively; the symbols do not depict the actual unit charges of the species involved. In practice, the form HA can correspond to a cation, an uncharged molecule, or an anion, and the same applies to A$^-$. The only essential point is that HA contains one dissociable hydrogen more than A$^-$.

where S is the solvent and P_1 and P_2 are the two products formed.

If the potential of the reduction of the base (E_2) is more negative than that of the acid form (E_1) at high pH values, the above-mentioned decrease in the rate of protonation with constant k_{-1} is shown by a decrease in wave i_{HA} of the acid form at potential E_1 with increasing pH. The change in limiting current with pH is discussed in detail in subsequent sections. It is useful to denote as pK_1' the pH value at which the current i_{HA} reaches one-half of the diffusion-controlled value (as recorded in sufficiently acidic media in which the height of the wave of HA is practically pH-independent).

The decrease in the height of wave i_{HA} is accompanied by a change in the shape or the slope of the plot $E_{1/2}$–pH of this wave. A simple case of this type of pH-dependence of the half-wave potential of the acid form HA is shown in Figure 3d.

In the pH region between $(pK_1 + 1)$ and $(pK_1' - 1)$, the half-wave potentials are shifted to more negative potentials, as described by Eq. 2a. At pH $> (pK_1' + 1)$, that is, at the alkaline end of the region studied, the half-wave potential of the acid form HA is pH-independent and is given by Eq. 4:

$$(E_{1/2})_{HA} = C - \frac{RT}{2\alpha nF} \ln K_1 - \frac{RT}{\alpha nF} \ln 0.886 \sqrt{k_1 t_1} - \frac{RT}{\alpha nF} \ln K_1' \quad (4)$$

The two linear sections of the pH-dependence which correspond to Eqs. 2a and 4, respectively, intersect at a pH value corresponding approximately to the value of pK_1' (Fig. 3d). Because the wave i_{HA} of the acid form HA becomes very small at pH $> (pK_1' + 1)$, it is usually impossible to measure the half-wave potential of the wave of HA at a pH more than about 1 pH unit higher than pK_1'. The characteristic feature of such systems is therefore that the linear pH-independent section corresponding to Eq. 4 does not extend for more than about 1 pH unit (Fig. 3d).

Even for compounds with acidic groupings, it is in some cases experimentally impossible to reach the region where a decrease in the wave height together with a change in the shape of the $E_{1/2}$–pH plot appears. This can happen when in the crucial pH range the wave i_{HA} is overlapped by another wave or when the compound undergoes cleavage at high pH values, and a shift in half-wave potential corresponding to Figure 3c is then observed.

Three types of $E_{1/2}$–pH and i–pH plots can be obtained according to the relative values of E_1 and E_2 at pH $> pK_1'$:

(a) If the reduction of the conjugate base occurs at a potential so negative so that wave i_A is overlapped by the current of the electrolysis of the supporting electrolyte, the wave of A^- is not observed. Only the wave of the acidic form HA is recorded; its height decreases with increasing pH in the shape of a dissociation curve and practically vanishes at pH > (pK'_1 + 1), and the half-wave potential of wave i_{HA} changes with pH in the form depicted in Figure 3d.

(b) If the reduction of the conjugate base occurs at potentials more positive than that of the electrolysis of the supporting electrolyte (so that a wave of the base A^- can be observed on polarographic current–voltage curves), but more negative than that of the acid form HA (so that separate waves result), the decrease in wave i_{HA} in the pH region (pK'_1 − 1) > pH > (pK'_1 + 1) is accompanied by an increase in wave i_A in the same pH region. At pH < (pK'_1 − 1) only wave i_{HA} is observed, and at pH > (pK'_1 + 1) practically only wave i_A. Provided that pH < pK_2 (where pK_2 corresponds to a further dissociation step of the conjugate base in the range studied, the half-wave potential of wave i_A remains pH-independent and the pattern of the pH-dependence corresponds to that depicted in Figure 3e.

(c) If the reduction of the conjugate base takes place at such positive potentials that it is not only more positive than the final rise in current, but that the potential E_2 at pH > pK'_1 is so little different from the potential E_1 of the acid form HA in this pH range that the two waves overlap, no obvious separation of two waves occurs. Furthermore, if in Eq. 3 $n_1 = n_2$, the wave height remains unchanged; if $n_1 > n_2$, one wave of decreasing height is observed; and if $n_1 < n_2$, one wave of increasing height is observed. The last-mentioned two cases can thus be easily recognized, but for $n_1 = n_2$ the wave heights offer no information about the existence of an acid–base reaction in which with increasing pH the rate of protonation decreases. In some cases in which no separation of waves is observed on current–voltage curves, derivative techniques or logarithmic analysis can reveal the existence of two waves and their change with pH. An example of this type was observed for p-diacetylbenzene (Fig. 4), showing a single wave of constant height on i–E curves, but for which logarithmic analysis reveals the presence of two sections of the wave, the ratio of which changes with increasing pH in the shape of a dissociation curve. However, even in cases in which attempts to separate and measure the two waves of forms HA and A are unsuccessful, the shape of the pH-dependence of half-wave potentials (Fig. 3f) indicates the change in the electroactive species from HA to A^-. The characteristic feature that helps to determine whether or not studied systems belong to this category is the fact that the half-wave potential of the "single" measured wave after becoming pH-independent remains so for a pH range corresponding to several pH units (more than one).

A more complex situation arises when the half-wave potential of the acid form, E_1, is more negative than the half-wave potential of the base, E_2. Fortunately, such a situation has not yet been observed because it was found empirically that the more protonated the species the more positive the

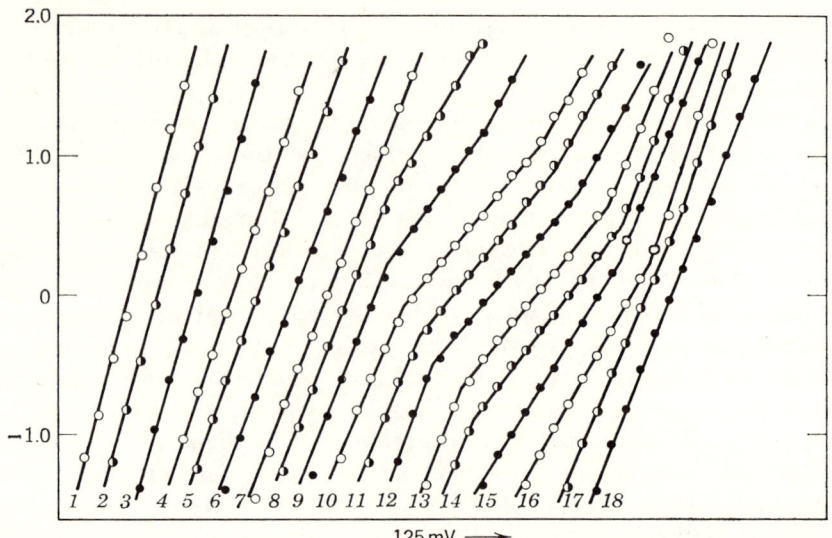

Fig. 4. Logarithmic analysis. Dependence of $\log i/(i_d - i)$ on potential for different pH values; 2×10^{-4} M p-diacetylbenzene; 30% ethanol. (*1*) $5N$ H_2SO_4; (*2*) $1N$ H_2SO_4; all others Britton–Robinson buffers of pH: (*3*) 1.95; (*4*) 2.8; (*5*) 3.8; (*6*) 4.8; (*7*) 5.5; (*8*) 6.1; (*9*) 6.5; (*10*) 7.0; (*11*) 7.2; (*12*) 7.4; (*13*) 7.7; (*14*) 7.9; (*15*) 8.65; (*16*) 9.15; (*17*) 9.8; (*18*) 10.35.

potential at which its reduction occurs. Biprotonated species are reduced at more positive potentials than univalent cations, and these in turn are reduced at more positive potentials than uncharged molecules. Similarly, the reductions of monoanions occur at potentials more negative than those of molecules but more positive than those of dianions.

Plots of the type shown in Figure 3d–f allow the following conclusions to be drawn: (*a*) that either only the acid (Fig. 3d) or both acid and base (Fig. 3e and f) forms are reduced, the acidic form being electroactive in wave i_{HA}, and the basic form in wave i_A; (*b*) that the basic form predominates in the bulk of the solution; (*c*) that the rate of protonation of the basic form in the pH range given by $(pK'_1 - 1) < pH < (pK'_1 + 1)$ producing the acidic form is decreased; (*d*) that the reduction of the basic form A^- does not involve an acid–base equilibrium antecedent to the electrode process proper.

When such a wide range is accessible that both the region $pH < (pK_1 + 1)$ and $pH > (pK'_1 - 1)$ can be studied, the overall picture corresponds to one of those shown in Figure 3g, h, or j. These figures are simple combinations of the curve in Figure 3b with those of Figure 3d, e, and f. Interpretation of the three or four linear sections remains the same as in previously discussed systems, and the deductions from such a graph are identical with those for Figure 3b and 3d–f when applied to the particular pH regions. It is possible to summarize that for irreversible systems the slope of an $E_{1/2}$–pH plot increases at a pH value close to the equilibrium pK_1 and decreases at a pH value close to the polarographic pK'_1.

The use of the shape of the pH-dependence of limiting currents in the interpretation of organic electrode processes is practically demonstrated in following sections. From a general point of view, it is noted here that most changes in wave heights with pH either have the shape of a dissociation curve or a part of it, are similar to it, or can be split into several overlapping parts of dissociation curves.

Every decrease in wave height with increasing pH in the shape of a dissociation curve or a part of it corresponds either to an acid–base equilibrium (with the acidic form electroactive), to a chemical reaction in which hydrogen ions are consumed, to a chemical reaction in which only a conjugate acid participates, or to a generally acid-catalyzed chemical reaction. However, every increase in wave height with increasing pH indicates the presence of an acid–base equilibrium with the basic form as the electroactive species, of a chemical reaction in which hydroxyl ions are consumed, of a chemical reaction in which only the conjugate base participates, or of a generally base-catalyzed reaction.

Even if the main aim of an investigation of courses or mechanisms of organic electrode processes is not the determination of rate constants of the fast chemical reactions accompanying the electrode process proper, it is useful to understand at least semiquantitatively how fast the reactions involved are.

The presence of a single wave the height of which is pH-independent but the half-wave potential of which is regularly shifted with pH indicates that the acid–base equilibrium involved is established rapidly over the entire pH range studied. The change in wave height with pH indicates that the establishment of the equilibrium involved in the pH region where the change occurs may be slow or relatively fast but is not extremely fast. The reactions reflected in waves the heights of which change with pH can be divided into two large subgroups: systems in which the wave heights are proportional to the equilibrium concentrations of the species in the bulk of the solution, and systems in which the heights are governed by the rate of formation of the electroactive species and do not reflect the equilibrium concentrations in the bulk of the solution. In the first case equilibria can be considered as slowly established, and consequently a component of the equilibrium mixture removed at the electrode surface by the electrolytic process is not renewed in any considerable degree during the life of a single drop (about 3 sec). In the second case the equilibrium perturbed in the vicinity of the electrode has the tendency to restore itself. The rate of this readjustment of the equilibrium governs polarographic currents. Diffusion-limited currents are observed in the former case and kinetic ones in the latter.

To distinguish between equilibria that are established relatively slowly and those that are established within an interval comparable with the drop time, it is necessary to determine whether the current is diffusion or kinetically controlled. This can be done by following the effect of mercury pressure at a pH at which the limiting current is 15% or less of the maximum value, by means of i–t curves, and by other techniques. Another proof is possible if the equilibrium under investigation can be examined in homogeneous media by methods that do not disturb the equilibrium (e.g., spectrophotometrically). If pK_1' obtained polarographically is practically identical with pK_1 obtained spectrophotometrically, it seems probable that equilibrium concentrations are measured polarographically and diffusion currents can be expected.* If, however, $pK_1' > pK_1$, the current is probably kinetic.

The next question to be discussed is what second-order rate constant must be involved in the recombination reaction in order to govern the limiting current and to give kinetic waves. At present this can be described in simple terms only for volume reactions, that is, reactions taking place in a thin layer of solution close to the surface of the electrode. For surface reactions, which to a greater or lesser degree take place at the electrode surface, the situation is more complex. A simple way to distinguish a volume reaction is to follow the shape of the limiting current; if the limiting current remains practically constant† over several hundred millivolts (Fig. 5a), it can be assumed that the reaction involved behaves as a volume reaction. If dips are observed on limiting currents (Fig. 5b), a surface reaction probably takes place.

For an electrode process accompanied by a volume reaction, the range of rates of recombination reactions over which the reactions are slow and over which they are comparably fast depends for acid–base equilibria on the value of the equilibrium dissociation constant pK_1. To distinguish a kinetic component in practice, the difference between polarographic pK_1' and equilibrium pK_1 must be at least 0.5 pH unit. To illustrate the ranges it is assumed that the drop time is 3 sec. Under such conditions for a compound with $pK_1 = 4$, reactions with $k_1 < 4 \times 10^4$ liter mole^{-1} sec^{-1} are slow and with $4 \times 10^4 < k_1 < 10^{11}$ liter mole^{-1} sec^{-1} are comparably fast. For compounds with $pK_1 = 7$, reactions with $k_1 < 4 \times 10^7$ liter mole^{-1} sec^{-1} are slow and those with $4 \times 10^7 < k_1 < 10^{11}$ liter mole^{-1} sec^{-1} are comparably fast. Finally, for compounds with $pK_1 = 10$, only reactions with

* Care must be taken to compare corresponding equilibrium processes. It sometimes happens that pK_1' possesses a numerical value close to pK_2 (rather than pK_1).

† It proved best to use a capillary with a regulated drop time to avoid effects of changes in the drop time with potential.

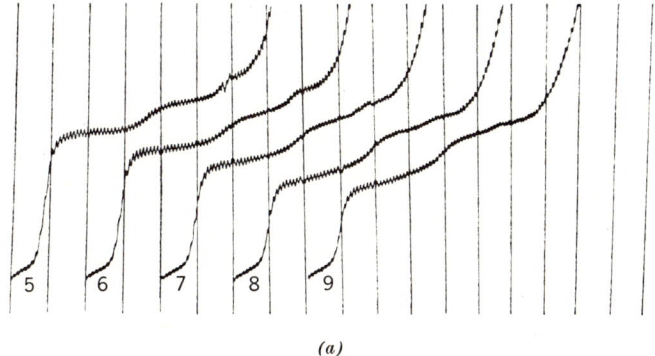

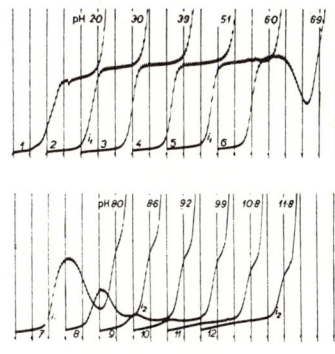

Fig 5. Polarographic curves showing a limiting current corresponding to a volume reactions (a) and surface reaction (b). (a) 2×10^{-4} M tropylium bromide, Britton-Robinson buffers, pH: 5— 4.5; 6— 4.8; 7— 5.0; 8— 5.2; 9— 5.4, SCE, 204 mV/absc., $h = 60$ cm, full-scale sens.: 0.5 μA. (b) 1×10^{-4} M nitrone $C_6H_5CH = N(O)C(CH_3)_3$, Britton-Robinson buffers, pH given on the polarogram, 1% ethanol. Curves starting at: 1–3 -0.2 V; 4–5 -0.4 V; 6–7 -0.6 V; 8–12 -0.8 V, SCE, 200 mV/absc., $h = 80$ cm, full-scale sens.: 6.6 μA.

$4 \times 10^{10} < k_1 < 10^{11}$ liter mole^{-1} sec^{-1} are comparably fast, whereas all those with $k_1 < 4 \times 10^{10}$ liter mole^{-1} sec^{-1} show behavior characteristic of slowly established equilibria. Hence even rather fast reactions do not perturb equilibria at the electrode surface provided that pK_1 is sufficiently high.

IV. Systems Showing One Reduction Wave

Examples of some more frequently observed types of pH-dependences of limiting currents and half-wave potentials and the corresponding reaction schemes are shown in Figures 6–13. Selected examples of chemical systems believed to correspond to such schemes are discussed below.

An example of a system for which the height of the single observed wave remains pH-independent and the half-wave potentials at pH > pK_1 are shifted to more negative values (Fig. 6b), is the reduction of methyl butyl phenacyl sulfonium ion (7, 8).

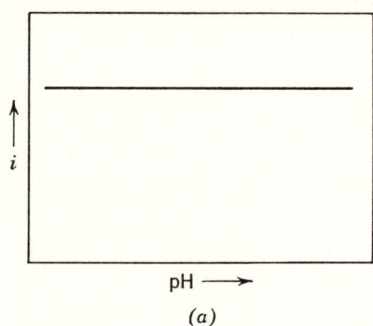

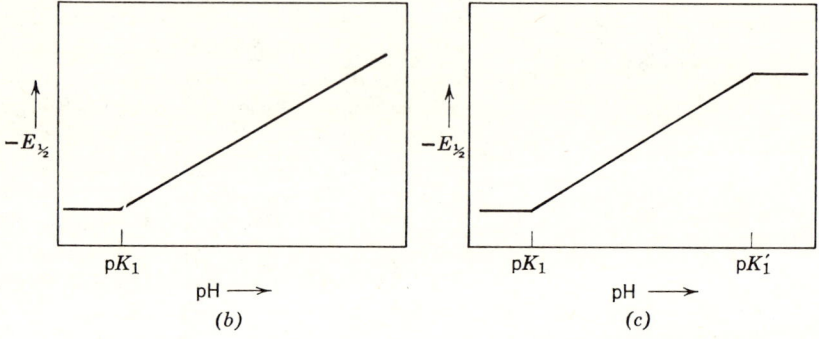

Fig. 6. pH-dependences of wave heights (a) and half-wave potentials (b, c) corresponding to the scheme:

(b) $\quad\quad$ HA $\rightleftharpoons$ A$^-$ + H$^+$ $\quad\quad$ fast

$\quad\quad\quad\quad$ HA + $n_1 e \xrightarrow{E_1}$ P$_1$ $\quad\quad\quad\quad$ pK_1

$\quad\quad\quad\quad$ A$^-$ $\quad\quad$ electroinactive

(c) $\quad\quad$ HA $\rightleftharpoons$ A$^-$ + H$^+$ $\quad\quad$ fast $\to$ comparably fast

$\quad\quad\quad\quad$ HA + $n_1 e \xrightarrow{E_1}$ $\quad\quad$ P$_1$ $\quad\quad$ pK_1 $\quad$ pK_1'

$\quad\quad\quad\quad$ A$^-$ + $n_2 e \xrightarrow{E_2}$ $\quad\quad$ P$_2$

$\quad\quad\quad\quad n_1 = n_2$; $E_1(\text{pH} > \text{p}K_1') \approx E_2$

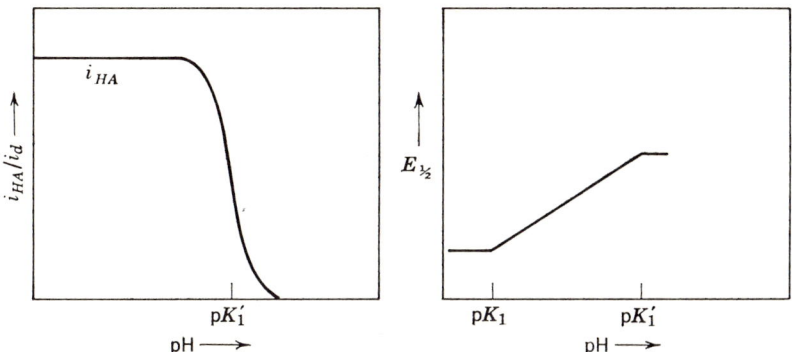

FIG. 7. pH-dependences of wave heights and half-wave potentials corresponding to the scheme:

$$HA \rightleftharpoons A^- + H^+ \quad \text{fast} \to \text{comparably fast}$$
$$HA + n_1 e \xrightarrow{E_1} P_1 \quad pK_1 \quad pK_1'$$
$$A^- \quad \text{electroinactive}$$

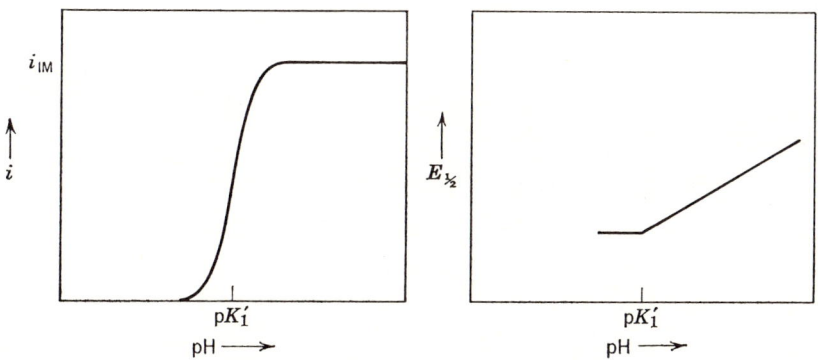

FIG. 8. pH-dependences of wave heights and half-wave potentials corresponding to the scheme:

$$A + OH^- \rightleftharpoons IM \quad \text{comparably fast}$$
$$IM + n_1 e \xrightarrow{E_1} P_1 \quad pK_1'$$
$$A \quad \text{electroinactive}$$

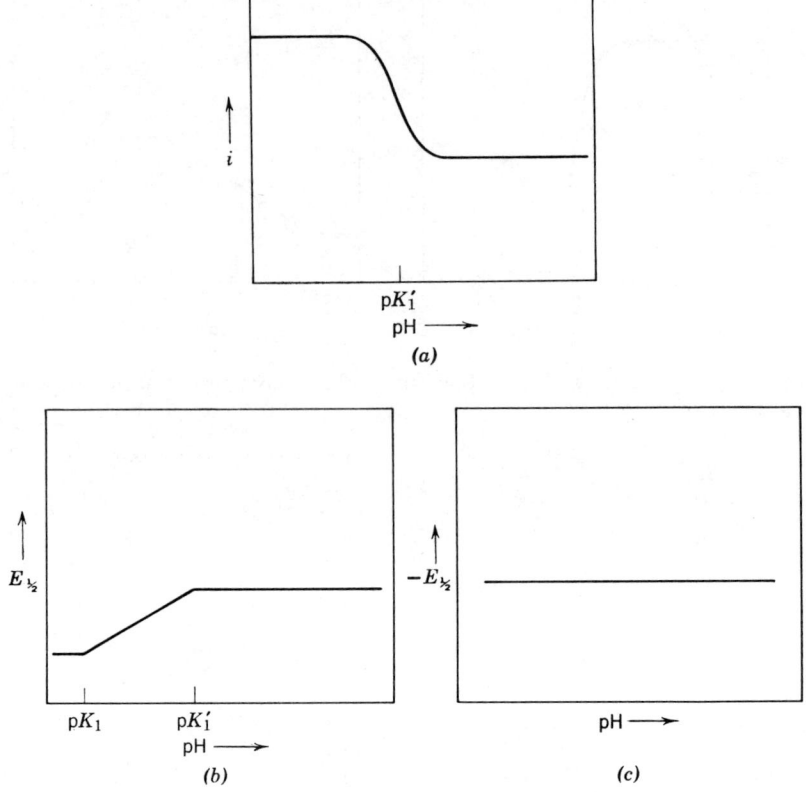

FIG. 9. pH-dependences of wave heights (a) and half-wave potentials (b, c) corresponding to the scheme:

(b) $\quad$ HA $\rightleftharpoons$ A$^-$ + H$^+$ $\quad$ fast → comparably fast
$\quad$ HA + $n_1 e \xrightarrow{E_1}$ P$_1$ $\quad\quad$ pK_1 $\quad$ pK_1'
$\quad$ A$^-$ + $n_2 e \xrightarrow{E_2}$ P$_2$
$\quad$ $n_1 > n_2$; $E_1(\text{pH} > \text{p}K_1') \approx E_2$

(c) $\quad$ A + $n_1 e \xrightarrow{E_1}$ IM
$\quad$ IM H$^+$ $\rightleftharpoons$ IM + H$^+$ $\quad$ fast → slow
$\quad$ IM H$^+$ + $n_2 e \xrightarrow{E_2}$ P
$\quad$ IM $\quad\quad$ electroinactive

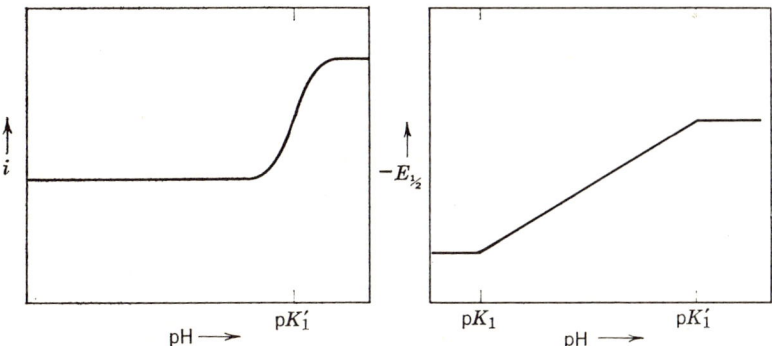

FIG. 10. pH-dependences of wave heights and half-wave potentials corresponding to the scheme:

$$HA \rightleftharpoons A^- + H^+ \quad \text{fast} \rightarrow \text{comparably fast}$$
$$HA + n_1 e \xrightarrow{E_1} P_1 \quad pK_1 \quad pK_1'$$
$$A^- + n_2 e \xrightarrow{E_2} P_2$$
$$n_1 < n_2; \quad E_1(pH > pK_1') \approx E_2$$

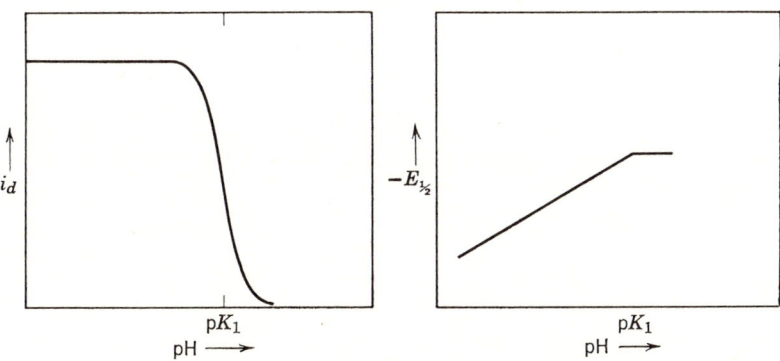

FIG. 11. pH-dependences of wave heights and half-wave potentials corresponding to the scheme:

$$HA \rightleftharpoons A^- + H^+ \quad \text{slow}$$
$$HA + n_1 e \xrightarrow{E_1} P_1 \quad pK_1$$
$$A^- \quad \text{electroinactive}$$

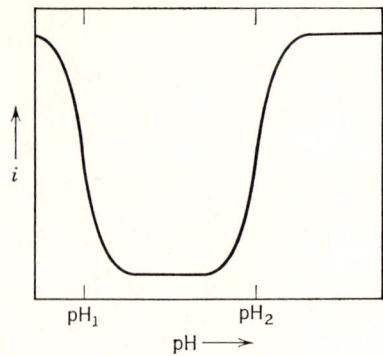

FIG. 12. pH-dependences of wave heights and half-wave potentials corresponding to the scheme:

(a)
$$A + \text{base} \longrightarrow IM \quad pH > (pH_2 - 1)$$
$$A + \text{acid} \longrightarrow IM \quad pH < (pH_1 + 1)$$
$$IM + n_1 e \xrightarrow{E_1} P_1$$
A   electroinactive

(b)
$$AH_2^+ \rightleftharpoons AH + H^+$$
$$AH \rightleftharpoons A^- + H^+$$
$$AH_2^+ \xrightarrow{k_1} IM$$
$$AH \xrightarrow{k_2} IM$$
$$A^- \xrightarrow{k_3} IM$$
$$IM + n_1 e \xrightarrow{E_1} P_1$$
$$k_2 < k_1, k_3$$

(c)
$$A + n_1 e \xrightarrow{E_1} B$$
$$B + \text{base} \longrightarrow IM$$
$$B + \text{acid} \longrightarrow IM$$
$$IM + n_1 e \xrightarrow{E_1} P_1$$
$$|E_1| \geqq |E_2|$$

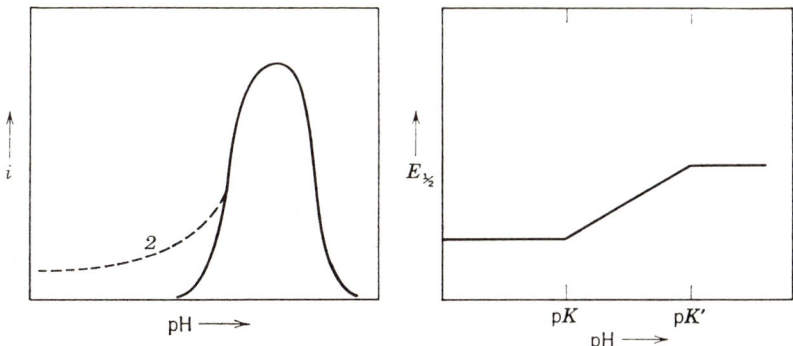

FIG. 13. pH-dependences of wave heights and half-wave potentials corresponding to the scheme:

(a)
$$A + \text{base} \rightleftharpoons \text{IMOH} + \text{base } H^+$$
$$\text{IMOH} \rightleftharpoons \text{IM} + OH^-$$
$$\text{IM} + n_1 e \xrightarrow{E_1} P_1$$

(b)
$$AH \rightleftharpoons A^- + H^+$$
$$AH + \text{base} \longrightarrow \text{IM}$$
$$\text{IM} + n_1 e \xrightarrow{E_1} P_1$$
$$A^- \quad \text{inactive}$$

This reduction follows Eq. 5:

$$C_6H_5COCH_2\overset{(+)}{S}(CH_3)C_4H_9 \rightleftharpoons C_6H_5CO\overset{(-)(+)}{CH}\ \overset{}{S}(CH_3)C_4H_9 + H^+ \quad (5a)$$

$$C_6H_5CO\overset{(-)(+)}{CH}\ S(CH_3)C_4H_9 \leftrightarrow C_6H_5\underset{\underset{O^{(-)}}{|}}{C}=CH\overset{(+)}{S}(CH_3)C_4H_9 \leftrightarrow$$

$$\leftrightarrow C_6H_5COCH=S(CH_3)C_4H_9 \quad (5b)$$

$$C_6H_5COCH_2\overset{(+)}{S}(CH_3)C_4H_9 + 2e \rightarrow C_6H_5COCH_2^{(-)} + CH_3SC_4H_9 \quad (5c)$$

The carbanion enolate or ylid might be stabilized by d-orbital resonance (Eq. 5b). The second carbanion enolate ($C_6H_5COCH_2^{(-)}$), formed in reaction 5c, can be transformed into acetophenone and further reduced in a more negative step. The observed intersection on the $E_{1/2}$–pH plot corresponds approximately to the value of pK_1 determined potentiometrically (8) or spectrophotometrically (7). A pH-independent limiting current and a shift in half-wave potentials corresponding to Figure 6c was reported (9) for the reduction of some 2-bromo-n-alkanoic acids and can be interpreted by scheme 7:

$$\text{RCH(Br)COOH} \underset{k_{-1}}{\overset{k_1}{\rightleftharpoons}} \text{RCH(Br)COO}^- + \text{H}^+ \qquad (7a)$$

$$\text{RCH(Br)COOH} + 2e \xrightarrow[\text{H}^+]{E_1} \text{RCH}_2\text{COOH} + \text{Br}^- \qquad (7b)$$

$$\text{RCH(Br)COO}^- + 2e \xrightarrow[\text{H}^+]{E_2} \text{RCH}_2\text{COO}^- + \text{Br}^- \qquad (7c)$$

in which the carbanions formed take up the proton in a consecutive reaction from the solvent. Because in most cases the pH-independent value of the half-wave potential remains constant for more than one pH unit [exceptions are the waves of 2-bromopropionic and 2-bromobutyric acids, for which separation of two waves and formation of a more negative wave of the anion were reported (9)], it can be assumed that at pH > pK'_1 the potential of the free acid (E_1) is so little different from that of the anion (E_2) that only one wave is observed. No special attention has been paid to the resolution of the two waves (e.g., by logarithmic analysis), but because of the strongly irreversible character of these drawn-out waves this might be difficult.

The approximate values for pK_1 and pK'_1 listed in the accompanying tabulation were obtained from the graph in reference (9) for compounds RCH(Br)COOH.

	pK_1		pK'_1
R	$E_{1/2}$	Equilibrium	$E_{1/2}$
H	2.9$_5$	2.86	5.7
CH$_3$	2.4	2.98	6.7
C$_2$H$_5$	2.5$_5$	2.99	7.5
C$_3$H$_7$	2.7	—	8.0$_5$
C$_4$H$_9$	3.4	—	7.6$_5$
C$_5$H$_{11}$	3.7	—	7.2
C$_6$H$_{13}$	3.7	—	7.4

(The accuracy of the values from polarographic ($E_{1/2}$) data is rarely better than ± 0.5 pH unit.)

Numerous examples are known that follow the pattern of pH-dependence shown in Figure 7. The simplest case is the reduction of tropylium ion (10) which follows scheme 8:

$$\text{[tropylium]}^+ + 2\text{H}_2\text{O} \underset{k_{-1}}{\overset{k_1}{\rightleftharpoons}} \text{[cycloheptatriene-OH-H]} + \text{H}_3\text{O}^+ \tag{8a}$$

$$\text{[tropylium]}^+ + e \longrightarrow \text{[tropyl radical]}\cdot \tag{8b}$$

The radical formed in reaction 8b can undergo dimerization (11) or interact with the surface of the mercury electrode.

The dependence on pH of the ratio of the limiting current i_{HA} of the acidic form HA (i.e., tropylium ion) to the diffusion current i_d has the shape of a dissociation curve (Fig. 7) and can be described by Eq. 9 (12):

$$\frac{i_{HA}}{i_d} = \frac{0.886 \left(\frac{k_{-1}t_1}{K_1}\right)^{1/2} [\text{H}^+]}{1 + 0.886 \left(\frac{k_{-1}t_1}{K_1}\right)^{1/2} [\text{H}^+]} \tag{9}$$

Because the limiting currents do not decrease as the potential becomes more negative and because the polarographic dissociation curve does not change on varying the buffer composition and ionic strength, reaction 8a takes place as a volume reaction (i.e., as a homogeneous reaction in a volume of the solution surrounding the dropping electrode) rather than as a surface reaction (13) (i.e., as a heterogeneous reaction in the immediate vicinity of the electrode surface). It is therefore possible to use Eq. 9 for calculation of the rate constant k_{-1}. The value of this constant ($k_{-1} = 2 \times 10^6$ liter mole^{-1} sec^{-1}) is in reasonable agreement with the value obtained from relaxation techniques (14).

Similar behavior is also shown by O-alkyl oximes (15) of the type $\text{C}_6\text{H}_5(\text{R})\text{C}=\text{NOCH}_3$ for R = H, CH$_3$, and C$_6$H$_5$, and the reactions can be described by Eq. 10:

$$\underset{\underset{\text{OCH}_3}{|}}{\text{C}_6\text{H}_5(\text{R})\text{C}=\overset{(+)}{\text{N}}-\text{H}} \underset{k_{-1}}{\overset{k_1}{\rightleftharpoons}} \text{C}_6\text{H}_5(\text{R})\text{C}=\text{NOCH}_3 + \text{H}^+ \tag{10a}$$

$$\underset{\underset{\text{OCH}_3}{|}}{\text{C}_6\text{H}_5(\text{R})\text{C}=\overset{(+)}{\text{N}}\text{H}} + 4e \xrightarrow{E_1} \text{P}_1 \tag{10b}$$

The free base $C_6H_5(R)C{=}NOCH_3$ is electroinactive, which makes it possible to distinguish the O-alkyl oxime derivatives from the corresponding N-alkyl oxime derivatives (nitrones)

$$C_6H_5(R)\underset{\underset{O}{|}}{C}{=}NCH_3$$

in which the free base is also reducible. Only for nitrones with all alkyl substituents such as N-cyclohexyl acetaldoximine

$$(CH_3\underset{\underset{O}{|}}{CH}{=}N{-}cyclo\text{-}C_6H_{11})$$

do reductions of the free base occur at potentials too negative to give a measurable wave, and the observed pH-dependences correspond to Figure 7.

The O- and N-substituted oximes differ from tropylium ion in two respects; because of acid hydrolysis the investigation cannot be carried out in media sufficiently strongly acidic to reach the value of pK_1 and the region of pH-independent potentials. The shape of the waves furthermore indicates, by a dip on the limiting current in the region $(pK'_1 - 1) < pH < (pK'_1 + 1)$, that the protonation occurs as a surface reaction. This is further supported by the large difference between pK'_1 (between 7 and 8) and pK_1 (smaller than 1). Hence even though it is possible from the pH-dependence of the limiting current and half-wave potentials to deduce and support scheme 10, the data cannot be used for calculation of the rate constants by means of Eq. 9.

A similar situation has been found for the reduction of semicarbazones (16). The recorded pH-dependences correspond to Figure 7, the study being restricted to pH > 0 because of streaming maxima in more strongly acid solutions, and the protonation occurred as a surface reaction, as shown by dips in limiting currents at $(pK'_1 - 1) < pH < (pK'_1 + 1)$. However, for semicarbazones it has been proved (16) that the shape of the dissociation curve depends on the buffer composition and that in the range $pH > (pK'_1 \pm 1)$ the limiting current increases with increasing buffer concentration. This indicates that acids other than hydronium ion can participate in the protonation reaction, as shown in scheme 11*:

$$[R^1R^2C{=}NNHCONH_2]H^+ + H_2O \underset{k_{-1}}{\overset{k_1}{\rightleftarrows}} R^1R^2C{=}NNHCONH_2 + H_3O^+ \tag{11a}$$

* Protons indicated below an arrow (e.g., in Eq. 11c) are transferred subsequently to the potential-determining step and have no influence on wave heights or potentials.

$$[R^1R^2C=NNHCONH_2]H^+ + A^- \underset{k_{-A}}{\overset{k_A}{\rightleftarrows}} R^1R^2C=NNHCONH_2 + HA \tag{11b}$$

$$[R^1R^2C=NNHCONH_2]H^+ + 4e \xrightarrow[4H^+]{E_1} R^1R^2CH_2NH_3^+ + NH_2CONH_2 \tag{11c}$$

One wave decreasing with increasing pH was also observed (17) for iodobenzoic acids, iodophenols, and iodoanilines, but shifts in half-wave potentials were not examined.

The pH-dependence of the limiting current shown in Figure 8 can be observed for electrode processes involving a base-catalyzed antecedent reaction. To reach a limiting value of the wave height at a sufficiently high pH value, the reaction must become sufficiently fast so that at pH > ($pK_1' + 1$) diffusion becomes the slowest step. To show the shape of a simple dissociation curve (Fig. 8), it is further necessary that the reaction catalyzed by hydroxyl ions predominates over reactions catalyzed by other bases. An example of this type is the reduction of acetaldehyde (18), following scheme 12:

$$CH_3C(OH)_2(OH) + OH^- \underset{k_{-1}}{\overset{k_1}{\rightleftarrows}} CH_3C(O^-)(OH) + H_2O \tag{12a}$$

$$CH_3C(OH)(OH) + B \underset{k_{-B}}{\overset{k_B}{\rightleftarrows}} CH_3C(O^-)(OH) + BH^+ \tag{12b}$$

$$CH_3C(O^-)(OH) \underset{k_{-2}}{\overset{k_2}{\rightleftarrows}} CH_3CH=O + OH^- \tag{12c}$$

$$CH_3CHO + 2e \xrightarrow{2H^+} CH_3CH_2OH \tag{12d}$$

The limiting current is predominantly governed by the reaction with rate constant k_1. For acetaldehyde in the accessible pH range in aqueous solutions at 20°C, the reverse reaction with rate constant k_{-2} is not fast enough to transform a significant part of the aldehyde into the electroinactive hydrated monoanion, and therefore no decrease in the limiting current was observed up to pH 14. The observed shift of the half-wave potentials corresponds to the effect of the antecedent acid–base equilibrium of Eq. 12c.

The behavior depicted in Figure 9 can be caused by two types of systems: by an antecedent protonation, when the number of electrons by which the acid form is reduced is greater than the number of electrons consumed in the electroreduction of the conjugate base, or by an acid–base reaction interposed between two electron transfers. A condition for the former is that for $pH > pK'_1$ the half-wave potentials of the acid and base form are so close that the waves overlap, for the latter that the unprotonated intermediate (IM) is electroinactive in the accessible potential range.

An example from the first subgroup is the reduction of sydnones (19), which follows scheme 13:

$$\underbrace{\begin{matrix}R_1-N\!-\!\!-\!\!-\!R_2\\ \mid\ \oplus\ \mid\\ N\!-\!\!-\!O^{(-)}\\ \diagdown O \diagup\end{matrix}}_{H^+} \underset{k_{-1}}{\overset{k_1}{\rightleftharpoons}} \begin{matrix}R_1-N\!-\!\!-\!\!-\!R_2\\ \mid\ \oplus\ \mid\\ N\!-\!\!-\!O^{(-)}\\ \diagdown O \diagup\end{matrix} + H^+ \qquad (13a)$$

$$\underbrace{\begin{matrix}R_1-N\!-\!\!-\!\!-\!R_2\\ \mid\ \oplus\ \mid\\ N\!-\!\!-\!O^{(-)}\\ \diagdown O \diagup\end{matrix}}_{H^+} + 6e \xrightarrow[4H^+]{E_1} \begin{matrix}R_1-N\!-\!\!-\!\!-\!CH\!-\!COOH\\ \mid\qquad\qquad\mid\\ NH_3^+\quad\ R_2\end{matrix} \qquad (13b)$$

$$\begin{matrix}R_1-N\!-\!\!-\!\!-\!R_2\\ \mid\ \oplus\ \mid\\ N\!-\!\!-\!O^{(+)}\\ \diagdown O \diagup\end{matrix} + 4e \xrightarrow[3H^+]{E_2} \begin{matrix}R_1-N\!-\!\!-\!\!-\!CH\!-\!COO^-\\ \mid\qquad\qquad\mid\\ NH_2\quad\ R_2\end{matrix} \qquad (13c)$$

In acidic media the reduction corresponds to a six-electron process (Eq. 13b); at higher pH values it takes place in a four-electron reduction step (Eq. 13c). The height of the wave of the protonated form decreases at $pH > (pK'_1 - 1)$ in the shape of a dissociation curve (Fig. 9), but because the potentials E_1 and E_2 in this region differ only very little for most sydnone derivatives, a separation into two waves was not observed. The region corresponding to $pH < (pK_1 + 1)$ is experimentally inaccessible because in strongly acidic media a cleavage of the sydnone molecule occurs.

An example from the second subgroup is the reduction of some aromatic aldehydes and ketones in alkaline media (20, 21), which follows scheme 14:

$$C_6H_5COR + e \xrightarrow{E_1} C_6H_5\overset{(\cdot)}{C}OR \qquad (14a)$$

$$C_6H_5\dot{C}(OH)R \underset{k_{-1}}{\overset{k_1}{\rightleftharpoons}} C_6H_5\overset{(\cdot)}{C}OR + H^+ \qquad (14b)$$

$$C_6H_5\dot{C}(OH)R + e \xrightarrow{E_2} C_6H_5\overset{(-)}{C}(OH)R \qquad (14c)$$

$$C_6H_5CH(OH)R \rightleftharpoons C_6H_5\overset{(-)}{C}(OH)R + H^+ \qquad (14d)$$

For these compounds, in which the reduction of the radical anion $C_6H_5\overset{(-)}{C}OR$ occurs at too negative a potential to be observed, polarograms show only one reduction wave. Its height corresponds to $n = 2$ at medium pH values, decreases with increasing pH, and corresponds to $n = 1$ at higher pH values. This indicates that the potential E_2, at which the radical is reduced, is equal to or more positive than the potential E_1, at which the first electron uptake occurs.

Because the first reduction step (Eq. 14a) is potential-determining, the half-wave potential remains pH-independent (Fig. 9c).

The shape of the plots shown in Figure 10 is less common. It can be observed for some carbonyl compounds in the acid and medium pH range, when at $pH < (pK_1' - 1)$ the protonated form is reduced in a one-electron step and the second one-electron process is overlapped by the current of the supporting electrolyte. A further condition for the observation of this type of dependence is that at $pH > pK'$ the half-wave potentials of the protonated and of the unprotonated form lie so close together that only a single wave is observed.

To distinguish systems that correspond to Figure 7 from those that correspond to Figure 11, it is essential to determine the character of the limiting current. For systems that correspond to Figure 11, the limiting current remains diffusion-controlled even when the wave height is only a small fraction of the value observed at lower pH. However, systems corresponding to Figure 7 produce kinetic currents when examined under such conditions. Furthermore, for systems described by the plot in Figure 11, both the pH at the inflection point on the i–pH curve and that at the intersection of the two linear parts on the $E_{1/2}$–pH curve are practically equal numerically to the "true" equilibrium constant pK_1 as determined potentiometrically or spectrophotometrically. For systems represented by Figure 7, the observed inflection and intersection points (pK_1') correspond to "polarographic dissociation constants" and occur at pH values several units larger than the true constant pK_1.

The behavior shown in Figure 11 is observed when a chemical equilibrium, established in the bulk of the solution, is practically not restored when one of the components is removed by electrolysis. This implies that the equilibrium appears to be either relatively slowly or rapidly established and depends not only on the value of the rate constant involved but also on the pK_1 value of the investigated acid. From the examples discussed in Section III, it follows that the higher the pK_1 value, the larger the rate constant of the faster reaction may be without affecting the diffusion-controlled current.

The two known examples of acid–base reactions of this type are both cases of C-acids involving formation of a carbanion enolate, 3-thianaphthenone (Eq. 15) (22):

[Structure: 3-thianaphthenone with CH$_2$] $\underset{k_{-1}}{\overset{k_1}{\rightleftarrows}}$ [Structure: carbanion with (−)] + H$^+$ (15a)

[Structure: carbanion] $\longleftrightarrow$ [Structure: enolate O$^{(-)}$] (15b)

[Structure: enol OH] $\underset{k_{-2}}{\overset{k_2}{\rightleftarrows}}$ [Structure: enolate O$^{(-)}$] + H$^+$ (15c)

[Structure: 3-thianaphthenone] + 2e $\underset{2H^+}{\overset{E_1}{\longrightarrow}}$ Products (15d)

and ethylbenzoyl benzoate (Eq. 16) (23) at pH > 8:

$$C_6H_5COCH_2COOC_2H_5 \underset{k_{-1}}{\overset{k_1}{\rightleftarrows}} C_6H_5CO\overset{(-)}{C}HCOOC_2H_5 + H^+ \quad (16a)$$

$$\overset{(-)}{C_6H_5COCHCOOC_2H_5} \longleftrightarrow C_6H_5\underset{\underset{O^{(-)}}{|}}{C}=CHCOOC_2H_5 \quad (16b)$$

$$C_6H_5\underset{\underset{OH}{|}}{C}CHCOOC_2H_5 \underset{k_{-2}}{\overset{k_2}{\rightleftarrows}} C_6H_5\underset{\underset{O^{(-)}}{|}}{C}=CHCOOC_2H_5 + H^+ \quad (16c)$$

$$C_6H_5COCH_2COOC_2H_5 + 2e \underset{2H^+}{\overset{E_1}{\longrightarrow}} C_6H_5CHOHCH_2COOC_2H_5 \quad (16d)$$

In neither of these systems does the carbanion enolate give a reduction wave in the accessible potential range, but for 3-thianaphthenone it undergoes oxidation. For benzoylbenzoate at lower pH values, the protonated form is reduced and the whole system is discussed from this point of view in Section V.

The ratio of wave heights i_{HA}/i_T, where i_T is the total value observed at pH < (pK_1 − 2), follows Eq. 17:

$$\frac{i_{HA}}{i_T} = \frac{[H^+]}{K_1 + [H^+]} \quad (17)$$

and the value of K_1 found from the pH-dependence of wave heights or half-wave potentials was in each case (22, 23) in reasonable agreement with values of the equilibrium constant obtained potentiometrically or spectrophotometrically.

To the same category belong slowly established (in relation to the drop time) equilibria other than acid–base which can be followed polarographically because one or more of the components are electroactive. Among the systems that show a pH-dependence of the limiting current corresponding to Figure 11 are the equilibria between 2-hydroxylchalcones and chromanones (25) and those between α, β-unsaturated ketones and mercaptans (26).

Several types of systems that show behavior of this type involve antecedent or interposed acid–base-catalyzed reactions or a combination of acid–base properties with general catalysis.

The simplest systems corresponding to Figure 12a are probably the dehydration or ring-opening of pyridoxal (27, 28) and the dehydration of N-alkylpyridinium aldehydes (29). For pyridoxal the pH-dependence of the limiting current has been ascribed either to hemiacetal formation (Eq. 18) (27, 28), or to hydration (28) of the aldehydic grouping (Eq. 19):

Comparison of the behavior of pyridoxal with that of pyridoxal 5-phosphate (**1**) is of importance in deciding between these two possibilities. With the ester **1** the formation of hemiacetal is impossible; hence no decrease in limiting current over the critical pH range between pH 0 and 10 is expected for hemiacetal formation. Experimentally, it has been observed (28) that whereas in phosphate and trimethylamine buffers the wave height remains constant and is the same as in acidic solutions, in barbital buffers the limiting current is approximately 10–20% lower and in borate buffers the decrease is even more pronounced. If these results are considered significant and not the effect of some uncontrolled factor, the interpretation based on dehydration (Eq. 19) seems more probable for the decrease in current of pyridoxal. The pronounced effect of boric acid can be explained by complex formation of the hydrated form of pyridoxal 5-phosphate. The smaller decrease observed for pyridoxal 5-phosphate when compared with pyridoxal can be interpreted as an effect of the phosphoric acid residue in position 5, either a shift in the equilibrium toward the hydrated form or an increase in the rate of dehydration.

$$\underset{1}{\text{HO}\underset{\text{CH}_3}{\diagdown}\overset{\text{CHO}}{\underset{\text{N}}{\bigcirc}}\diagup\text{CH}_2\text{OPO(OH)}_2}$$

In scheme 19 only the simplest mechanism for acid- and base-catalyzed dehydration is considered, and ionization of the pyridine ring or the phenolic grouping is not considered. The pH-dependence of half-wave potentials (28) does not show a sharp break and does not give any information. In considering scheme 19, the small wave at pH 0–6 is attributable to the dehydration catalyzed by solvent (Eq. 19b), whereas the increase in current in acid media is attributable to acid-catalyzed reaction 19a and that in alkaline media attributable to base-catalyzed reaction 19c.

The type of pH-dependence depicted in Figure 12 has been observed for quinazoline (29, 30) and was also attributed to hydration–dehydration equilibria in which either the dehydration of the conjugate base is faster than that of the conjugate acid or the conjugate base is less hydrated. The pH dependence of the limiting current shows an increase in current at pH < 0 and pH > 2. The increase in concentration or in rate of formation of the conjugate base accounts for the increase at pH > 2 with increasing pH. The other increase in current at pH < 0 with increasing acidity corresponds to an acid-catalyzed dehydration, probably involving a protonated form. If the hydration is prevented, for example, in 4-methylquinazoline (30), the decrease in current in acidic media is not observed. Shifts in half-wave potentials were not related to this interpretation.

A reduction scheme (Eq. 20) analogous to Eq. 19 can be proposed for the reduction of N-alkylpyridinium aldehydes (31) showing a similar type of pH-dependence (Fig. 14):

$$\text{CH(OH)}_2\text{H}^+\text{-Py-R} \underset{k_{-2}}{\overset{k_2}{\rightleftharpoons}} \text{CHO-Py-R} + \text{H}_3\text{O}^+ \quad (20a)$$

$$k_1 \updownarrow k_{-1} \qquad\qquad k_3 \updownarrow k_{-3}$$

$$\text{CH(OH)}_2\text{-Py-R} \underset{k_{-7}}{\overset{k_7}{\rightleftharpoons}} \text{CHO-Py} + \text{H}_2\text{O} \quad (20b)$$

$$k_6 \updownarrow k_{-6} \qquad\qquad k_4 \updownarrow k_{-4}$$

$$\text{CH(OH)O}^-\text{-Py-R} \underset{k_{-5}}{\overset{k_5}{\rightleftharpoons}} \text{CHO-Py-R} + \text{OH}^- \quad (20c)$$

$$\text{CHO-Py-R} + 2e \xrightarrow[2\text{H}^+]{E_1} \text{CH}_2\text{OH-Py-R} \quad (20d)$$

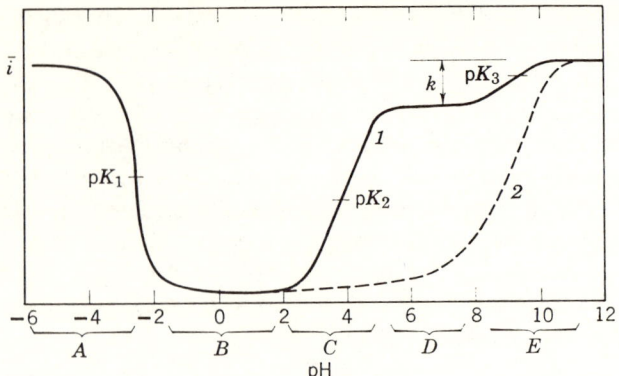

FIG. 14. pH-dependences of the limiting current $\bar{i}$ of a pyridine aldehyde (—) and N-methylpyridinium aldehyde (– – –). pK_1, pK_2, and pK_3 correspond to Eqs. 20 and 21 (schematically).

In region A, the acid-catalyzed reaction follows the path k_{-1}, k_2, k_3.

In regions B, C, and D the uncatalyzed reaction follows path k_7, while in region E the base-catalyzed reaction follows the path k_6, k_5, k_{-4}. Whether the limiting current depends solely on pH or also on buffer type and concentration remains to be determined. The pH-dependence of the half-wave potential for this system has not been reported.

The situation becomes more complicated when pyridine carboxaldehydes (27, 31) unsubstituted on the ring nitrogen are investigated, as in this case the ring can be also protonated. A comparison of the N-alkyl (curve 2, Fig. 14) and the unmethylated aldehyde (curve 1) reveals which part of the pH-dependence is affected by the protonation. The dehydration can be interpreted by scheme 21:

$$\text{4-CH(OH)}_2\text{-pyridinium} \underset{k_{-3}}{\overset{k_3}{\rightleftharpoons}} \text{4-CHO-pyridinium} + \text{H}_3\text{O}^+ \qquad (21a)$$

$$k_2 \updownarrow k_{-2}$$

SYSTEMS SHOWING ONE REDUCTION WAVE 111

$$\underset{\underset{H}{\overset{+}{N}}}{\text{CH(OH)}_2\text{-pyridinium}} \underset{k_{-9}}{\overset{k_9}{\rightleftharpoons}} \underset{\underset{H}{\overset{+}{N}}}{\text{CHO-pyridinium}} + H_2O \quad (21b)$$

$$k_1 \updownarrow k_{-1} \qquad\qquad k_4 \updownarrow k_{-4}$$

$$\underset{\underset{\text{\textcircled{3}}}{N}}{\text{CH(OH)}_2\text{-pyridine}} \underset{}{\overset{k_{10}}{\rightleftharpoons}} \underset{N}{\text{CHO-pyridine}} + H_2O \quad (21c)$$

$$k_8 \updownarrow k_{-8}$$

$$\underset{N}{\text{CH(OH)O}^-\text{-pyridine}} \overset{k_7}{\rightleftharpoons} \underset{N}{\text{CHO-pyridine}} + OH^- \quad (21d)$$

$$\underset{\underset{H}{\overset{+}{N}}}{\text{CHO}} + 2e \xrightarrow[2H^+]{E_2} \underset{\underset{H}{\overset{+}{N}}}{\text{CH}_2\text{OH}} \quad (21e)$$

$$\underset{N}{\text{CHO}} + 2e \xrightarrow[2H^+]{E_1} \underset{N}{\text{CH}_2\text{OH}} \quad (21f)$$

Only the simplest mechanism for the acid- and base-catalyzed dehydration is indicated in this scheme. In region A (Fig. 14), the dehydration of the protonated form (**2**) occurs by an acid-catalyzed reaction following the path

k_{-2}, k_3, k_4. In region B the protonated form (2) is dehydrated by an uncatalyzed reaction k_9. In region D the unprotonated form (3) undergoes a dehydration catalyzed only by the solvent (k_{10}), and in region E form 3 is dehydrated by a base-catalyzed reaction following the path k_8, k_7. The change in the limiting current in region C is based on comparison with the behavior of quaternized compounds and with changes in ultraviolet spectra with pH ascribed to the acid–base equilibrium between forms 2 and 3 with the equilibrium constant $k_1/k_{-1} = K_1$. The shifts in half-wave potentials (32) with pH resemble the plots in Figure 3f and do not exclude the possibility of separation of two waves in the region E, corresponding to pK_8.

The third subgroup includes systems in which there is a general catalyzed reaction interposed between two electron transfers and in which the reduction of the product of this chemical reduction occurs at the same potential as, or at more positive potentials than, the initial reduction step. An example of this type is the reduction of o- and p-nitrophenols (33) and nitroanilines, demonstrated for p-nitrophenol in scheme 22:

$$p\text{-}NO_2\text{-}C_6H_4\text{-}OH + 2e + 4H^+ \xrightarrow{E_1} p\text{-}NHOH\text{-}C_6H_4\text{-}OH \quad (22a)$$

$$p\text{-}NHOH\text{-}C_6H_4\text{-}OH \underset{\text{general catalysed}}{\rightleftarrows} p\text{-}NH\text{=}C_6H_4\text{=}O + H_2O \quad (22b)$$

$$p\text{-}NH\text{=}C_6H_4\text{=}O + 2e + 2H^+ \underset{E_2}{\rightleftarrows} p\text{-}NH_2\text{-}C_6H_4\text{-}OH \quad (22c)$$

The acid catalysis of the dehydration (Eq. 22b) results in an increase in the four-electron wave into a six-electron wave in acid media, and the base catalysis of the dehydration results in an analogous increase in alkaline solutions. Because potential E_2 is more positive than potential E_1, the reduction occurs in a single wave. Because the lifetime of the quinoneimine intermediate is short, its hydrolysis—sometimes considered a side reaction—can be neglected.

Application of the treatment (34, 35) derived for polarographic currents of systems in which a chemical reaction of a product of an irreversible electrode process is further reduced (ECE mechanism) permits calculation of rate constants. The computed values of these constants depend on the drop time (36), which indicates that the treatment is not complete.

The reduction of p-nitrosophenol (37, 39) follows a path similar to Eq. 22, except that the first step corresponds to a transfer of two rather than four electrons and is reversible. It is the latter property that has made theoretical treatments (38, 39) of this system possible.

Systems that show a pH-dependence similar to that in Figure 13 were observed for hydrated carbonyl compounds and belong to two subgroups. The first includes compounds for which a decrease in the i–pH plot in alkaline media is attributable only to the dehydration reaction, and the second includes those systems in which this decrease is caused by dissociation in another reaction center of the molecule. In both cases, nevertheless, the original increase in current observed at somewhat lower pH values is caused by dehydration catalyzed predominantly by hydroxyl ions.

An example of the first type is the reduction of formaldehyde (18, 40), which follows the path shown in (Eq. 23):

$$H_2C(OH)_2 + \text{base} \underset{k_{-1}}{\overset{k_1}{\rightleftarrows}} H_2C(O^-)(OH) + \text{base H}^+ \qquad (23a)$$

$$H_2C(O^-)(OH) \underset{k_{-2}}{\overset{k_2}{\rightleftarrows}} H_2C{=}O + OH^- \qquad (23b)$$

$$H_2C{=}O + 2e \xrightarrow[2H^+]{E_1} CH_3OH \qquad (23c)$$

The formation of an anion in alkaline solutions of formaldehyde has been proved by ultraviolet spectra and by conductivity measurements. The anion could in principle be derived from the hydrated form in Eq. 23a, or from the nonhydrated form as a carbanion $[HCO]^-$. This second possibility has been excluded on the basis of the absence of an absorption band at 300 nm and of the nonreactivity of formaldehyde in alkaline solutions in typical nucleophilic reactions of carbanions (18).

The formation of the anion of the hydrated form $H_2C(OH)O^-$ had earlier been considered (40) a side reaction relative to the dehydration, and the equilibrium Eq. 23a (40) or the rate of the anion formation (41) with rate constant k_1 was thought to be responsible for the decrease in the limiting current with increasing pH at pH $>$ 12. To explain the general base catalysis in accordance with other reactions of this type, it is nevertheless necessary to assume that the anion of the hydrated form $H_2C(OH)O^-$ is an intermediate rather than a side product. Scheme 23 is further supported by the fact that OH^- ions present in alkaline media are much stronger nucleophilic agents than water. Their addition in the reverse reaction (Eq. 23b) is therefore more probable, and scheme 23 is thus in full agreement with other nucleophilic reactions of carbonyl compounds (e.g., additions of amines or mercaptan anions), as well as with nucleophilic additions of hydroxyl ions to other double bonds. If the anion $H_2C(OH)O^-$ is thus considered an intermediate, its formation in reaction 23a cannot be an explanation of the decrease in the limiting current at pH $>$ 12.

Present (18) interpretation of the pH-dependence of the limiting current of formaldehyde in a shape corresponding to Figure 13 is that the increase in current in the region between pH 7 and 12 is attributable to an increase in the rate of reaction 23a with rate constant k_1 resulting from the increase in base concentration. The decrease in the current at pH $>$ 12 is explained by an increase in the rate of the reverse reaction (Eq. 23b) with rate constant k_{-2}. With an increase in the rate of nucleophilic addition of hydroxyl ions as a result an increase in the concentration of hydroxyl ion as nucleophilic reagent, the surface concentration of the electroactive form H_2CO decreases at the expense of the electroinactive anion $H_2C(OH)O^-$.

Whereas most aliphatic aldehydes correspond to scheme 12 and show at 20°C a limiting value rather than a decrease in alkaline media, aldehydes bearing a phenyl group in the position alpha to the aldehydic group show a pH-dependence of the limiting current corresponding to Figure 13. Contrary to formaldehyde, however, these compounds (e.g., mono- and diphenylacetaldehyde or 2-phenylpropionaldehyde) (18) show in ultraviolet spectra the characteristic absorption band of carbanions at 300 nm. Furthermore, these aldehydes show reactions characteristic of carbanions, for example, with molecular oxygen. For these reasons it is assumed that the observed dependence can be interpreted by scheme 24:

$$R^1R^2CHCH(C_6H_5)(OH)(OH) + \text{base} \underset{k_{-1}}{\overset{k_1}{\rightleftharpoons}} R^1R^2CHCH(C_6H_5)(O^-)(OH) + \text{base} + H^+ \quad (24a)$$

$$R^1R^2CHCH\genfrac{}{}{0pt}{}{O^-}{\diagdown OH}\genfrac{}{}{0pt}{}{}{C_6H_5} \underset{k_{-2}}{\overset{k_2}{\rightleftharpoons}} R^1R^2CHCH\!=\!O + OH^- \quad (24b)$$
$$\phantom{R^1R^2CHCH\genfrac{}{}{0pt}{}{O^-}{\diagdown OH}}\underset{C_6H_5}{|}$$

$$R^1R^2CHCHO \underset{k_{-3}}{\overset{k_3}{\rightleftharpoons}} R^1R^2\overset{(-)}{C}CH\!=\!O + H^+ \quad (24c)$$
$$\underset{C_6H_5}{|} \phantom{\underset{k_{-3}}{\overset{k_3}{\rightleftharpoons}}} \underset{C_6H_5}{|}$$

$$R^1R^2\overset{(-)}{C}CH\!=\!O \longleftrightarrow R^1R^2C\!=\!CH\!-\!O^{(-)} \quad (24d)$$
$$\underset{C_6H_5}{|} \underset{C_6H_5}{|}$$

$$R^1R^2CHCH\!=\!O + 2e \xrightarrow[2H^+]{E_1} R^1R^2CHCH_2OH \quad (24e)$$
$$\underset{C_6H_5}{|} \phantom{+ 2e \xrightarrow[2H^+]{E_1}} \underset{C_6H_5}{|}$$

The increase in current at pH 7–11 is attributable to an increase in the rate of formation of the anion of the hydrate (Eq. 24a), while the decrease at higher pH values can result either from an increase in the rate of the nucleophilic reaction (Eq. 24b) with a constant k_{-2}, or from an increase in the rate of dissociation (Eq. 24c) with rate constant k_3. Because of spectroscopic evidence and, because α-phenylacetaldehydes are considerably less hydrated than formaldehyde, and in agreement with the increase in acidity of the CHCHO grouping resulting from the introduction of phenyl, the formation of the carbanion enolate (Eq. 24c) seems to be a more plausible interpretation.

The principal role in distinguishing between schemes 23 and 24, apart from structural deductions, is played by ultraviolet spectra at pH values several units lower than the pH region in which the decrease in current occurs.

A similar type of pH-dependence has been observed (42, 43) for some N-alkyl-γ-piperidones (4). As N-dialkyl-γ-piperidonium salts (5) have shown an increase in current at pH 6–10, but no decrease at higher pH values, the decrease observed for the N-alkyl derivatives seems properly attributed to dissociation of hydrogen ion from an N-alkyl-γ-piperidinium ion (6). The nitrogen-protonated γ-piperidone derivative (6) seems to be the electroactive form. However, the interpretation suggesting that the increase in current at pH 6–10 is attributable to base-catalyzed dehydration is doubtful. The original claim based on spectroscopic evidence (42) could not be confirmed (43), as the spectra change with pH much less than expected for the degree of hydration necessary to bring about the observed decrease in current. Either the grouping responsible for polarographic reduction is not the same as that responsible for the ultraviolet absorption at 290 nm, or another group affecting the polarographic reducibility but not the ultraviolet absorption is hydrated.

V. Systems Showing Two Reduction Waves

Changes in polarographic waves with pH that result in the formation of two separate waves on current–voltage curves but that do not affect the overall height are rather frequently observed. For most of the systems, nevertheless, the region of sufficiently low pH values, where the half-wave potential remains pH-independent, is experimentally inaccessible, either because of cleavage, because of a change in the reduction scheme (a further protonation), or because of an overlapping of the observed waves by the current of hydrogen ion evolution. Thus for α-keto acids for which this type of phenomena was first observed (44) and correctly interpreted (45, 46), protonation, hydration (47, 48), and possibly even enol formation (47, 48) affect the behavior at lower pH values. Moreover, it has been shown (49, 50) that acids other than hydronium ions can participate in the protonation reaction. For pyridinecarboxylic acids (51, 52), the reaction scheme, hence the shift in half-wave potentials, is complicated by the protonation of the pyridine ring; for 1,3-diketones and related compounds (23), it is complicated by protonation of the carbonyl group in addition to the dissociation of the methylene grouping; for α, β-unsaturated alicyclic ketones (53, 54), the reduction waves overlap the current of hydrogen evolution in strongly acidic media, in which the protonated form exists in equilibrium. For nitrones derived from benzaldehyde, acetophenone, and benzophenone (15), sufficiently acidic media cannot be reached because of hydrolyses.

Thus one of the simplest systems showing behavior corresponding to Figure 15 has been observed (31, 54, 55) for the reduction of halopyridines.

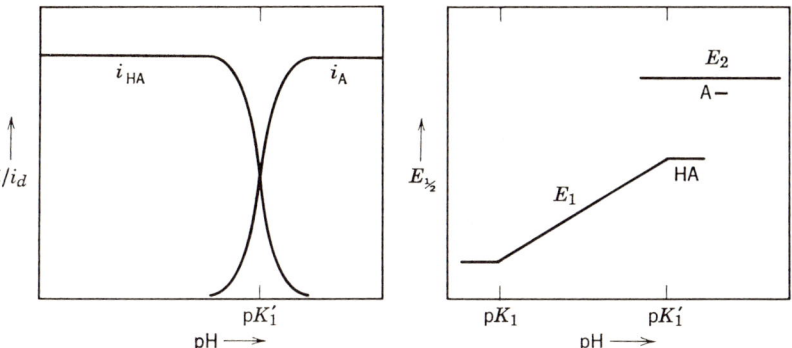

FIG. 15. pH-dependences of wave heights and half-wave potentials corresponding to the scheme:

$$HA \rightleftharpoons A^- + H^+ \quad \text{fast} \to \text{comparably fast}$$
$$HA + n_1 e \xrightarrow{E_1} P_1 \quad pK_1 \quad pK_1'$$
$$A^- + n_2 e \xrightarrow{E_2} P_2$$
$$n_1 = n_2; \ |E_1| < |E_2|$$

The best developed waves and fewest complications from catalysis were observed for iodopyridines, which follow scheme 25:

(25a) $\text{pyridinium-I} \underset{k_{-1}}{\overset{k_1}{\rightleftharpoons}} \text{pyridine-I} + H^+$

(25b) $\text{pyridinium-I} + 2e \xrightarrow[H^+]{E_1} \text{pyridinium} + I^-$

(25c) $\text{pyridine-I} + 2e \xrightarrow[H^+]{E_2} \text{pyridine} + I^-$

With increasing pH the height of the wave of the protonated form decreases at the expense of that of the more negative wave of the free base. The potential (E_2) of the reduction of free base is pH-independent. A proof for the above scheme is the behavior of corresponding N-alkyliodopyridinium salts (31, 54, 55). Neither the wave heights nor the half-wave potentials of these alkylated compounds depends on pH. Formation of the halogenides has been proved by means of the commutator technique (55).

The values in the accompanying tabulation have been found in 50% ethanol (55) for iodopyridines [from graphs in reference (55)].

	pK		pK'	
	Spectra	$E_{1/2}$–pH	i–pH	$E_{1/2}$–pH
2-I	1.11	1.2	8.4	7.8
3-I	2.32	—	9.05	8.5
4-I	2.93	2.6	9.75	9.4

It has been stated (55) that "the positions of the bends on the $E_{1/2}$–pH plots are not strictly identical with pK and pK'." It is important to check whether the difference between pK values obtained from spectra and those from $E_{1/2}$–pH plots is significant, or reflects the inaccuracy in determination of the latter. Similarly, the difference in the pK' values is of the order predicted by Saveant (7), and the system would be useful for checking equations derived in his paper (7). For the sake of deduction about mechanism, in which an approximate coincidence of pK values obtained by polarographic and other methods and of pK' values obtained from polarographic currents and potentials can be considered sufficient, the above data and for pK' the identical sequence of both pairs of values can be assumed to support scheme 25.

Because of the dips on limiting currents of the protonated form at pH ≈ pK', because of the large difference between pK and pK' (which corresponds to rate constants of the order of 10^{15} liter mole^{-1} sec^{-1}, and because of the dependence of the pK' value on halopyridine concentration, it can be deduced that reaction 25a occurs as a surface reaction (5, 13). The dependence of pK' on halopyridine concentration can be explained by adsorption of the electroactive compound or the reduction product. The adsorbed pyridine derivatives can also act at the surface as proton donors.

Calculation of the rate constants of the protonation reaction is possible for some C-acids, such as ω-cyanoacetophenone or benzoylacetone (23), as in these cases protonation takes place as a homogeneous, volume reaction. These systems are at lower pH values complicated by protonation of the carbonyl group, and are discussed in full detail in Section VI. Here we restrict ourselves to the behavior at pH values greater than about 7, where only reaction scheme 26 can be considered (R = CN, COC_6H_5, and so on):

$$C_6H_5COCH_2R \underset{k_{-1}}{\overset{k_1}{\rightleftharpoons}} C_6H_5\overset{(-)}{C}OCHR + H^+ \qquad (26a)$$

$$C_6H_5\overset{(-)}{C}OCHR \longleftrightarrow \underset{\underset{O^{(-)}}{|}}{C_6H_5C}=CHR \qquad (26b)$$

$$\underset{\underset{OH}{|}}{C_6H_5C}=CHR \underset{k_{-2}}{\overset{k_2}{\rightleftharpoons}} \underset{\underset{O^{(-)}}{|}}{C_6H_5C}=CHR + H^+ \qquad (26c)$$

$$C_6H_5COCH_2R + 2e \xrightarrow[2H^+]{E_1} C_6H_5CHOHCH_2R \qquad (26d)$$

$$C_6H_5\overset{(-)}{C}OCHR + 2e \xrightarrow[3H^+]{E_2} C_6H_5CHOHCH_2R \qquad (26e)$$

As the rate with constant k_2 is considerably greater than that with constant k_{-1}, it is the rate of the latter reaction that governs the rate of formation of the electroactive form $C_6H_5COCH_2R$. To calculate the value of the rate constant k_{-1}, it is necessary to determine the value of $K_1 = k_1/k_{-1}$, which is not accessible experimentally but can be calculated from the overall acid–base constant K_T determined spectrophotometrically and from the ratio $[C_6H_5COCH_2R]/[C_6H_5C(OH)=CHR]$. According to the nature of group R, the value of the rate constant k_{-1} was found (23) to be 10^8–10^{10} liter mole^{-1} sec^{-1}, the calculation being made with Eq. 27a or b:

$$\frac{i_{HA}}{i_{HA} + i_A} = \frac{0.886[H_3O^+](k_{-1}t_1/K_1)^{1/2}}{1 + 0.886[H_3O^+](k_{-1}t_1/K_1)^{1/2}} \qquad (27a)$$

$$\log k_{-1} = 2pK' - pK_1 - 2\log 0.886 - \log t_1 \qquad (27b)$$

where i_{HA} refers to the limiting current of the form $C_6H_5COCH_2R$ and $i_{HA} + i_A$ to the total diffusion-controlled two-electron wave.

Experimental data fitting Figure 15 were obtained for some aryl ketones over a pH range such that, in acidic media, two one-electron waves coalesce before their height decreases, while at higher pH values the separation of the one-electron step does not occur (20, 21). For these compounds scheme 28 is in agreement with experimental data:

$$\text{ArCOHR} \overset{(+)}{\underset{k_{-1}}{\overset{k_1}{\rightleftarrows}}} \text{ArCOR} + \text{H}^+ \qquad (28\text{a})$$

$$\text{ArCOHR}^{(+)} + 2e \underset{\text{H}^+}{\overset{E_1}{\rightleftarrows}} \text{ArCHOHR} \qquad (28\text{b})$$

$$\text{ArCOR} + 2e \underset{2\text{H}^+}{\overset{E_2}{\rightleftarrows}} \text{ArCHOHR} \qquad (28\text{c})$$

For these systems it has been proved (56) that in addition to hydronium ion other proton-donors can participate in the protonation reaction with rate constant k_{-1}.

The way in which buffers are prepared is of great importance in studies of the pH-dependence of the limiting current in such systems. When buffers are prepared in the most conventional way by mixing the acidic and basic buffer components in varying ratios while keeping the total analytical concentration (i.e., the sum of the concentrations of the acidic and basic forms of each buffer component) constant, the resulting polarographic curves are difficult to interpret quantitatively. This is because three variables (the pH and the concentrations of both the acidic and the basic components of the buffer) are all changed simultaneously. The treatment of results obtained with such solutions is unnecessarily complicated.

Even if the concentration of the acidic buffer component $[\text{BH}^+]$ is kept constant, the shape of the resulting i–pH curve varies with changes in buffer composition and differs from the theoretical dissociation curve, corresponding to Eq. 27. This can be deduced from Eq. 29, which describes the pH-dependence of the ratio i/i_d in buffers containing a constant concentration of the acidic buffer component $[\text{BH}^+]$ but having different pH values because they contain different concentrations of the basic buffer component $[\text{B}]$.

$$\frac{\bar{i}}{\bar{i}_d} = \frac{0.886(t_1/K_1)^{1/2}[\text{H}_3\text{O}^+]\{k_{\text{H}_3\text{O}^+} + k_{\text{BH}^+}[\text{BH}^+]/[\text{H}_3\text{O}^+]\}^{1/2}}{1 + 0.886(t_1/K_1)^{1/2}[\text{H}_3\text{O}^+]\{k_{\text{H}_3\text{O}^+} + k_{\text{BH}^+}[\text{BH}^+]/[\text{H}_3\text{O}^+]\}^{1/2}} \qquad (29)$$

With changes in pH the value of $k_{\text{BH}^+}[\text{BH}^+]/[\text{H}_3\text{O}^+]$ at constant $[\text{BH}^+]$ changes, and therefore the resulting shape of the dissociation curve may differ from that corresponding to Eq. 27. For the shape of the dissociation curve, the ratio $k_{\text{BH}^+}[\text{BH}^+]/k_{\text{H}_3\text{O}^+}$ is decisive (Fig. 16). When the value of this ratio is large (i.e., when $k_{\text{H}_3\text{O}^+} \ll k_{\text{BH}^+}[\text{BH}]$, as is most likely to be true at high concentrations of the acidic buffer component), the slope of the dissociation curve is one-half that predicted by Eq. 27.

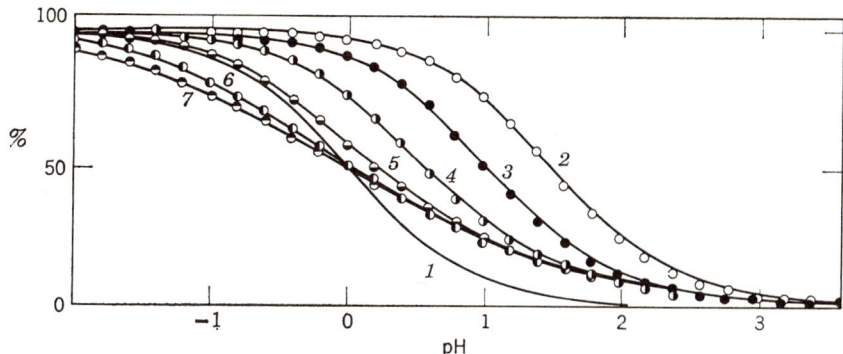

FIG. 16. pH-dependence of the ratio i/i_d for various values of $k_{BH^+}[BH^+]/k_{H_3O^+}$ in buffers with acid component concentration constant. Theoretical curves for: 1 $k_{H_3O^+} \gg k_{BH^+}[BH^+]/[H_3O^+]$; ratio $k_{BH^+}[BH^+]/k_{H_3O^+}$ is: 2 10^{-3}; 3 10^{-2}; 4 10^{-1}; 5 1; 6 10; 7 100.

Only when the concentration of the basic buffer component [B] is kept constant, the pH being changed by varying the concentration of the acidic buffer component [BH$^+$], is the shape of the dissociation curve the same as that predicted by Eq. 27. Under these conditions Eq. 30 is valid:

$$\frac{i}{i_d} = \frac{0.886(t_1/K_1)^{1/2}[H_3O^+]\{k_{H_3O^+} + k_{BH^+}[B]/K_{BH^+}\}^{1/2}}{1 + 0.886(t_1/K_1)^{1/2}[H_3O^+]\{k_{H_3O^+} + k_{BH^+}[B]/K_{BH^+}\}^{1/2}} \quad (30)$$

It can be seen from this equation that when [B] is kept constant the quantity within the braces is constant and Eq. 30 assumes the same form as Eq. 27. Nevertheless, it can be shown (56) that, although the shape of the dissociation curve is identical with that for the reaction of a monobasic acid anion with a proton, it is shifted along the pH axis according to the value of [B]. The value of pK' is therefore a function of [B]. From the dependence of pK' on [B], it is possible to determine both $k_{H_3O^+}$ and k_{BH^+} if the value of K_{BH^+} (the dissociation constant of the buffer acid) is known.

A separation of two waves and a decrease in the height of the more positive one in the shape of a dissociation curve have also been described for iodobenzoic acids (17, 57), iodoanilines (17), nitrobenzoic acids (58, 59), benzaldehydes bearing an amino or phenolic grouping (60), formylbenzoic acids (61), oximes (62, 63), thiosemicarbazones (16, 64), and numerous other examples for which the data are not detailed enough for a thorough discussion. Generally speaking, a dependence of the type shown in Figure 15 can be obtained whenever the organic molecule contains a group that can undergo protonation and the reduction process of the conjugate base involves the same number of electrons as the corresponding acid. The acidic group can be either the electroactive group that undergoes an electrochemical change in the course of reduction, or a substituent that is not chemically affected in the reduction process.

Other cases in which two waves have been observed on polarographic curves have been reported much less frequently. A summary of some of the observed types is presented in Figures 17–24, and examples are discussed briefly.

The dependence of wave heights and half-wave potentials depicted in Figure 17 has been observed for some sydnones (19) and corresponds to

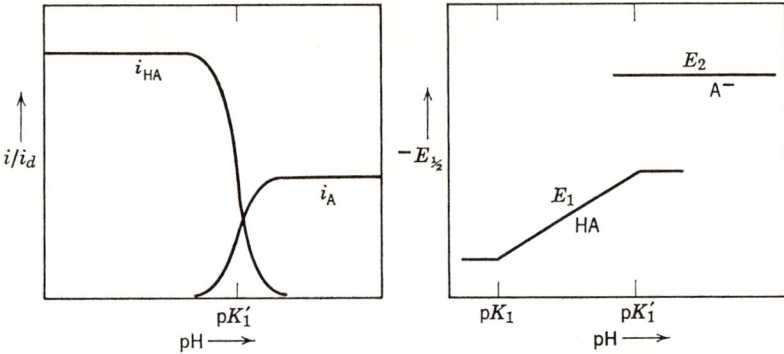

FIG. 17. pH-dependences of wave heights and half-wave potentials corresponding to the scheme:

$$HA \rightleftharpoons A^- + H^+ \quad \text{fast} \to \text{comparably fast}$$
$$HA + n_1 e \xrightarrow{E_1} P_1 \quad pK_1 \quad pK_1'$$
$$A^- + n_2 e \xrightarrow{E_2} P_2$$
$$n_1 > n_2; \quad |E_1| < |E_2|$$

scheme 13 under the condition that at pH = $(pK' \pm 1)$ the potential for the reduction of the protonated form (E_1) is more positive than that for the reduction of the free base (E_2), and two separate waves result.

The plots in Figure 18 are shown (20, 21) by aryl ketones and substituted benzaldehydes for which the second one-electron wave, corresponding to

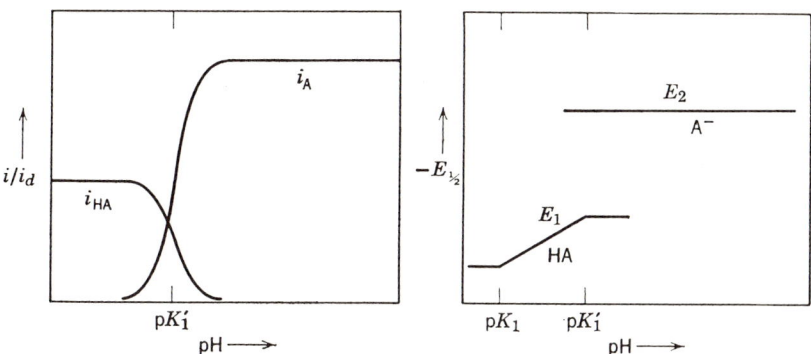

FIG. 18. pH-dependences of wave heights and half-wave potentials corresponding to the scheme:

$$\text{HA} \rightleftharpoons \text{A}^- + \text{H}^+$$
$$\text{HA} + n_1 e \xrightarrow{E_1} \text{P}_1$$
$$\text{A}^- + n_2 e \xrightarrow{E_2} \text{P}_2$$
$$n_1 < n_2; \quad |E_1| > |E_2|$$

reduction of the radical, is superimposed on the current resulting from hydrogen evolution in acidic media. Scheme 28 applies with the exception that the two-electron reduction (Eq. 28b) is replaced by a one-electron step (Eq. 28d):

$$\text{Ar}\overset{(+)}{\text{C}}\text{OHR} + e \xrightarrow{E_1} \text{Ar}\dot{\text{C}}\text{OHR} \tag{28d}$$

For systems corresponding to both Figures 17 and 18, it is usually impossible to reach the acid pH range in which potential E_1 is pH-independent.

Two types of systems can manifest themselves in polarographic curves the heights of which decrease with increasing pH until they reach one-half the original value with a simultaneous increase in a new, more negative wave i_2. The overall height ($i_1 + i_2$) remains constant. The first system involves an interposed proton transfer, the second an antecedent one.

An example of the first type (Fig. 19b) is the reduction of numerous carbonyl compounds in alkaline media, as interpreted by scheme 31:

$$\text{RCOR}' + e \xrightarrow{E_1} \overset{(-)}{\text{R}\dot{\text{C}}\text{OR}'} \tag{31a}$$

$$\dot{\text{R}}\text{COHR}' \underset{k_{-1}}{\overset{k_1}{\rightleftharpoons}} \overset{(-)}{\text{R}\dot{\text{C}}\text{OR}'} + \text{H}^+ \tag{31b}$$

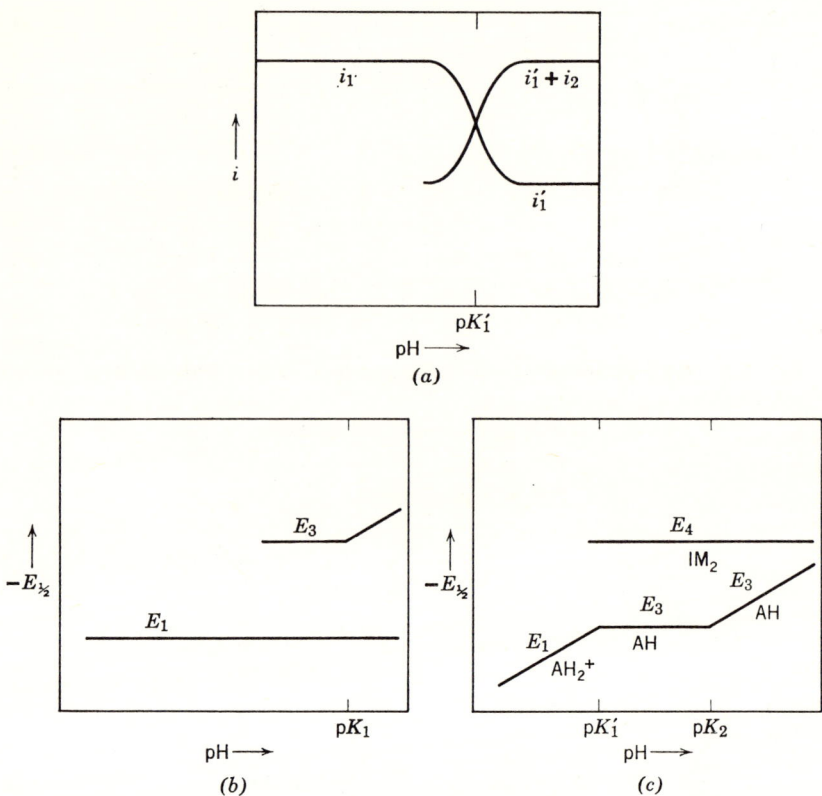

FIG. 19. pH-dependences of wave heights (a) and half-wave potentials (b, c) corresponding to the scheme:

(b) $AH + n_1 e \xrightarrow{E_1} IM$
          $IMH^+ \rightleftharpoons IM + H^+$ fast → comparably fast
          $IMH^+ + n_2 e \xrightarrow{E_2} P_1$ pK_1 pK_1'
          $IM + n_3 e \xrightarrow{E_3} P_2$
          $|E_1| \approx |E_2| < |E_3|$

(c) $AH_2^+ \rightleftharpoons AH + H^+$ fast → comparably fast
          $AH_2^+ + n_1 e \xrightarrow{E_1} IM_1$ pK_1'
          $IM_1 + n_2 e \xrightarrow{E_2} P_1$
          $AH + n_3 e \xrightarrow{E_3} IM_2$
          $IM_2 + n_4 e \xrightarrow{E_4} P_2$
          $AH \rightleftharpoons A^- + H^+$ fast
          $|E_1| \gtrapprox |E_2|$; $|E_3| \approx |E_1|(pH > pK_1')$; $|E_3| < |E_4|$

$$\text{R}\overset{\cdot}{\text{C}}\text{OHR}' + e \xrightarrow{E_2} \overset{(-)}{\text{RCOHR}'} \tag{31c}$$

$$\overset{(\div)}{\text{R}\overset{\cdot}{\text{C}}\text{OR}'} + e \xrightarrow{E_3} \overset{(2-)}{\text{RCOR}'} \tag{31d}$$

The products of reaction 31c and d are protonated in a subsequent step to give an alcohol. The radicals $\text{R}\overset{\cdot}{\text{C}}\text{OHR}'$ and radical anions $\overset{(\div)}{\text{R}\overset{\cdot}{\text{C}}\text{OR}'}$ can undergo side reactions, such as dimerization or interaction with electrode material, solvent, or other molecules of the carbonyl compound. Reduction of the radical takes place at potential E_2 which differs little from, or is more positive than, that of the first electron uptake E_1. As long as the rate of protonation (k_{-1}) is fast enough to transform all the radical anion $\overset{(\div)}{\text{R}\overset{\cdot}{\text{C}}\text{OR}'}$ into radical, one two-electron wave is observed. Because no proton transfer takes place before the potential-determining first irreversible electron uptake, the half-wave potential E_1 remains pH-independent (Fig. 19b). Decreasing the rate of protonation (Eq. 31b) results in a decrease in wave i_1, the unprotonated radical anion being reduced at a more negative potential E_3 in wave i_2. The half-wave potential of wave i_2 is pH-dependent, as the potential-determining step (Eq. 31d) is preceded by the acid–base equilibrium of Eq. 31b. The intercept on the E_3–pH plot corresponds approximately to pK_1.

A similar change in wave heights with pH has been observed for phthalimide (65) and was also attributed to an interposed protonation of the radical formed in the first step. A recent reinvestigation (66), with particular attention paid to the shift in half-wave potentials of the first wave, indicated that an antecedent protonation must be involved and that the reduction follows a more complex scheme (Eq. 32):

(32a)

(32b)

(32c)

(32d)

$$\underset{(\rightleftharpoons)}{\text{phthalimide-CO-NH-CO}} + e \xrightarrow[2\text{H}^+]{E_4} \text{phthalimide-CHOH-NH-CO} \quad (32\text{e})$$

$$\text{phthalimide-CO-NH-CO} \underset{k_{-2}}{\overset{k_2}{\rightleftharpoons}} \text{phthalimide-CO-N}^{(-)}\text{-CO} + \text{H}^+ \quad (32\text{f})$$

This scheme can be complicated by side reactions of radical anions even though its contribution does not seem essential, as the concentration dependence of the wave heights of the two-electron wave in acidic and of the two one-electron waves in alkaline media is linear and does not indicate a second-order reaction. The observed (65) dependence of the polarographic dissociation curve on phthalimide concentration can be attributed to the surface character of protonation reaction 32a, indicated in the shape of polarographic waves at pH $\approx$ pK_1'. If the adsorbed protonated phthalimide can act as a proton donor, the observed dependences of the wave height i_1 at (pK_1' − 1) < pH < (pK_1' + 1) and of the half-wave potential of i_1 on phthalimide concentration can be understood, as well as the pH-independence of the half-wave potential of i_1 at pH > (pK_1' + 1).

The potential E_2 for the reduction of the radical (Eq. 32c) is similar to, or more positive than, the potential for the first one-electron transfer (E_1), and therefore only one two-electron wave i_1 is observed in acidic media. Because the acid–base equilibrium (Eq. 32a) is antecedent to the potential-determining step (Eq. 32b), the half-wave potential of the first wave (E_1) in this pH range [at pH < (pK_1' − 1)] is pH-dependent. When the rate of protonation (k_{-1}) is no longer fast enough to transform all the phthalimide into the protonated form, a decrease in wave i_1 and the formation of a more negative wave i_2 are observed at pH = (pK_1' ± 1). At (pK_1' + 1) < pH < (pK_2 − 1), reduction occurs in two pH-independent steps (Eq. 32d and e). The pH-independence of the half-wave potential E_4 of the wave i_2 indicates that the radical anion formed in the first step (Eq. 32d) is not protonated to any considerable degree before being reduced in Eq. 32e. The half-wave potential of the first wave E_3 is pH-independent, as the uncharged phthalimide predominating in the bulk of the solution is reduced. At pH = (pK_1' + 1) the potentials E_1 and E_3 must be similar, as no separation of two waves is observed. When pH > (pK_2 − 1), equilibrium (Eq. 32f) becomes important and because of its rapid establishment the potential E_3 starts to shift to more negative potentials. The pH of the intersection of the two linear portions on the $E_{1/2}$–pH plot (9.6) is in reasonable agreement with the reported (65) value of pK_2 (9.80).

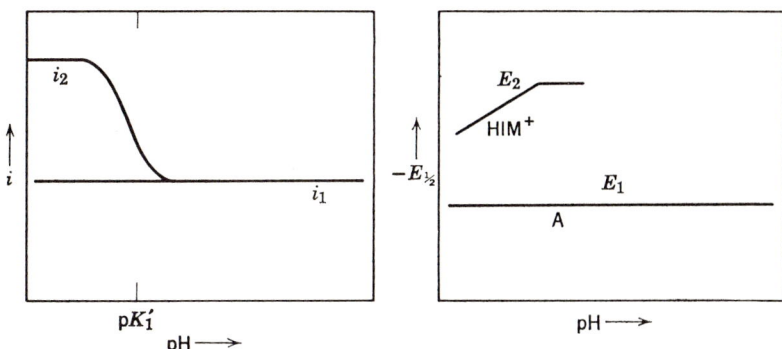

FIG. 20. pH-dependences of wave heights and half-wave potentials corresponding to the scheme:

$$A + n_1 e \xrightarrow{E_1} IM_1$$
$$IM_1 + \text{acid} \rightleftharpoons IMH^+$$
$$IMH^+ + n_2 e \xrightarrow{E_2} P_1$$

The essential factor in determining whether a given compound follows a scheme analogous to Eq. 31 or 32 is the effect of pH on the half-wave potential of the first wave. pH-independence indicates interposed reaction 31; pH-dependence indicates preceding reaction 32.

The most frequently studied systems corresponding to Figure 20 are reductions of aliphatic and aromatic nitro compounds in which the hydroxylamino derivative formed in the first four-electron process is reduced only in the protonated form (Eq. 33):

$$R\text{—}NO_2 + 4e \xrightarrow[2H^+]{E_1} R\text{—}NHOH \qquad (33a)$$

$$R\text{—}NHOH_2^+ \underset{k_{-1}}{\overset{k_1}{\rightleftharpoons}} R\text{—}NHOH + H^+ \qquad (33b)$$

$$R\text{—}NHOH_2^+ + 2e \xrightarrow[2H^+]{E_2} RNH_3^+ + H_2O \qquad (33c)$$

The pH-dependence of the half-wave potential of the first wave is more complex than indicated in Figure 20, as the four-electron reduction involves hydronium ion transfer prior to electron transfer. At higher pH values the decreasing rate of protonation (Eq. 33b) with rate constant k_{-1} leads to a decrease in the height of the more negative wave.

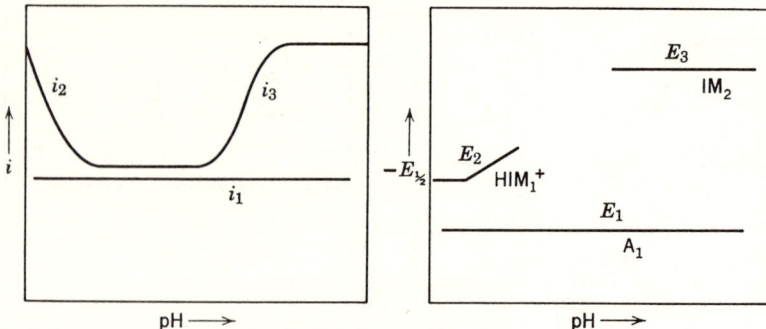

FIG. 21. pH-dependences of wave heights and half-wave potentials corresponding to the scheme:

$$A + n_1 e \xrightarrow{E_1} IM_1$$
$$IM_1 + acid \longrightarrow HIM_1^+$$
$$IM_1 + base \longrightarrow IM_2$$
$$HIM_1^+ + n_2 e \xrightarrow{E_2} P_1$$
$$IM + n_2 e \xrightarrow{E_3} P_2$$

Systems having plots as shown in Figure 21 correspond, similarly to Eq. 33, to an interposed chemical reaction, but in this case the reaction rate increases not only with the concentrations of acids but also of bases. An example of this type is the reduction of p-diacetylbenzene (Eq. 34) (67):

$$\underset{\underset{\text{CHOH}}{|}}{\overset{\underset{\text{CH}_3}{|}}{\text{C}=\text{OH}^+}} + e \xrightarrow{E_2} \underset{\underset{\text{CHOH}}{|}}{\overset{\underset{\text{CH}_3}{|}}{\cdot\text{C}-\text{OH}}} \quad (34\text{c})$$

$$\underset{\underset{\text{CH}_3}{|}}{\overset{\underset{\text{CH}_3}{|}}{\cdot\text{COH}\cdots\cdot\text{COH}}} \xrightarrow{\text{Base catalysis}} \underset{\underset{\text{CHOH}}{|}}{\overset{\underset{\text{CH}_3}{|}}{\text{C}=\text{O}\cdots\text{CHOH}}} \quad (34\text{d})$$

$$\underset{\underset{\text{CHOH}}{|}}{\overset{\underset{\text{CH}_3}{|}}{\text{CO}}} + 2e \xrightarrow[2\text{H}^+]{E_3} \underset{\underset{\text{CHOH}}{|}}{\overset{\underset{\text{CH}_3}{|}}{\text{CHOH}}} \quad (34\text{e})$$

At pH approximately 2–4 a two-electron reversible wave is observed; the product of this process gives an anodic wave with a Kalousek commutator. When the pH is either decreased or increased, the height of the anodic wave decreases, and simultaneously the height of another cathodic wave increases. At higher pH values this new cathodic wave corresponds to a further two-electron reduction. The reversible process (Eq. 34a) is followed by an acid- (Eq. 34b) or base-catalyzed (Eq. 34d) reaction in which the biradical or quinoid form, which is not reducible, is transformed into a ketol, which can undergo further reduction (Eqs. 34c and e) to the dialcohol. When controlled-potential electrolysis is carried out at pH 2–4, where only the first two-electron wave is observed, the reduction yields the ketol; even though reactions 34b and d are too slow under these conditions to give rise to a polarographic wave corresponding to the reduction of acetophenone, they are fast enough to form the ketol during an electrolysis carried out during

1-2 hr. Chronopotentiometric measurements (68) verified that reactions 34b and d are general catalyzed.

When the mathematical treatment for interposed (ECE) reactions (34, 35) was applied (36) to the currents i_2, it was found that calculated values of rate constants were a function of the drop time and differed by two to three orders of magnitude from values obtained from chronopotentiometric measurements (68) and from an estimate based on controlled-potential electrolysis (67). The treatment (34, 35) seems to be incomplete.

Similar results were obtained for other benzene derivatives with two carbonyl groups in the para position (69). The lifetime of the primary product of two-electron reduction of $RCOC_6H_5COR$ increases in the sequence $R = H < CH_3 < C_6H_5$. Exchange of one or two of the carbonyl groups for an oxime or carboxyl group prevents this type of process (67, 69).

Another example of this type are reductions of α-substituted ketones of the type $RCOCH_2X$, which contain a group X that undergoes reduction at potentials more positive than the potential at which the reduction of the carbonyl group occurs. The C—X bond is sufficiently activated if $X = NR_2$ (70–72), $NR_3^{(+)}$ (70), SR (73), $SR_2^{(+)}$ (7, 8), OR (73), $PR_3^{(+)}$ (71), or a halogen (74). Compounds of this type are reduced in acidic media according to scheme 35:

$$RCOCH_2X + 2e \xrightarrow{E_1} RCOCH_2^{(-)} + X^- \qquad (35a)$$

$$RCOCH_2^{(-)} \longleftrightarrow \underset{O^{(-)}}{RC\!\!=\!\!CH_2} \qquad (35b)$$

$$RCOCH_3 \underset{k_{-1}}{\overset{k_1}{\rightleftarrows}} RCOCH_2^{(-)} + H^{(+)} \qquad (35c)$$

$$\underset{OH}{RC\!\!=\!\!CH_2} \underset{k_{-2}}{\overset{k_2}{\rightleftarrows}} \underset{O^{(-)}}{RC\!\!=\!\!CH_2} + H^{(+)} \qquad (35d)$$

$$RCO\overset{(+)}{H}CH_3 \underset{k_{-3}}{\overset{k_3}{\rightleftarrows}} RCOCH_3 + H^{(+)} \qquad (35e)$$

$$RCO\overset{(+)}{H}CH_3 + e \xrightarrow{E_2} R\dot{C}OHCH_3 \qquad (35f)$$

$$RCOCH_3 + 2e \xrightarrow[2H^+]{E_3} RCHOHCH_3 \qquad (35g)$$

Polarograms of such compounds show a two-electron reduction wave i_1 at potential E_1; the height of this wave is independent of pH. At a more negative potential E_2, there is another wave i_2 (Fig. 21). The half-wave potentials of the wave at potential E_2 are identical, over a range of pH values from 0 to 6, with the half-wave potentials obtained with the parent ketone. Hence this wave can be attributed to the reduction of the product of the first reduction step. Nevertheless, the height of the more negative wave i_2 at pH 2–5 is smaller than would correspond to a one-electron reduction. This is attributed to the fact that the enolate

$$\text{ArC}\!\cdots\!\text{CH}_2^{(-)}$$
$$\underset{\text{O}}{|\!:}$$

which is the product of the first reduction step i_1, is electroinactive. To be further reduced this intermediate must be first transformed into the electroactive form. This transformation (Eq. 35c) is acid–base-catalyzed, and its rate controls the height of the more negative wave. As the rate of reaction 35c also depends on the structure of the compound, the height of the more negative wave depends on the structure of the ketone involved.

A similar reduction scheme is also followed for α,β-unsaturated compounds, but because the system is further complicated by acid–base equilibria, it is discussed in Section VI.

In most cases discussed so far, a wave at more negative potentials increases with pH, and the height of the more positive of the two waves either remains unchanged or decreases. An interesting type of polarographic behavior has been observed for some α-dicarbonyl compounds (Fig. 22) in which, on the

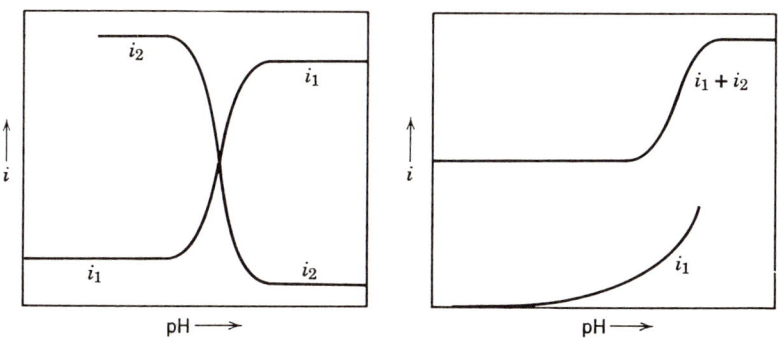

FIG. 22. pH-dependences of wave heights corresponding to the scheme:

$$\text{A} + \text{base} \rightleftharpoons \text{IM}_1$$
$$\text{IM}_1 + n_2 e \xrightarrow{E_2} \text{P}_1$$
$$\text{IM}_1 + \text{base} \rightleftharpoons \text{IM}_2$$
$$\text{IM}_2 + n_1 e \xrightarrow{E_1} \text{P}_2$$

contrary, the height of the more positive wave increases with increasing pH while that of the more negative one either remains almost constant or decreases (75). This is caused by the fact that these dicarbonyl compounds can be hydrated on both carbonyl groups, in which case they are electroinactive (Eq. 36); they can be hydrated on one group, giving rise to a negative reduction wave at potential E_2; or they can be nonhydrated, in which case the two conjugated carbonyl groups can interact and cause the reduction to occur at the more positive potential E_1.

$$RC(OH)_2C(OH)_2CH_3 + \text{base} \underset{k_{-1}}{\overset{k_1}{\rightleftharpoons}} RC\overset{O^-}{\underset{OH}{-}}C(OH)_2CH_3 + \text{base H}^+ \quad (36a)$$

$$RC\overset{O^-}{\underset{OH}{-}}C(OH)_2CH_3 \underset{k_{-2}}{\overset{k_2}{\rightleftharpoons}} RCOC(OH)_2CH_3 + OH^- \quad (36b)$$

$$RCOC(OH)_2CH_3 + 2e \xrightarrow[2H^+]{E_2} P_2 \quad (36c)$$

$$RCOC(OH)_2CH_3 + \text{base} \underset{k_{-3}}{\overset{k_3}{\rightleftharpoons}} RCOC\overset{O^-}{\underset{OH}{-}}CH_3 + \text{base H}^+ \quad (36d)$$

$$RCOC\overset{O^-}{\underset{OH}{-}}CH_3 \underset{k_{-4}}{\overset{k_4}{\rightleftharpoons}} RCOCOCH_3 + OH^- \quad (36e)$$

$$RCOCOCH_3 + 2e \xrightarrow[2H^+]{E_1} P_1 \quad (36f)$$

The height of the wave at the more positive potential E_1 depends on the positions of the equilibria of Eq. 36a, b, d, and e and on the rates at which they are established, while the height of the wave at the more negative potential E_2 depends on the positions and rates of establishment of the equilibria of Eq. 36a and b. The increase in the more negative wave to a limiting value for methylglyoxal (Fig. 22b) and decrease for diacetyl (Fig. 22a) are caused by the effect of the nature of the group R on the equilibrium and rate constants of reactions 36a, b, d, and e.

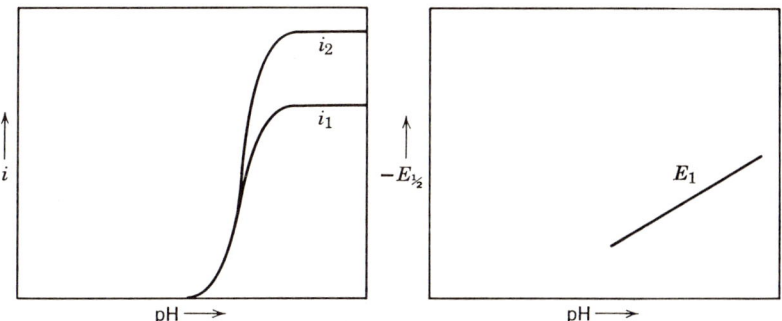

FIG. 23. pH-dependences of wave heights and half-wave potentials corresponding to the scheme:

$$A + \text{base} \rightleftharpoons IM_1$$
$$IM_1 + n_1 e \xrightarrow{E_1} IM_2$$
$$B + \text{base} \rightleftharpoons IM_2$$
$$IM_2 + n_2 e \xrightarrow{E_2} P$$

Simple α-substituted aliphatic aldehydes that carry, in the position vicinal to the carbonyl group a substituent that is not ionized (e.g., OH, NR_3^+, halogen) follow the reduction pattern shown in Figure 23. This type of dependence can be explained (7, 18) by scheme 37:

$$\underset{X}{\text{RCHCH}}\diagup\overset{OH}{\diagdown}_{OH} + \text{base} \underset{k_{-1}}{\overset{k_1}{\rightleftharpoons}} \underset{X}{\text{RCHCH}}\diagup\overset{O^-}{\diagdown}_{OH} + \text{base H}^+ \quad (37a)$$

$$\underset{X}{\text{RCHCH}}\diagup\overset{O^-}{\diagdown}_{OH} \underset{k_{-2}}{\overset{k_2}{\rightleftharpoons}} \underset{X}{\text{RCHCHO}} + OH^- \quad (37b)$$

$$\underset{X}{\text{RCHCHO}} + 2e \xrightarrow{E_1} \overset{(-)}{\text{RCHCHO}} + X^- \quad (37c)$$

$$\overset{(-)}{\text{RCHCHO}} \longleftrightarrow \text{RCH}=\text{CH}-O^{(-)} \quad (37d)$$

$$\text{RCH}_2\text{CHO} \underset{k_{-3}}{\overset{k_3}{\rightleftharpoons}} \overset{(-)}{\text{RCHCHO}} + \text{H}^+ \qquad (37e)$$

$$\text{RCH}=\text{CH}-\text{OH} \underset{k_{-4}}{\overset{k_4}{\rightleftharpoons}} \text{RCH}=\text{CH}-\text{O}^{(-)} + \text{H}^+ \qquad (37f)$$

$$\text{RCH}_2\text{CH}\begin{array}{c}\text{OH}\\ \\ \text{OH}\end{array} + \text{base} \underset{k_{-5}}{\overset{k_5}{\rightleftharpoons}} \text{RCH}_2\text{CH}\begin{array}{c}\text{O}^-\\ \\ \text{OH}\end{array} + \text{base H}^+ \qquad (37g)$$

$$\text{RCH}_2\text{CH}\begin{array}{c}\text{O}^-\\ \\ \text{OH}\end{array} \underset{k_{-6}}{\overset{k_6}{\rightleftharpoons}} \text{RCH}_2\text{CHO} + \text{OH}^- \qquad (37h)$$

$$\text{RCH}_2\text{CHO} + 2e \xrightarrow[2\text{H}^+]{E_2} \text{RCH}_2\text{CH}_2\text{OH} \qquad (37j)$$

The increase in wave i_1 at potential E_1, which corresponds to cleavage of the C—X bond, is governed by the rate of dehydration with constant k_1. Only in the unhydrated form can the CO group exert an activating influence on the neighboring C—X bond, making it polarizable enough to give a reduction wave in the accessible potential range. If in the general base-catalyzed reaction 37a the effect of the hydroxyl ions predominates, the shape of the pH-dependence of wave i_1 resembles a dissociation curve (Fig. 23). The dependence of the half-wave potential E_1 on pH is caused by the antecedent acid–base equilibria of Eq. 37a and b.

The second wave i_2 at the more negative potential E_2 corresponds to the reduction of the saturated aldehyde RCH_2CHO. This has been proved by comparison of the half-wave potentials E_2 with those of the authentic aldehyde RCH_2CHO, and by proof of formation of this aldehyde and of X$^-$ in controlled-potential electrolysis.

The shape of the pH-dependence of the limiting current of wave i_2 is similar to that of the aldehyde RCH_2CHO, but the currents are considerably smaller (by about 50%) than those of an equimolar solution of the saturated aldehyde. The pH-dependence of the wave heights i_2 cannot thus be explained only by the hydration of the primary electrolysis product of reaction 37c. If the hydration equilibria of Eq. 37g and h were the only side reactions of the consecutive reduction processes of Eq. 37c and j, it would be expected that the height of wave i_2 for RCH(X)CHO would be the same as that of an equimolar solution of RCH_2CHO at a given pH. The observed difference was explained by the interposed reaction of Eq. 37e with the rate constant k_3, by which the carbanion enolate formed by the first two-electron uptake (Eq. 37c) is transformed into the electroactive aldehyde RCH_2CHO. This interpretation is supported by findings on the polarographic reduction of cinnamaldehyde in alkaline media (76), in which the saturated aldehyde formed electrolytically in a process not involving formation of a carbanion behaves quantitatively identically to the authentic saturated aldehyde. It is possible that the more negative waves observed at higher pH values on the current–voltage curves of erythrose (77) and of some other sugars correspond to an analogous process.

If the electroactive group X undergoes ionization (7) (as for $X = NR_2$), the system involves another acid–base equilibrium.

Finally, two waves can be distinguished on the current–voltage curves of glyoxalic acid (78) (Fig. 24). The pH-dependence of the limiting current can be explained by Eq. 38:

$$\begin{array}{c}
\underset{COOH}{C(OH)_2H^+} \underset{k_{-3}}{\overset{k_3}{\rightleftharpoons}} \underset{COOH}{CHO} + H_3O^+ \quad (38a) \\
k_2 \updownarrow k_2 \\
\underset{COOH}{C(OH)_2} \underset{k_{-7}}{\overset{k_7}{\rightleftharpoons}} \underset{COOH}{CHO} + H_2O \xrightarrow{E_1} \quad (38b) \\
K_1 = k_1/k_{-1} \quad k_1 \updownarrow k_{-1} \qquad k_5 \updownarrow k_{-5} \\
\underset{COO^-}{C(OH)_2} \underset{k_{-8}}{\overset{k_8}{\rightleftharpoons}} \underset{COO^-}{CHO} + H_2O \xrightarrow{E_2} \quad (38c) \\
k_{10} \updownarrow k_{-10} \\
\underset{COO^-}{C(OH)O^-} \underset{k_{-9}}{\overset{k_9}{\rightleftharpoons}} \underset{COO^-}{CHO} + OH^- \quad (38d)
\end{array}$$

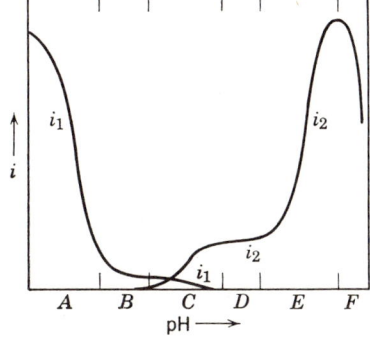

FIG. 24. pH-dependences of the limiting currents of waves of undissociated (i_1) glyoxalic acid and its anion (i_2).

$$\begin{array}{lll}
AH & \rightleftharpoons A^- + H^+ & (C) \\
AH + acid & \longrightarrow IM_1 & (A) \\
AH + H_2O & \longrightarrow IM_1 & (B) \\
IM_1 + n_1 e & \xrightarrow{E_1} P_1 & \\
A + H_2O & \longrightarrow IM_2 & (D) \\
A + base & \longrightarrow IM_2 & (E) \\
A + OH^- & \rightleftharpoons IM_2OH & (F)
\end{array}$$

Region A corresponds to the acid-catalyzed dehydration of the free acid following path k_2, k_3, region B to the uncatalyzed dehydration of the free acid k_7, and region D to the uncatalyzed dehydration of the anion k_8. Region C corresponds to the establishment of the acid–base equilibrium with the constant K_1. The decrease in current in region F at pH > 12 is interpreted as being attributable to the increase in the rate of the reverse reaction (Eq. 38d) (constant k_{-9}) with increasing hydroxyl ion concentration. Region E corresponds to the base-catalyzed dehydration of the anion 7 following path k_{10}, k_9.

VI. Systems Showing Three or More Reduction Waves

Apart from the reduction of dibasic acids, which shows a pattern that is rather general, reductions of more complex systems often show a pattern that is not frequently repeated in other systems. For this reason a few types of polarographic behavior are illustrated here with randomly chosen examples.

A. Dibasic Acids

One of the more frequently observed types of i–pH and $E_{1/2}$–pH plots is that shown in Figure 25, which indicates that the electroactive compound

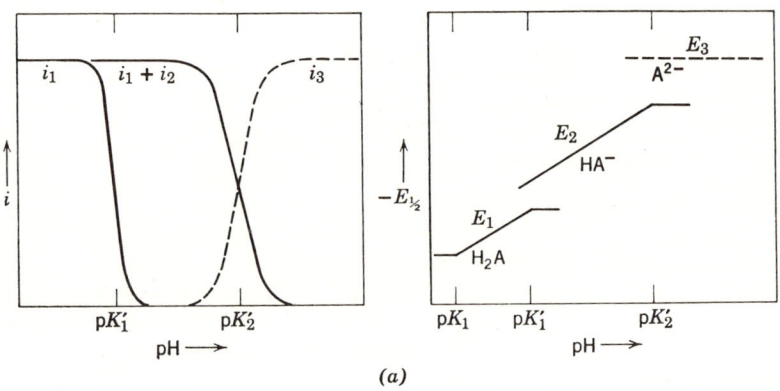

FIG. 25a. pH-dependences of wave heights and half-wave potentials corresponding to the scheme:

$H_2'A \rightleftharpoons HA^- + H^+$ fast → comparably fast $H_2A + n_1e \xrightarrow{E_1} P_1$
   pK_1 pK_1'

$HA^- \rightleftharpoons A^{2-} + H^+$ fast → comparably fast $HA^- + n_2e \xrightarrow{E_2} P_2$
      pK_2' $A^{2-} + n_3e \xrightarrow{E_3} P_3$

$pK_2 < pK_1'$; $|E_1| < |E_2| < |E_3|$ $n_1 = n_2 = n_3$

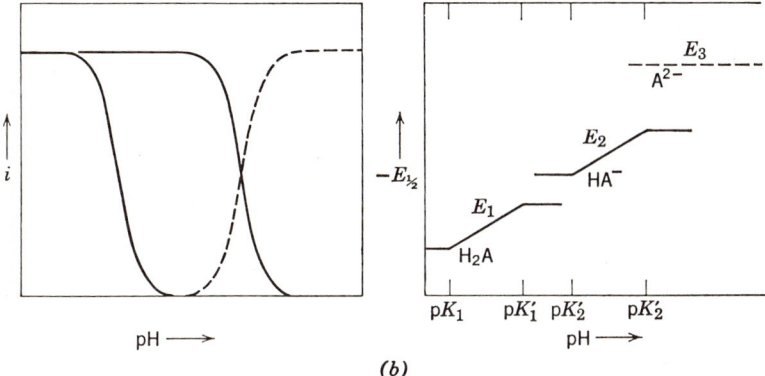

FIG. 25b. pH-dependences of wave heights and half-wave potentials corresponding to the scheme:

$H_2A \rightleftharpoons HA^- + H^+$ fast → comparably fast $H_2A + n_1 e \xrightarrow{E_1} P_1$
                                   pK_1 pK'_1

$HA^- \rightleftharpoons A^{2-} + H$ fast → comparably fast $HA^- + n_2 e \xrightarrow{E_2} P_2$
                                   pK_2 pK'_2

                                                         $A^{2-} + n_3 e \xrightarrow{E_3} P_3$

$pK'_1 < pK_2$; $|E_1| < |E_2| < |E_3|$ $n_1 = n_2 = n_3$

bears two groups with acidic properties. Three waves, i_1, i_2, and i_3, are observed at potentials E_1, E_2, and E_3 respectively, changing their height with pH, and forming two polarographic dissociation curves. When we restrict ourselves to a discussion of systems with a constant number of electrons transferred, it is possible to distinguish two types of systems (Fig. 25a and b) which differ in the shape of the first dissociation curve. Systems belong to the first or second category according to their relative equilibrium values (K_1, K_2) and polarographic dissociation constants (K'_1, K'_2).

The second dissociation curve has in all cases the same shape as for a monobasic acid, described by Eq. 9, and $pK'_2 \geqq pK_2$. The shape of the first dissociation curve depends on the ratio K'_1/K_2. If $pK'_1 > pK_2$ the first dissociation curve, which is observed at lower pH values, has a greater slope (Fig. 25a) than the second dissociation curve and corresponds to Eq. 39 (79, 80):

$$\frac{\bar{i}}{\bar{i}_d} = \frac{0.886 \left(\dfrac{t_1 k_1}{K_1}\right)^{1/2} \dfrac{[H^+]^2}{K_2 + [H^+]}}{1 + 0.886 \left(\dfrac{t_1 k_1}{K_1}\right)^{1/2} \dfrac{[H^+]^2}{K_2 + [H^+]}} \qquad (39)$$

If, however, $pK_1' \ll pK_2$, then both dissociation curves have the same shape (Fig. 25b). It is possible to show that under these conditions Eq. 39 degenerates into Eq. 9. Hence both dissociation curves are described by the same equation (Eq. 9). The two dissociation steps differ, however, in the sets of values k_1 and K_1 (for the first dissociation curve) and k_2 and K_2 (for the second dissociation curve).

The treatment has been derived for maleic, fumaric, and citraconic acids (79), but it has been recognized (13, 79) that these acids behave rather more as tribasic than dibasic acids, and thus they are discussed in the next section.

An example of the first type (Fig. 25a), showing a steeper first dissociation curve, is the reduction of pyridoxaloxime (28), which follows scheme 40:

$$\text{[protonated pyridoxaloxime cation]} \underset{k_{-1}}{\overset{k_1}{\rightleftarrows}} \text{[pyridoxaloxime, N-protonated]} + \text{H}^+ \quad (40a)$$

$$\text{[oxime-protonated form]} \underset{k_{-2}}{\overset{k_2}{\rightleftarrows}} \text{[neutral pyridoxaloxime]} + \text{H}^+ \quad (40b)$$

$$\text{[oxime-H}^+\text{, N-H}^+\text{]} + 4e \xrightarrow[4\text{H}^+]{E_1} \text{[pyridoxamine, N-H}^+\text{]} + \text{H}_3\text{O}^+ \quad (40c)$$

$$\text{[oxime-H}^+\text{]} + 4e \xrightarrow[4\text{H}^+]{E_2} \text{[pyridoxamine]} + \text{H}_3\text{O}^+ \quad (40d)$$

$$\text{[neutral oxime]} + 2e \xrightarrow{E_3} P_3 \quad (40e)$$

The products of the two-electron reduction (Eq. 40e) of the species unprotonated on the oxime group have not been identified. In accordance with Figure 25a are not only the shapes of the pH-dependences of the limiting currents i_1, i_2, and i_3, but also those of the half-wave potentials. The half-wave potentials E_1 of wave i_1 are pH-independent at pH > pK_1' [Fig. 18 and Table I in reference (28)]; the half-wave potentials E_2 of wave i_2 at pH > pK_2' and those (E_3) of wave i_3 are pH-independent at all pH values. The $E_{1/2}$–pH plot for wave i_1 shows a change in slope at about pH 6. The role of dissociation of the phenolic group—not considered in scheme 40—cannot be excluded.

The shape of the first dissociation curve indicates that $pK_1' > pK_2$. The value of $pK_1' = 10.3$ indicates that from the three pK values reported (28) for pyridoxaloxime (4.10, 8.10, and 10) it can be concluded that $pK_1 = 4.10$ and $pK_2 = 8.10$. Protonation of the oxime grouping, which is considered (24, 62–64) a necessary condition for polarographic reducibility of an oxime, was not considered in earlier assessments of dissociation constants. The attribution of pK_1 to Eq. 40a and pK_2 to Eq. 40b can nevertheless be affected by conditions at the surface of the mercury electrode, where owing to adsorption phenomena the sequence of the protonated centers can differ from those observed in the bulk of the solution. The proton transfers accompanying the reduction of pyridoxaloxime are surface reactions (5, 13), and it is therefore impossible to calculate the rate constants.

An example of a dibasic acid showing two dissociation curves of the same slope (Fig. 25b), identical with that observed for monobasic acids, is the reduction of pyridoxal 5-phosphate (28). Half-wave potentials of wave i_1 are shifted at pH < pK_1' to more negative potentials and at pH > pK_1' become pH-independent [Fig. 16 and Table IV in reference (28)]. Because of the small difference between pK_1' (10.9) and pK_2' (11.9), the half-wave potentials of wave i_2 remain practically pH-independent, as do those of wave i_3. When $pK_1' < pK_2$, it is necessary to assume $pK_2 > 10.9$. Potentiometrically, nevertheless, only values of $pK = 3.65, 6.20$, and 8.69 were found, and from kinetic measurements another value of $pK = 0.9$. Since the formation of waves i_2 and i_3 was not observed for pyridoxal, it is assumed that the phosphoric acid portion is responsible for the observed acid–base reaction but, without further spectroscopic evidence for behavior in alkaline media, it is impossible to attribute structures to the individual dissociation steps.

Another system that gives two dissociation curves of identical slope (Fig. 25b) is ω-cyanoacetophenone (23), the reduction of which follows scheme 41:

$$C_6H_5CO\overset{(+)}{H}CH_2CN \underset{k_{-1}}{\overset{k_1}{\rightleftharpoons}} C_6H_5COCH_2CN + H^+ \qquad (41a)$$

$$C_6H_5COCH_2CN \underset{k_{-2}}{\overset{k_2}{\rightleftharpoons}} C_6H_5CO\overset{(-)}{C}HCN + H^+ \qquad (41b)$$

$$\text{C}_6\text{H}_5\text{CO}\overset{(-)}{\text{CH}}\text{CN} \longleftrightarrow \text{C}_6\text{H}_5\underset{\underset{\text{O}(-)}{|}}{\text{C}}=\text{CHCN} \qquad (41\text{c})$$

$$\text{C}_6\text{H}_5\underset{\underset{\text{OH}}{|}}{\text{C}}=\text{CHCN} \underset{k_{-3}}{\overset{k_3}{\rightleftharpoons}} \text{C}_6\text{H}_5\underset{\underset{\text{O}^-}{|}}{\text{C}}=\text{CHCN} + \text{H}^+ \qquad (41\text{d})$$

$$\text{C}_6\text{H}_5\overset{(+)}{\text{CO}}\text{HCH}_2\text{CH} + 2e \xrightarrow{E_2} \text{C}_6\text{H}_5\text{CHOHCH}_2\text{CN} \qquad (41\text{e})$$

$$\text{C}_6\text{H}_5\text{COCH}_2\text{CN} + 2e \xrightarrow[2\text{H}^+]{E_2} \text{C}_6\text{H}_5\text{CHOHCH}_2\text{CN} \qquad (41\text{f})$$

The experimental data correspond to Figure 25b, except that wave i_3 of the carbanion enolate (Eq. 41c) is at a potential so negative that it is superimposed on the current of the supporting electrolyte. The intersection of the half-wave potentials of wave i_2 at pH 7.4 corresponds to the spectrophotometrically determined value $pK_2 = 7.6$, but the other intersection at 9.0 is somewhat larger than the value $pK'_2 = 8.4$ obtained from the pH-dependence of limiting currents.

In this case it is possible to show that the condition $pK'_1 < pK_2$ needed for the formation of two dissociation curves of the same slope is fulfilled when $pK'_1 = 6.35$ and $pK_2 = 7.6$.

Whereas reaction 41a governing at pH > $(pK'_1 - 1)$ the height of the first wave i_1 takes place as a surface reaction, protonation of the carbanion enolate in reaction 41b with rate constant k_{-2} was proved (23) to be a volume reaction. To calculate the rate constant k_{-2}, it is nevertheless necessary to calculate the value $K_2^0 = k_2/k_{-2}$, as the experimentally accessible value K_2 depends on the ratio [enol]/[keto] by the relation $K_2 = K_2^0\{(1 + [\text{enol}])/[\text{keto}]\}$. By taking the keto–enol equilibrium into account, the value $k_{-2} = 2 \times 10^8$ liter mole^{-1} sec^{-1} was found.

Finally, the reduction of ethyl benzoylacetate (23) also shows two dissociation curves of identical slope (Fig. 25b) and corresponds to scheme 42:

$$\text{C}_6\text{H}_5\overset{+}{\text{CO}}\text{HCH}_2\text{COOC}_2\text{H}_5 \underset{k_{-1}}{\overset{k_1}{\rightleftharpoons}} \text{C}_6\text{H}_5\text{COCH}_2\text{COOC}_2\text{H}_5 + \text{H}^+ \qquad (42\text{a})$$

$$\text{C}_6\text{H}_5\text{COCH}_2\text{COOC}_2\text{H}_5 \underset{k_{-2}}{\overset{k_2}{\rightleftharpoons}} \text{C}_6\text{H}_5\text{CO}\overset{-}{\text{CH}}\text{COOC}_2\text{H}_5 + \text{H}^+ \qquad (42\text{b})$$

$$\text{C}_6\text{H}_5\text{CO}\overset{(-)}{\text{CH}}\text{COOC}_2\text{H}_5 \longleftrightarrow \text{C}_6\text{H}_5\underset{\underset{\text{O}(-)}{|}}{\text{C}}=\text{CHCOOC}_2\text{H}_5 \qquad (42\text{c})$$

$$\underset{\mathrm{OH}}{\mathrm{C_6H_5C}}\!=\!\mathrm{CHCOOC_2H_5} \underset{k_{-3}}{\overset{k_3}{\rightleftharpoons}} \underset{\mathrm{O}^{(-)}}{\mathrm{C_6H_5C}}\!=\!\mathrm{CHCOOC_2H_5} + \mathrm{H^+} \qquad (42\mathrm{d})$$

$$\mathrm{C_6H_5CO\overset{+}{H}CH_2COOC_2H_5} + e \xrightarrow{E_1{}^a} \mathrm{C_6H_5\dot{C}OHCH_2COOC_2H_5} \qquad (42\mathrm{e})$$

$$\mathrm{C_6H_5\dot{C}OHCH_2COOC_2H_5} + e \xrightarrow[\mathrm{H^+}]{E_1{}^b} \mathrm{C_6H_5CHOHCH_2COOC_2H_5} \qquad (42\mathrm{f})$$

$$\mathrm{C_6H_5COCH_2COOC_2H_5} + 2e \xrightarrow[2\mathrm{H^+}]{E_2} \mathrm{C_6H_5CHOHCH_2COOC_2H_5} \qquad (42\mathrm{g})$$

As regards ω-cyanoacetophenone, the wave of carbanion enolate (Eq. 42c) is too negative to be observed. The wave i_1 is split into two one-electron steps at pH < 5, but the significant difference between ethyl benzoylacetate and ω-cyanoacetophenone is attributable to the fact that the wave i_2 is diffusion controlled even when it is only a small fraction of the two-electron overall wave. The inflection point of the i_2–pH dependence at pH 10.4, the intersection of the two linear parts on the $E_{1/2}$–pH plot of this wave at pH 10.5, and the spectrophotometrically obtained value $pK_2 = 10.64$ also indicate that the decrease in wave i_2 is attributable to the shift in the equilibrium of Eq. 42b rather than to the rate of its establishment. (The effect of Eq. 42c and d can be neglected, as the contribution of the enol form in aqueous solutions is less than 1%). The protonation of the carbonyl governing the height of wave i_1 has characteristics of a surface reaction and is thus unsuitable for calculation of k_{-1}. The condition that pK_1' (6.5) is smaller than pK_2 (10.5) is fulfilled in this case. The diffusion control of the current i_2 is not as much caused by the fact that the protonation reaction with constant k_{-2} is slow, but rather that the value of pK_2 is high (pp. 92–3). The diffusion character of wave i_2 allows only the deduction that $k_{-2} < 10^{10}$ liter mole^{-1} sec^{-1}.

Among further systems that can behave as dibasic acids, polarographic reduction has been reported, for example, for oxalic (81) and terephthalic (82) acids and for 2-ethyl-4-thiocarbamoyl pyridine (83).

B. Tribasic Acids

Reduction processes accompanied by three proton transfers can be identified when the i–pH plot shows three dissociation curves (Fig. 26).

The most thoroughly studied system of this type is phthalic acid (84, 85) in which—in accordance with theory (84)—the steepness of the dissociation curves decreases in the sequence $i_1 > i_2 > i_3$ (Fig. 26a), the change of i_2

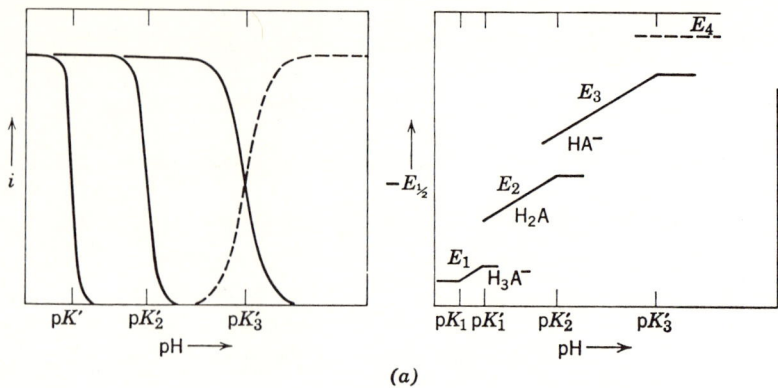

FIG. 26a. pH-dependences of wave heights and half-wave potentials corresponding to the scheme:

$$H_3A^+ \rightleftharpoons H_2A + H^+ \quad \text{fast} \to \text{comparably fast}$$
$$ pK_1 \quad pK_1'$$
$$H_2A \rightleftharpoons HA^- + H^+ \quad pK_2'$$
$$HA^- \rightleftharpoons A^{2-} + H^+ \quad pK_3'$$
$$H_3A^+ + n_1 e \xrightarrow{E_1} P_1$$
$$H_2A + n_2 e \xrightarrow{E_2} P_2$$
$$HA^- + n_3 e \xrightarrow{E_3} P_3 \quad |E_1| < |E_2| < |E_3| < |E_4|$$
$$A^{2-} + n_4 e \xrightarrow{E_4} P_4$$
$$n_1 = n_2 = n_3 = n_4$$

with pH being given by Eq. 39 and that of i_3 by Eq. 9. The third, most negative, wave i_3 hence behaves as a wave of a monobasic acid. This behavior was explained on the basis of scheme 43:

$$C_6H_4(COOH)_2H^+ + \text{base} \underset{k_{-1}}{\overset{k_1}{\rightleftharpoons}} C_6H_4(COOH)_2 + \text{base } H^+ \tag{43a}$$

$$C_6H_4(COOH)_2 + \text{base} \underset{k_{-2}}{\overset{k_2}{\rightleftharpoons}} C_6H_4(COOH)COO^- + \text{base } H^+ \tag{43b}$$

$$C_6H_4(COOH)COO^- + \text{base} \underset{k_{-3}}{\overset{k_3}{\rightleftharpoons}} C_6H_4(COO^-)_2 + \text{base } H^+ \tag{43c}$$

$$C_6H_4(COOH)_2H^+ + 2e \xrightarrow{E_1} P_1 \tag{43d}$$

$$C_6H_4(COOH)_2 + 2e \xrightarrow{E_2} P_2 \tag{43e}$$

$$C_6H_4(COOH)COO^- + 2e \xrightarrow{E_3} P_3 \tag{43f}$$

$$C_6H_4(COO^-)_2 + 2e \xrightarrow{E_4} P_4 \tag{43g}$$

The existence of the protonated form $C_6H_4(COOH)_2H^+$ has been proved (85) in acid media spectrophotometrically, and it has been found that equal equilibrium concentrations of $C_6H_4(COOH)_2H^+$ and $C_6H_4(COOH)_2$ are attained in 24.4 N sulfuric acid. By using Eqs. 9 and 39, the value $k_{-3} = 2.2 \times 10^8$ liter mole^{-1} sec^{-1} and $k_{-2} = 1.4 \times 10^7$ liter mole^{-1} sec^{-1} were obtained (84) under the assumption that the acid–base equilibria are volume reactions.

A completely analogous scheme was deduced for the reduction of maleic acid (79, 80). It has been proved that the third dissociation curve has a slope corresponding to that of the monobasic acid (Eq. 9), while the second dissociation curve has a slope corresponding to that of a dibasic acid for $pK_1' > pK_2$ ($pK_1' \approx 6$, $pK_2 = 5.5$). The waves i_1 and i_2 are ill separated, and therefore it is difficult to express the slope of the first dissociation curve quantitatively. It was concluded (13) that the first two proton transfers take place as surface reactions, but that the recombination of the dianion, which corresponds to the third dissociation curve, occurs as a volume reaction (Eq. 44) (numbering analogous to Eq. 43):

$$\begin{array}{c} CHCOO^- \\ \| \\ CHCOOH \end{array} + \text{base} \underset{k_{-3}}{\overset{k_3}{\rightleftharpoons}} \begin{array}{c} CHCOO^- \\ \| \\ CHCOO^- \end{array} + \text{base H}^+ \qquad (44c)$$

$$\begin{array}{c} CHCOO^- \\ \| \\ CHCOOH \end{array} + 2e \xrightarrow[2H^+]{E_3} \begin{array}{c} CH_2COO^- \\ | \\ CH_2COOH \end{array} \qquad (44f)$$

$$\begin{array}{c} CHCOO^- \\ \| \\ CHCOO^- \end{array} + 2e \xrightarrow[2H^+]{E_4} \begin{array}{c} CH_2COO^- \\ | \\ CH_2COO^- \end{array} \qquad (44g)$$

Rate constants k_{-3} were calculated (13) for the reactions with various acids (base H$^+$) from the dependences of the current i_3 on buffer concentration. It was assumed that the reason reaction 44c takes place as a volume rather than a surface reaction is that at the negative potentials at which the reduction wave i_3 is observed the negatively charged mono- and dianions of maleic acid are no longer adsorbed.

Three waves corresponding to different forms of the electroactive species and three polarographic dissociation curves as in Figure 26 were observed (86) in the reduction of *trans*-urocanic acid. The slope of the third dissociation curve corresponds to that observed for a monobasic acid with $pK_3' \approx 8.5$. The second dissociation curve ($pK_2' = 7.1$) is steeper as corresponds to Eq. 39. The first dissociation curve ($pK_1' \approx 5.85$) is steep as shown by the narrow pH range (pH 5.2–6.1) in which the change in waves was observed. However, the waves i_1 and i_2 are so ill separated that it is impossible to

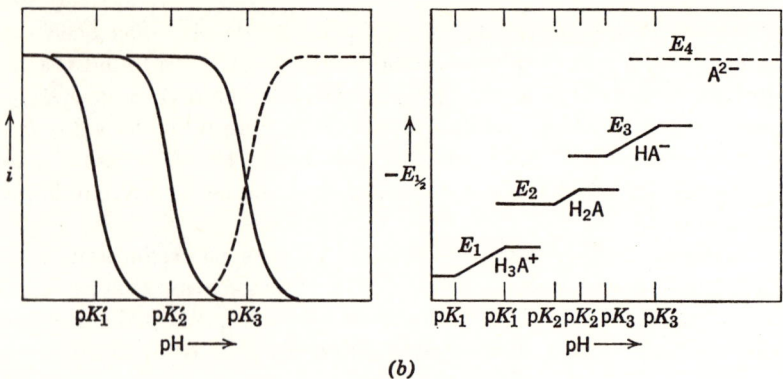

FIG. 26b. pH-dependences of wave heights and half-wave potentials corresponding to the scheme:

$H_3A^+ \rightleftharpoons H_2A + H^+$ fast → comparably fast

                                  pK_1 pK_1'

$H_2A \rightleftharpoons HA^- + H^+$ pK_2 pK_2'

$HA^- \rightleftharpoons A^{2-} + H^+$ pK_3 pK_3'

$H_3A^+ + n_1e \xrightarrow{E_1} P_1$

$H_2A + n_2e \xrightarrow{E_2} P_2$

$HA^- + n_3e \xrightarrow{E_3} P_3$ $|E_1| < |E_2| < |E_3| < |E_4|$

$A^{2-} + n_4e \xrightarrow{E_4} P_4$

$n_1 = n_2 = n_3 = n_4$

determine whether the steepness corresponds to Eq. 39 or is even greater. The shifts in half-wave potentials in principle correspond to the $E_{1/2}$–pH plot shown in Figure 26.

Finally, four waves changing with pH were observed for benzoylacetone. The i–pH plot for this compound shows three dissociation curves with practically identical slopes (Fig. 26b). This behavior was interpreted by scheme 45:

$$C_6H_5\overset{(+)}{C}OHCH_2\overset{(+)}{C}OHCH_3 \underset{k_{-1}}{\overset{k_1}{\rightleftharpoons}} C_6H_5\overset{(+)}{C}OHCH_2COCH_3 + H^+ \quad (45a)$$

$$C_6H_5\overset{(+)}{C}OHCH_2COCH_3 \underset{k_{-2}}{\overset{k_2}{\rightleftharpoons}} C_6H_5COCH_2COCH_3 + H^+ \quad (45b)$$

$$C_6H_5COCH_2COCH_3 \underset{k_{-3}}{\overset{k_3}{\rightleftharpoons}} C_6H_5CO\overset{(-)}{C}HCOCH_3 + H^+ \quad (45c)$$

$$C_6H_5CO\overset{(-)}{C}HCOCH_3 \longleftrightarrow C_6H_5C=CHCOCH_3 \quad (45d)$$
$$\qquad\qquad\qquad\qquad\qquad\qquad\qquad |$$
$$\qquad\qquad\qquad\qquad\qquad\qquad\qquad O^{(-)}$$

$$\underset{\underset{OH}{|}}{C_6H_5C=CHCOCH_3} \underset{k_{-4}}{\overset{k_4}{\rightleftharpoons}} \underset{\underset{O^{(-)}}{|}}{C_6H_5C=CHCOCH_3} + H^+ \quad (45e)$$

$$C_6H_5\overset{(+)}{CO}HCH_2\overset{(+)}{CO}HCH_3 + 2e \xrightarrow[H^+]{E_1} C_6H_5CHOHCH_2\overset{(+)}{CO}HCH_3 \quad (45f)$$

$$C_6H_5\overset{(+)}{CO}HCH_2COCH_3 + 2e \xrightarrow[H^+]{E_2} C_6H_5CHOHCH_2COCH_3 \quad (45g)$$

$$C_6H_5COCH_2COCH_3 + 2e \xrightarrow[2H^+]{E_2} C_6H_5CHOHCH_2COCH_3 \quad (45h)$$

$$\left.\begin{array}{c} C_6H_5CO\overset{(-)}{C}HCOCH_3 \\ \updownarrow \\ C_6H_5C=CHCOCH_3 \\ | \\ O^{(-)} \end{array}\right\} + 2e \xrightarrow[3H^+]{E_4} C_6H_5CHOHCH_2COCH_3 \quad (45j)$$

The inflection points on the i–pH plots ($pK'_1 = 6.8$, $pK'_2 = 7.95$, $pK'_3 = 9.5$) roughly corresponded to intersections of linear portions on $E_{1/2}$–pH plots (at pH approximately 7.3, 7.7, and 9.5). The protonations of the carbonyl groups (Eqs. 45a and b) took place as surface reactions, while dissociation of the 1,3-diketone (Eq. 45c–e) took place as a volume reaction. After separating the value of pK_3 from the experimentally accessible quantity $K_T [= K_3(1 + [enol]/[keto])]$, the value 1×10^{10} liter mole^{-1} sec^{-1} was calculated for the rate constant k_{-3}.

C. Aryl Ketones

Various aspects of the reduction of carbonyl compounds have already been discussed. The most general overall scheme (Eq. 46) involves formation of four waves (Fig. 27):

$$\overset{(+)}{ArCOHR} \underset{k_{-1}}{\overset{k_1}{\rightleftharpoons}} ArCOR + H^+ \quad (46a)$$

$$\overset{(+)}{ArCOHR} + e \xrightarrow{E_1} Ar\dot{C}OHR \quad (46b)$$

$$Ar\dot{C}OHR + e \xrightarrow[H^+]{E_2} ArCHOHR \quad (46c)$$

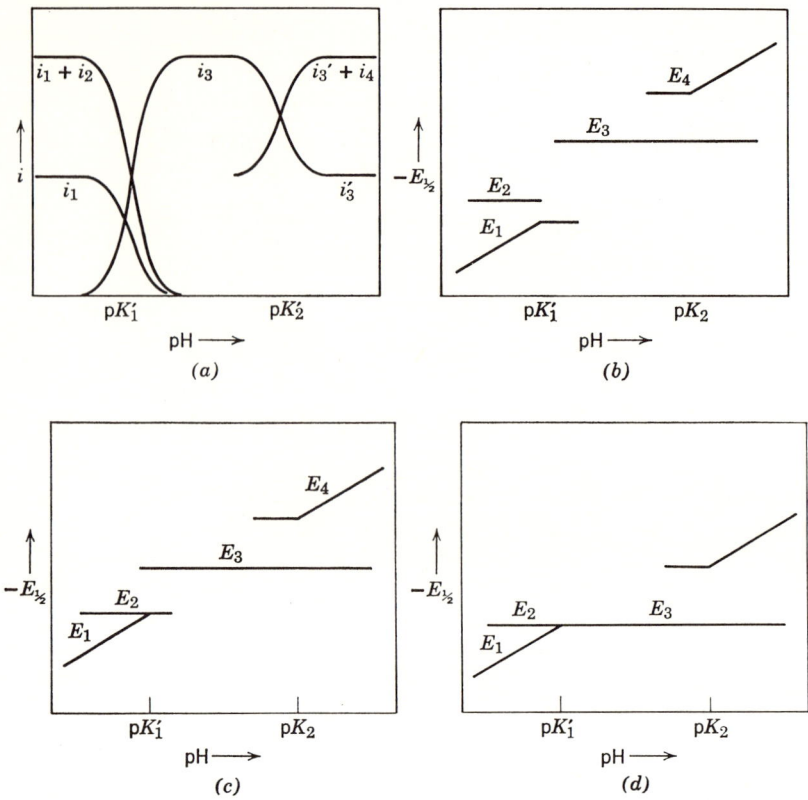

Fig. 27. pH-dependences of limiting currents (a) and half-wave potentials (b–d) of carbonyl compounds of the type ArCOR for wave i_1 (E_1) of the protonated form i_2 (E_2) of the radical i_3 and i'_3 (E_3) of the unprotonated form and i_4 (E_4) of the carbanion. (b) $|E_1|(\text{pH} > \text{p}K'_1) < |E_2| < |E_3|$; (b) $|E_1|(\text{pH} > \text{p}K'_1) \approx |E_2| < |E_3|$; (c) $|E_1|(\text{pH} > \text{p}K'_1) \approx |E_2| \approx |E_3|$.

$$\text{ArCOR} + e \xrightarrow{E_3} \overset{(-)}{\text{ArCOR}} \qquad (46d)$$

$$\text{Ar}\dot{\text{C}}\text{OHR} \underset{k_{-2}}{\overset{k_2}{\rightleftharpoons}} \overset{(-)}{\text{ArCOR}} + \text{H}^+ \qquad (46e)$$

$$\text{Ar}\dot{\text{C}}\text{OH} + e \xrightarrow[\text{H}^+]{E_2} \text{ArCHOHR} \qquad (46c)$$

$$\text{Ar}\dot{\text{C}}\text{OMeR} \underset{k_{-M}}{\overset{k_M}{\rightleftharpoons}} \overset{(-)}{\text{ArCOR}} + \text{Me}^+ \qquad (46f)$$

$$\text{Ar}\dot{\text{C}}\text{OMeR} + e \xrightarrow[2\text{H}^+]{E_M} \text{ArCHOHR} + \text{Me}^+ \qquad (46g)$$

$$\overset{(-)}{\text{ArCOR}} + e \xrightarrow[2\text{H}^+]{E_4} \text{ArCHOHR} \qquad (46h)$$

The decrease in the rate of protonation (Eq. 46a) with constant k_{-1} causes the heights of waves i_1 and i_2 to decrease; the reaction usually takes place as a surface reaction, and polarographic curves do not allow calculation of the value of rate constant k_{-1}. The rate of protonation of the radical anion (Eq. 46e) governs the decrease in wave i_3 to i'_3 and the increase in wave i_4 if it occurs at sufficiently positive potentials. The equilibrium constant corresponding to the dissociation of the radical (Eq. 46e) is usually not available for calculation of the rate constant k_{-2}; an approximate value of K_2 can be obtained from $E_{1/2}$–pH plots. The study of reaction 46e is often complicated by the fact that the alkali metal ions of the supporting electrolyte react with the radical anion $\overset{(-)}{\text{ArCOR}}$ in a way (Eq. 46f) similar to that in which hydronium ion does. Moreover, the potential E_M of the reduction of the product of the interaction of the radical anion with metal cations is usually similar to the potential E_2 of the reduction of the radical $\text{Ar}\dot{\text{C}}\text{OHR}$. Hence increasing metal ion concentration increases wave i_3 in the same way as decreasing pH.

Even though quantitative interpretation is usually impossible, a plot of the type shown in Figure 27 is usually sufficient to elucidate the course of the process. In practice, some of the waves may overlap. For instance, if potential E_2 is little different from E_1, only one two-electron wave appears in acidic media, and its height decreases as $i_1 + i_2$ (Fig. 27c). If, in addition, at pH > $(pK'_1 + 1)$ E_3 is close to E_2, only the one-electron wave i_1 decrease and is replaced by a two-electron wave $i_2 + i_3$ (Fig. 27d). Finally, if i_1 and i_2 coalesce at pH < $(pK'_1 - 1)$ and E_2 is close to E_3, the total wave height of $i_1 + i_2 + i_3$ does not change up to pH > $(pK_2 - 1)$. The feature to look for in such a case is the presence of two one-electron waves in acidic media merging into one two-electron wave, the half-wave potential of which remains pH-independent for more than 1 pH unit. In alkaline media the height of the two-electron wave decreases until it reaches the value corresponding to a one-electron process, with possible formation of another more negative one-electron wave which might be superimposed on the current of the supporting electrolyte.

The radical formed in reactions 46b and e and the radical anion produced in reaction 46d can undergo side reactions—with another radical, with the original compound, with solvent, or with the material of the electrode.

D. α, β-Unsaturated Aldehydes and Ketones

The reductions of α,β-unsaturated ketones (21, 87, 88) and aldehydes (76, 89) that carry a phenyl group attached to the conjugated system take place principally in two two-electron steps; the first step (split in acid media into two one-electron steps) is followed by another one at more negative potentials (Fig. 28). Although the half-wave potential of the second wave is

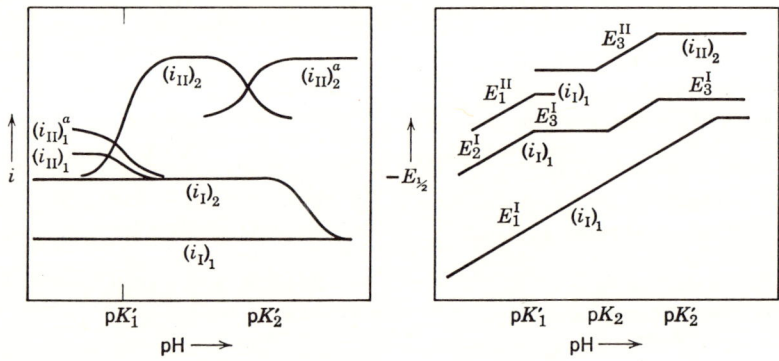

FIG. 28. pH-dependences of limiting currents and half-wave potentials of the individual waves of chalcone (schematically) $(i_I)_1$ and $(i_I)_2$ correspond to reduction of the double bond, $(i_{II})_1$, $(i_{II})_2$ to the reduction of ω-benzylacetophenone formed in the first two-electron step.

identical with that of the corresponding saturated ketone under the same conditions, the height of the second wave is much lower than predicted and reaches the theoretical value corresponding to a two-electron process only

SYSTEMS SHOWING THREE OR MORE REDUCTION WAVES 149

at pH > 8. Reduction scheme 47 can be suggested for the reduction of ketones (87, 88).

$$[\text{ArCOCH}=\text{CHR}]\text{H}^+ \underset{k_{-1}}{\overset{k_1}{\rightleftharpoons}} \text{ArCOCH}=\text{CHR} + \text{H}^+ \quad (47\text{a})$$

$$[\text{ArCOCH}=\text{CHR}]\text{H}^+ + e \xrightarrow{E_1^\text{I}} \text{ArCO}\dot{\text{C}}\text{HCH}_2\text{R} \quad (47\text{b})$$

$$\overset{(+)}{\text{ArCOH}}\dot{\text{C}}\text{HCH}_2\text{R} \underset{k_{-2}}{\overset{k_2}{\rightleftharpoons}} \text{ArCO}\dot{\text{C}}\text{HCH}_2\text{R} + \text{H}^+ \quad (47\text{c})$$

$$\overset{(+)}{\text{ArCOH}}\dot{\text{C}}\text{HCH}_2\text{R} + e \xrightarrow[\text{H}^+]{E_2^\text{I}} \overset{(+)}{\text{ArCOH}}\text{CH}_2\text{CH}_2\text{R} \quad (47\text{d})$$

$$\text{ArCO}\dot{\text{C}}\text{HCH}_2\text{R} + e \xrightarrow{E_3^\text{I}} \text{Ar}\overset{(-)}{\overbrace{\text{C}\cdots\text{CH}}}\text{CH}_2\text{CH}_2\text{R} \quad (47\text{e})$$
$$\underset{\text{O}}{\vdots}$$

$$\text{ArCOCH}_2\text{CH}_2\text{R} \underset{k_{-3}}{\overset{k_3}{\rightleftharpoons}} \text{Ar}\overset{(-)}{\overbrace{\text{C}\cdots\text{CH}}}\text{CH}_2\text{R} + \text{H}^+ \quad (47\text{f})$$
$$\underset{\text{O}}{\vdots}$$

$$\text{ArCOCH}=\text{CHR} + e \xrightarrow{E_4^\text{I}} \overset{(-)}{\overbrace{\text{ArCOCHCHR}}} \quad (47\text{g})$$

$$\text{ArCO}\dot{\text{C}}\text{HCH}_2\text{R} \underset{k_{-4}}{\overset{k_4}{\rightleftharpoons}} \overset{(-)}{\overbrace{\text{ArCOCHCHR}}} + \text{H}^+ \quad (47\text{h})$$

$$\overset{(+)}{\text{ArCOH}}\text{CH}_2\text{CH}_2\text{R} \underset{k_{-5}}{\overset{k_5}{\rightleftharpoons}} \text{ArCOCH}_2\text{CH}_2\text{R} + \text{H}^+ \quad (47\text{j})$$

$$\overset{(+)}{\text{ArCOH}}\text{CH}_2\text{CH}_2\text{R} + e \xrightarrow{E_1^\text{II}} \text{Ar}\dot{\text{C}}\text{OHCH}_2\text{CH}_2\text{R} \quad (47\text{k})$$

$$\text{Ar}\dot{\text{C}}\text{OHCH}_2\text{CH}_2\text{R} + e \xrightarrow[\text{H}^+]{E_2^\text{II}} \text{ArCHOHCH}_2\text{CH}_2\text{R} \quad (47\text{l})$$

$$\text{ArCOCH}_2\text{CH}_2\text{R} + e \xrightarrow{E_3^\text{II}} \overset{(-)}{\text{Ar}\dot{\text{C}}\text{OCH}_2\text{CH}_2\text{R}} \quad (47\text{m})$$

$$\text{Ar}\dot{\text{C}}\text{OHCH}_2\text{CH}_2\text{R} \underset{k_{-6}}{\overset{k_6}{\rightleftharpoons}} \overset{(-)}{\text{Ar}\dot{\text{C}}\text{OCH}_2\text{CH}_2\text{R}} + \text{H}^+ \quad (47\text{n})$$

$$\text{Ar}\dot{\text{C}}\text{OHCH}_2\text{CH}_2\text{R} + e \xrightarrow[\text{H}^+]{E_4^\text{II}} \text{ArCHOHCH}_2\text{CH}_2\text{R} \quad (47\text{p})$$

$$\overset{(-)}{\text{Ar}\dot{\text{C}}\text{OCH}_2\text{CH}_2\text{R}} + e \xrightarrow[2\text{H}^+]{E^\text{III}} \text{ArCHOHCH}_2\text{CH}_2\text{R} \quad (47\text{r})$$

In the first two-electron step (Eq. 47a–h), a saturated ketone $\text{ArCOCH}_2\text{CH}_2\text{R}$ is produced as the final product. The rate of reaction 47f governs the transformation of the carbanion enolate into the reducible saturated ketone, which is subsequently reduced in Eq. 47j–r in a sequence of reactions analogous to scheme 46.

In the first one-electron step, the protonated form of the unsaturated ketone is reduced up to pH $\approx$ pK_1'. Because the difference between $E_1^{\rm I}$ and $E_4^{\rm I}$ is small in this pH range, the presence of two waves can be observed only on logarithmic analysis. The change in the reduction of protonated to unprotonated form is clearly marked on the $E_{1/2}$–pH plot. The second one-electron uptake in strongly acidic media must take place on the protonated radical according to Eq. 47d to result in the observed shift in $E_2^{\rm I}$. Only at pH > 6, that is, at pH > pK_2', does the half-wave potential become pH-independent; the equilibrium of Eq. 47c is no longer shifted to the left and the radical is reduced according to Eq. 47e. The potentials $E_2^{\rm I}$ and $E_3^{\rm I}$ at pH > (pK_2' + 1) are so close that only one wave $(i_{\rm I})_2$ is observed. At pH > pK_1' (i.e., 10.6), a radical anion is formed in the reduction of the unprotonated chalcone (Eq. 47g). The protonation of this radical anion with pK_4 about 10.6 is responsible for the observed shift in wave $(i_{\rm I})_2$ in this region. This shift continues until pK_4' is reached, after which protonation is no longer fast enough, half-wave potential $E_3^{\rm I}$ becomes pH-independent, and a decrease in current $(i_{\rm I})_2$ with increasing pH is observed.

The pH-dependences of waves $i_{\rm II}$ correspond to those shown in Figure 27 and are interpreted by scheme 46.

The radicals and radical anions formed in reaction 47b, g, k, and m can react with mercury or with the solvent. Because the wave heights are linearly proportional to the concentration of unsaturated ketone and their ratio remains concentration-independent, we may conclude that if reactions of higher order, for example, dimerization or addition to the original compound, do occur, they are not fast enough to result in significant conversion during the drop life.

Scheme 47 has been fully verified for chalcone (88) and in principle for phenyl vinyl ketone also (87). The behavior of the unsaturated aldehyde with a phenyl ring conjugated with the unsaturated system (Fig. 29) differs in certain aspects from that observed for ketones (Fig. 28). The two one-electron waves merge, the height of the two-electron step decreases in alkaline media, no wave beyond the two-electron reduction is observed in acidic media, and a single two-electron wave of the saturated aldehyde is formed in alkaline media. Scheme 48 is in accordance with experimental data (Fig. 29):

$$[{\rm C_6H_5CH{=}CHCHO}]{\rm H}^+ \underset{k_{-1}}{\overset{k_1}{\rightleftharpoons}} {\rm C_6H_5CH{=}CHCHO} + {\rm H}^+ \qquad (48{\rm a})$$

$$[{\rm C_6H_5CH{=}CHCHO}]{\rm H}^+ + e \xrightarrow{E_1^{\rm I}} {\rm C_6H_5CH_2\dot{C}HCHO} \qquad (48{\rm b})$$

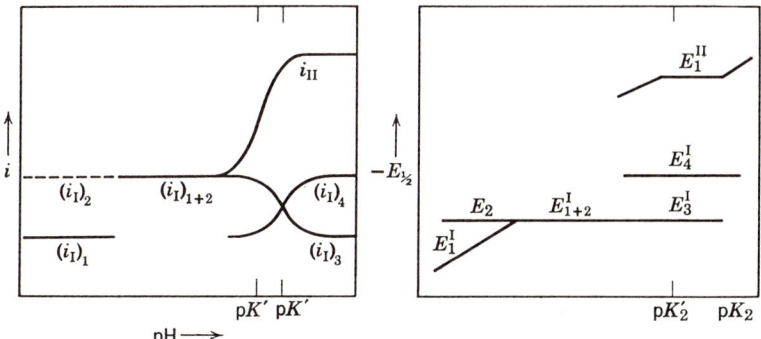

FIG. 29. pH-dependences of limiting currents and half-wave potentials of the individual waves of cinnamaldehyde (schematically). $(i_I)_1$, $(i_I)_2$, $(i_I)_3$, and $(i_I)_4$ correspond to the reduction of the double bond, i_{II} to the reduction of 3-phenyl-propionaldehyde formed in the first two-electron step.

$$C_6H_5CH_2\dot{C}HCHO + Me^+ \underset{k_{-N}}{\overset{k_N}{\rightleftarrows}} [C_6H_5CH\dot{C}HCHO]Me + H^+ \qquad (48c)$$

$$[C_6H_5CH\dot{C}HCHO]Me + e \xrightarrow[2H^+]{E_N} C_6H_5CH_2CH_2CHO + Me^+ \qquad (48d)$$

$$C_6H_5CH_2\dot{C}HCHO + e \xrightarrow[H^+]{E_2^I} C_6H_5CH_2CH_2CHO \qquad (48e)$$

$$C_6H_5CH=CHCHO + e \xrightarrow{E_3^I} [C_6H_5CHCHCHO]^{(\cdot -)} \qquad (48f)$$

$$C_6H_5CH_2\dot{C}HCHO \underset{k_{-2}}{\overset{k_2}{\rightleftarrows}} [C_6H_5CHCHCHO]^{(\cdot -)} + H^+ \qquad (48g)$$

$$C_6H_5CH_2\dot{C}HCHO + e \xrightarrow[H^+]{E_2^I} C_6H_5CH_2CH_2CHO \qquad (48e)$$

$$[C_6H_5CHCHCHO]^{(\cdot -)} + Me^+ \underset{k_{-M}}{\overset{k_M}{\rightleftarrows}} [C_6H_5CH\dot{C}HCHO]Me \qquad (48f)$$

$$[C_6H_5CH\dot{C}HCHO]Me + e \xrightarrow[2H^+]{E_M} C_6H_5CH_2CH_2CHO + Me^+ \qquad (48g)$$

$$[C_6H_5CHCHCHO]^{(\cdot -)} + e \xrightarrow[2H^+]{E_4} C_6H_5CH_2CH_2CHO \qquad (48h)$$

$$C_6H_5CH_2CH_2C\begin{array}{c}OH\\ \diagup\\ \diagdown\\ OH\end{array} + \text{base} \underset{k_{-3}}{\overset{k_3}{\rightleftarrows}} C_6H_5CH_2CH_2C\begin{array}{c}O^-\\ \diagup\\ \diagdown\\ OH\end{array} + \text{base } H^+$$

$$(48j)$$

$$C_6H_5CH_2CH_2C\diagup_{OH}^{O^-} \underset{k_{-4}}{\overset{k_4}{\rightleftharpoons}} C_6H_5CH_2CH_2CH{=}O + OH^-$$

(48k)

$$C_6H_5CH_2CH_2CHO + 2e \xrightarrow[2H^+]{E^{II}} C_6H_5CH_2CH_2CH_2OH \qquad (48l)$$

Protonation (Eq. 48a) causes the shift in the potential E_1^I, but the potential of wave E_2^I remains pH-independent because the radical is not protonated. The potentials of the combined wave $(i_l)_1$ and $(i_l)_2$ of the protonated form are so close to the potential E_3^I of the reduction of the unprotonated form (Eq. 48f) that no separation of two waves is observed. When protonation (Eq. 48g) of the radical anion formed in the reduction of the unprotonated cinnamaldehyde (Eq. 48f) becomes slow, the overall height of the combined wave $(i_l)_1$ and $(i_l)_2$ decreases and a new wave of the reduction of the radical anion (Eq. 48h) is formed at more negative potentials E_4^I. The potential of this wave is pH-independent for pH $>$ pK_2.

The 3-phenylpropionaldehyde formed in the first two-electron step, is reduced in wave i_{II}. For this wave the half-wave potentials, the wave heights, and the pH-dependence are all quantitatively identical with values obtained for an authentic compound. This indicates that whereas for ketones the protonation of $[ArCOCHCH_2R]^-$ is a relatively slow reaction interposed between two electron transfers, for cinnamaldehyde the protonation of $[C_6H_5CH_2CHCHO]^-$ is fast when compared with subsequent reduction and a side reaction, hydration. Shifts in the half-wave potential E^{II} are caused by the two acid–base equilibria of Eq. 48j and k.

VII. Conclusions

Combinations of experimental techniques and modifications in structure of the organic compounds are essential, in connection with the studies of effects of the composition of reaction mixture, for the elucidation of organic electrode processes. Even with the development of new, more sophisticated experimental techniques, however, classical dc polarography still offers a large amount of information when suitable systems are chosen, really informative experiments are set up, and the results are properly evaluated. Even though the use of non-aqueous solvents makes it possible to obtain information on some thermodynamic parameters impossible to deduce from experiments carried out in aqueous media, there is still interest in aqueous chemistry and in the development of methods in this area, among others, of polarography. One of the advantages that classical dc polarography has in common with studies of organic reactions in water-containing solutions is that in both these fields deductions and interpretations of experimental data are based on a relatively large amount of material available for comparison.

In polarography this material was nevertheless frequently classified according to the type of chemical system involved. This sometimes made it difficult to see the relation between experimental data and the course of the reaction operating. It has been the aim of this chapter to show in a systematic way how the type of the chemical and electrochemical process involved can be deduced from experimental data, in particular as a function of the acidity of the investigated solution.

Developments during the past decade show that our ability to deduce from experimental results enables us to elucidate more and more details of the organic electrode process. Many systems discussed underwent almost continuous assessment and adjustment of the proposed mechanism. It can be assumed that in the future numerous systems discussed here will undergo refinement, and perhaps even alteration in the present interpretation of mechanisms. At present, the simultaneous and combined use of $E_{1/2}$–pH and i–pH plots proves to be much more informative than either of these plots separately.

There is no doubt that one of the final aims of such studies is to describe quantitatively all chemical and electrochemical steps involved in the organic electrode process and determine the values of all rate and equilibrium constants involved. In this article it has been shown that even in those cases in which it is not possible to apply an exhaustive quantitative treatment it is possible to achieve some progress based on qualitative information. The other extreme—that is, the use of complex mathematical apparatus for evaluation of systems not fully qualitatively understood—should perhaps be avoided.

If we learn to interpret polarographic curves and understand their changes with pH, interesting information on electrode mechanisms and on fast chemical reactions can be obtained.

References

1. Zuman, P., in *Progress in Physical Organic Chemistry* (A. Streitwieser, Jr. and R. W. Taft, Eds.), Vol. 5, Interscience, New York, 1967, p. 81.
2. Zuman, P., *J. Polarograph. Soc.,* **13**, 53 (1967).
3. Zuman, P., *Substituent Effects in Organic Polarography*. Plenum Press, New York, 1967.

4. Zuman, P., *The Elucidation of Organic Electrode Processes*, Academic Press, New York, 1969.
5. Majranovskij, S. G., *Catalytic and Kinetic Waves in Polarography*, Plenum Press, New York, 1968.
6. Gardner, H. J., and L. E. Lyons, *Rev. Pure Appl. Chem.*, **3**, 134 (1953).
7. Saveant, J. M., *Bull. Soc. Chim. France*, 481, 486, 493 (1967).
8. Zuman, P., and S. Tang, *Collection Czech. Chem. Commun.*, **28**, 829 (1963).
9. Rosenthal, T., C. H. Albright, and P. J. Elving, *J. Electrochem. Soc.*, **99**, 227 (1952).
10. Zuman, P., J. Chodkowski, and F. Šantavý, *Collection Czech. Chem. Commun.*, **26**, 380 (1961).
11. Zhdanov, S. I., and A. N. Frumkin, *Dokl. Akad. Nauk USSR*, **122**, 412 (1958).
12. Koutecký, J., *Collection Czech. Chem. Commun.*, **18**, 587 (1953).
13. Majranovskij, S. G., and S. C. Lishczeta, *Collection Czech. Chem. Commun.*, **25**, 3025 (1960).
14. Eigen, M., *Angew. Chem. (Engl. Ed.)*, **3**, 1 (1964).
15. Zuman, P., and O. Exner, *Collection Czech. Chem. Commun.*, **30**, 1832 (1965).
16. Fleet, B., and P. Zuman, *Collection Czech. Chem. Commun.*, **32**, 2066 (1967).
17. Zuman, P., *Chem. Listy*, **56**, 219 (1962).
18. Barnes, D., and P. Zuman, *J. Chem. Soc., B*, 1118 (1971).
19. Zuman, P., *Collection Czech. Chem. Commun.*, **25**, 3245 (1960).
20. Zuman, P., *Collection Czech. Chem. Commun.*, **33**, 2548 (1968).
21. Zuman, P., D, Barnes, and A. Ryvolová-Kejharová, *Discussions Faraday Soc.*, **45**, 202 (1968).
22. Kucharczyk, N., M. Adamovský, V. Horák, and P. Zuman, *J. Electroanal. Chem.*, **10**, 503 (1965).
23. Nișli, G., D. Barnes, and P. Zuman, *J. Chem. Soc., B*, 764, 770, 778 (1970).
24. Zuman, P., *Collection Czech. Chem. Commun.*, **15**, 839 (1950).
25. Tirouflet, J., and A. Corvaisier, *Bull. Soc. Chim. France*, 540 (1962).
26. Zuman, P., and I. Šestáková, unpublished results.
27. Volke, J., *Z. Physik. Chem. (Leipzig)*, Sonderheft, 268 (1958).
28. Manoušek, O., and P. Zuman, *Collection Czech. Chem. Commun.*, **29**, 1432 (1964).
29. Lund, H., *Nature*, **204**, 1087 (1964).
30. Lund, H., *Acta Chem. Scand.*, **18**, 1984 (1964).
31. Tirouflet, J., and E. Laviron, *Ric. Sci.*, **29**, *Contributi Teor. Sper. Polarografia*, **4**, 189 (1959).

32. Fornasari, E., G. Giacometti, and G. Rigatti, *Ric. Sci.*, **30**, *Contributi Teor. Sper. Polarografia*, **5**, 261 (1960).
33. Stočesová, D., *Collection Czech. Chem. Commun.*, **14**, 615 (1949).
34. Kastening, B., and L. Holleck, *Z. Elektrochem.*, **63**, 166 (1959).
35. Nicholson, R. S., J. M. Wilson, and M. L. Olmstead, *Anal. Chem.*, **38**, 542 (1966).
36. Barnes, D., and P. Zuman, unpublished results.
37. Tachi, I., and M. Senda, *Advances in Polarography*, Vol. II, Pergamon Press, Oxford, 1960, p. 454.
38. Testa, A. C., and W. H. Reinmuth, *J. Am. Chem. Soc.*, **83**, 784 (1961).
39. Alberts, G. S., and I. Shain, *Anal. Chem.*, **35**, 1859 (1963).
40. Veselý, K., and R. Brdička, *Collection Czech. Chem. Commun.*, **12**, 313 (1947).
41. Kůta, J., *Collection Czech. Chem. Commun.*, **24**, 2532 (1959).
42. Majranovskij, S. G., D. I. Dzhaparidze, and O. I. Sorokin, *Izv. Akad. Nauk SSSR, Ser. Khim.*, No. 5, 795 (1964).
43. Traini, A., and P. Zuman, unpublished results.
44. Müller, O. H., and J. P. Baumberger, *J. Am. Chem. Soc.*, **61**, 590 (1939).
45. Brdička, R., and K. Wiesner, *Collection Czech. Chem. Commun.*, **12**, 138 (1947).
46. Brdička, R., *Collection Czech. Chem. Commun.*, **12**, 212 (1947).
47. Ono, S., M. Takagi, and T. Wasa, *Collection Czech. Chem. Commun.*, **26**, 141 (1961).
48. Takagi, M., S. Ono, and T. Wasa, *Rev. Polarog. (Kyoto)*, **11**, 210 (1963).
49. Wiesner, K., M. Wheatley, and J. Los, *J. Am. Chem. Soc.*, **76**, 4858 (1954).
50. Becker, M., and H. Strehlow, *Z. Elektrochem.*, **42**, 818 (1960).
51. Volke, J., and V. Volková, *Collection Czech. Chem. Commun.*, **20**, 1332 (1955).
52. Volke, J., *Collection Czech. Chem. Commun.*, **22**, 1777 (1957).
53. Zuman, P., J. Tenygl, and M. Březina, *Collection Czech. Chem. Commun.*, **19**, 46 (1954); see also *J. Electrochem. Soc.*, **105**, 758 (1958).
54. Laviron, E., *Compt. Rend.*, **250**, 3671 (1960).
55. Holubek, J., and J. Volke, *Collection Czech. Chem. Commun.*, **27**, 680 (1962).
56. Zuman, P., and B. Turcsanyi, *Collection Czech. Chem. Commun.*, **33**, 3090 (1968).
57. Gergely, E., and T. Iredale, *J, Chem. Soc.*, 2638 (1951).
58. Gergely, E., and T. Iredale, *J. Chem. Soc.*, 3226 (1953).
59. Holleck, L., and H. J. Exner, *Z. Elektrochem.*, **56**, 46, 677 (1952).
60. Kemula, W., Z. R. Grabowski, and E. Bartel, *Roczniki Chem.*, **33**, 1125 (1959).
61. Holleck, L., and H. Marsen, *Z. Elektrochem.*, **57**, 301, 944 (1953).

62. Souchay, P., and S. Ser, *J. Chim. Phys.*, **49**, C172 (1952).
63. Gardner, H. J., and W. P. Georgans, *J. Chem. Soc.*, 4180 (1956).
64. Lund, H., *Acta Chem. Scand.*, **13**, 249 (1959).
65. Ryvolová, A., *Collection Czech. Chem. Commun.*, **25**, 420 (1960).
66. Kejharová-Ryvolová, A., and P. Zuman, *Collection Czech. Chem. Commun.*, **36**, 1019 (1971).
67. Kargin, Yu., O. Manoušek, and P. Zuman, *J. Electroanal. Chem.*, **12**, 443 (1966).
68. Fischer, O., L. Kišová, and J. Stěpánek, *J. Electroanal. Chem.*, **17**, 233 (1968).
69. Zuman, P., O. Manoušek, and S. K. Vig, *J. Electroanal. Chem.*, **19**, 147 (1968).
70. Zuman, P., and V. Horák, *Collection Czech. Chem. Commun.*, **26**, 176 (1961).
71. Saveant, J. M., *Compt. Rend.*, **257**, 448 (1963); **258**, 585 (1964).
72. Micheel, F., and E. Heiskel, *Chem. Ber.*, **94**, 143 (1961).
73. Lund, H., *Acta Chem. Scand.*, **14**, 1927 (1960).
74. Delaroff, V., M. Bolla, and M. Legrand, *Bull. Soc. Chim. France*, 1912 (1961).
75. Fedoreňko, M., J. Königstein, and K. Linek, *Collection Czech. Chem. Commun.*, **32**, 1497 (1967).
76. Barnes, D., and P. Zuman, *Trans. Faraday Soc.*, **65**, 1668, 1681 (1969).
77. Zuman, P., and H. Zinner, *Chem. Ber.*, **95**, 2089 (1962).
78. Kůta, J., *Collection Czech. Chem. Commun.*, **24**, 2532 (1959).
79. Hanuš, V., and R. Brdička, *Chem. Listy*, **44**, 291 (1950); *Khimiya*, **1**, 28 (1951).
80. Koutecký, J., *Collection Czech. Chem. Commun.*, **19**, 1093 (1954).
81. Kůta, J., *Collection Czech. Chem. Commun.*, **21**, 697 (1956).
82. Bezuglyj, V. D., and E. Yu. Novik, *Zavodsk. Lab.*, **27**, 544 (1961).
83. Kane, P. O., *J. Electroanal. Chem.*, **2**, 152 (1961).
84. Ryvolová, A., and V. Hanuš, *Collection Czech. Chem. Commun.*, **21**, 853 (1956).
85. Buck, R. P., *Anal. Chem.*, **35**, 1853 (1963).
86. Kůta, J., and E. Krejčí, *Collection Czech. Chem. Commun.*, **24**, 258 (1959).
87. Zuman, P., and J. Michl, *Nature*, **192**, 655 (1961).
88. Ryvolová-Kejharová, A., and P. Zuman, *J. Electroanal. Chem.*, **21**, 197 (1969).
89. Barnes, D., and P. Zuman, *J. Electroanal. Chem.*, **16**, 575 (1968).

CHAPTER III

INTERMETALLIC COMPOUNDS IN AMALGAMS*

Michail Kozlovsky and Alexandra Zebreva

The Kirov Kazakh State University, Alma-Ata, U.S.S.R.

Contents

I. Introduction	157
II. The Classification of Amalgams. Intermetallic Compounds in Amalgams	158
III. Electrochemical Properties of Simple Amalgams	159
A. The equilibrium potential	159
B. Current potential curves (polarographic waves)	170
IV. Intermetallic Compound Formation in Complex Amalgams	175
A. Effect on the potential of a complex amalgam	175
B. Effect on cathodic polarographic curves	180
C. Effect on anodic polarographic curves	181
V. Intermetallic Compounds in Anodic Stripping Voltammetry	183
VI. Intermetallic Compounds in Phase Exchange (Cementation)	187
VII. Conclusion	188
References	189

I. Introduction

In voltammetric studies, as well as in electrolysis with mercury cathodes, one must consider the interaction of metals in amalgams, which leads to the formation of intermetallic compounds. This interaction may take place both in simple and complex (multicomponent) amalgams, between metal and mercury in the former case, and among all the metals present in the latter case. The formation of intermetallic compounds is especially important in amalgam voltammetry, particularly in anodic stripping voltammetry [amalgam polarography with preliminary accumulation (APPA)], in which the amount of one or another metal is calculated from the magnitude of its anodic limiting current. The interaction of metals in amalgams must also be taken into consideration in methods of dividing and refining metals that involve the use of amalgams.

* Manuscript received July, 1968.

This chapter is concerned with the formation of intermetallic compounds in amalgams and their effects on the electrochemical properties of amalgams, that is, their equilibrium potentials and their behavior in cathodic and anodic processes.

II. The Classification of Amalgams. Intermetallic Compounds in Amalgams

We define an amalgam as a metallic system that has mercury as one of its components.

Most metal–mercury systems have already been investigated more or less sufficiently (1). Complete diagrams have been obtained for many systems. For other systems only partial diagrams are available. The solubilities of some metals in mercury at certain temperatures, and the compositions of metal–mercury compounds and their melting or decomposition points, have been determined as well.

Different metals may behave toward mercury in very different ways; they may react with it to form intermetallic compounds, they may form solutions (particularly solid solutions) with it, or they may not mix with it at all. Table I lists the characteristics of several metal–mercury systems.

According to the classification of metallic alloys suggested by Sauerwald (2, 3), amalgams can be divided into three groups:

Group I: Metals that react to form compounds with considerable evolution of heat; interaction between metals is also observed in liquid alloys.

Group II: Metals that form only solutions, both in the solid and in the liquid state.

Group III: An intermediate group between groups I and II; in the solid state the metals react to form compounds with the evolution of small quantities of heat, while in the liquid state the properties of the amalgams are similar to those of group II.

To the first group belong amalgams of the alkali, alkaline earth, and probably rare earth metals. [Phase diagrams of rare earth metal amalgams have not yet been studied over wide ranges of concentration and temperature. The formation of solid and liquid amalgams of metals of this group are accompanied by considerable evolution of heat (Table I)].

The second group includes amalgams of bismuth, lead, tin, and zinc, in which no compounds are observed at temperatures above 0°C. The heats of solution of these metals in mercury to give the liquid amalgams are very low. In the resulting liquid amalgams, these metals are present as atoms distributed in the general mass in statistical disorder (4, 5). Antimony and aluminum can also be assigned to this group of metals; their solubilities in mercury are very low at room temperature but increase notably on heating.

Amalgams of thallium, indium, copper, gold, silver, manganese, and cadmium can be assigned to the third group. In solid alloys of these metals with mercury, intermetallic compounds are formed with the evolution of small quantities of heat. These compounds as a rule decompose on heating, either before melting or at the melting point.

Too few data are as yet available on the alloys of mercury with nickel, platinum, palladium, and the other metals of the platinum group to permit them to be assigned definitely to one or another group. The compounds of these metals with mercury that exist at room temperature in the solid phase are so slightly soluble in mercury that there is no possibility of studying the liquid phase by any of the methods known at present. As the temperature increases, the solubilities of these compounds in mercury rise very slowly to the temperatures at which they decompose (32).

In the solid state, chromium, cobalt, and iron form neither compounds nor solid solutions with mercury and are very slightly soluble in mercury at all temperatures up to the boiling point (32). Amalgams of these metals are heterogeneous (in liquid amalgams the metals are suspended in mercury) at any metal concentration above $10^{-5}\%$.

Liquid amalgams are usually employed in electrochemical methods of dividing, obtaining, refining, and determining metals. Therefore we consider the electrochemical characteristics of only liquid amalgams, both homogeneous and heterogeneous.

III. Electrochemical Properties of Simple Amalgams

A. The Equilibrium Potential

The main electrochemical characteristic of an amalgam is its equilibrium potential. The equilibrium potential of a liquid amalgam is described by Eq. 1 (42):

$$E = E_M^0 - \frac{\Delta G^0}{nF} + \frac{RT}{nF} \ln {}^*a_{M(Hg)} {}^*a_{Hg}^y + \frac{RT}{nF} \ln \frac{a_{M^{n+}}}{a_{M(Hg)}} \quad (1)$$

where E_M^0 is the standard potential of the metal, ΔG^0 the standard change in free energy accompanying the formation of a solid phase in equilibrium with a liquid phase at a given temperature, ${}^*a_{M(Hg)}$ the activity of the metal in a saturated amalgam, ${}^*a_{Hg}$ the activity of mercury in a saturated amalgam, y the number of mercury atoms associated with one atom of metal in the

TABLE I. Characteristics of Alloys of Different Metals with Mercury

Metal	Melting point of metal, °C	Formula of intermetallic compound	Melting point of compound, °C	Reference	Heat of formation of solid compound, kcal/g-atom of metal	Heat of formation of liquid alloy, kcal/g-atom of metal	Reference
Mercury	−38.9	—	—	—	—	—	—
Lithium	179	LiHg	590	6–9	10.45	—	10
		LiHg$_2$	Decomp. 339	6–9	8.3	—	11
		Li$_3$Hg	375	6–9	—	—	—
Sodium	97.5	NaHg$_2$	353	14	6.1–6.8	5.5	16
		NaHg$_4$	Decomp. 157	14	4.0–4.4	—	20
Potassium	63.5	KHg	180	12	5.5–6.5	4.0	15
		KHg$_2$	270	13	6.1–6.6	5.0	16
		KHg	Decomp. 170	14	2.5	—	11
Cesium	28.4	CsHg$_2$	208	17	—	—	—
		CsHg$_4$	164	17	—	—	—
		CsHg	158	17	—	—	—
Beryllium	1284	BeHg	—	18	—	—	—
Magnesium	620	MgHg	627	19	—	—	—
		Mg$_5$Hg$_3$	562	19	3.5	—	20
		MgHg$_4$	580	19	—	—	—
Calcium	850	Ca$_3$Hg$_2$	—	21	—	—	—
Strontium	770	SrHg	—	22	—	SrHg: 59.5 kcal/mole	20

Element		Compound					
Barium	710	BaHg	—	22	—	—	—
Lanthanum	810	LaHg	Decomp. 250	23	—	—	—
		LaHg$_4$	—	23	—	—	—
		LaHg$_3$	—	23	—	—	—
Europium	1150	Eu$_3$Hg$_2$	Stable at red heat	24	—	—	—
Cerium	815	CeHg$_4$	—	24	4.2–4.5	—	10, 11
Uranium	1700	UHg$_2$	—	25	3.9	—	10, 11
		UHg$_4$	Decomp. 200	25	4.1	—	26
		UHg$_3$	—	25	4.4	—	26
Indium	156	InHg$_5$	−13.2	29	—	0.22	30
		InHg	−18.5	29	—	0.17	30
Thallium	302	Tl$_2$Hg$_5$	14.5	27	0.12	−0.06	16, 28
Zinc	419.5	?	Decomp. 20	37	—	−0.13	11
Cadmium	321	Cd$_3$Hg	Decomp. 188	1	0.52	0.5	11, 16
Manganese	1244	Mn$_2$Hg$_5$	Decomp. 75	31	1.1	—	26
		MnHg	Decomp. 265	31	1.0	—	26
Nickel	1455	NiHg$_4$	Decomp. 220	32	2.4	—	26
Copper	1083	Cu$_4$Hg$_3$	Decomp. 96	32	—	—	—
Silver	960	Ag$_5$Hg$_8$	Decomp. 276	34	0.8	—	20
		Ag$_3$Hg$_4$	Decomp. 127	34	0.1	—	20
Gold	1063	Au$_3$Hg	Decomp. 420	33	0.2	0.03	1, 10
Palladium	1555	PdHg	—	35	11.3	—	26
Platinum	1773	PtHg	Decomp. 480	36	4.6	—	26
		PtHg$_2$	Decomp. 245	36	5.5	—	26
		PtHg$_4$	Decomp. 160	36	9.1	—	26

TABLE I. Characteristics of Alloys of Different Metals with Mercury

Metal	Melting point of metal, °C	Formula of intermetallic compound	Melting point of compound, °C	Reference	Heat of formation of solid compound, kcal/g-atom of metal	Heat of formation of liquid alloy, kcal/g-atom of metal	Reference
Thorium	1700	$ThHg_3$	?	40	—	—	—
Bismuth	271.3	No compounds; only a little mercury dissolves in solid bismuth				−0.2	15
Tin	231.9	$SnHg_4$ exists below −35°; at and above 0° a solid solution of mercury in tin is in equilibrium with a liquid amalgam				−0.1	20
Lead	327.3	No compounds; a solid solution of mercury in lead is in equilibrium with a liquid amalgam (1)				−0.3 −0.025	15 25
Aluminum	660	No compounds; mercury is virtually insoluble in solid aluminum (39)					
Antimony	627	No compounds; mercury is virtually insoluble in antimony (32)	—	41	—	—	—
Plutonium	639	$PuHg_3$ $PuHg_4$	—	—	—	—	—
Chromium	1875	No compounds; chromium is virtually insoluble in mercury, and mercury is insoluble in chromium (32)					
Iron	1539	No compounds; iron is virtually insoluble in mercury, and mercury is insoluble in iron (32)					
Cobalt	1492	No compounds; cobalt is virtually insoluble in mercury, and mercury is insoluble in cobalt (32)					

solid phase, $a_{M(Hg)}$ the activity of the metal in the liquid phase of the amalgam, and $a_{M^{n+}}$ the activity of the ions of the same metal in the solution. All the other symbols are conventional.

If the solid phase in equilibrium with a saturated amalgam is a pure metal, the potential of a heterogeneous amalgam is equal to the potential of this metal, since in this case both y and ΔG^0 are equal to zero, so that

$$E = E_M^0 + \frac{RT}{nF} \ln a_{M^{n+}} \qquad (2)$$

If the liquid phase of a heterogeneous amalgam is in equilibrium with a compound MHg_y, which is weakly dissociated in the liquid phase, then, according to Jangg and Kirchmayr (42), the potential of the heterogeneous amalgam is described by the equation

$$E = E_M^0 - \frac{\Delta G^0}{nF} + \frac{RT}{nF} \ln {^*a_{Hg}^y} + \frac{RT}{nF} \ln a_{M^{n+}} \qquad (3)$$

If the liquid phase of a heterogeneous amalgam is in equilibrium with a solid solution of mercury in the metal or an intermetallic compound fully dissociating in the liquid phase, then, according to Jangg and Kirchmayr (42, 69), the potential of the heterogeneous amalgam is equal to that of the pure metal.

However, experimental data do not conform to the latter assertion. As seen from Table II, the difference in potential E_s between a heterogeneous amalgam of cadmium, thallium, lead, or tin and the corresponding pure metal is not equal to zero even though it is impossible to presume the existence of nondissociated compounds in the liquid phases of these amalgams. In our opinion, Eq. 3 can be deduced without assuming, as Jangg and Kirchmayr did, that a nondissociated compound is present in the liquid phase (166). The experimentally observed finite values of E_s for cadmium, thallium, lead, and tin amalgams are theoretically justified and are not an experimental error as Kirchmayr suggested (69).

It follows from Eqs. 1–3 that the potentiometric method can be successfully applied for determining the solubilities of metals in mercury. For this purpose it is most effective to measure the emf of a concentration cell of the type

| M(Hg) | M^{n+} | M(Hg) |
| Electrode 1 | | Electrode 2 |

The concentration of the metal in electrode 1 is kept constant, while that

TABLE II. Potential Differences between Saturated Amalgams and the Corresponding Pure Metals

System	Composition of the solid phase in equilibrium with a saturated amalgam (at 20°C)	E_s, mV	Reference	Heat of formation of the solid phase, kcal/g-atom	Reference	Comment
$Zn/Zn^{2+}/Zn(Hg)$	Pure Zn	0 ± 1	—	—	—	—
$Bi/Bi^{3+}/Bi(Hg)$	Pure Bi	0.0	42	—	—	—
		0.0	43	—	—	—
		0.0	45	—	—	—
$Pb/Pb^{2+}/Pb(Hg)$	Solution of Hg in Pb	6.4	—	—	—	—
	Up to 20 at. % Hg	0 ± 1	42	-0.28	16	—
	Up to 33 at. % Hg	6.3	45	—	—	—
		6.0	44	0.017	—	$T = 330°C$
$Sn/Sn^{2+}/Sn(Hg)$	Solution of Hg in Sn; 10 at. % Hg	0.9	46	—	—	—
		1.0	47	-0.23	16	$T = 325°C$
		1.2	45	—	—	—
$In/In^{3+}/In(Hg)$	Solution of Hg in In; 12 at. % Hg; compound In_7Hg	0.1	48	0.17	16	$T = 160°C$
$Tl/Tl^{+}/Tl(Hg)$	Solution of Hg in Tl; 15 at. % Hg	0.8	50	-0.14	28	$T = 25°C$
		2.1	49	—	—	—
$Cd/Cd^{2+}/Cd(Hg)$	Content of Hg = 76.5 at. % at 25°C	51.0	48	—	—	—
		49.0	51	0.52	16	$T = 25°C$
		51.4	45	—	—	—
$Cu/Cu^{2+}/Cu(Hg)$	Cu_4Hg_3	0 ± 1	42	—	—	—
$Mn/Mn^{2+}/Mn(Hg)$	Mn_2Hg_5	40.0	4	1.1	52	—
$Na/Na^{+}/Na(Hg)$	$NaHg_4$	780.0	53	4.0	—	$T = 25°C$
$K/K^{+}/K(Hg)$	KHg_{10}	1001.0	54	3.3	11	$T = 25°C$
	KHg_{11}	—	—	—	—	—

in electrode 2 is varied over a wide range. The emf of such a cell is given by the equation

$$E = \frac{RT}{nF} \ln \frac{a_{M(Hg),\,1}}{a_{M(Hg),\,2}} \qquad (4)$$

If the activity coefficient of the metal in the amalgam is equal to unity, or if it does not change as the concentration of the metal changes, then the metal activity in Eq. 4 can be replaced by its concentration. Then the dependence of E on $\log \{[M(Hg)]_1/[M(Hg)]_2\}$ will be linear. Upon reaching the limit of solubility of the metal in mercury, there is a break in the straight line, after which the potential of the cell does not depend on the concentration of the metal present in electrode 2. The break corresponds to the solubility of the metal in mercury (Fig. 1). As seen from Table III, the potentiometrically determined solubilities of a number of metals in mercury coincide with those obtained by other methods.

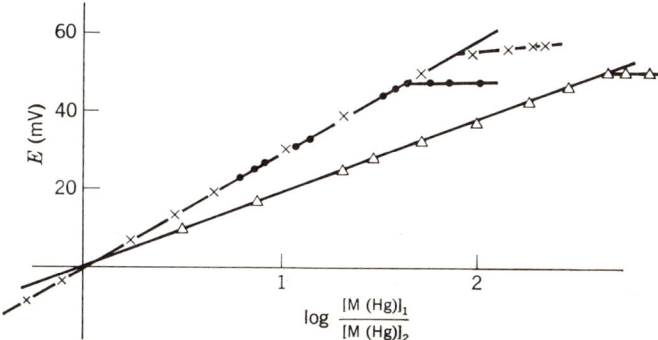

FIG. 1. EMF as a function of $\log \{[M(Hg)]_1/[M(Hg)]_2\}$ for various metals. ($\triangle$) Bismuth, $[Bi(Hg)]_1 = 1.7 \times 10^{-3}$ g-atom/liter; ($\bullet$) lead $[Pb(Hg)]_1 = 1.5 \times 10^{-2}$ g-atom/liter; ($\times$) tin $[Sn(Hg)]_1 = 6.3 \times 10^{-3}$ g-atom/liter.

However, in some potentiometric determinations of the solubilities of metal in mercury, namely, those by Tammann and co-workers (58, 68), there were admitted errors. Thus Tammann, to evaluate the solubilities of chromium and metals of the iron group in mercury, obtained experimental values of the potentials of saturated amalgams of those metals and simply substituted them in the Nernst equation:

$$E = E_M^0 + \frac{RT}{nF} \ln \frac{a_{M^{n+}}}{a_{M(Hg)}} \qquad (5)$$

TABLE III. Comparison of Values Found by Different Methods for the Solubilities of Some Metals in Mercury

Metal	Solubility found potentiometrically, at. %	Reference	Solubility found by amalgam polarography, at. %	Reference	Solubility found by other methods at. %	Reference
Indium	68.3	55	—	—	70.3	67
	68.0	56	—	—	—	—
	69.0	57	—	—	—	—
Thallium	42.4	58	—	—	—	—
	42.6	59	—	—	43.7	54
Zinc	5.6	42	—	—	—	—
	6.4	57	—	—	6.4	48
Tin	1.26	47	—	—	—	—
	1.27	38	—	—	1.2	60
Lead	1.2	42	—	—	—	—
	1.1	61	—	—	1.6	62
	1.3	58	—	—	—	—
Bismuth	1.2	58	—	—	—	—
	1.1	45	—	—	1.4	48
Copper	8×10^{-3}	42	1.0×10^{-2}	64	6.4×10^{-3}	48
Manganese	6.6×10^{-3}	42	9.5×10^{-3}	65	6.5×10^{-3}	66
Antimony	3.8×10^{-4}	67	3.8×10^{-4}	67	—	—
	4×10^{-4}	56	—	—	—	—
	4.8×10^{-5}	58	—	—	—	—
Nickel	2.1×10^{-3}	68	—	—	27.6×10^{-6}	66
	1.8×10^{-5}	56	—	—	—	—
Cobalt	3×10^{-4}	68	—	—	3.4×10^{-6}	—

He then calculated the concentration of the metal in the saturated amalgam, that is, the solubility of the metal in mercury. As shown above, however, the potential of a saturated amalgam is given not by the usual Nernst equation but by Eqs. 2 and 3 which do not contain the concentration of metal at all.

In other cases, to evaluate the solubility of a metal in mercury, Tammann dealt with amalgams containing two metals in addition to mercury. The formation of intermetallic compounds between these metals brought about changes in the amalgam potential which in turn led to erroneous values of the solubility. Thus, for example, to obtain the solubility of antimony in mercury, Tammann worked with an amalgam containing both copper and antimony (58). As shown earlier (67), the formation of a compound between antimony and copper in an amalgam results in a large change in the potential.

If similar errors in theory or technique can be avoided, the potentiometric method is a most accurate one for evaluating the solubilities of metals in mercury (42, 43, 63).

In studying very dilute amalgams, it has been found that the Nernst–Turin equation for concentration cells of the above-mentioned type has a lower limit of applicability (42, 56, 70, 71). On reaching this limit the potential of the amalgam quickly shifts toward more positive values up to the potential of pure mercury. As seen from Table IV, the deviations begin at concentrations around 10^{-5} g-atom of metal per liter of mercury. However, various authors give different values for this limit even for the same metal.

TABLE IV. The Lower Limit of Applicability of the Nernst Equation to Dilute Amalgams

Metal	Lower limit of applicability of the Nernst equation, g-atom of metal per liter of mercury			
	According to Liebl (56)	According to Tammann and Kollmann (68)	According to Jangg and Kirchmayr (42)	Our data
Manganese	—	—	1×10^{-4}	—
Zinc	—	2×10^{-6}	8×10^{-6}	2×10^{-5}
Cadmium	—	—	—	2×10^{-5}
Indium	10^{-4}	—	—	4×10^{-5}
Tin	—	—	—	2×10^{-5}
Lead	—	6×10^{-5}	3×10^{-5}	1×10^{-5}
Antimony	6×10^{-5}	—	—	3×10^{-5}
Copper	8×10^{-6}	1×10^{-6}	4×10^{-6}	—

Comparing different concepts concerning the behavior of metals in very dilute amalgams (42, 70) with our own observations leads us to believe that a very low concentration of a metal in an amalgam gives rise to a stationary or "mixed" rather than an equilibrium potential. This is because, when the concentration of the metal is very low, the limiting current for the reaction $M \rightarrow M^{n+} + ne$ becomes comparable with the total limiting current for the reductions of more easily reducible substances present in the solution $Ox + ne \rightarrow Red$. The value of such a stationary potential is governed by the concentration of metal in the amalgam and also by the identities and concentrations of the easily reducible impurities in the solution. One of these impurities is oxygen, of which traces are present in a solution even after prolonged deaeration by a specially purified inert gas (72). Distilled water may be another source of impurities. Although doubly distilled water is ordinarily used in electrochemical studies, special studies carried out in our laboratory and in many others have shown how difficult it is to eliminate all traces of heavy metals such as copper and zinc from water.

Recent studies employing radioactive indicators (74, 75) have shown that the potentials of amalgams obey the Nernst equation even when the concentration of metal is as small as approximately 10^{-7}–10^{-9} g-atom/liter, well below that usually found in measuring the potentials of dilute amalgams.

Of great interest is the relationship between the potential of an amalgam and the concentration of metal in it in the range between the lower limit of applicability of the Nernst equation and the limit of solubility of the metal in mercury. In several cases this relationship provides information about the form in which the metals are present in mercury.

In their earlier studies, Richards and co-workers (76–78), followed by other investigators (79–81), discovered that the potentials of amalgams of zinc, lead, tin, or bismuth vary much less rapidly with increasing concentration of dissolved metal than they should according to the Nernst equation. That is, the activity of the metal in the amalgam becomes lower than its concentration.

Hildebrand (82) assumed that the metals in such amalgams are associated into di- and triatomic molecules. Hildebrand used Richards' experimental data to calculate the percentage of associated atoms in amalgams of various concentrations. Close consideration of these data, however, reveals that the calculated percentage of associated atoms is large only when the solubility of the metal in mercury is exceeded. For example, according to Hildebrand, 26.2% of the zinc atoms are associated in an amalgam containing 3.16 wt % of zinc, which is 1.5 times the solubility of zinc in mercury. Thus deviations from the Nernst equation in such cases are not caused by the association of atoms but merely by the separation of a solid phase from the amalgam. The data obtained in our laboratory (41, 61, 63, 67) and by Jangg and Kirchmayr (42) allow us to assert that for amalgams of lead, zinc, copper, or bismuth the potential of the amalgam varies linearly with the logarithm of the concentration of metal in it up to the limit of solubility of the metal in mercury. The slope of the straight line is equal to the theoretical value predicted by the Nernst equation.

Hence in these amalgams the activity of the metal is equal to its concentration, and therefore there can be no association of atoms.

In amalgams of thallium, indium, sodium, or potassium, different investigators (55–57, 76–78, 83, 85, 90) have established that the activity of the metal is higher than its concentration. Hildebrand (82) supposed this to be connected with the formation of molecular metal–mercury compounds in the amalgams. Such a process would bind part of the mercury and increase the concentration of the potential-determining particles. With only earlier data on the potentials of amalgams at his disposal, Hildebrand concluded that indium amalgams contain the compound $InHg_4$, thallium amalgams the compound $TlHg_6$, and sodium amalgams the compounds $NaHg_2$, $NaHg_4$, and $NaHg_6$ (91).

However, more recent data covering wider ranges of concentration do not confirm this conclusion. Bent and Swift (90) were the first to conclude that the deviations of sodium amalgams from ideal behavior, as expressed by Raoult's law, among others, cannot be explained quantitatively by the formation of sodium–mercury compounds of only these formulas. We have made some calculations, based on more recent investigations, of the number of mercury atoms bound to one atom of indium, thallium, or sodium. These calculations show that the number of atoms gradually decreases as the concentration of metal in the amalgam increases (Fig. 2). This fact permits us to conclude that the liquid phases of such amalgams do not contain molecules of a definite composition but rather only solvates of variable composition.

This conclusion is also confirmed by x-ray investigations of thallium amalgams (92), by the absence of extrema on the curve showing the dependence of the conductivity of a liquid indium amalgam on its composition (93, 94), and by the results of surface tension measurements for amalgams of the alkali metals (95).

However, it is necessary to note the difference between amalgams of the alkali metals on the one hand, and amalgams of indium on the other. Liquid amalgams of the alkali metals are characterized by large heats of formation (about -5 kcal/g-atom). This allows us to attribute a significant fraction of ionic (heteropolar) character to the bonds in the solvates of alkali metals with mercury. The heats of formation of amalgams of indium and thallium

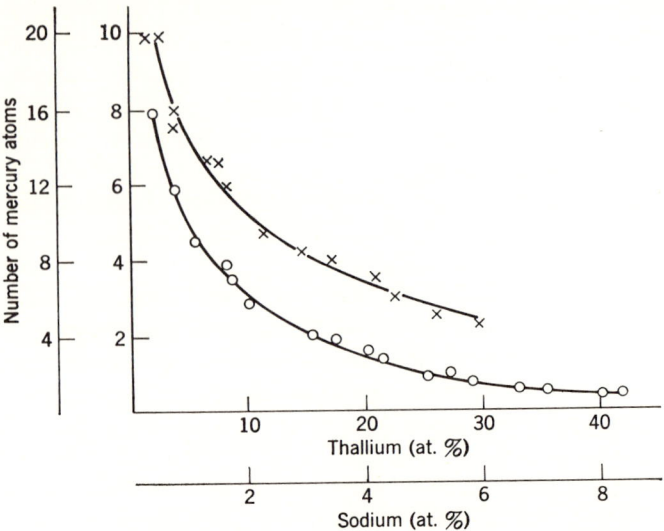

Fig. 2. Number of mercury atoms bound with one atom of thallium (○), and sodium (×) for various concentrations of these metals in amalgams.

are much smaller (less than 0.5 kcal/g-atom), so that the bonds between these metals and mercury in the solvates must be mainly metallic.

B. Current–Potential Curves (Polarographic Waves)

In addition to the potentiometric method, polarography is also of great value for studying the interaction of metals in amalgams. The most important polarographic characteristic—the half-wave potential—of a reversible half-reaction can be described by Eq. 6 (96):

$$E_{1/2} = E^0_{M(Hg)} + \frac{RT}{nF} \ln \left(\frac{D'}{D}\right)^{1/2} \qquad (6)$$

where D' is the diffusion coefficient of the metal atoms in the amalgam, D the diffusion coefficient of the metal ions in the solution, and $E^0_{M(Hg)}$ the standard potential of the amalgam described by the equation

$$E^0_{M(Hg)} = E^0_M - \frac{\Delta G^0}{nF} + \frac{RT}{nF} \ln {}^*a_{M(Hg)} {}^*a_{Hg}^y \qquad (7)$$

For an irreversible polarographic wave, the half-wave potential is given by Eq. 8 (97):

$$E_{1/2} = E^0_{M(Hg)} + \frac{RT}{\alpha nF} \ln 0.886\, k_s \left(\frac{t}{D}\right)^{1/2} \qquad (8)$$

where α is the transfer coefficient, k_s the rate constant of the electrode reaction at the standard potential, and t the drop time at the half-wave potential.

Equation 8 contains both the standard potential of the amalgam and the rate constant. The difference between the standard potential of the amalgam and that of the pure metal depends mainly on the free energy of formation of the amalgam. Consequently, the difference between the half-wave potential and the standard potential of the metal depends on the free energy of formation of the amalgam as well as on the rate constant of the electrode half-reaction.

Table V lists values of cathodic and anodic half-wave potentials, standard potentials, heat of formation of the amalgam, and the rate constant of the electrode half-reaction for several common metals.

It is seen from Table V that for the alkali metals, cadmium, thallium, and lead in the absence of surface-active substances the electrode reaction is reversible ($k_s > 10^{-2}$ cm/sec). The formation of the amalgams of these metals are accompanied by the evolution of different quantities of heat. For the alkali metals these quantities are significant, and therefore the half-wave potentials of both the cathodic and the anodic processes are displaced toward values much more positive than the standard potentials.

For cadmium and thallium the heat of formation of the amalgams is not high, and the difference between the half-wave potentials and the standard potential is therefore much less. For lead there is no difference between the half-wave potentials and the standard potential.

For zinc, tin, copper, and bismuth in absence of surface-active substances, the half-wave potentials are only slightly more negative than the standard potentials. This is because the heats of formation of amalgams of these metals are low, while the rate constants of the electrode reactions characterize them as quasireversible.

Quite an interesting picture can be observed for indium. The heat of formation of indium amalgams is low, and therefore this factor has only a very small effect upon the half-wave potential. In perchloric acid solutions, in which the rate constant for the half-reaction is very low, the half-wave potential of reduction is strongly displaced to the negative side, whereas the half-wave potential of oxidation is removed to the positive side. In hydrochloric acid solutions the rate constant is high, and the half-wave potential

TABLE V. Heats of Formation and Polarographic Data for Metal Amalgams

Metal	$-\Delta H$, kcal/g-atom[a]	k_s, cm/sec	E_M^0, V	Supporting electrolyte	$E_{1/2}$ cathodic, V	$E_{1/2}$ anodic, V
Lithium	4.0	0.09	−3.05	$1\,M$ $(CH_3)_4NCl$ + $0.1\,M$ $(CH_3)_4NOH$	−2.10	—
Sodium	3.1	0.4	−2.71	$1\,M$ $(CH_3)_4NCl$ + $0.1\,M$ $(CH_3)_4NOH$	−1.85	−1.85
Potassium	3.3	0.1	−2.93	$1\,M$ $(CH_3)_4NCl$ + $0.1\,M$ $(CH_3)_4NOH$	−1.88	−1.88
Rubidium	—	0.2	−2.92	$1\,M$ $(CH_3)_4NCl$ + $0.1\,M$ $(CH_3)_4NOH$	−1.88	—
Cesium	—	0.09	−2.92	$1\,M$ $(CH_3)_4NCl$ + $0.1\,M$ $(CH_3)_4NOH$	−1.88	—
Zinc	—	$(3-6) \times 10^{-3}$	−0.76	$1\,M$ KCl	−0.78	−0.75
	—	10^{-7}	—	$1\,M$ KCl + camphor	−0.90	−0.20
Cadmium	0.5	0.6	−0.40	$1\,M$ KNO$_3$	−0.36	−0.35
	—	2.9	—	$1\,M$ KCl	−0.35	−0.35
	—	6.7×10^{-8}	—	$1\,M$ KCl + camphor	−1.04	−0.18
Thallium	0.08	0.3–1.0	—	$1\,M$ KNO$_3$	−0.22	−0.21
Indium	0.2	10^{-5}	−0.34	$1\,M$ HClO$_4$	−0.75	−0.30
	—	1	—	$1\,M$ HCl	−0.36	−0.34
Tin	−0.05	—	−0.14	$1\,M$ NaClO$_4$ + $1\,M$ HClO$_4$	−0.18	—
Lead	—	1	−0.13	$1\,M$ NaClO$_4$ + HClO$_4$	−0.13	−0.12
Bismuth	0.05	3×10^{-3}	+0.16	$1\,M$ HCl	+0.14	+0.12
Copper	—	5×10^{-4}	+0.34	$1\,M$ HClO$_4$	+0.23	—
Manganese	1.1 (Mn$_2$Hg$_5$)	—	−1.18	$1\,M$ KCl	−1.26	−1.12
Iron	—	10^{-6}	−0.44	$1\,M$ HClO$_4$	−1.02	—
Cobalt	—	10^{-6}	−0.28	$1\,M$ K$_2$SO$_4$	−1.18	—
Nickel	2.4 (NiHg)	10^{-6}	−0.25	$1\,M$ HClO$_4$	−0.85	—

[a] The values of ΔH pertain to amalgams in which the mole fraction of mercury is 0.9. All potentials are referred to the NHE.

of the cathodic wave is only slightly more negative than the standard potential of indium.

The cathodic half-wave potential is also displaced toward more negative values, while the anodic one is displaced toward more positive values, by the addition of surface-active substances which slow down the electrode reaction. Table V shows the effects of camphor on the half-wave potentials for cadmium and zinc.

For metals of the iron group, the rate constants of the electrode half-reactions are very small (98, 101), and the half-wave potentials of reduction are strongly displaced to the negative side. This displacement is nearly the same for iron as for nickel, despite the fact that nickel and mercury form a compound ($NiHg_4$), slightly soluble in mercury with significant evolution of heat, while iron does not form any compounds with mercury. This fact proves that during the electrolytic reductions of metals on mercury a solution of the metal in mercury is formed first, the discharge of an intermetallic compound as a solid phase being a secondary process (102).

Stromberg (103, 104), Furman and Cooper (105, 106), and von Stackelberg (107) have established that the Ilkovič equation can also be applied to the anodic limiting currents for metal oxidation from amalgams, that is, that the diffusion current for the oxidation of the metal is directly proportional to the concentration of metal in the amalgam. However, working with amalgams of metals slightly soluble in mercury (e.g., copper, manganese, and antimony), we have found that the limiting current for oxidation of the metal increases in proportion to the concentration of metal in the amalgam up to a definite value (64, 65, 67). At higher concentrations of metal in the amalgam, this proportionality ceases, and a break is seen on the calibrating straight line. As shown by Figures 3 and 4, proportionality between the anodic limiting current and the concentration of metal in the amalgam is observed for copper at concentrations up to 6.8×10^{-3} g-atom/liter (9.0×10^{-3} at. %), for manganese up to 6.5×10^{-3} g-atom/liter (9.0×10^{-3} at. %), and for antimony up to 2.6×10^{-4} g-atom/liter (3.8×10^{-4} at. %). Each of these values is very close to the solubility of the metal in mercury as found by potentiometric and other methods (Table III). The presence of the break on the calibrating straight line is explained by the fact that the anodic limiting current depends on the diffusion of only that part of the metal present in the liquid phase of the amalgam in the form of a true solution. The part of the metal present in a solid phase cannot immediately take part in the electrode reaction, since as a result of the surface tension of mercury there are no solid particles on the surface of the amalgam (108). The solid phase present within the amalgam affects the limiting current for the oxidation only because the concentration of the metal in the liquid

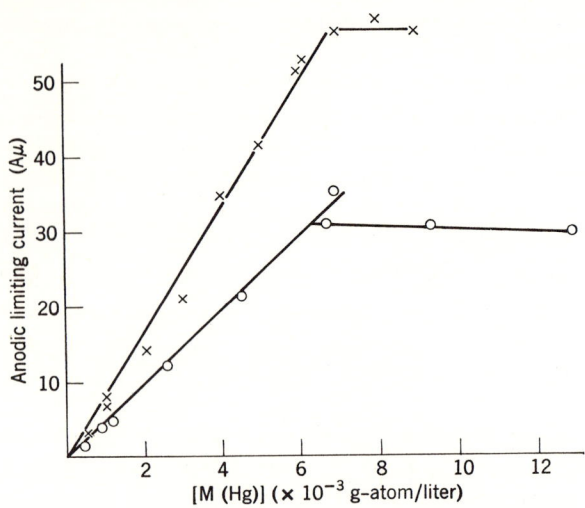

FIG. 3. Dependence of anodic limiting current on the concentration of amalgams, manganese (×), copper (○).

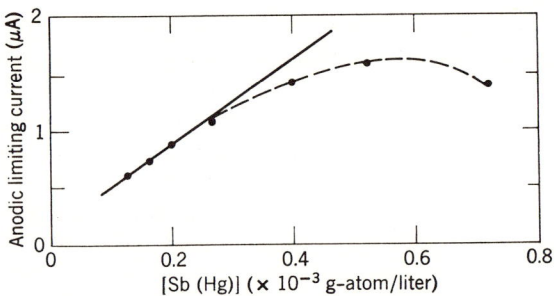

FIG. 4. Dependence of anodic limiting current on the concentration of antimony amalgam.

phase decreases in the polarographic process. This shifts the equilibrium between the solid and liquid phases in favor of the dissolution of the solid phase.

However, this influence cannot be great, for the rate of solution of the solid phase is much lower than the rate of diffusion.

It should be noted that the break in proportionality between the anodic limiting current and the concentration of metal in the amalgam is connected with the limit of solubility of the metal in mercury no matter whether it is the pure metal (antimony) or a compound with mercury (Cu_3Hg_4, Mn_2Hg_5) that is discharged in the solid phase.

On a dropping amalgam electrode containing cobalt, iron, or nickel, it is impossible to obtain oxidation currents of these metals before the beginning of the oxidation of mercury (109–111). This is because the concentrations of iron, cobalt, and nickel in the liquid phases of these amalgams are very low (about 10^{-5} g-atom/liter), which is even lower than the sensitivity of the polarographic method.

IV. Intermetallic Compound Formation in Complex Amalgams

A. Effect on the Potential of a Complex Amalgam

The introduction of two metals into an amalgam may bring about the formation of compounds between these metals. Tammann (112) and, later, Russel and co-workers (113, 114) were the first to report the formation of intermetallic compounds in multicomponent amalgams.

In principle, metals in amalgams may form both binary compounds of the types A_mB_n, A_mHg_n, and B_mHg_n, and ternary compounds of the type $A_mB_nHg_p$.

Numerous investigations of complex amalgams, carried out by Lihl et al. (115) by the x-ray method, showed that from such amalgams there precipitate, as a rule, intermetallic compounds that are typical of binary alloys (A_mB_n) and do not contain mercury. Of 32 cases studied by Lihl, only in two (Cr/Zn/Hg and Mn/Zn/Hg) was mercury also present in the solid phase discharged from the amalgams.

Thus mercury behaves as a common solvent as regards the metals and intermetallic compounds in complex amalgams. Similarly to any solvent, it can interact with certain components or be a component part of the solid phase. Very often mercury helps in the preparation, at room temperature, of intermetallic alloys free of mercury and characteristic of the corresponding binary alloys obtained at high temperatures.

The formation of intermetallic compounds in complex amalgams has a great influence upon the electric properties of the amalgams.

The mutual influence of metals on the potential of the amalgam can be studied by measuring the potentials of concentration cells of the type

A(Hg)	A^{n+}	A, B(Hg)
Electrode 1		Electrode 2

where A is the most easily oxidized component of the system. In the course of the experiment, electrode 1 remains unchanged and acts as a reference electrode. In electrode 2 the concentration of metal B is also kept constant, while that of metal A is varied. The measured values of E are plotted against log $[A]_1/[A]_2$. As noted above, for simple amalgams of metals that do not interact with mercury, the dependence of E on log $[A]_1/[A]_2$ is linear. The linear dependence is also preserved in the absence of interaction in the complex amalgam.

If metals A and B form a compound in electrode 2, the potential of the concentration cell changes owing to the decrease in the concentration of free A. The potential of the compound itself, as shown by numerous studies carried out by Pushin (116), is usually intermediate between the potentials of metals A and B.

Analyzing plots of E against log $([A]_1/[A]_2)$ for different cases of metal interaction in amalgams allows us to make the following general conclusions:

(a) If an intermetallic compound is present in the liquid phase of the amalgam, the equilibrium in the system $A_mB_n(Hg) \rightleftharpoons mA(Hg) + nB(Hg)$ is defined by the constant $K = a_A^m a_B^n / a_{A_mB_n}$, where a_A and a_B are the activities (or concentrations) of the free metals A and B in the complex amalgam and $a_{A_mB_n}$ is the activity of the compound.

The dependence of E on log $([A]_1/[A]_2)$ in this case is described by curve 2 (Fig. 5).

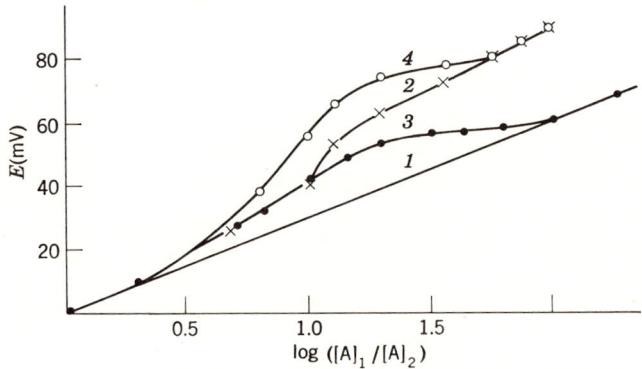

FIG. 5. Dependence of E on log $([A]_1/[A]_2)$ calculated theoretically. (*1*) Metals do not interact in amalgam. (*2*) An intermetallic compound is present and partially dissociated in the liquid phase of the amalgam. $K = 10^{-3}$. (*3*) An intermetallic compound precipitates as a solid phase. In the liquid phase of the amalgam, the compound is fully dissociated. $S_p = 10^{-5}$. (*4*) An intermetallic compound precipitates as a solid phase. In the liquid phase of the amalgam, the compound is partially dissociated. $S_p = 2 \times 10^{-6}$; $K = 10^{-3}$.

(b) If the intermetallic compound separates as a solid phase, the equilibrium in the system is defined by the solubility product of the compound:

$$S_p = a_A^m a_B^n \qquad (9)$$

The resulting dependence of E on log $([A]_1/[A]_2)$ may be described by either curve 3 or curve 4 of Figure 5. Curve 3 is obtained when the compound A_mB_n is fully dissociated in the liquid phase of the amalgam; curve 4 is obtained when the intermetallic compound is only partially dissociated in the liquid phase of the amalgam.

The experimental data that are available on the potentials of such cells show that in amalgams containing antimony and tin, or silver and tin (118), silver and indium (119), and copper and indium, or nickel and indium (120) interaction between metals is not observed at concentrations up to 10^{-2}–10^{-1} g-atom per liter of mercury, that is, at concentrations notably exceeding the solubility of the less soluble metal in the system.

With amalgams containing gold and zinc, or gold and cadmium (121), gold and tin (118), gold and indium (122), copper and zinc (117), copper and tin (123), antimony and zinc (124), antimony and cadmium (124), antimony and indium (125), and nickel and tin (126), the deviations of E from the straight lines computed for simple amalgams are attributable to the separation of intermetallic compounds into the solid phase (Figs. 6 and 7).

From Eq. 9 it follows that a plot of log a_A against log a_B must be linear for amalgams saturated with the compound A_mB_n, and this is confirmed experimentally (Fig. 8). The value of the ratio m/n is easily deduced from the slope of the straight line. Thus, for example, Figure 8 shows that the compound AuCd separates as a solid phase from amalgams containing gold and cadmium, while indium and antimony, and gold and tin, react to give InSb and Sn_2Au, respectively.

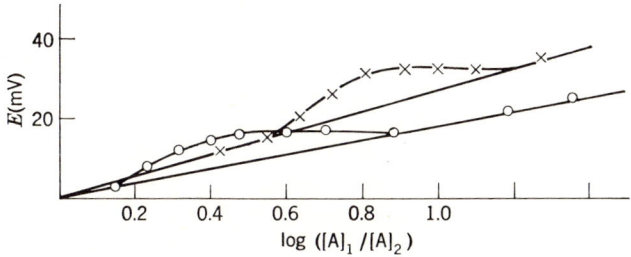

FIG. 6. Dependence of E on log $([A]_1/[A]_2)$ for zinc–antimony ($\times$) and indium–antimony ($\circ$) amalgams.

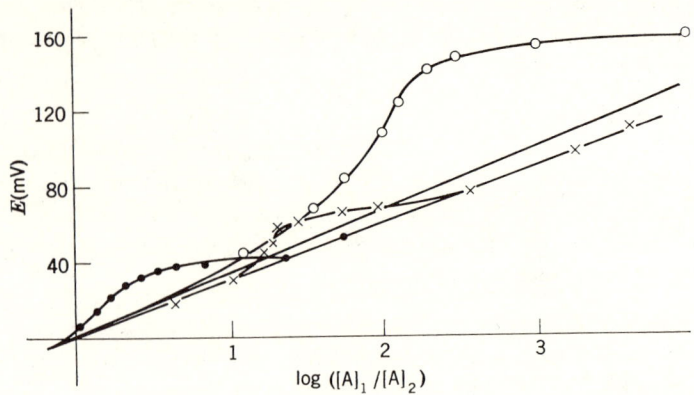

Fig. 7. Dependence of E on log $([A]_1/[A]_2)$ for gold-containing amalgams. (○) Gold–zinc; (×) gold–cadmium; (●) gold–tin.

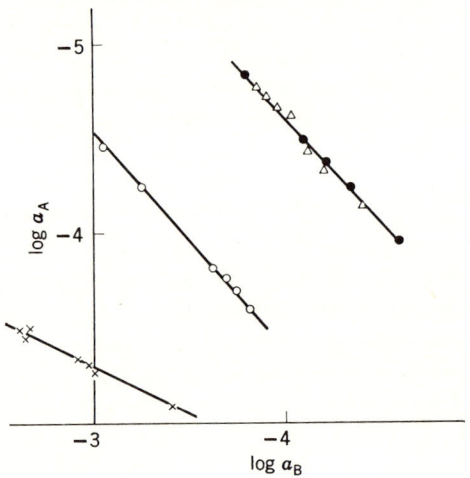

Fig. 8. The dependence of log a_A on log a_B. (●) Gold–cadmium; (×) gold–tin; (○) indium–antimony; (△) zinc–antimony.

Values of the solubility products of several intermetallic compounds in mercury obtained in our laboratory are shown in Table VI.

TABLE VI. Solubility Products of Several Intermetallic Compounds at 20–25°C

Compound	Type of compound	Solubility product
AuZn	β-Phase of the Hume-Rothery type	2.5×10^{-12} (at 90°C)
AuCd	β-Phase of the Hume-Rothery type	2.5×10^{-10}
AuSn$_2$	β-Phase of the Hume-Rothery type	2.8×10^{-10}
AuIn	β-Phase of the Hume-Rothery type	1.8×10^{-6}
Au$_3$In	β-Phase of the Hume-Rothery type	10^{-12}
AgZn	β-Phase of the Hume-Rothery type	3×10^{-6a}
AgCd	β-Phase of the Hume-Rothery type	7×10^{-6a}
CuZn	β-Phase of the Hume-Rothery type	4×10^{-6}
Cu$_3$Sn	ε-Phase of the Hume-Rothery type	3×10^{-12}
CuSn	Nickel arsenide	4×10^{-6}
Ni$_3$Sn$_4$(NiSn)	Nickel arsenide	2.5×10^{-12} (computed for NiSn)
SbZn	Nickel arsenide	2×10^{-9}
SbCd	Nickel arsenide	10^{-8}
SbIn	Nickel arsenide	2×10^{-8}

[a] Calculated from polarographic data.

Figures 6 and 7 show that several amalgams have been studied potentiometrically in concentration ranges in which the value of the solubility product has not been reached. In all these cases the measured potentials lie on the straight line that corresponds to simple amalgams of the most easily oxidizable metal. Since the precision of measuring the potential is usually within ±0.5 mV, the concentration of the undissociated compound (if it is present in the liquid phase of the amalgam) cannot exceed a few percent of the concentration of the most easily oxidizable component.

Table VI shows that the solubilities in mercury of all of the intermetallic compounds are of the order of 10^{-5} g-atom/liter and lower. It follows that the concentration of the undissociated intermetallic compound in the liquid phase of the amalgam is always lower than 10^{-5} mole per liter of mercury. So we can assert that all the intermetallic compounds listed in Table VI are virtually completely dissociated into free metals when dissolved in mercury.

B. Effect on Cathodic Polarographic Curves

Interactions of metals can manifest themselves not only on the potentials of amalgams but also on current–potential curves. For example (127), the reduction waves of bismuth, antimony, and zinc appear at more positive potentials on a dropping copper–amalgam electrode than on a dropping mercury electrode (DME). For bismuth and antimony this shift is insignificant, but for zinc it reaches 0.14 V. Baimakov (128, 129) observed an analogous phenomenon in the simultaneous reduction of cobalt and copper on a DME. In the reduction of zinc at a large copper–amalgam cathode, the displacement of the reduction potential of zinc strongly depends on the current density; the higher the current density, the less the displacement (130).

This phenomenon can be explained in the following way. At the start of the electrolysis, some quantity of metal is accumulated in the layer at the surface of the mercury and definite rates of the cathodic and anodic processes, which depend on the potential of the electrodes, are established (Fig. 9). To reduce metal A on the mercury cathode at a rate corresponding to the current i_1, E_1 is required, while to reduce it at a rate corresponding to i_2, the potential E_2 is required. If the reduction of metal A takes place not on pure mercury but on an amalgam of metal B (or if both metals A and B are reduced on mercury), then, because some of the atoms of A are bound as a compound

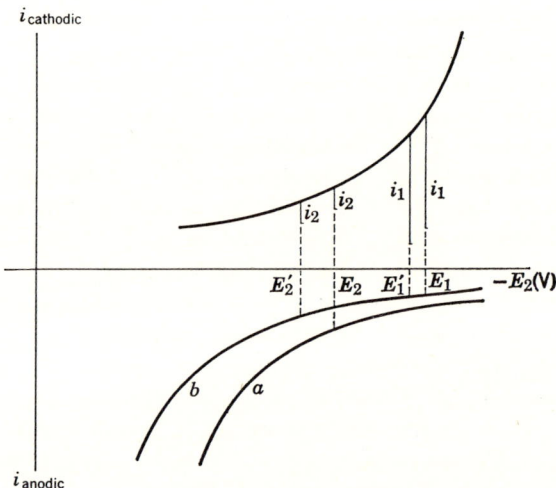

FIG. 9. Schematic plot of the dependence of i on E for metal reduction processes on pure mercury (a) and on an amalgam of another metal (b).

with B, the curve describing the true rate of the anodic process will shift from position a to position b. Then to reduce metal A at rates corresponding to i_1 and i_2, the potentials E'_1 and E'_2, respectively, are required. As seen from Figure 9, E'_1 and E'_2 are more positive than the original values E_1 and E_2. This shift increases as the reduction rate decreases.

The interaction of metals does not affect the values of the limiting reduction currents on an amalgam since the limiting current is governed by the rate of transport of the metal ions to the surface of the electrode.

C. Effect on Anodic Polarographic Curves

The behavior of metals during the oxidation of their amalgams has been studied in detail in our laboratory (118–120, 123, 125, 126, 131–134). In cases in which the metals do not interact in mercury-free binary alloys, their oxidation waves are observed separately on anodic polarograms of the complex amalgam. The wave heights and the half-wave potentials remain practically the same as those for the corresponding simple amalgams of the same concentration. Thus, for instance, the presence of zinc, lead, or cadmium in an amalgam of indium in no way affects the anodic waves of indium, and the indium in turn does not affect the anodic polarographic wave of the other metal present (132, 133).

Nor is there any mutual influence of metals upon their oxidation currents from complex amalgams in cases in which metals in mercury-free binary alloys do form compounds among themselves. In such cases, as a rule, potentiometric investigations show that compounds are not formed in the amalgams. For example, the presence of silver in an amalgam does not exert any influence upon the oxidation wave of indium or tin (118, 119). Similarly, for amalgams containing copper together with either indium or cadmium, separate oxidation waves are observed for the copper and the other metal, and the heights of these waves are the same as for the corresponding simple amalgams (120–134). Potentiometric studies of these amalgams also indicate the absence of interaction between the metals in them.

In amalgams for which the results of potentiometric measurements indicate the separation of an intermetallic compound as a solid phase, the interaction of metals can also be observed by polarographic studies.

We have found mutual influence of metals in polarographic studies of the following amalgams: gold and indium (122), gold and tin (118), silver and zinc (108), silver and cadmium (108), copper and tin (123), copper and zinc (130), indium and antimony (125), and nickel and tin (126).

The interaction of metals in amalgams causes the anodic waves of these metals to decrease, and in some instances to disappear completely (Fig. 10).

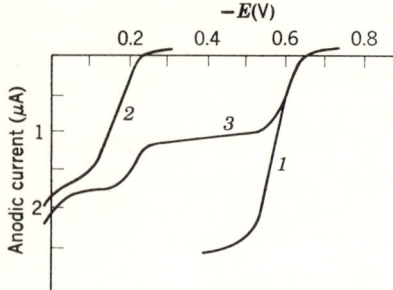

Fig. 10. Anodic polarograms, with 0.5 M hydrochloric acid as the aqueous phase, of amalgams of (1) indium, (2) antimony, and (3) indium–antimony amalgams.

The formula of the compound can be determined from a plot of the height of the anodic wave of one metal (usually the more easily oxidizable one) against the concentration of the other in the amalgam. Figure 11 shows this dependence for gold–tin amalgams. The curve in this figure represents in fact an amperometric titration curve for the titration of tin with gold, the measured current being the oxidation current of tin. The point of intersection of the straight lines is the equivalence point. As seen from Figure 11, the formula of the gold–tin compound proves to be $AuSn_2$, which is the same formula found by potentiometric measurements.

In all the cases mentioned above, the formulas of compounds determined by amalgam polarography are the same as those found potentiometrically.

It is necessary to stress that the interaction of metals in amalgams leads only to decreases in their individual wave heights. On the anodic polarograms of amalgams, no new waves which might be attributed to the oxidation of intermetallic compounds were ever observed. As mentioned before, the values of potentials of intermetallic compounds are as a rule intermediate

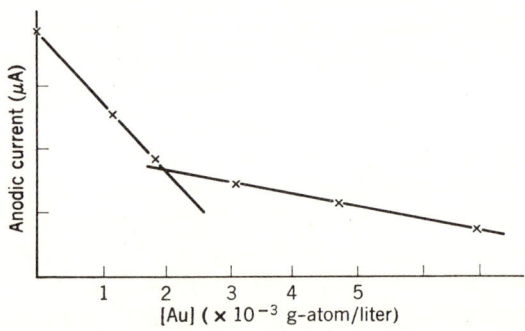

Fig. 11. The dependence of the anodic limiting current of tin on the concentration of gold in an amalgam. $[Sn] = 4.07 \times 10^{-3}$ g-atom/liter.

between the potentials of the corresponding metals. If the heights of anodic waves of the metals in an amalgam decrease because of the formation of soluble but slightly dissociated compounds, the oxidation waves of the compounds should be observed on the anodic polarogram of the complex amalgams. The absence of such waves proves that the intermetallic compounds are only sparingly soluble in mercury.

When the concentration of the metals in an amalgam are small (so that the solubility product is not attained), the separate waves of all the metals in the amalgam appear on the anodic polarograms. The heights of these waves are the same as they would be in the corresponding simple amalgams. This is the case, for example, with amalgams containing copper and tin, zinc and silver, or indium and silver.

Thus the method of amalgam polarography with a dropping electrode once more confirms the thesis that intermetallic compounds that form in an amalgam separate into a solid phase. In the liquid phase of the amalgam, the compounds are actually completely dissociated, at least within the limits of sensibility of potentiometric and polarographic methods.

V. Intermetallic Compounds in Anodic Stripping Voltammetry

Different variations of the method of anodic stripping voltammetry, or APPA, on a stationary or hanging mercury drop have recently been widely applied to the determinations of extremely small (10^{-5}–10^{-9} M) concentrations of metal ions (135–138). The method has been used not only for analytical purposes but also to determine the solubilities of some metals in mercury (139, 140) and to study the interaction of metals in amalgams (141–143).

The APPA method has certain experimental advantages over amalgam polarography with a dropping electrode. Since the anodic curves are obtained immediately after the accumulation of metal in the drop and no transfer of the amalgam to another vessel is required, the APPA method excludes the possibility of errors connected with uncontrolled oxidation of metal in the amalgam. Moreover, since the charging current can be made much smaller on a stationary electrode than on an electrode with a continuously varying surface area, it is possible by the APPA method to obtain distinct oxidation peaks for metals even at concentrations one or two orders of magnitude lower than the limit of detection with a dropping electrode.

However, there are two factors that complicate the use of the APPA method for determining the solubilities of metals in mercury and for studying intermetallic compounds:

(a) In studying amalgams by the APPA method, there is a substantial danger of obtaining supersaturated amalgams. As a rule, one proceeds with oxidation of an amalgam immediately after it is obtained, while the drop of the amalgam is at rest. Especially when the supersaturation is small, the solid phase may separate only very slowly.

Hickling and Maxwell (144) have shown that supersaturated amalgams are readily formed in electrolytic reductions of metal ions at stationary macroelectrodes. The formation of supersaturated amalgams may be responsible for the fact that the solubilities of some metals [such as antimony (145) and nickel (146)] in mercury are found to be higher by the APPA method than by other methods. The slow establishment of equilibrium in complex amalgams is mentioned by Ficker and Meites (120) and was also observed in our laboratory. Aging an amalgam by allowing it to stand in contact with a solution for some time to equilibrate before obtaining its anodic current–potential curve entails some latent danger of complications of another kind. Thus, for example, when aged manganese amalgams obtained by electrolysis on stationary mercury drops are polarized anodically, the original peak at -1.2 V versus SCE disappears completely. It is at first replaced by peaks at -0.35 and -0.9 V but on longer standing only the one at -0.35 V remains (147). This fact was explained (147) as follows. During the electrolysis a solution of manganese in mercury is obtained at first; then, as the amalgam ages, an intermetallic compound of manganese with mercury is formed, and this compound is oxidized at a different potential from manganese. This explanation, however, is untenable. As shown by Jangg (52), mercury and manganese form only one compound, Mn_2Hg_5, at room temperature, and the potential of this compound is only 0.041 V more positive than that of manganese itself. It can easily be seen therefore that the peaks mentioned above cannot correspond to this intermetallic compound. However, the appearance of new anodic peaks on the voltammogram can be easily explained by the presence of contaminants in the solution with which the amalgam is in contact. Kemula and Galus (147) used reagent grade chemicals, but these compounds may contain 10^{-3}–$10^{-4}\%$ lead and zinc. In typical 0.1 M solutions of such chemicals, the concentrations of lead and zinc ions are about 10^{-6}–10^{-7} M. These concentrations are so small that no appreciable amounts of these metals could accumulate in the amalgam drop during the 30-sec deposition periods employed, and peaks attributable to them could therefore not appear on the current–potential curve of a fresh manganese amalgam. However, when such an amalgam is allowed to stand in contact with the same solution for 15 min or more, the manganese reduces the more easily reducible metal ions from the solution, and the amalgam becomes more and more contaminated with lead and zinc. Therefore, the height of the manganese peak decreases on aging and the peak may eventually disappear altogether, and at the same time there will appear the peaks of the more difficultly oxidizable metals—zinc at -0.9 V and lead at -0.35 V.

This theory was verified experimentally; no new peaks appeared when manganese amalgams were allowed to stand in contact with sufficiently pure solutions, while on aging in solutions to which small concentrations (about 10^{-5}–10^{-6} M) of zinc and lead ions had been added, anodic peaks were observed to appear and disappear in the manner just described (148). Analogous behavior is obtained on aging a zinc amalgam; new peaks appear although zinc is known not to form compounds with mercury at room temperature.

When the interaction of metals in complex amalgams is studied by the APPA method, the solution from which the metals are accumulated contains at least two metals. As the complex amalgam stands in contact with this solution, the more easily oxidizable metal in the amalgam "cements" out the more easily reducible metal ion from the solution, and as a result the amalgam becomes enriched in the metal giving a peak at the more positive potential. We have observed this phenomenon with stationary copper–cadmium amalgam macroelectrodes in contact with solutions containing both copper and cadmium salts (149).

(b) The other factor complicating the use of the APPA method in studying the behavior of metals in amalgams is the possibility that metals or intermetallic compounds may deposit onto the surface of the mercury drop.

During electrolysis the primary process is the reduction of metal ions on the surface of the electrode. The diffusion of the atoms of metal from the surface into the interior of the mercury drop is a secondary process. Preelectrolysis from a sufficiently diluted solution yields an amalgam that is far from saturated, and in this case (150, 151) the rate of diffusion of metal atoms into the interior of the mercury drop is sufficiently high, by comparison with the rate of their generation on the surface of the amalgam, that their concentration quickly becomes uniform throughout the drop. If the rate at which atoms are generated on the surface is comparable with the rate at which they diffuse into the interior of the drop, then the solubility of the metal or of an intermetallic compound in mercury can be reached on the surface of the drop before the concentration becomes uniform throughout the bulk of the drop. In this event the metal or alloy will begin to deposit as a solid phase on the surface of the electrode. This has even been observed with thallium (152), whose solubility in mercury is very high.

Conditions are especially favorable for the deposition of a solid metal on the surface of a mercury drop when the amalgam is nearly saturated, as it is likely to be in determining the solubility of a metal in mercury or in dealing with a metal sparingly soluble in mercury. A solid metal or an intermetallic compound deposited on the surface of an electrode is easily oxidized before the oxidation of mercury begins. For metals of the iron group, anodic currents have been observed as the result of their oxidation on the electrode surface (135, 153, 154).

In studying nickel amalgams with the help of a specially devised method, we separated the current corresponding to the oxidation of nickel on the surface of a mercury drop from that corresponding to the oxidation of nickel from the bulk of the amalgam (140, 146).

Studies of interactions between metals in amalgams by the APPA method have been described in several papers (135, 141–143, 155–158). Many of these state that the interaction manifests itself in the form of a decrease in or disappearance of the oxidation currents of the corresponding metals (135, 142, 155, 156, 158), which is analogous to the observations, as described above, from the method of amalgam polarography with a dropping electrode. Some authors describe the appearance of new peaks which are attributed to the oxidation of intermetallic compounds from the bulk of the drop (159–161). We believe, however, that these peaks correspond to the oxidation of intermetallic compounds accumulated on the surface of the electrode. In agreement with this assumption is the fact that when such peaks are observed the oxidation potential of mercury itself is usually shifted toward positive potentials (Fig. 12). Moreover, studies of numerous complex amalgams [among them those containing nickel together with either tin (126) or zinc (162)] by polarography with dropping amalgam electrodes (that is, under conditions that exclude the presence of intermetallic compounds on the surface of the electrode) have revealed the existence of no new waves; only those resulting from the oxidation of the individual free metals are observed.

All the phenomena described above—namely, the possibility of obtaining supersaturated amalgams, the possibility of cementation of more easily deposited metals from the solution during the aging of the amalgam, and the possibility of depositing metals and intermetallic compounds on the surface of the mercury drop—force us to admit that the APPA method should be applied only with great care in studying metal interaction in amalgams and especially in studying the main characteristics of intermetallic compounds.

VI. Intermetallic Compounds in Phase Exchange (Cementation)

The formation of solid intermetallic compounds in amalgams must be taken into account not only in separations of metals by anodic oxidations

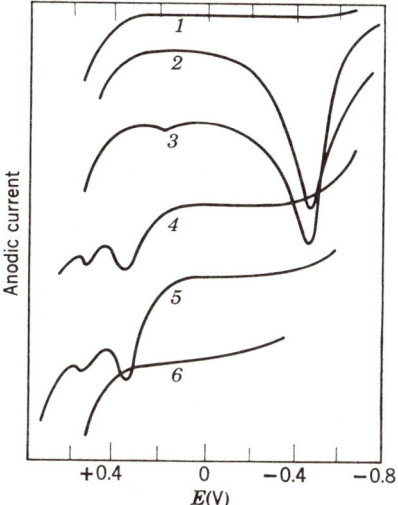

Fig. 12. Anodic polarization curves of cadmium–silver amalgams. The amalgams were produced by preelectrolysis from solutions of various concentrations (159). (1) Mercury alone; (2) a cadmium amalgam prepared from $10^{-3}\ N\ Cd(NO_3)_2$; (3) to (5) cadmium–silver amalgams prepared from: (3) $10^{-3}\ N\ Cd(NO_3)_2$, $3 \times 10^{-3}\ N\ AgNO_3$; (4) $10^{-3}\ N\ Cd(NO_3)_2$, $6 \times 10^{-4}\ N\ AgNO_3$; (5) $10^{-3}\ N$ $Cd(NO_3)_2$, $9 \times 10^{-4}\ N\ AgNO_3$; and (6) a silver amalgam prepared from $9 \times 10^{-4}\ N\ AgNO_3$.

of complex amalgams but also in separations by the amalgam cementation method in which easily reducible metal ions are displaced from the solution by reacting with amalgams of more easily oxidizable metals.

Cementation is a combination of two mutually connected electrochemical processes: the anodic dissolution of the "cementing" metal and the cathodic reduction of the ion of the metal being cemented (163). The rate of the cementation process is governed by the current that corresponds to the point of intersection of two current–potential curves: the cathodic curve for the solution containing the ions of the metal being cemented, and the anodic curve of the cementing metal (164). During the cementation process, as the concentrations of the metal ions in the solution and of the metal atoms in the amalgam change, there are corresponding changes in the slopes of the current–potential curves, and the position of the point of intersection changes as a result. The ordinate of the point of intersection becomes smaller, that is, the rate of the cementation process decreases.

All the factors that influence the shapes of current–potential curves affect the cementation process as well. Especially important among these factors are the solubilities of the metals and intermetallic compounds in mercury. If the solubility of the cementing metal is very low, the rate of the cementation process will be very low, since the limiting anodic current will be very small. Thus, for example, the cementation of copper by an iron amalgam is extremely slow because of the very low solubility of iron in mercury.

The effect of the formation of an intermetallic compound in the amalgam on the course of a cementation process can be illustrated by the cementation of copper with a zinc amalgam. At first the cementation of copper by zinc is very fast, but the copper passing into the amalgam binds the remaining zinc into an intermetallic compound sparingly soluble in mercury. Zinc bound with copper does not take part in the cementation process, and with the disappearance of zinc from the liquid phase of the amalgam the cementation process stops or, to be more exact, it continues at an extremely low rate. However, the cementation of copper by a cadmium amalgam proceeds very rapidly until the cadmium is completely removed from the amalgam, since copper–cadmium compounds are quite soluble in mercury and do not separate from it during the cementation process (149).

Of great interest is the interaction of a zinc amalgam with a solution containing both copper and cadmium. At first both metals—copper and cadmium—are cemented by zinc. As the concentration of copper in the amalgam increases, however more and more of the zinc remaining in it is bound by copper and prevented from taking part in the cementation process. Then the copper remaining in the solution begins to be cemented by the cadmium amalgam formed at the beginning of the process (162).

VII. Conclusion

Investigations of the electrochemical properties of liquid amalgams allow us to state that the difficulties that arise in separating metals by oxidizing complex amalgams or in polarographic determinations of metals are connected with the formation of a solid phase, either of the metal or of an intermetallic compound, both in the bulk and on the surface of the amalgam. From this point of view, controversies among various investigators concerning the mutual influence of metals in complex amalgams become quite explicable. Until the concentrations of the metals in the amalgam become high enough to satisfy the solubility product of the corresponding intermetallic compound in mercury, these metals have no visible influence on the electrochemical properties of the amalgam. At concentrations that surpass the solubility product of the intermetallic compound, however, the solid phase begins to separate and difficulties in the separation and determination of metals begin to arise.

While the formation of intermetallic compounds in amalgams usually causes difficulties in separating metals, it can also be used in masking certain metals that disturb polarographic determinations. Thus, Vinogradova, Vasilyeva, and Iobst (165) suspended a mercury drop on a gold electrode in order to mask cadmium when determining lead by the APPA method. As already pointed out, gold binds cadmium in an intermetallic compound but does not combine with lead. This makes it possible to determine lead in the presence of even 1000 times as much cadmium.

References

1. Hansen, M., and K. Anderko, *Constitution of Binary Alloys*, McGraw-Hill, New York, 1958.
2. Sauerwald, F., *Z. Metall.*, **41**, 97, 214 (1950).
3. Sauerwald, F., P. Brand, and W. Menz, *Z. Metall.*, **57**, 103 (1966).
4. Karlikov, D. M., *Ukr. Fiz. Zh.*, **3**, 370 (1958).
5. Evseev, A. M., and G. F. Voronin, *Termodinamika i Struktura Zhidkikh Metallicheskikh Splavov* (Thermodynamics and Structure of Liquid Metallic Alloys), M.G.U., Moscow, 1966.
6. Grube, G., and W. Wolf, *Z. Elektrochem.*, **41**, 675 (1935).
7. Zhukovskii, G. Ya., *Z. Anorg. Allgem. Chem.*, **71**, 403 (1911).
8. Zintl, E., and A. Schneider, *Z. Elektrochem.*, **41**, 771 (1935).
9. Zintl, E., and G. Brauer, *Z. Physik. Chem.*, **B20**, 243 (1933).
10. *Selected Values of Chemical Thermodynamic Properties*, Part 1, Circular, National Bureau of Standards, Washington D.C., 1952, p. 500.
11. Kubaschewski, O., and E. Evans, *Metallurgische Thermochemie*, VEB Verlag Technik, Berlin, 1959.
12. Jänecke, E., *Z. Metall.*, **20**, 113 (1928).
13. Vierk, A. L., and K. Hauffe, *Z. Elektrochem.*, **46**, 348 (1940).
14. Kurnakov, N. S., *Z. Anorg. Allgem. Chem.*, **23**, 441 (1900).
15. Weibe, F., and O. Kubaschewski, *Thermochemie der Legierungen*, Springer, Berlin, 1943.
16. Sauerwald, F., *Z. Metall.*, **35**, 105 (1943).
17. Friedman, H. L., and M. Kahlweit, *J. Am. Chem. Soc.*, **78**, 4243 (1956).
18. Rolls, M. C., K. B. Holden, and C. J. Whitman, *J. Am. Chem. Soc.*, **79**, 3925 (1957).
19. Nowotny, H., *Z. Metall.*, **37**, 130 (1946).
20. Britske, E. V., A. F. Kapustinskii, B. K. Veselovskii, et al., *Termicheskie Konstanty Neorganicheskikh Veshchestv* (Thermal Constants of Inorganic Substances), Akad. Nauk SSSR, Moscow, 1949.

21. Kraus, C. A., and H. F. Kurtz, *J. Am. Chem. Soc.*, **47**, 43 (1925).
22. Ferro, R., *Acta Cryst.*, **7**, 781 (1954).
23. Daniel'chenko, P. T., *Zh. Obshch. Khim.*, **1**, 467 (1931).
24. McCoy, H. N., *J. Am. Chem. Soc.*, **63**, 1662 (1941).
25. Jangg, G., *Metall*, **13**, 407 (1959).
26. Jangg, G., *Metall*, **19**, 442 (1965).
27. Sauerwald, F., and E. Osswald, *Z. Anorg. Allgem. Chem.*, **257**, 195 (1948).
28. Claire, J., H. Tachoire, and M. Laffitte, *Bull. Soc. Chim. France*, 1613 (1967).
29. Jangg, G., *Z. Metall.*, **53**, 612 (1962).
30. Kleppa, O., and M. Kaplan, *J. Phys. Chem.*, **61**, 1120 (1957).
31. Lihl, F., *Monatsh. Chem.*, **86**, 186 (1955); *Chem. Abstr.*, **49**, 9998 (1955).
32. Jangg, G., and H. Palman, *Z. Metall.*, **54**, 364 (1963).
33. Shishakov, I. A., *Izv. Akad. Nauk SSSR, Otd. Khim. Nauk*, 683 (1941).
34. Hume-Rothery, W., J. O. Batterton, and J. Regnolds, *J. Inst. Metals*, **80**, 609 (1951).
35. Bittner, H., and H. Novotny, *Monatsh. Chem.*, **83**, 287, 308 (1952).
36. Plaksin, I. P., and K. A. Suvorovskaya, *Izv. Sektora Platiny Drug. Blagorod. Metal. Inst. Obshch. Neorg. Khim. Akad. Nauk SSSR*, **18**, 67 (1945).
37. Simson, C. V., *Z. Physik. Chem.*, **109**, 192 (1924).
38. Bonnier, E., P. Desré, and G. Petot-Ervas, *Compt. Rend.*, **255**, 2432 (1961).
39. Müller, K., *Z. Elektrochem.*, **31**, 304 (1925).
40. Boenziger, N. C., *Acta Cryst.*, **9**, 93 (1956).
41. Begli, K., *Plutonii i evo splavy* (Plutonium and Its Alloys), Atomizdat, Moscow, 1958.
42. Jangg, G., and H. Kirchmayr, *Z. Chem.*, **3**, 41 (1963).
43. Nigmatullina, A. A., and A. I. Zebreva, *Izv. Akad. Nauk Kaz. SSR, Ser. Khim.*, 18 (1964).
44. Gercke, R. H., *J. Am. Chem. Soc.*, **44**, 1684 (1922).
45. Pushin, N. A., *Zh. Russk. Fiz. Khim. Obshchestva*, **34**, 856 (1902).
46. Van Heteren, W. J., *Z. Anorg. Allgem. Chem.*, **24**, 129 (1902).
47. Haring, M. M., and J. C. White, *Trans. Electrochem. Soc.*, **73**, 211, (1938).
48. Pushin, N. A., *Z. Anorg. Allgem. Chem.*, **31**, 201 (1903).
49. Richards, T. W., and C. P. Smyth, *J. Am. Chem. Soc.*, **44**, 524 (1922).
50. Kozin, L. F., *Tr. Inst. Khim. Nauk Akad. Nauk Kaz. SSR*, **9**, 93 (1962).
51. Hildebrand, J. H., *J. Electrochem. Soc.*, **22**, 319 (1912).
52. Jangg, G., *Monatsh. Chem.*, **95**, 1103 (1964).
53. Bent, H. E., and A. F. Vorziatti, *J. Am. Chem. Soc.*, **58**, 2220 (1936).
54. Roos, G. D., *Z. Anorg. Allgem. Chem.*, **94**, 358 (1916).

55. Sunden, N., *Z. Elektrochem.*, **57**, 100 (1953).
56. Liebl, G., *Über die Amalgammetallurgische Herstellung von Reinsten Indium und Antimon*, Diss., Bucholz. F. X. Seitz, Munich, 1956.
57. Kozin, L. F., *Tr. Inst. Khim. Nauk Akad. Nauk Kaz. SSR*, **9**, 71 (1962).
58. Tammann, G., and J. Hinnüber, *Z. Anorg. Allgem. Chem.*, **160**, 249 (1927).
59. Hohn, H., *Wien. Chem. Ztg.*, **49**, 15 (1948).
60. Joyner, R. A., *J. Chem. Soc.*, **99**, 195 (1911).
61. Nigmatullina, A. A., and A. I. Zebreva, *Izv. Akad. Nauk Kaz. SSR, Ser. Khim.*, 20 (1956).
62. Thompson, H. E., *J. Phys. Chem.*, **36**, 201 (1903).
63. Levitskaya, S. A., and A. I. Zebreva, *Tr. Inst. Khim. Nauk Akad. Nauk Kaz. SSR*, **15**, 66 (1967).
64. Sagadieva, K. Zh., and M. T. Kozlovskii, *Izv. Akad. Nauk Kaz. SSR, Ser. Khim.*, **15**, 22 (1959).
65. Sagadieva, K. Zh., and M. T. Kozlovskii, *Vestn. Akad. Nauk Kaz. SSR*, 85 (1963).
66. De Vet, J. F., and R. A. W. Haul, *Z. Anorg. Allgem. Chem.*, **277**, 96 (1954).
67. Zebreva, A. I., and M. T. Kozlovskii, *Collection Czech. Chem. Commun.*, **25**, 3188 (1960).
68. Tammann, G., and K. Kollmann, *Z. Anorg. Allgem. Chem.*, **160**, 244 (1927).
69. Kirchmayr, H. R., *Electrochim. Acta*, **9**, 459 (1964).
70. Erdey-Gruz, T., and A. Vazsonyi-Zilahy, *Z. Physik. Chem.*, **A177**, 292 (1936).
71. Hartmann, H., and K. Schölzel, *Z. Physik. Chem. N.F.*, **9**, 106 (1956).
72. Chao, F., and M. Costa, *Compt. Rend.*, **262C**, 1357 (1966).
73. Marpl, T. L., and L. B. Rogers, *Anal. Chim. Acta*, **11**, 574 (1954).
74. Losev, V. V., E. V. Leontovich, O. K. Kudra, and V. G. Prikhodchenko. *Tezisy Dokladov na Vsesoyuznoi Konferentsii Teoriya i Praktika Amal'gamnykh Protsessov* (Abstracts of the Conference on Theory and Practice of Amalgam Reactions), Alma-Ata, 1966, Izd. Nauka, p. 52.
75. Antropov, L. I., and V. P. Chviruk, *Elektrokhim. Margantsa, Akad. Nauk Gruz. SSR*, **3**, 1351 (1967).
76. Richards, T. W., and G. S. Forbes, *Z. Physik. Chem.*, **58**, 683 (1907).
77. Richards, T. W., and J. H. Wilson, *Z. Physik. Chem.*, **72**, 128 (1910).
78. Richards, T. W., and R. N. Garrod-Thomas, *Z. Physik. Chem.*, **72**, 179 (1910).
79. Pearce, J. N., and J. F. Eversole, *J. Phys. Chem.*, **32**, 209 (1928).
80. Yoshimura, C., *J. Chem. Soc. Japan, Pure Chem. Sect.*, **77**, 1672 (1956).
81. Liebhafsky, H. A., *J. Am. Chem. Soc.*, **57**, 2657 (1935).
82. Hildebrand, J. H., *J. Am. Chem. Soc.*, **35**, 501 (1913).

83. Richards, T. W., and M. Daniels, *J. Am. Chem. Soc.*, **41**, 1732 (1919).
84. Irvin, N. M., and A. S. Russel, *J. Chem. Soc.*, 891 (1932).
85. Richards, T. W., and J. B. Conant, *J. Am. Chem. Soc.*, **44**, 601 (1922).
86. Trümpler, G., and D. Schuler, *Helv. Chim. Acta,* **33**, 790 (1950); **32**, 1940 (1949).
87. Hauffe, K., *Z. Elektrochem.*, **46**, 348 (1940).
88. Armbroster, M. H., and J. Grenschau, *J. Am. Chem. Soc.*, **56**, 2525 (1934).
89. Vierk, A. L., and K. Hauffe, *Z. Elektrochem.*, **54**, 383 (1950).
90. Bent, H. E., and E. Swift, *J. Am. Chem. Soc.*, **58**, 2216 (1936).
91. Bent, H. E., and J. H. Hildebrand, *J. Am. Chem. Soc.*, **49**, 3011 (1927).
92. Smallmann, R., and B. Frost, *Acta Met.*, **4**, 611 (1956).
93. Cusack, N., P. Kendall, and M. Fielder, *Phil. Mag.*, **10**, 87 (1964).
94. Schulz, L. G., *Advan. Phys.*, **6**, 102 (1957).
95. Pugachevich, P. P., and O. A. Timofeicheva, *Zh. Neorgan. Khim.*, **1**, 1387 (1956).
96. Lingane, J. J., *J. Am. Chem. Soc.*, **61**, 2099 (1939).
97. Heyrovský, J., and J. Kůta, *Principles of Polarography,* Academic Press, New York, 1965.
98. Roiter, V. A., V. A. Yuza, and E. S. Poluyan, *Zh. Fiz. Khim.*, **13**, 305 (1939); *Acta Physicochim. URSS,* **10**, 845 (1939).
99. Roiter, V. A., E. S. Poluyan, and V. A. Yuza, *Zh. Fiz. Khim.*, **13**, 805 (1939).
100. Yuza, V. A., and L. D. Kopyl, *Zh. Fiz. Khim.*, **14**, 1074 (1940).
101. Solov'eva, Z. A., L. A. Uvarov, and A. T. Vagramyan, *Zh. Prikl. Khim.,* **5**, 1185 (1960).
102. Chao, F., and M. Costa, *Compt. Rend.*, **261**, 3812 (1965).
103. Stromberg, A. G., *Dokl. Akad. Nauk SSSR,* **85**, 831 (1952).
104. Stromberg, A. G., and A. I. Zelyanskaya, *Tr. Komis. Analit. Khim. Akad. Nauk SSSR,* **4**, 1 (1952).
105. Furman, N. H., and W. C. Cooper, *J. Am. Chem. Soc.*, **72**, 5667 (1950); **74**, 6183 (1952).
106. Cooper, W. C., *J. Am. Chem. Soc.*, **77**, 2047 (1955).
107. von Stackelberg, M., and V. Toome, *Z. Elektrochem.*, **58**, 226 (1954).
108. Zebreva, A. I., *Tr. Inst. Khim. Nauk Akad. Nauk Kaz. SSR,* **9**, 55 (1962).
109. Coriou, H., J. Huré, and N. Mennier, *Anal. Chim. Acta,* **9**, 171 (1953).
110. Porter, J. T., and W. D. Cooke, *J. Am. Chem. Soc.*, **77**, 1481 (1955).
111. Nigmatullina, A. A., and A. I. Zebreva, *Izv. Akad. Nauk Kaz. SSR, Ser. Khim.,* 19 (1967).
112. Tammann, G., and W. Jander, *Z. Anorg. Allgem. Chem.*, **124**, 105 (1922).
113. Russel, A. S., P. V. F. Cazelet, and M. Irvin, *J. Chem. Soc.*, 837, 841 (1932).

114. Russel, A. S., and A. M. Luoys, *J. Chem. Soc.*, 852 (1932).
115. Lihl, F., and H. Kirrbauer, *Z. Metall.*, **48**, 9, 62 (1957).
116. Pushin, N. A., *Zh. Russk. Fiz. Khim. Obshchestva*, **39**, 353, 528, 921 (1907).
117. Zebreva, A. I., *Tr. Inst. Khim. Nauk Akad. Nauk Kaz. SSR*, **15**, 54 (1967).
118. Kovaleva, L. M., and A. I. Zebreva, *Elektrokhimia*, **1**, 1084 (1965).
119. Val'ko, A. V., A. I. Zebreva, S. A. Levitskaya, and B. K. Toibaev, *Zh. Fiz. Khim.*, **38**, 1839 (1964).
120. Ficker, H. K., and L. Meites, *Anal. Chim. Acta*, **26**, 172 (1962).
121. Hartmann, H., and K. Scholzel, *Z. Phys. Chem. (Frankfurt)*, **9**, 106 (1956).
122. Zebreva, A. I., and S. A. Levitskaya, *Zh. Fiz. Khim.*, **36**, 2799 (1962).
123. Kovaleva, L. M., and A. I. Zebreva, *Zh. Fiz. Khim.*, **38**, 1162 (1964).
124. Zebreva, A. I., *Zh. Fiz. Khim.*, **36**, 1822 (1962).
125. Levitskaya, S. A., and A. I. Zebreva, *Elektrokhimia*, **2**, 92 (1966).
126. Zebreva, A. I., and L. M. Kovaleva, *Zh. Fiz. Khim.*, **39**, 855 (1965).
127. Kozlovskii, M. T., and S. P. Bukhman, *Izv. Akad. Nauk Kaz. SSR, Ser. Khim.*, 14 (1953).
128. Baimakov, Yu. V., *Tr. IV. Vses. Sovesch. Elektrokhim.*, Akad. Nauk SSSR, Moscow, 1959, p. 427.
129. Baimakov, Yu. V., *Tr. Leningr. Politekhn. Inst.*, 162 (1955).
130. Zebreva, A. I., and M. T. Kozlovskii, *Zh. Fiz. Khim.*, **30**, 1553 (1956).
131. Babkin, G. N., and M. T. Kozlovskii, *Izv. Vysshikh Uchebn. Zavedenii, Khim. Khim. Tekhnol.*, **1**, 129 (1959).
132. Ilyushchenko, V. M., and M. T. Kozlovskii, *Izv. Akad. Nauk SSSR, Ser. Khim.*, 23 (1958).
133. Sagadieva, K. Zh., and M. T. Kozlovskii, *Izv. Akad. Nauk SSSR, Ser. Khim.*, 3 (1965).
134. Zebreva, A. I., E. F. Speranskaya, and M. T. Kozlovskii, *Zh. Fiz. Khim.*, **3**, 2715 (1959).
135. Kemula, W., *Proceedings of the 2nd International Congress of Polarography*, Cambridge, 1959, p. 105.
136. Stromberg, A. G., and E. A. Stromberg, *Zavodsk. Lab.*, **27**, 3 (1961); **30**, 261 (1964).
137. Nikelly, J. G., and W. D. Cooke, *Anal. Chem.*, **29**, 933 (1957).
138. Sinyakova, S. I., et al., *Sovremennye Metody Analiza* (Contemporary Methods of Analysis), Akad. Nauk SSSR, Moscow, 1965, p. 192.
139. Stepanova, O. S., and M. S. Zakharov, *Izv. Vysshikh Uchebn. Zavedenii, Khim. Tekhnol.*, **7**, 184 (1964).
140. Krasnova, I. E., and A. I. Zebreva, *Elektrochim. Margantsa Akad. Nauk Gruz. SSR*, **2**, 247 (1966).
141. Kemula, W., Z. Galus, and Z. Kublik. *Nature*, **36**, 1223 (1962).

142. Stromberg, A. G., and V. E. Gorodovykh, *Zh. Neorgan. Khim.*, **8**, 2355 (1963).
143. Zakharov, M. S., L. F. Zaichko, N. A. Mesyats, and L. G. Baletskaya, *Izv. Vysshikh Uchebn. Zavedenii, Khim. Khim. Tekhnol.*, **9**, 355 (1966).
144. Hickling, A., and J. Maxwell, *Trans. Faraday Soc.*, **51**, 44 (1955).
145. Zaichko, L. F., and M. S. Zakharov, *Zh. Analit. Khim.*, **21**, 65 (1966).
146. Krasnova, I. E., A. I. Zebreva, and M. T. Kozlovskii, *Dokl. Akad. Nauk SSSR*, **156**, 415 (1964).
147. Kemula, W., and Z. Galus, *Roczniki Chem.*, **36**, 1223 (1962).
148. Krasnova, I. E., and A. I. Zebreva, *Zh. Fiz. Khim.*, **38**, 1675 (1964).
149. Ilyushchenko, V. M., and M. T. Kozlovskii, *Izv. Akad. Nauk Kaz. SSR, Ser. Khim.*, 49 (1961).
150. Vasil'eva, L. N., and E. N. Vinogradova, *Zavodsk. Lab.*, **27**, 1079 (1961).
151. Vinogradova, E. N., and L. N. Basil'eva, *Zh. Analit. Khim.*, **17**, 579 (1962).
152. Kůta, J., and I. Smoler, *Collection Czech. Chem. Commun.*, **26**, 76 (1961).
153. Ivanov, V. F., and Z. A. Iofa, *Dokl. Akad. Nauk SSSR*, **140**, 1368 (1961).
154. Ivanov, V. F., and Z. A. Iofa, *Zh. Fiz. Khim.*, **36**, 1080 (1962).
155. Kalvoda, R., *Collection Czech. Chem. Commun.*, **22**, 1390 (1957).
156. Zakharov, M. S., O. S. Stepanova, and V. I. Aparina, *Izv. Tomsk. Politekhn. Inst.*, **128**, 36 (1965).
157. Zakharov, M. S., and L. F. Zaitsko, *Izv. Tomsk. Politekhn. Inst.*, **164**, 183 (1967).
158. Kemula, W., and Z. Galus, *Bull. Acad. Polon. Sci. (Chim.)*, **7**, 553 (1959).
159. Kemula, W., and Z. Galus, *Bull. Acad. Polon. Sci. (Chim.)*, **7**, 613 (1959).
160. Kemula, W., Z. Galus, and Z. Kublik, *Bull. Acad. Polon. Sci. (Chim.)*, **6**, 661 (1958); **7**, 729 (1959).
161. Stepanova, O. S., *Izv. Tomsk. Politekhn. Inst.*, **151**, 14 (1966).
162. Zebreva, A. I., and M. T. Kozlovskii, *Sovremennye Metody Analiza* (Contemporary Methods of Analysis), Izd. Akad. Nauk SSSR, Moscow, 1965, p. 214.
163. Kozlovskii, M. T., *Rtyt i Amal'gamy v Elektrokhimicheskikh Metodakh Analiza* (Mercury and Amalgams in Electrochemical Methods), Izd. Akad. Nauk Kaz. SSR, Alma-Ata, 1956.
164. Kozlovskii, M. T., *Tr. IV. Vses Soveshch. po Elektrokhim.*, Akad. Nauk SSSR, Moscow, 1959, p. 704.
165. Vinogradova, E. N., L. N. Vasil'eva, and K. Iobst, *Zavodsk. Lab.*, **27**, 525 (1961).
166. Kozlovskii, M. T., A. I. Zebreva, and V. P. Gladyschev, *Amal'gamy i ikh Primenenie* (Amalgams and their Application), Izd. "Nauka", Alma-Ata, 1971.

CHAPTER IV

FREE RADICALS IN ORGANIC POLAROGRAPHY*

B. Kastening

Forschungsabteilung Angewandte Elektrochemie der Kernforschungsanlage Jülich, Jülich, Germany

To Prof. Dr. Ludwig Holleck on his 65th Birthday

Contents

I. Introduction	195
II. General Mechanisms	199
A. Wave height	199
B. Reversibility	200
C. Slow charge transfer	202
D. Stepwise electron transfer	203
E. Proton transfer	206
F. Dimerization and dismutation	208
III. Special Mechanisms	212
A. Simple unsaturated compounds (C=C)	212
B. Compounds containing C=N or N=N	214
C. Aromatic, polycyclic, and heterocyclic compounds	217
D. Quinones and related compounds	227
E. Carbonyl compounds	230
F. Unsaturated carbonyl compounds	242
G. Halogen-containing and related compounds	244
H. Compounds containing the groups NO, PO, SO, and SO_2	250
I. Onium compounds	252
J. Sulfur compounds	255
K. Organometallic compounds	257
L. Nitro and nitroso compounds	259
References	267

I. Introduction

In dealing with electrode reactions of organic compounds, it can as a rule be presumed that a simultaneous transfer of two or more electrons from the

* Manuscript received February, 1969.

electrode to the depolarizing species, or vice versa, is rather exceptional. Hence the elementary step involved must in general be written

$$R^z \pm e^- \rightarrow R^{z \mp 1} \qquad (1)$$

Consequently, one of the two species R^z or $R^{z \mp 1}$ must have an odd number of electrons and represents a free radical (or radical ion). It is apparent from this that free radicals play an important role in organic polarography. The free radical involved may of course be rather unstable. In numerous cases the radical cannot escape from the electrode surface without stabilization, for instance, by transfer of another electron, and it may become difficult to ascertain whether or not radicals are involved at all. Moreover, depolarizer molecules adsorbed at the electrode are sometimes subject to strong electronic interaction with the metal, for instance, aromatic molecules with an orientation of the ring system parallel to the interface [cf. investigations with aniline (91)]. In this case we cannot distinguish between R^z and $R^{z \mp 1}$ in the adsorbed state, the latter being intermediate and characterized by a partial charge transfer number λ, with $0 < \lambda < 1$ (296, 371). Equation 1 is better rewritten as follows (for a cathodic process, and, for simplification, setting $z = 0$):

$$R_{sol} + e^-_{met} \rightarrow (R \cdot e^-)_{ad} \rightarrow R^-_{sol} \qquad (2)$$

The charge transfer virtually proceeds in two steps: adsorption of R and desorption of R^-. If R^- denotes a radical, the process can be considered to proceed through a radical intermediate if R^- is actually desorbed from the electrode as indicated in Eq. 2; the formation of a radical intermediate, however, becomes questionable if only dimers are desorbed, for example, according to

$$2(R \cdot e^-)_{ad} \rightarrow R_2^{-2} \qquad (3)$$

In other cases, because of low stability, special techniques and/or thorough consideration of several facts are necessary to prove whether or not radicals participate in the electrode process. However, a large number of instances exist in which evidence for the generation of radicals during polarographic reduction (oxidation) can be derived simply from inspection of current–voltage curves.

The literature related to this subject is rather voluminous and is growing nearly as fast as that of organic polarography as a whole. At present, the majority of mechanisms proposed involve radicals, either with or without adequate justification. It is the aim of this chapter to outline the more important features of this subject, mainly guided by more recent papers,*

---

* It must be noted that this chapter was written in 1968. More recent developments are, therefore, not considered. This applies also to the quoted literature.

rather than to cover the whole literature which may comprise a multiple of that cited here. For further information the reader is referred to books (493), comprehensive articles (362, 495), and current reviews on organic polarography (368) as well as on polarographic theory and methodology (211).

In examining the matter from a historical point of view, the 45 years of polarography can be divided into three periods, each of them lasting for about 15 years. In the first period, up to the end of the 1930s, no considerable work seems to have been done in the field with which we are concerned. During the second period, up to about 1955, much pioneering work was done with different groups of compounds. The last period has seen remarkable development, for several reasons some of which are mentioned here. Growing use has been made of organic solvents, in which numerous radicals show marked stability; a review on polarography in non-aqueous media published in 1958 was mainly devoted to inorganic depolarizers (167), whereas 5 years later nearly 250 references were listed in a review on organic polarography in such solvents (289); some fundamentals have been discussed in references (269) and (469). Increasing application has been made of modified electrochemical techniques, such as cyclic voltammetry, by which radical mechanisms can often be elucidated by virtue of the reversibility of one-electron processes. The results obtained from combining electron-spin resonance (ESR) techniques with electrochemical research, which were reviewed several years ago (3) and again more recently (247) [cf. also reference (397)], should convince the very last doubter that radicals are involved. Two convenient techniques developed at the beginning of the 1960s, the generation of radicals at electrodes situated inside the microwave cavity (146, 313–315, 369, 370) and the pumping of the electrolyzed solution through an ESR spectrometer (381), have been applied in numerous studies since then. Finally, new aspects of the subject were discovered by irradiation of solutions and/or electrodes; the reader is referred to reviews of this field (38, 183, 278).

As a tribute to the early studies, we briefly review several more important investigations made during the second period mentioned above.

The redox system quinone/semiquinone/hydroquinone, well established by the work of Michaelis (1932–1938), has been dealt with polarographically and the appropriate theoretical relations have been derived (47, 334, 389); well-separated one-electron waves were observed in aprotic media (105) and quinonelike systems (actually α,β-diketones) were shown to be reduced reversibly in aqueous alkaline solutions (330).

Early investigations also involved aromatic and polycyclic hydrocarbons (79, 280, 458, 460), as well as heterocyclic compounds (250, 251, 286, 446, 478), and relevant theories concerning correlations with molecular orbital calculations and the effect of protonation on the stability of such radicals were developed (185).

A radical mechanism for the reduction of carbonyl compounds (aldehydes and ketones) was established (10, 191, 346); it explained the essential features of this process and has often since been confirmed. Preprotonation in acid solutions was shown to play a role (108). Half-wave potentials were correlated with molecular orbital considerations (87, 88). Systematic investigations were performed dealing with the influence of substituents on this process (192), and the resulting data were correlated with Hammett's rule (160).

For unsaturated carbonyl compounds it was shown that a somewhat different mechanism is valid (123, 142, 176, 272, 346, 376), inasmuch as the products of secondary reduction or dimerization of the radicals are saturated carbonyl and dicarbonyl compounds, respectively, instead of alcohols and pinacols. This also holds true for unsaturated esters (109) including coumarin (66, 174, 347) and other derivatives of acids the radicals of which undergo polymerization (103, 261, 345). However, for tropolone, which is also an unsaturated ketone, the situation is somewhat different (219, 337), hydroxy compounds being the products of either reduction or dimerization of the corresponding radical.

Organic halogen compounds were found to produce radicals by reduction-induced halide ion abstraction (110, 421), and the mechanisms were elucidated (111).

Although an analysis of current–voltage curves suggested that less than one electron was consumed in reaching the transition state during the reduction of nitro groups (351), it was not before the application of surface-active inhibitors that the corresponding anion radical was shown to be the primary product of electron transfer (186–189). Whereas the first investigations were performed in aqueous solutions, the radical showed considerably improved stability in aprotic solvents and became, a decade later, one of the systems most frequently studied by joint electrochemical and ESR techniques.

Further work involving radical mechanisms concerned the reduction of C=N (375) and N=N double bonds (127), mercury compounds (25) and other organometallics (476), sulfur compounds such as cystine (268), and diazonium (385) as well as iodonium compounds (82).

Certainly, this list is not complete. Furthermore, in the following sections earlier studies are not mentioned in each case, and the reader is referred to the literature cited and quotations therein.

As regards the scope of the subjects presented, we have not adopted the use of the term polarography to denote voltammetric techniques irrespective of the type of electrode applied. Instead, polarography is understood in its original sense, namely, the utilization of the dropping mercury electrode

(DME) (and, eventually, the streaming and hanging drop mercury electrodes). Although results obtained with other electrode types and materials are occasionally referred to, they are essentially disregarded. Thus certain groups of compounds are not considered since they are oxidized only at potentials too positive to be accessible with mercury electrodes. These include, for example, the oxidations of aromatic and aliphatic amines, which have been investigated at platinum (104, 326, 342, 470) and carbon paste electrodes (139, 288).

Results obtained by controlled-potential electrolysis, usually with mercury pool electrodes, for the purpose of identifying products and elucidating mechanisms, are frequently given without special reference to the method; reviews on these techniques have been published recently (300, 496).

II. General Mechanisms

Although the application of further methods may be necessary for conclusive elucidation of electrode mechanisms, a simple polarographic current–voltage curve in general provides some pertinent information. We therefore consider in detail the relevant features of these curves.

A. Wave Height

A first suggestion that radicals are involved comes from the wave height. If the height corresponds to the transfer of one electron, a radical mechanism is rather probable (irrespective of the fate of the radicals, for example, stabilization or rapid dimerization). As a rule of thumb, it may be noted that, with the usual DME (drop time about 3 sec, height of mercury column about 60 cm) and with 10^{-3} M depolarizer, a mean current of about 3 μA corresponds to a single-electron process (in water, ethanol, and dimethylformamide; 4 μA in methanol; 5 μA in acetonitrile). This criterion, however, is not conclusive. For instance, in an aprotic solvent a substance containing an acid hydrogen (symbolized by RH^+, although not necessarily a cation) may react as:

$$RH^+ + 2e^- \rightarrow RH^- \qquad (4)$$

$$RH^+ + RH^- \rightarrow R + RH_2 \qquad (5)$$

A two-electron mechanism is in operation, but for each molecule being reduced another molecule is deactivated, R being reducible only at more negative potentials; thus judging from the wave height, the curve seems to indicate the transfer of one electron. Some unusual mechanisms resembling those of Eqs. 4 and 5 are mentioned in Section III. (Quite similar observations can be made when a proton donor is present at a concentration comparable to that of the depolarizer.) However, radicals may well be involved in processes with a wave height corresponding to two or more electrons if the intermediate undergoes further reduction (oxidation) at a sufficient rate.

B. Reversibility

Further conclusions as to the nature of the mechanism may be drawn from the shape of the wave, namely, the dependence of the current on potential. Here we are rather intimately concerned with the problem of reversibility.

Until about 15 years ago, the majority of polarographic investigations with organic compounds were carried out in aqueous or aqueous-alcoholic solutions. Since most electrode reactions proceed irreversibly in these media, the contention seemed to be justified in that only a limited number of organic electrode processes show reversibility. Since then, an increasing amount of work has been done in non-aqueous, and especially in aprotic, solvents, and it has developed that reversibility is a rather frequently observed phenomenon in organic polarography and that radicals are involved in many or even most cases.

Polarographic reversibility is characterized by the fact that all the reactions involved (electrochemical as well as chemical) proceed reversibly and are fast in comparison with mass transfer. For a reversible process, therefore, the equation of the current–voltage curve includes no rate constants except diffusion coefficients. The most simple case is characterized by

$$i = i_d \frac{1}{1 + \lambda} \qquad (6)$$

(provided that $D_{ox} = D_{red}$), where i_d, involving the diffusion coefficient, is given by the Ilkovič equation, and λ, defined by

$$\lambda = \exp\left[\frac{F(E - E_0)}{RT}\right] \qquad (7)$$

stands for the thermodynamic equilibrium constant (E_0 being the standard redox potential) of the process considered:

$$R + e^- \rightleftharpoons R^- \qquad (8)$$

(Here and in the following discussion we deal with reduction processes, although the equations are also valid for oxidations if the signs are properly chosen.) R^-, or occasionally R, is then a radical. (λ^z must be inserted in place of λ if a reversible transfer of z electrons occurs.) Several simple methods have been used to analyze the curves with respect to Eq. 6, the most usual of which are: (a) determination of the potential difference $E_{1/4} - E_{3/4}$ for one and three quarters of the limiting current; (b) measurement of the difference between the potentials at which the tangent at $E_{1/2}$ intersects the background and limiting currents; and (c) evaluation of the slope of the plot of E versus log $(i_d - i)/i$. At 25°C Eq. 6 requires: (a) 56.5; (b) 103; and (c) 59 mV/log unit. The most accurate method is (c); since, however, more involved mechanisms reveal slopes not far from 59 mV/log unit, it may be difficult to draw the correct conclusion, especially if the current–voltage curve is distorted by experimental problems, such as errors resulting from an iR drop across the cell, an unfavorably low ratio of the diffusion current to the background current at low concentration, and complications arising from adsorption phenomena. Further proofs of reversibility according to Eq. 8 are: failure of $E_{1/2}$ to depend on either the concentration of depolarizer or the drop time (height of mercury reservoir); current–time curves for single drops following the relation

$$i = \text{constant} \times t^{1/6} \tag{9}$$

even on the ascending part of the wave (in practice, 0.2 is observed instead of the theoretical value of 1/6 because of deviations from the simplified theory); and equality of the cathodic and anodic half-wave potentials for the starting material and product.

Beyond the inspection of simple current–voltage curves, advantage may be taken of other methods applying periodic or pulsed potential changes and indicating the reoxidizability of the product of reduction. Among these methods the most frequently applied are: the Kalousek commutator (applied square-wave voltage, mostly 5–50 Hz, the potential being kept constant during one half-cycle but changing linearly with time during the other half-cycle, while in general the mean current during the periods of changing potential is recorded versus this potential); oscillographic polarography according to Heyrovský (applied sine wave current, usually 50 Hz, of such an amplitude as to cover the accessible range of potentials, the function dE/dt versus E being usually recorded); ac polarography according to Breyer (a sine wave voltage of low amplitude, about 10 mV, being superimposed on a linearly changing dc voltage, the amplitude of the ac current being recorded as a function of the dc voltage); and cyclic voltammetry (applied triangular voltage, the sweep rate varying from 0.1 V/min to several hundred volts per second and the current being recorded versus potential). The theory of this method has been developed for the observation of unstable intermediates applying the hanging drop electrode (291). All these methods show characteristic features according to the reversibility of the electrode reaction. More recently, cyclic voltammetry has become the most frequently applied method. It should, however, be emphasized that at rapid voltage sweeps reversibility of the electrode process may be indicated, although the overall polarographic process is irreversible; the intermediate radical may undergo subsequent irreversible changes (including also adsorption or desorption), the rates of which are fast as compared with the drop time but slow as compared with the sweep time.

Another method frequently applied is that of controlled-potential electrolysis at electrodes of large area by means of potentiostats. In this case the product of the electrode process is generated as a constituent of the bulk solution; polarograms recorded during, or even at the end of, exhaustive electrolysis show an anodic wave having the same $E_{1/2}$ as the original cathodic wave if the product is stable and reversibly oxidized. Since the time constants of electrolysis are usually of the order of several minutes or more (although they can be made as small as 10–15 sec by special techniques), there is again some restriction as to the indication of polarographic reversibility; the product, although stable during the drop life, may well undergo irreversible changes during the period of electrolysis. Thus under unfavorable conditions only combined application of different methods may help the experimenter to judge whether or not the process is reversible and to elucidate its mechanism.

C. Slow Charge Transfer

After this digression on the methods employed supplementary to ordinary polarography, we return to the radical-producing reaction (Eq. 8). Thus far electron transfer has been considered fast as compared with mass transfer. If this is not true at the potential E_0, then $E_{1/2} \neq E_0$, and the current–voltage relation differs from Eq. 6. The rigorous solution of the relevant differential equation of mass transfer with the proper boundary conditions, including the rate equation of electron transfer, yields

$$i = i_d F(\chi) \tag{10}$$

which may be approximated, within the limits of experimental error, by an equation analogous to Eq. 6:

$$i = i_d \frac{1}{1 + \rho\lambda^x} \tag{11}$$

Whereas $F(\chi)$ is Koutecký's well-known function, χ and ρ are given by

$$\chi = k_{e0}\lambda^{-\alpha}\left(\frac{12\tau}{7D}\right)^{1/2} \tag{12}$$

$$\rho = (0.87 k_{e0}\sqrt{\tau/D})^{-1} \tag{13}$$

these parameters being dependent on the drop time τ, the diffusion coefficient D, and the standard rate constant k_{e0} of electron transfer, α being the transfer coefficient. Since $\rho\lambda^{\alpha} = 1$ at the half-wave potential, it follows that

$$E_{1/2} = E_0 - \frac{RT}{\alpha F}\ln\rho \tag{14}$$

Again, the current–voltage curve may be analyzed by the methods mentioned in Section II-B; however, as can be seen from Eq. 11, since $0 < \alpha < 1$ and frequently $\alpha \sim 0.5$, the curve is flatter than the reversible one, and the potential differences or slopes are greater by a factor of $1/\alpha$. Techniques employing rapid change of potentials mentioned above may help in this case, since their response is sensitive to the degree of reversibility.

If the transfer of the first electron is rate-determining, then the shape of the current–voltage curve is invariably given by Eq. 11, irrespective of what happens to the (anion) radical produced; it may be stable, undergo dimerization or dismutation or, with or without intermediate protonation, accept another electron. However, with dismutation or subsequent electron transfer, the height i_d of the wave is twice that for single-electron transfer. (If dismutation occurs, ρ will have twice the value given by Eq. 13.) If the (anion) radical is stable, the half-wave potential will be more negative than E_0. However, if subsequent changes occur, the irreversibility in general is attributable to the removal of R^-, preventing it from being reoxidized, and $E_{1/2}$ will shift toward more positive values.

The current–voltage curve will be identical with that of a reversible process if $\alpha = 1$ for a barrierless electron transfer (116), although $E_{1/2} \neq E_0$. The same is true for an irreversible two-electron transfer with $\alpha \sim 0.5$, the wave height, however, being twice as large (except for the special mechanism shown in Eqs. 4 and 5). In both cases techniques employing rapid change of potentials indicate irreversibility.

D. Stepwise Electron Transfer

Thus far we have dealt with mechanisms giving rise to a single wave on which the potential is exclusively governed by the transfer of the first electron, whether this is reversible or not. We now consider mechanisms in which the (anion) radical undergoes another charge transfer reaction. The influence of any other reaction being absent, we can distinguish the mechanisms, according to the reversibility of both charge transfer reactions:

$$R \underset{E_{0\text{I}}}{\overset{+e^-}{\rightleftarrows}} R^- \underset{E_{0\text{II}}}{\overset{+e^-}{\rightleftarrows}} R^{-2} \qquad (15)$$

$$R \underset{E_{0\text{I}}}{\overset{+e^-}{\rightleftarrows}} R^- \underset{E_{1/2\text{II}}}{\overset{+e^-}{\longrightarrow}} R^{-2} \qquad (16)$$

$$R \underset{E_{1/2\text{I}}}{\overset{+e^-}{\longrightarrow}} R^- \underset{E_{1/2\text{II}}}{\overset{+e^-}{\longrightarrow}} R^{-2} \qquad (17)$$

(A mechanism in which the first step is irreversible, while the second proceeds reversibly, occurs only under unusual conditions; e.g., if an inhibitor is adsorbed at the electrode, which renders the first step irreversible, but is desorbed prior to the second wave.)

In the case of overall reversibility (Eq. 15), the two E_0 values reflect the thermodynamic formation constant of the anion radical from the species R and R^{-2}:

$$K = \frac{[R^-]^2}{[R][R^{-2}]} = \exp\left[\frac{F(E_{0\text{I}} - E_{0\text{II}})}{RT}\right] \qquad (18)$$

This situation is realized by the system quinone/semiquinone/hydroquinone, which was the first polarographic process involving radicals to be studied in detail (47, 334). The theory (47) describing consecutive reversible electron transfer (also including dimerization of the intermediate radical) has recently been refined and extended for the application of chronopotentiometry (168). Separate waves, however, each of which is described by Eq. 6 with its correct E_0 value, occur only when K is sufficiently large. Otherwise, the waves merge and K can be evaluated only by careful analysis of the curve (which may easily be mistaken for irreversible two-electron transfer), until for small values of K this constant cannot be determined since the wave assumes the shape for a reversible two-electron transfer: $i = i_d/(1 + \lambda^2)$. For a more detailed consideration, see references (168) and (181).

Separate waves with stable anion radicals and large values of K have been observed with numerous compounds in aprotic solvents and for some systems in aqueous solution. With most compounds in aqueous solution, however, the behavior is different from Eq. 15, first, because of protonation reactions accompanying electron transfer, which are considered in Section II-E. However, even when Eq. 15 is followed in aqueous solution (as it is for the system p-benzoquinone/hydroquinone dianion at pH > 12), the formation constant K is in general small and a single two-electron wave occurs.

This results from increases in the importance of solvation phenomena with increasing dielectric constant and/or molecular dipole moment of the solvent molecules, as well as with increasing charge of the solvated species. Hence the dianion R^{-2} is thermodynamically favored as compared with R^- when passing from organic solvents to water, even if the anion radical R^- is stabilized by the odd electrons occupying a molecular orbital of low energy in a highly resonant π-electron system, since the repulsion energy of the two electrons is attenuated by solvation. Apart from electrostatic phenomena, other effects, such as hydrogen bonding, play a role.

The mechanism represented by Eq. 16 is often observed in organic solvents as well as in water. The current–voltage curve is described by

$$i = \frac{i_d}{2} \frac{2 + \rho}{1 + \rho + \lambda\rho} \tag{19}$$

where ρ is given by Eq. 13 if, for simplicity, the standard rate constant k_{e0} is referred to the potential E_{0I} of the first step. There will be two separate waves if $\rho \gg 1$ (k_{e0} small), the first wave being represented (cf. Eq. 6) by

$$i = \frac{i_d}{2} \frac{1}{1 + \lambda} \tag{20}$$

and the second wave [with $\lambda \to 0$, including the limiting current of the first wave (cf. Eq. 11)] by

$$i = \frac{i_d}{2} + \frac{i_d}{2} \frac{1}{1 + \rho\lambda^\alpha} \tag{21}$$

However, a single wave occurs at potentials more positive than E_{0I} if $\rho \ll 1$ (k_{e0} large); then

$$i = i_d \frac{1}{1 + \rho\lambda^{(1+\alpha)}} \tag{22}$$

In this case, since $\alpha \sim 0.5$, a plot of E versus $\log (i_d - i)/i$ yields a slope of approximately 40 mV/log unit.

Apart from slow electron transfer to R^-, there are other processes that may be responsible for the irreversibility of the second step in Eq. 16, including fast and irreversible chemical changes of R^{-2}. Rather frequently, R^{-2} is a strong proton acceptor that readily abstracts protons even from the molecules of an "aprotic" solvent such as dimethylformamide or acetonitrile. Proton donors other than the solvent may of course be operative as well. In this case, however, only Eqs. 19–22 hold true and the process is independent of the donor concentration only as long as the second charge transfer step is exclusively rate determining (apart from eventual effects of the solution composition on the rate constant k_{e0}, such as by double-layer and other effects). In the next section, therefore, we are concerned with the effect of proton transfer.

E. Proton Transfer

The shapes of the waves and the values of $E_{1/2}$ depend on the prevailing mechanism when proton transfer is involved. Some frequently observed mechanisms proceed as:

$$R \xrightleftharpoons[E_{0I}]{+e^-} R^- \xrightarrow{+e^-}_{k_e} R^{-2} \xrightarrow[\text{Very fast}]{+H^+} RH^- \tag{23}$$

$$R \xrightleftharpoons[E_{0I}]{+e^-} R^- \xrightarrow{+H^+}_{k_r} RH \xrightarrow[\text{Very fast}]{+e^-} RH^- \tag{24}$$

$$R + H^+ + e^- \xrightleftharpoons[E_{0I}]{} RH \xrightarrow{+e^-}_{k_e} RH^- \tag{25}$$

The mechanism represented by Eq. 23 was considered at the end of the preceding section.

If protons are readily available (e.g., from water or proton donors added to aprotic solvents), R^- may undergo rapid protonation, and the radical RH is usually rather unstable toward subsequent chemical and/or electrochemical attack. Dismutation and dimerization reactions may serve for stabilization (cf. Section II-F), as well as further electron transfer as indicated in Eq. 24. The latter is attributable to the fact that in general the E_0 value of RH, in contrast to R^-, is more positive than that of R itself. This has been shown for alternant hydrocarbons by comparing the energies of the lowest vacant molecular orbitals of R and RH (185). Hence, at potentials at which R^- is generated, a second electron is immediately transferred subsequent to rapid protonation. In this situation the current is given by

$$i = i_d \frac{1}{1 + \rho_r \lambda} \tag{26}$$

where

$$\rho_r = (0.87\sqrt{k_r \tau})^{-1} \tag{27}$$

and the half-wave potential ($\rho_r \lambda = 1$)

$$E_{1/2} = E_{0I} + \frac{RT}{F} \ln \frac{1}{\rho_r} \tag{28}$$

is shifted to a potential more positive than E_{0I}. Equation 26 is applicable only when $\rho_r \ll 1$, that is, when the chemical reaction (rate constant k_r) is fast as compared with mass transfer but slow as compared with transfer of the second electron. Since k_r is in general proportional to the concentration of proton donors c_D,

$$k_r = k_H c_D \qquad (29)$$

we find that, at 25°

$$\frac{dE_{1/2}}{d \log c_D} = \frac{RT}{2F} \ln 10 = 29.5 \text{ mV/log unit} \qquad (30)$$

When the rate of proton transfer becomes comparable with that of mass transfer, then Koutecký's function and the corresponding approximate solution (Eq. 26) are no longer applicable. The part of the limiting current i_1 that exceeds i^*, the value corresponding to the diffusion-controlled one-electron process, will be a kinetic rather than a diffusion-controlled current. A solution to this problem has been given (232) [cf. also references (338) and (246)]:

$$i_1 = i^*[1 + \phi(k_r\tau)] \qquad (31)$$

and the function $\phi(k_r\tau)$ tabulated. For example, $i_1 = 1.5i^*$ for $k_r\tau = 4$. The shape of the curve resembles that for a reversible one-electron transfer and $E_{1/2}$ does not differ markedly from E_{0I}. A somewhat similar situation occurs if the rate of proton transfer is high but the concentration of the proton donor not sufficient to protonate all the anion radicals R^- generated; the limiting current will then be the sum of diffusion currents attributable to R and the proton donor.

At this point it should be remembered that chemical reactions such as proton transfer may proceed by heterogeneous as well as by homogeneous mechanisms, the former involving species adsorbed at the electrode. It is beyond the scope of this chapter to cover the field of adsorption in any detail; this has been done elsewhere (239) and has included adsorbed radicals. It may, however, be noted that, especially in aqueous solutions, the mechanism represented by Eq. 24 may proceed with R^- being adsorbed, k_r then denoting the rate constant of heterogeneous proton transfer (244). Since the adsorption of R^- depends on the potential, k_r is also potential-dependent, and the dependence of half-wave potentials is a more involved function of proton donor concentrations (or pH values), moreover, potential-dependent limiting currents may be observed [see also reference (247b)].

The mechanism according to Eq. 25, involving simultaneous reversible transfer of a proton and an electron, indicates that the rates of both transfer reactions are too high to permit distinguishing which particle is transferred first; the charge transfer may involve the species RH^+ as well as R. Here, Eqs. 19–22 hold true, except that E_{0I} is pH-dependent; for pH $>$ pK_a,

$$E_{0I} = E_{0(RH^+/RH)} + 0.059\ \text{p}K_a - 0.059\ \text{pH} \tag{32}$$

where K_a is the acid dissociation constant of RH^+. (A corresponding relation can be written with $E_{0(R/R^-)}$ and the acid dissociation constant of RH.) Whereas, as noted above, the potential E_0 for the couple RH/RH^- is usually more positive than that of R/R^-, so that protonation of R^- leads to the immediate transfer of a second electron, the positive shift in E_{0I} caused by simultaneous participation of protons in the first step (Eqs. 25 and 32) may render E_{0I} more positive than E_0 for RH/RH^-, thus preventing the transfer of a second electron at potentials on the first wave. Charge transfer is then restricted to the first electron and, since RH is often unstable, dimerization occurs; this is discussed in the following section.

Another possible sequence of reactions involving the participation of protons, in which charge transfer proceeds with protons rather than with the organic molecule which in turn is reduced by intermediate hydrogen atoms:

$$H^+ + e^- \rightarrow H; \qquad R + H \rightarrow RH \tag{33}$$

has been ruled out by early investigations (422). Such a mechanism may, however, take place if reduction proceeds at potentials close to the final current rise resulting from hydrogen evolution (308); an example is given in Section III-E.

F. Dimerization and Dismutation

Apart from the fact that many radicals are stable at least in aprotic solvents (excluding oxygen and, in many cases, also water, the potential being insufficiently negative to effectuate a subsequent charge transfer), many radicals tend to be stabilized by subsequent chemical reactions that remove the odd number of electrons. This can take place through reactions with solvent molecules or with other constituents of the solution, thus producing new radicals which undergo further stabilization reactions. There are, however, two other ways occurring rather frequently by which an even number of electrons is realized, namely, the reaction of two radicals with one another in which a new chemical bonding is established (dimerization), or in which an electron is exchanged (dismutation)

$$R^- + R^- \xrightarrow{k_{\text{dim}}} (R\text{---}R)^{-2} \tag{34}$$

$$R^- + R^- \xrightarrow{k_{\text{dis}}} R + R^{-2} \tag{35}$$

Both reactions proceed irreversibly in most cases, especially dimerization, although certain examples of reversible reactions are known.

Furthermore, in the presence of proton donors, the anion radical R^- may be protonated prior to the reactions shown, which then take place according to, for example,

$$RH + R^- \rightarrow RH^- + R \tag{36}$$

In this case the preceding proton transfer reaction ($R^- + H^+ \rightarrow RH$) may be rate determining, the subsequent bimolecular reaction being fast, and first-order kinetics applies. However, equilibrium conditions may control the protonation, and the rate of the bimolecular reaction will therefore be pH dependent. Apart from these implications, we consider what happens if reaction 34 or 35 takes place.

In both reactions we are—in contrast to all the other mechanisms dealt with so far—concerned with second-order kinetics. A rigorous solution of the proper differential equations for mass transfer and second-order kinetics involves rather difficult mathematics, which has been overcome so far only under certain restricted conditions, namely, for reactions that are very rapid as compared with mass transfer (274) and for dismutation reactions that are very slow as compared with mass transfer (273). In the former case approximate solutions making use of the concept of "reaction layers" are applicable as well (173).

We can distinguish the following frequently observed mechanisms:

(*a*) Transfer of the first electron is reversible and dimerization is fast as compared with mass transfer; only at more negative potentials does the transfer of a second electron become fast enough to compete with dimerization:

$$R + e^- \rightleftharpoons R^- \begin{array}{c} \xrightarrow{k_{\text{dim}}} \tfrac{1}{2}(R\text{—}R)^{-2} \\ \xrightarrow[k_e]{+e^-} R^{-2} \end{array} \tag{37}$$

Theoretical relations pertaining to this sequence have been developed for different electrochemical techniques (140, 392). In ordinary polarography the first one-electron wave is shifted to potentials more positive than E_{01} because of the rapid removal of R^- by dimerization. The current cannot be given explicitly, but the current–voltage curve obeys the approximate relation

$$\frac{1 - (i/i_d)}{(i/i_d)^{2/3}} = 1.15 \frac{\lambda}{(k_{\text{dim}} c_0 \tau)^{1/3}} \tag{38}$$

In the original article (274), which gave a more rigorous solution to this problem,

$$i/i_d = f(\gamma) \tag{39}$$

with

$$\gamma = (\lambda^3 k_{\text{dim}} c_0 \tau)^{1/2} \tag{40}$$

and $f(\gamma)$ being tabulated, the approximate relation corresponding to Eq. 38 [Eq. 55 in reference (274)] was given with incorrect exponents for $1 - (i/i_d)$ and i/i_d, and this error has been transferred to subsequent publications (362). This is, incidentally, the reason for the rather large differences suggested to exist between the approximate and the rigorous solutions, which are largely eliminated by applying the correct approximation (Eq. 38). According to this equation, the half-wave potential is given by

$$E_{1/2} = E_{0I} + \frac{RT}{3F} \ln (0.33 k_{\text{dim}} c_0 \tau) \tag{41}$$

The second polarographic wave is governed by competition of charge transfer and dimerization; the current–voltage curve approximately corresponds to

$$\frac{[1 - (i/i_d)]^{2/3}}{i/i_d} = 1.07 \lambda^\alpha \frac{D^{1/2}(k_{\text{dim}} c_0)^{1/3}}{k_{e0} \tau^{1/6}} \tag{42}$$

k_{e0} is again referred to E_{0I}, and i only to the current exceeding i_d of the first wave. For the half-wave potential, it follows that

$$E_{1/2} = E_{0I} - \frac{RT}{\alpha F} \ln \left[0.85 \frac{D^{1/2}(k_{\text{dim}} c_0)^{1/3}}{k_{e0} \tau^{1/6}} \right] \tag{43}$$

Equations 38–41 and Eqs. 42 and 43 also hold true when one electron and one proton are reversibly transferred in the first step:

$$R + e^- + H^+ \rightleftharpoons RH \begin{array}{c} \xrightarrow{k_{\text{dim}}} \frac{1}{2}(HR\!-\!RH) \\ \xrightarrow{+e^-} RH^- \end{array} \tag{44}$$

However, E_{0I} is then pH-dependent according to Eq. 32.

Dimerization may also proceed as a heterogeneous reaction with radicals being adsorbed at the electrode (307); in this case the corresponding relations for the first wave (with k_{dim} being given in cm sec^{-1} l·mole^{-1}) are

$$\frac{1 - (i/i_d)}{(i/i_d)^{1/2}} = 1.11 \frac{\lambda}{(k_{\text{dim}} c_0 \sqrt{\tau/D})^{1/2}} \tag{45}$$

$$E_{1/2} = E_{01} + \frac{RT}{2F} \ln (0.41 k_{\text{dim}} c_0 \sqrt{\tau/D}) \tag{46}$$

(b) In certain systems transfer of the second electron competing with dimerization is governed by the rate of an intermediate chemical reaction rather than by the rate of electron transfer:

$$R + e^- \rightleftharpoons R^- \begin{array}{c} \xrightarrow{k_{\text{dim}}} \frac{1}{2}(R\text{—}R)^{-2} \\ \xrightarrow[k_r]{+H^+} RH \xrightarrow[\text{Very fast}]{+e^-} RH^- \end{array} \tag{47}$$

If the first- and second-order reactions proceed at comparable rates, the limiting current of the first wave will be intermediate between the values corresponding to the transfer of one and two electrons; when both chemical reactions are fast as compared with mass transfer, the limiting current i_1 of the first wave will exceed that corresponding to one-electron transfer i_d according to

$$\frac{[2 - (i_1/i_d)]^2}{[(i_1/i_d) - 1]^3} = 1.23 \frac{k_{\text{dim}} c_0}{k_r^{3/2} \tau^{1/2}} \tag{48}$$

As long as $i_1 < 2i_d$, the half-wave potential does not differ markedly from that given by Eq. 41.

(c) If a dismutation reaction according to Eq. 35 and fast as compared with mass transfer follows the reversible transfer of the first electron, the limiting current i_d corresponds to the transfer of two electrons as a result of the regeneration of the depolarizer R. The current–voltage curve obeys the relation

$$\frac{1 - (i/i_d)}{(i/i_d)^{2/3}} = 0.73 \frac{\lambda}{(k_{\text{dis}} c_0 \tau)^{1/3}} \tag{49}$$

and the half-wave potential is given by

$$E_{1/2} = E_0 + \frac{RT}{3F} \ln (1.28 k_{\text{dis}} c_0 \tau) \tag{50}$$

If the rate of dismutation is comparable with that of mass transfer, a reduced limiting current $i_d > i_1 > i_d/2$ will occur. However, the approximate method of reaction layers is not applicable in this case, and the rigorous solution involves serious mathematical difficulties. The solution seems to have been given only for rather low rates of dismutation (273).

III. Special Mechanisms

A classification of special mechanisms according to the different classes of chemical compounds would be thoroughly justified, if a given type of substance (for example, carbonyl compounds) would, more or less invariably, follow a certain sequence of reactions. From polarographic and related electrochemical observations, however, it is evident that such a close correlation between the chemical nature of a substance and its reduction or oxidation mechanism does not exist. Mechanisms are in fact rather sensitive to changes of the experimental conditions (solvent, proton availability, depolarizer concentration, and so on). Moreover, the actual path followed in a branched mechanism depends also on the employed technique because, for example, the distribution among competing branches depends on the characteristic time constant of the method. From a more theoretical point of view, it would be preferable to discuss the present subject according to the behavior and/or mechanism of formation of radicals. One could, for instance, collect all depolarizers whose reductions (or oxidations) result in radicals in which the odd electron belonging to the π-electron system extends throughout the molecule, whereas other characteristics might be chosen in order to subdivide radicals in which the odd electron is localized at certain groups or atoms. Actually, such a classification would reveal some surprising features that quite different compounds have in common. However, there are several reasons why such a procedure is inadequate. These include the fact that we are often uncertain what mechanism takes place and what the actual structure of the radical is, as well as the fact that our chemical sense is programmed in terms of active groups rather than in terms of, for instance, electronic structures. In the following sections we therefore essentially retain the customary classic classification. As an alternative, an attempt has been made by the present author more recently (247a) to classify mechanisms in terms of the characteristic individual steps involved.

A. Simple Unsaturated Compounds (C=C)

Olefins and their derivatives often show two separate one-electron reduction waves in aprotic media, whereas a two-electron wave occurs when protons are available, especially in aqueous solutions, owing to immediate subsequent reduction of the protonated radical RH. This is in accord with theoretical considerations, based on molecular orbital calculations, of the relative reducibilities of R^- and RH as compared with R (185). This explanation, although restricted by its theoretical foundations to alternant hydrocarbons, seems to apply to rather different types of compounds; some restrictions are mentioned in Section III-C. Here, we refer only to some selected observations on compounds containing C=C double bonds.

The reduction of polymethine compounds such as $(CH_3)_2N—CH=CH—CH=CH—CH=\overset{+}{N}(CH_3)_2$ is in accord with the comments made above; separate one-electron waves are observed in dimethylformamide, whereas an immediate two-electron reduction proceeds in aqueous solutions (209). Similarly, stepwise one-electron oxidation is observed in aprotic solvents with tetra(dimethylamino)ethylene, the cation radical being indicated by ESR (279). Esters of maleic and fumaric acids show two separate one-electron waves in anhydrous pyridine, whereas the molecules of the free acids transfer protons to solvent molecules giving rise to the reduction of pyridinium cations (439). With the esters one-electron reductions, resulting in the corresponding dimers, have been said to occur even in aqueous solutions of higher pH values, whereas these and most similar compounds more commonly undergo two-electron reductions (109). However, anion radicals generated by the reduction of conjugated double bonds may abstract protons even from anhydrous dimethylformamide at a rate sufficient for immediate two-electron reduction (311, 312). Proton abstraction by anion radicals (or dianions?) may even proceed with depolarizer molecules arriving in the vicinity of the electrode from the bulk of the solution, thus deactivating these molecules and giving rise to limiting currents lower than expected, as has been shown with ethyl methylacrylate in dimethyl sulfoxide, for which the limiting current is tripled by the addition of phenol as a proton donor (365).

For unsaturated nitriles, R—CH=CH—CN, in dimethylformamide, the concentration and the substituent R determine whether two-electron reduction to the saturated nitrile, or one-electron reduction followed by dimerization of the anion radical and a second wave at more negative potentials (405), is observed. With R = H (acrylonitrile) propionitrile and adipodinitrile are generated simultaneously, and currents lower than those for two-electron reduction are observed; preprotonation and reduction of $CH_2=CH—C\overset{+}{N}H$ to the corresponding radical has been suggested to proceed in acidic media (6), whereas for generation of the dimer in alkaline media both radical

$$R + e^- \rightarrow R^-; \qquad 2R^- \rightarrow (R—R)^{-2} \qquad (51)$$

and anionic mechanisms

$$R + 2e^- \rightarrow R^{-2}; \qquad R^{-2} + R \rightarrow (R—R)^{-2} \qquad (52)$$

have been discussed (6, 483). The same is true for the polymerization of acrylamide, $CH_2=CH-CO-NH_2$, induced by cathodic reduction in water (407).

The two-electron wave observed with certain dipyridylethylenes splits into two one-electron waves at higher depolarizer concentrations; hence dimerization is fast and competes with the transfer of the second electron (454). Reversible charge transfer observed in one case allowed the determination of the radical formation constant.

The triple bond in dicarboxyacetylene, $HOOC-C\equiv C-COOH$, shows a peculiar reduction mechanism; the radical generated in the first step immediately decarboxylates to give $CH_2=\overset{\cdot}{C}-COOH$ with subsequent dimerization and complete reduction of the two double bonds of the dimer (384).

The behavior of compounds containing double bonds in resonance with aromatic ring systems resembles that of polycyclic compounds. The ease of reducibility in these cases is attributable to the extended π-electronic system and consequently a low energy of the lowest vacant molecular orbital, as has been calculated for stilbenes (132).

B. Compounds Containing C=N or N=N

Hydrazones such as $R_2C=N-NHR'$ (375), as well as numerous azomethine compounds of different types (488), were shown to be reduced with the transfer of two electrons, although in the former case a potential-determining one-electron step was suggested. A rate-determining one-electron transfer and a radical intermediate have also been suggested for the reduction of dioximes in aqueous solutions (143, 419).

Two separate one-electron waves, however, were observed with benzalaniline, $\phi-CH=N-\phi$, in acidic aqueous solutions, and the generation of a radical during the first wave was postulated (194). Some similarities in the reduction mechanism to that of benzaldehyde have been noted; after merging of the two waves at higher pH values, however, the behavior is different inasmuch as no decrease in limiting currents to one-electron reduction occurs. This was explained in terms of ready protonation of nitrogen as compared with oxygen atoms in the respective radicals, and the subsequent transfer of a second electron at a rate much higher than that of the dimerization. Hydrolysis of benzalaniline and related compounds (resulting in the appearance of the corresponding benzaldehyde waves) must be taken into account since hydrolysis proceeds too rapidly to permit polarographic investigation at pH < 6; it is not quite clear whether or not this has been properly considered in the work cited hereafter.

The behavior of $\phi_2C{=}N{-}\phi$ resembles that of benzalaniline described above (298). The production of radicals has been confirmed by controlled-potential electrolysis at potentials corresponding to the first wave with benzalaniline and analogous compounds substituted in the aromatic rings (448). The benzalaniline radical is also generated by a three-electron reduction of the corresponding nitrone (494),

$$\phi-CH{=}N{-}\phi$$
$$\downarrow$$
$$O$$

In dimethylformamide, benzalaniline and numerous similar Schiff bases show two one-electron waves, the first of which is almost reversible, although a stable ESR signal was detected only with one particular compound (derived from the condensation of naphthaldehyde and naphthylamine) (403); a correlation of half-wave potentials with molecular orbital calculations was also carried out in this study. Two separate one-electron waves were observed with phenylisocyanate only at high depolarizer concentrations, whereas the two-electron wave at lower concentrations was interpreted in terms of sufficient proton supply from the solvent (408). Substituted ω-diazoacetophenones show reversible one-electron reduction in dimethylformamide, the half-wave potentials being correlated with Hammett's equation (442).

Stepwise transfer of the two electrons during reduction of azobenzene even in aqueous solutions was discussed as early as 1952 (127). More detailed investigations, however, of the radical intermediates have more recently been restricted to non-aqueous solutions since in water, at least in more acidic solutions, reduction to hydrazobenzene proceeds reversibly in one two-electron wave. In dimethylformamide two one-electron waves were observed; the first is reversible, while the second, producing stable dianions, is governed by slow electron transfer (14). The ESR spectrum observed at potentials of the first wave showed more than 180 lines and could not be analyzed, whereas perdeuterated azobenzene shows a simple five-line spectrum as expected because of coupling with two equivalent nitrogen nuclei. A detailed study of different azo compounds employing different techniques has been reported for the same solvent (388); stable anion radicals and dianions were obtained at potentials of the two separate waves. On the addition of a proton donor (hydroquinone), the height of the first wave increased at the expense of the second because of direct subsequent reduction of the anion radical upon protonation; in this case the product of two-electron reduction is reoxidizable to the parent azo compound. The reduction of azopyridine dioxide (387) is more complex since the oxygen atoms are reduced first and slow protonation of the reduction products, anion radical and dianion, precedes the reduction to azopyridine and subsequent reduction of the latter to give the corresponding anion radical and dianion. ESR spectra recorded at potentials of the first polarographic wave were significantly non-symmetrical, which was explained as being attributable to the fact that radicals of the dioxide and of azopyridine are present simultaneously. Radicals are also obtained by oxidation of hydrazo compounds such as 9,9′-hydrazoacridine in acetonitrile (74); the isolated ESR-active, brown-black crystals consist of the uncharged free radicals formed according to

$$RH_2 \to RH\cdot + e^- + H^+ \tag{53}$$

whereas the anion radical is formed in alkaline methanol:

$$RH_2 \to R^- + e^- + 2H^+ \tag{54}$$

The uncharged free radical is subject to a dismutation reaction (slow in the solid state and more rapid in dimethyl sulfoxide solution) to give the corresponding azo and hydrazo compounds.

Azines, being the bases of dicationic hydrazo compounds, undergo oxidation in two separate reversible one-electron waves in aqueous solutions, producing rather stable radical intermediates (212, 213). Over a certain pH range, the first oxidation wave is pH-dependent owing to the preceding dissociation equilibria:

(55)

Certain sulfonylhydrazones, $=\overline{N}-\overline{N}-SO_2-\phi$, constitute similar systems which have intermediate uncharged free radicals with rather large formation constants (215). The stable radical diphenylpicrylhydrazyl (DPPH) was shown to give one-electron oxidation and reduction waves in aprotic media with different electrode materials (138, 416); it can also be used for solvent purity tests in polarography (170). Reference is made to further work (159, 403), also noting that radicals are involved in the electrochemistry of azo compounds.

C. Aromatic, Polycyclic, and Heterocyclic Compounds

There are many compounds, very different in chemical structure and behavior, which show remarkable similarities in electrochemical response as far as the generation of radicals is concerned. As prototypes we can consider unsaturated polycyclic hydrocarbons and heterocyclic compounds, especially those of the aromatic type. The common feature of their electrochemical behavior can be ascribed to the fact that the electron transferred upon reduction is not fixed to a certain atom or group but becomes part of the π-electron system of the whole molecule. The same is true for oxidation; the electron is released from the π-electron system. In both cases there is a certain change in electron densities throughout the molecule causing, for instance, facilitated transmission of para substituents in aromatic anion radicals because of a greater statistical weight of quinoidal structures (53). In general, however, no significant change takes place in structure or in groups of atoms responsible for chemical behavior. This, as already noted, includes even substances whose reduction (or oxidation) is usually viewed as affecting a certain electroactive group. Spin density calculations, for instance, and their correlation with corresponding coupling constants observed by ESR with the benzophenone negative ion in dimethylformamide solution (380), show that the electroactive keto group contains less than one-half of the unpaired electron which is rather spread over the two aromatic rings. The nitrobenzene negative ion generated in aprotic solvents (133, 147) may serve as another example, although there is an enhanced probability of finding the odd electron at the nitro group; the sensitivity of charge distribution to substitution in the ring and, as a result of solvation effects, to the kind of solvent as well as to the cations of supporting electrolytes reminds us that the electroactive group may well play a role and that we are dealing with charged species. Furthermore, if transfer of protons and/or transfer of another electron occurs, then the peculiarities attributable to the electroactive group become conspicuous, since protons are usually bonded to these groups and cause considerable change in the chemical conditions. This section is therefore restricted to radicals derived from aromatic and unsaturated polycyclic hydrocarbons and the heterocyclic analogs, whereas radicals derived from compounds containing characteristic electroactive groups such as C=O, NO_2, and so on, are dealt with in separate sections.

As early as 1942, it was suggested that the reductions of polycyclic hydrocarbons proceed by a radical mechanism (280, 458). Almost at the same time, a similar conclusion was drawn with respect to heterocyclic systems (446). A recently published review (358) notes some of the more important aspects in this field, including the close relationship between polarographic half-wave potentials and theoretically calculated energies of lowest vacant molecular orbitals (185, 344, 479). Another correlation has been established for polycyclic hydrocarbons between polarographic behavior and the ability to act as organic semiconductors (359).

As mentioned above, a significant feature of the corresponding radicals is the delocalization of the odd electron, from which a certain stability results, for instance, toward dimerization which requires localization of the electron (161), as well as toward protonation and subsequent charge transfer. The stability of cation radicals is enhanced by electron-donating substituents (159), while electron-withdrawing groups may produce more stable anion radicals, both effects being the result of a reduction in the free energies. However, certain substituents, as well as the replacement of carbon atoms by hetero atoms or, in general, even a decrease in molecular symmetry, may reduce the uniform distribution of electrons and facilitate stabilization reactions. Briefly, stability is closely connected, next to proton availability, to the particular structure. Thus the cation and anion radicals of 9,10-diphenylanthracene (357, 412) are relatively stable, whereas the cation radicals of anthracene and some similar hydrocarbons show lifetimes only of the order of several milliseconds, as has been found from cyclic voltammetry at different scan rates (357).

Dimerization has been observed with triphenylmethyl (113, 462) and, at even higher rates, with triphenylcyclopropenyl (49) radicals. Furthermore, there are quite different reasons for dimerization, for instance, the insolubility of the dimer, as has been suggested in the case of phenanthrene anion radicals (358, 463). Dimerization also takes place with the cation radicals formed during oxidation (at platinum electrodes in acetonitrile) of 9-arylaminoanthracenes if the 10-position is not substituted, whereas the corresponding 10-phenyl derivatives form cation radicals stable toward dimerization (73); ESR spectra show that the proton at the amino group is retained and cation radicals are formed. Dimerization of reduction and oxidation products of heterocyclic compounds has been observed, for instance, with pyridine (420) and its derivatives (286, 482), pyrimidine (414), pyrylium cations (118) and, presumably, isobenzpyrylium cations (450).

In aprotic solvents and under such conditions that dimerization is absent (or sufficiently slow, or the equilibrium is shifted largely to the monomeric radicals), a reversible one-electron transfer is often observed resulting in relatively stable (ion) radicals; in numerous cases even the second electron is transferred reversibly according to Eq. 15. The current–voltage curves correspond to Eq. 6, the half-wave potentials giving the standard redox potentials which in turn constitute an electrochemical series similar to that for metal/metal cation couples. A close correlation therefore exists between these potentials and the equilibrium constants of reactions such as

$$X + Y^- \rightleftharpoons X^- + Y \qquad (56)$$

where X and Y stand for two different hydrocarbons (348, 358). Interactions of both the hydrocarbon and its corresponding radical with constituents of the solution, namely, solvation, ion pair formation, and formation of charge transfer complexes, may significantly affect these potentials (358). Close correlations have repeatedly been observed between such changes in reduction potentials and ESR characteristics, indicating changes in electron distribution in the radicals.

Two reversible one-electron waves without participation of protons in a solvent of medium proton availability are scarcely observable [cf. fulvene (441) in 1:3 water–dioxane], since the availability of protons in general significantly affects electrochemical behavior. Some phenomena frequently observed upon addition of proton donors (such as phenol, benzoic acid, and water) to aprotic solvents, differing from one another because of different mechanisms, have been listed (358). If protonation follows the transfer of the second electron, removing R^{-2} from the equilibrium to give RH^- or even RH_2, the second wave is in general rendered irreversible and shifted to more positive potentials with increasing donor concentration (460); cyclic voltammetry in this case indicates irreversibility of the second wave, but a new peak may be observed during the anodic half-cycle corresponding to the couple RH^-/RH rather than R^{-2}/R^- (101). Theoretically well established is the protonation of the anion radical R^- with immediate subsequent reduction of RH, which results in an increase in the first wave at the expense of the second; a shift to more positive potentials upon further addition of proton donor occurs once the first wave has attained the full height of the former two waves. This is in accordance with molecular orbital calculations

(185) of the relative potentials of the couples R/R^- and RH/RH^- as mentioned in Section II-E. This theory, derived for alternant hydrocarbons, also applies to other compounds; there are, however, marked restrictions as to a more general application. This may be attributable (155) to the nonvalidity of the implied relative amounts of energies of lowest vacant molecular orbitals as a result of structural factors, as well as to pecularities in the prevailing mechanisms. The latter applies, for instance, when kinetics rather than thermodynamics predominate; a strongly limited rate of protonation of R^- may prevent RH from being generated even if protons are available. However, rapid dimerization may compete with proton transfer and prevent further reduction. Another source of the nonapplicability of this theory involves the preprotonation of the species R to give RH^+ which is reduced at more positive potentials. Hence a new wave at less negative potentials occurs. Whether this is a one-electron wave or implies the transfer of more electrons depends upon the particular conditions. If the potential is too positive to allow the reduction of RH, the radical may either be stable or dimerize; the radical may also undergo other chemical reactions (e.g., dismutation) which result in the transfer of further electrons. Apart from other systems such as carbonyl compounds or quinones, preprotonation is more common with basic heterocyclics than with hydrocarbons; however, the reductions of tropyl alcohol and azulene proceed with tropylium and azulenium ions, respectively, as discussed below. Moreover, proton transfer producing RH_2 instead of R^{-2} may give rise to a reduction exceeding the two-electron stage, reflected by additional waves or increased wave heights, as has been observed with azulene (79, 154).

For organic polarography in aqueous solutions, it has long been agreed that buffered solutions are in general preferable in order to avoid complications resulting from depletion of H^+ ions frequently observed in earlier polarographic work. With aprotic solvents, however, this problem seems to have been recognized only recently and has been handled (360) by using acid/base couples as buffers down to proton activities as low as pH 30, instead of the more customary addition of different amounts of proton donors. In consequence, a more systematic investigation of processes involving proton transfer has been effected, for example, for the reduction of naphthacene in buffered tetrahydrofuran solutions. Current–voltage curves show that, even at pH 16, protons are readily supplied by the buffer acid to protonate the anion radical and effect immediate reduction to the dihydro compound, whereas at pH 30 the anion radical and dianion are generated in separated waves. From the plot of half-wave potentials versus pH values, the relative stabilities of the different species are illustrated. It is, inter alia, concluded that the anion radical would not be stable in solutions with pH < 28 if thermodynamic equilibria were established, but would evolve hydrogen according to

$$R^- + H^+ \rightarrow R + \tfrac{1}{2}H_2 \tag{57}$$

After these more general considerations, we now examine more recent results with some particular compounds. Cyclooctatetraene was found in earlier studies to be reduced with the transfer of two electrons in organic solvents containing water, as well as by reduction with alkali metals in aprotic solvents (248, 435). Apart from proton transfer facilitating the transfer of a second electron, this was explained on the basis of a dismutation equilibrium (Eq. 35) largely shifted to the right, although ESR spectra of the anion radical have been obtained. In a more recent investigation (4) applying different techniques, namely, temperature dependence of ordinary polarograms, ac polarography, cyclic voltammetry, and ESR, it was shown that ion pairing with alkali metal cations of the dianion is probably the driving force in dismutation, whereas in the absence of these cations the anion radical is stable toward dismutation (in dimethylformamide). Peculiarities in the electrochemical behavior (e.g., a transfer coefficient significantly smaller than 0.5) combined with energy considerations (similarity of free energy of activation to the energy needed to flatten the molecule) suggested that the conformation of the parent molecule plays an important role in the first electron transfer reaction; the conformation of the transition state presumably resembles that of the flat anion radical. The likewise nonplanar dibenzcyclooctatetraene also shows two separate one-electron waves in aprotic solvents, and ESR investigations indicate delocalization of the odd electron in the anion radical (249).

Azulene was shown to be reduced in an aqueous-organic solvent (79) and in dimethylformamide (154) in the first wave to give the anion radical, which is reduced at more negative potentials with the transfer of one or more electrons according to proton availability. At very low pH values in water, a new one-electron wave is observed at rather positive potentials, which has been ascribed to the reduction of the azulenium ion as a result of preprotonation (373, 489):

$$\underset{C_{10}H_8}{\text{[structure]}} \longleftrightarrow \underset{}{\text{[structure]}} \overset{+H^+}{\rightleftharpoons} \underset{C_{10}H_9^+}{\text{[structure]}} \overset{+e^-}{\longrightarrow} \underset{C_{10}H_9}{\text{[structure]}} \tag{58}$$

Preprotonation is also involved in the reduction of tropylium ions representing protonated tropyl alcohol (480, 481, 486, 491); the radical dimerizes to give ditropyl:

$$\text{tropyl alcohol} + H^+ \rightleftharpoons \text{tropylium}^+ + H_2O \quad (59)$$

$$\xrightarrow{+e^-} \text{tropyl radical} \longrightarrow \tfrac{1}{2}\,\text{ditropyl}$$

Reduction of tropylium cations ($E_{1/2} = -0.27$ V versus SCE) proceeds at potentials almost 2 V more positive than that of the alcohol (-2.09 V). The height of the wave is determined by the kinetics of preprotonation and decreases at pH > 4 in the form of a dissociation curve; the corresponding rate constants have been determined (491). Instead of the mechanism shown above, anionic dimerization according to Eq. 52 (with R corresponding to the cation) has also been discussed. This involves a two-electron transfer, but the wave height corresponds to one electron per molecule; experimental data, however, seem to favor the radical mechanism, although a reoxidation current of the radicals was observed neither with cyclic voltammetry nor with the Kalousek technique, indicating rapid dimerization with lifetimes smaller than 1 msec (264). Complications in current–voltage curves arise from an inhibiting effect resulting from the adsorption of products, namely, radicals or, more probably, the dimer (480). At higher concentrations of sulfuric acid, complications (cathodic shifts) arise from complexation of tropylium cations with sulfate anions (210). A comprehensive review of the electrochemistry of nonbenzenoid aromatic compounds has been published in Russian (486).

Preprotonation frequently takes place with basic heterocyclic compounds; however, the corresponding cations may show rather different behavior under different conditions. Thus the pyridinium ion in a solvent that is essentially water (e.g., pyridine with >50% water) only catalyzes the reduction of H^+ ions (via the discharge of the cation and liberation of hydrogen from the adsorbed radical), whereas at low concentrations of water cleavage of the ring takes place (126). In pure pyridine, with only a low concentration of proton donors to give the pyridinium cation, the latter reduces to the radical which subsequently dimerizes in ortho or para position (420, 439). In the absence of proton donors but in the presence of aluminum chloride, the radical with seven electrons in the π-electron system is also generated and stabilized by complexation with aluminum(III); subsequent hydrolysis in the presence of water is followed by opening of the ring and production of polymer products (81). The formation of a radical and its reaction with mercury presumably proceeds during the reduction of 3-cyanopyridine, in contrast with the behavior of isomeric compounds (230).

N-alkylpyridinium cations, not capable of catalyzing H^+-ion reduction, are reduced to the corresponding radicals even in aqueous solution (446); similarly, uncharged radicals form during the reduction of diphosphopyridine nucleotide (cozymase) and model compounds derived from N-substituted nicotinamide cations in aqueous solution (40). The N, N'-dialkylbipyridinium dication is reversibly reduced to give the corresponding cation radical, which is then irreversibly reduced with the transfer of a second electron at more negative potentials (107); the same also applies to bipyridine (478).

Studies of compounds containing two pyridinium rings with systematic variation in the structure (separation of the two pyridine rings by condensed benzene rings, by the group CH=CH, and so on) have been made in non-aqueous solvents and in aqueous media (214); transfer of two electrons in two reversible waves (corresponding to the system dication/cation radical/reduced quinoid molecule) separated enough for the evaluation of formation constants (Eq. 18) of the cation radicals has been found in several cases.

The behavior of N-cyclopentadienylpyridinium ion resembles that of tropylium and azulenium ions (482). In acidic solutions one electron is transferred to give the uncharged radical ($E_{1/2} = -0.91$ V versus SCE); this wave decreases in height at higher pH values, since protonation of the uncharged species $C_5H_5N-C_5H_4$ becomes rate determining. The latter species is formed by releasing H^+ from the five-membered ring of the cation; it is itself capable of reduction as well as oxidation with the transfer of one electron to give the corresponding anion radical (-1.66 V) and cation radical (-0.05 V), respectively, both of which undergo dimerization.

Pyrazine, in a mixture of water and dimethylformamide (50:50) buffered with acetic acid and sodium acetate, shows two moderately separated waves; each involves the reversible transfer of one electron and one proton, and they give rise to the uncharged radical and then the dihydro compound. The radical exhibits considerable stability in this medium, the formation constant (cf. Eq. 18 with RH and RH_2 instead of R^- and R^{-2}) amounting to about 10^3 at room temperature (168). The corresponding radical of pyrimidine was shown to dimerize in acidic aqueous solutions (414).

Porphyrin complexes in non-aqueous solvents show two one-electron reduction waves, the monoanion produced at the first wave showing an ESR spectrum (119); evidence indicates that both electrons are transferred to orbitals belonging to the porphyrin ring system rather than to a metal-centered orbital, except for the cobalt compound. Decay kinetics of radicals generated by oxidation of 1-phenyl-3-pyrazolidone were followed with ESR (69).

Substituted quinolines have been studied in dimethylformamide. The 8-hydroxy compound (135) shows a first wave attributable to hydrogen evolution resulting from the acid hydrogen. For the second wave, resulting from reduction of the pyridine ring of the remaining phenolate anion, a somewhat doubtful mechanism including the participation of hydrogen atoms as intermediates has been suggested:

$$RH + e^- \rightarrow R^- + H$$
$$R^- + e^- \rightleftharpoons R^{-2} \qquad (60)$$
$$R^{-2} + H \rightarrow RH^{-2} \rightarrow Products$$

The first wave of the 6-chloro compound in anhydrous dimethylformamide (136, 137) is ascribed to two-electron reduction with the elimination of chloride ion, the resulting anion abstracting a proton from the solvent to give quinoline; the latter is reduced in a second wave to the anion radical which may—by proton transfer from traces of water—undergo immediate subsequent reduction or, by direct electron transfer, show a third wave to give the corresponding dianion. In the presence of water or proton donors the mechanism is more involved and no chloride ion elimination occurs.

In aqueous solution the reduction of N-phenylisoindoline, substituted in the five-membered ring (409), is interpreted in terms of electron transfer to the isoindolinium cation with subsequent transfer of another proton and electron. Protonation preceding the first electron transfer also takes place with phthalimide in aqueous solutions (386) [cf. also reference (444)]; a second proton is transferred to the uncharged radical, followed by immediate subsequent reduction, thus showing a single two-electron wave. At pH > 4 the transfer of the second proton becomes rate-determining, and the limiting current decreases to one-electron reduction. There is, however, a certain influence of the depolarizer concentration, suggesting that dimerization is also involved. In dimethylformamide, phthalimide exhibits a reversible one-electron wave (413); the corresponding anion radical has been investigated with the ESR technique. In this work an ESR spectrum reported earlier (379), which was composed of contributions of more than one radical, was interpreted in terms of proton rearrangement within the anion radical (with the participation of an unreduced depolarizer molecule as proton donor):

(61)

More familiar is the occurrence of different ESR spectra when a radical is generated in different stages of protonation. Thus phenazine in acetonitrile is reduced to the anion radical R^-, whereas the dication RH_2^{2+} is reduced to the cation radical RH_2^+ which is also produced by oxidation of dihydrophenazine RH_2 (177). The anion radical has also been generated in dimethylformamide, in which two one-electron waves occur for phenazine and several derivatives (158); the corresponding ESR spectra have been investigated (54). The two well-separated one-electron waves observed in aqueous acidic solutions have also been interpreted in terms of a semiquinonelike structure as the product of the first wave (251); this explanation has been rejected in favor of the generation of a molecular compound similar to quinhydrone in a more recent study based upon results of controlled-potential electrolysis (336); the argumentation is, however, not quite clear, partly because of grammatical difficulties. Acridicinium cations also show two one-electron waves in acidic aqueous medium, the first of which is reversible (129); because of the quaternary structure of the nitrogen atom, there is no chance of protonation prior to the transfer of the second electron, both waves consequently being pH independent. However, the corresponding waves of acridine exhibit pH dependence since the tertiary bonded nitrogen is capable of protonation (250). The behavior of this compound in non-aqueous solvents with the possibility of obtaining, according to the conditions, uncharged or anionic radicals, was mentioned above (74, 75). Pyrylium ions, similar to acridicinium ions, exhibit a pH-independent wave in acidic aqueous solutions; the radical undergoes dimerization, but the dimer can be reoxidized to pyrylium ions (118). Benzo- and naphthofurazans undergo one-electron reductions in dimethylformamide; in aqueous solution the wave height corresponds to the transfer of six electrons, but the first one-electron step is probably rate determining (293). Analogous behavior has also been found with the sulfur and selenium isologs (piazthiazol, piazselenazol) (487); the formation of the anion radical in dimethylformamide was shown to be reversible by application of ac polarography, cyclic voltammetry, and the Kalousek method. On addition of water or sulfuric acid, the height of this wave increases as a result of protonation of the anion radical with subsequent reduction; sulfuric acid, moreover, produces a new wave at more positive potentials, which has been ascribed to the reduction of the protonated depolarizer.

Dibenzofuran as well as its sulfur and selenium isologs (dibenzothiophene, dibenzoselenophene) show one-electron waves in dimethylformamide; the ESR spectra of the anion radicals have been recorded, and correlations of shifts in half-wave potentials upon substitution of methyl groups in the six-membered ring with molecular orbital calculations have been shown to correspond to an inductive effect (145). Polarographic as well as ESR investigations have been performed with several N-heterocyclics (150), as well as with the compound

$$\underset{S\text{———}S}{\overset{CH=CH}{|}}\!\!\diagdown\!\!\underset{\diagup}{CH\text{—}SH} \xrightarrow{+H^+} \underset{S\text{———}S}{\overset{CH=CH}{|}}\!\!\diagdown\!\!\underset{\diagup}{CH\text{—}SH_2^+} \xrightarrow{+e^-} \quad (62)$$

and similar substances (12) in aprotic media; the results were correlated with molecular orbital parameters. Phenothiazine is oxidized in acetonitrile at platinum electrodes in two reversible one-electron waves to the cation radical and dication, the waves being separated by about 0.3 V (43); the ESR spectrum of the cation radical has been observed [cf. also reference (152)]. On addition of water the two waves merge because of a decreased formation constant of the radical and a consequent tendency to dismutate (44). In a highly acidic aqueous solution (3.5 M sulfuric acid), however, a reversible oxidation wave to give the cation radical is observed as well (260).

Reversible reduction to a radical has been observed with a dithiolium cation in nitromethane as a solvent (456). Trithion in aqueous solutions (400) is first reduced, presumably via a radical, to give the corresponding mercaptan:

$$\underset{S\text{———}S}{\overset{CH=CH}{|}}\!\!\diagdown\!\!\underset{\diagup}{CH\text{—}\overset{\cdot}{S}H_2} \longrightarrow \underset{S\text{———}S}{\overset{CH=CH}{|}}\!\!\diagdown\!\!\underset{\diagup}{CH\text{—}SH} + \tfrac{1}{2}H_2$$

which, in a second wave, catalyzes hydrogen evolution via an adsorbed radical:

$$\begin{array}{c}\diagup\!\!\!\diagdown\\\diagdown\!\!\!\diagup\end{array}\!\!\!\!\overset{N}{\underset{N}{\diagdown\!\!\diagup}}\!\!X \quad (X = O, S, Se) \quad\quad\quad (63)$$

Phenyl-substituted borazines (410) are reversibly reduced in dimethylformamide to the anion radicals which disappear according to first-order kinetics and with first-order rate constants of about 40 sec^{-1}. Because of extended distribution of the electron throughout the molecule, reduction takes place at potentials at which neither borazine nor benzene is reduced; half-wave potentials have been correlated with energies of the lowest vacant molecular orbitals.

D. Quinones and Related Compounds

Early investigations with quinone/hydroquinone systems, well known to show stable semiquinone radical intermediates, have been mentioned in Section I. As with other types of compounds, the first studies were mostly carried out in aqueous solutions. When the application of organic solvents became more familiar in polarography, well-separated one-electron waves were more frequently shown to correspond to the formation of semiquinone anion radicals and hydroquinone dianions, both waves often indicating reversible electrode reactions (105, 153, 461). The first application of the ESR method to electrochemically generated radicals also seems to have involved benzo- and anthrasemiquinones (13); electrolyzed dimethylformamide solutions were chilled with liquid nitrogen and portions of them transferred to the ESR cavity. The two more convenient techniques (cf. Section I) have also been applied to semiquinones. More recent ESR investigations concerned, for example: benzocyclobutadienoquinone which is reversibly reduced in acetonitrile to its semiquinone, the odd electron of the latter essentially assembling within the six-membered ring (148); anthraquinone sulfonic acids (51); quinones from vitamins K and E (128); as well as the application of the ^{17}O isotope in order to extend the conclusions being drawn from hyperfine coupling constants (58, 166).

Several further investigations in aprotic solvents can be mentioned here. Thus anthraquinones frequently show two reversible one-electron waves in dimethylformamide; the influence of substituents on half-wave potentials, occasionally attributable to intramolecular hydrogen bonding, has been discussed and correlated with Hammett's constants (52, 477). Quinone derivatives of triptycene have been studied in the same solvent, and semiquinone formation constants determined (316). With halogen-substituted anthraquinones, semi- and hydroquinone formation proceeds prior to the reduction of the halogen atom in dimethylformamide and buffered methanolic solutions (42). Homogeneous equilibria of the type shown in Eq. 56 were determined in acetonitrile with quinones and tetracyanoquinodimethane by light absorption measurements and correlated with the corresponding half-wave potentials (217).

The interaction of quinones with electron-donating molecules leads to shifts in half-wave potentials (356). Charge transfer between acceptor (A) and donor (D) molecules giving rise to the corresponding anion and cation radicals:

$$A + D \rightarrow A^- + D^+ \tag{64}$$

as well as to molecular complexes:

$$A + D \rightarrow A \cdot D \tag{65}$$

are reflected in the polarographic curves, and the corresponding equilibrium constants may be calculated. A correlation also exists between the reduction and oxidation potentials of the two molecules and the wavelength (or energy $h\nu$) of light absorption corresponding to the charge transfer:

$$A \cdot D \xrightarrow{h\nu} A^- + D^+ \tag{66}$$

Furthermore, ion pairing of semiquinone anions as well as of hydroquinone dianions is indicated by shifts in half-wave potentials; the second wave is markedly influenced by the presence of monovalent cations with small radii, whereas the first wave is sensitive to additions of divalent cations.

Through the availability of protons upon the addition of water or other proton donors to the organic solvent, effects are brought about which have been discussed in previous sections. In general, the second wave is shifted to more positive potentials because of the protonation of the hydroquinone dianion, until both waves merge, indicating reversible two-electron transfer which may also be considered in terms of an easy dismutation reaction of the semiquinone. This has been observed with different aromatic quinones in dimethylformamide and acetonitrile (461), with anthraquinones (292, 263) and benzoquinone (95) in dimethylformamide, with stilbenequinone (154), and with tetracyanoquinodimethane (1, 325) in comparisons of behavior in aprotic with that in acidic media. Liquid $LiNO_3 \cdot NH_3$ as a solvent seems to be an adequate source of protons since the formation of semiquinones is not indicated by the polarographic curves in this medium (498). However, the hydroquinones, QH_2, are not oxidizable at mercury electrodes (63) except in the presence of water (461) which facilitates dissociation to give the anion, QH^-. Occasionally, there is also an increase in the first reduction wave as a result of protonation of the semiquinone anion to give QH with fast subsequent reduction of the latter; such behavior was observed with anthraquinone in dimethylformamide containing more than 20% water (449).

The change in half-wave potentials of the first polarographic wave of benzoquinones (355) when acetonitrile is replaced by aqueous methanol as a solvent is more pronounced than the corresponding energy of light absorbed to effect a charge transfer according to Eq. 66, with hexamethylbenzene as an electron donor. This is explained by the fact that the difference in solvation and hydrogen bonding in these two solvents is entirely reflected in the half-wave potentials, which correspond to thermodynamic equilibrium, whereas only part of this difference affects the charge transfer (Eq. 66), which is too fast as to allow reorientation of solvent molecules. The influence of substituents is discussed in terms of changes in both electron density and solvation, both of which affect hydrogen bonding.

Several quinones and related compounds show one-electron waves even in aqueous solutions of higher pH values. Such behavior has been observed with, among others: naphthoquinone (389); substituted anthraquinones (in 50% aqueous ethanol, whereas one two-electron wave is observed in pure water) (332); halogen-substituted fluoresceins (the formation constants of which depend on the concentration of Na^+ ions present, as a result of an influence of the latter on the hydration of the semiquinone anion) (67); fluorescein itself (16, 157) [the half-wave potential of which is shifted to more positive values upon irradiation with light owing to the formation of excited molecules (35)]; and the α,β-diketone camphorquinone (330). The radical of the latter compound was shown by oscillopolarography to dimerize in alkaline solutions (45); a dismutation reaction has also been excluded by electrolytic production of the radical and subsequent homogeneous decay; no increase in the reduction wave of camphorquinone was observed during the decrease in the oxidation wave of its radical (240). Second-order decay kinetics of radicals of diketones generated by nonelectrochemical methods [namely, reduction with palladium sol (30) and by flash photolysis (36)] have also been studied using the decrease in the corresponding oxidation waves.

The radical generated by the reduction of benzil is in equilibrium with benzil and benzoin (27), the latter being formed through intermediate stilbenediol (22, 426, 453). The intermediate radical was shown by oscillopolarography to be reoxidizable even in strongly acidic media (227). In aprotic solvents the generation of the anion radical is obvious from the appearance of two separate one-electron waves and the observed ESR signal (366, 367), although when lithium ion is present the reduction involves two electrons, is reversible, and no ESR signal is observed, obviously because of ion pairing of the dianion.

The semiquinone anion of anthraquinone-1,5-disulfonic acid is less stable than the isomeric 2,6-disulfonic acid (28); for the former compound, moreover, the electrode reaction proceeds irreversibly for pH < 11, reduction and oxidation potentials of quinone and hydroquinone differing by several tenths of a volt. This is explained in terms of intramolecular hydrogen bonding with the sulfonate group.

For quinones, diketones, and simple ketones, there are correlations between the half-wave potentials, corresponding to the generation of radicals with the destruction of the C=O π-bonding, and the long-wavelength band of the electronic absorption spectra corresponding to an n–π^* transition and leading to an excited molecule with an electronic biradical structure (32), [cf. also reference (59)].

A review on the polarography of quinones was published several years ago (34); some results obtained in dimethyl sulfoxide have been summarized in a general review on polarography in this solvent (63).

E. Carbonyl Compounds

This section concerns saturated carbonyl compounds, especially aldehydes and ketones, including cyclic (mostly aromatic) ones. Unsaturated compounds, however, including those in which the carbon atom of the C=O group belongs to an unsaturated ring system (e.g., tropone), are considered in the following section. This accounts for a significant difference in mechanisms applying to these two groups of compounds, although a strict distinction is not invariably valid. As already seen in the preceding section, the behavior of α,β-diketones resembles that of quinones in many respects, but the former are also capable of undergoing reactions having more in common with simple carbonyl compounds. Similarly, the behavior of a compound such as p-diacetylbenzene, as seen later, resembles that of quinones or α,β-diketones rather than that of a bifunctional ketone with separated C=O groups.

The principles of the reduction mechanism of saturated aldehydes and ketones applying in most cases, at least for media of high proton availability such as water, were established in early investigations (10, 191, 346). Including more recent insights into the mechanism, this can be formulated as shown in the scheme on page 231. Preprotonation of the carbonyl group, in general being shifted largely in favor of C=O rather than $\overset{+}{\text{C=OH}}$, has been claimed to give rise to the pH-dependence of the first wave in early studies (108) and subsequently has been dealt with by several investigators (112, 155, 232, 304, 349, 373, 451, 438). Under these conditions the mechanism corresponds to that shown in Eq. 44 and the half-wave potentials of the two waves, E_I and E_{II}, are given by Eqs. 41 and 43, respectively, E_{0I} being pH-dependent according to Eq. 32; when dimerization proceeds as a heterogeneous reaction (307), the half-wave potential of the first wave is given by Eq. 46. At low pH values the second wave may be masked by the current rise due to hydrogen evolution. Even preprotonation itself, however, may proceed as a heterogeneous reaction rather than as a homogeneous equilibrium (112) [cf. also reference (233)]; in this case both reoxidation and dimerization of the free radical may

$$
\begin{array}{c}
\frac{1}{2}\begin{array}{c}\diagdown\\\diagup\end{array}\!\!C\!-\!OH\\|\\\begin{array}{c}\diagdown\\\diagup\end{array}\!\!C\!-\!OH\\\uparrow H^+\\\frac{1}{2}\begin{array}{c}\diagdown\\\diagup\end{array}\!\!C\!-\!O^-\\|\\\begin{array}{c}\diagdown\\\diagup\end{array}\!\!C\!-\!O^-\\\uparrow
\end{array}
\qquad
\begin{array}{c}\diagdown\\ \diagup\end{array}\!\!\overset{H}{C}\!-\!OH\\\uparrow 2H^+
$$

$$
\begin{array}{c}\diagdown\\\diagup\end{array}\!\!C\!=\!O \xrightarrow[E_{III}]{+e^-} \begin{array}{c}\diagdown\\\diagup\end{array}\!\!\overset{\bullet}{C}\!-\!O^- \xrightarrow[E_{IV}]{+e^-} \begin{array}{c}\diagdown\\\diagup\end{array}\!\!\overset{-}{C}\!-\!O^-
$$

$$H^+ \updownarrow \qquad\qquad\qquad H^+ \downarrow$$

$$
\begin{array}{c}\diagdown\\\diagup\end{array}\!\!C\!=\!\overset{+}{O}H \xrightarrow[E_I]{+e^-} \begin{array}{c}\diagdown\\\diagup\end{array}\!\!\overset{\bullet}{C}\!-\!OH \xrightarrow[E_{II}]{+e^-} \begin{array}{c}\diagdown\\\diagup\end{array}\!\!\overset{-}{C}\!-\!OH
$$

$$
\downarrow \qquad\qquad\qquad H^+ \downarrow
$$

$$
\frac{1}{2}\begin{array}{c}\diagdown\\\diagup\end{array}\!\!C\!-\!OH\\|\\\begin{array}{c}\diagdown\\\diagup\end{array}\!\!C\!-\!OH
\qquad
\begin{array}{c}\diagdown\\\diagup\end{array}\!\!\overset{H}{C}\!-\!OH
$$

(67)

proceed at comparable rates, and none of the equations that assume only one rate-determining step to be involved fit the experimental current–voltage characteristics (438). The heterogeneous protonation may also be rate determining and give rise to limiting currents controlled by its kinetics rather than by diffusion (304). Another step preceding charge transfer may be the dehydration of the carbonyl group, which has been observed with, for example, formaldehyde (48), ninhydrin (198), and other compounds (349). When the rate-determining step governing the potential of the first wave is dimerization and this reaction is not too fast, the reversibility of the preceding equilibrium [Eq. 44, as has been established with xanthone in acidic solutions (474)] can be verified by rapid methods, such as cyclic voltammetry (471) or faradaic impedance measurements (350), whereas the benzaldehyde radical was shown to have too short a lifetime to be detected by chronopotentiometry (117). Half-wave potentials, E_I, corresponding to pH = 0, have been compiled for a series of substituted benzaldehydes (192); the applicability of the theoretical relations for the half-wave potentials, as mentioned above, has been proved in several studies (281, 361, 428, 429, 471).

When the second wave, E_{II}, being masked at low pH values by hydrogen evolution, was observed to become just visible in more weakly acidic solutions close to the final current rise (which is attributable to: $H^+ + e^- \rightarrow H \rightarrow \frac{1}{2}H_2$), it was thought that reduction of the free radical was effected by hydrogen atoms rather than by electron transfer (308). This was concluded from the fact that the current of the second wave, corrected for the reduction of hydrogen ion, was too small and decreased with more negative potentials, at which the rate of reduction of hydrogen ion and accordingly the steady-state concentration of adsorbed hydrogen atoms increased. Since hydrogen ion reduction currents have been taken from those of the pure supporting electrolyte, it is not clear whether or not a possible influence on this reaction of depolarizer reduction products adsorbed at the electrode has been taken into account.

Adsorption plays an important role in the total mechanism. Both preprotonation and the dimerization of the free radical by heterogeneous mechanisms have already been mentioned. The negative shift in the second wave with increasing depolarizer concentration, which is observed at higher concentrations and is contrary to expectations based on Eq. 41, has been attributed to an autoinhibition effect resulting from product adsorption (281, 282). The same explanation accounts for the increase in reversibility when the alcohol content of the solvent is increased, although this is accompanied by a decrease in the rate of transfer of a second electron, resulting from radical desorption (472) [cf. also references (305), (399), and (438)]. Peculiarities observed with benzoylpyridines were also found to result from the adsorption of the depolarizer as well as that of the product (453).

As the pH value increases, preprotonation is eliminated and the reduction process begins with the transfer of an electron at E_{III} to give the anion radical which again can undergo dimerization. Protonation may also occur after the electron transfer step (108, 191) and, since E_{III} is less negative than E_{II}, immediate reduction of the free radical takes place; this protonation again proceeds as a heterogeneous reaction, as has been concluded from recent results showing that this reaction path is partly or totally inhibited in the presence of surface-active agents (244). The rate of protonation decreases in more alkaline solutions and, since dimerization becomes predominant, the current drops to one-electron reduction (10, 41, 191, 346, 361).

Since dimerization is often very rapid, fast methods may be necessary to prove the reversibility of electron transfer at E_{III} (19). Half-wave potentials E_{III}, becoming constant in alkaline solutions in which dimerization predominates, have been determined for a series of substituted benzaldehydes (192) and correlated with Hammett's substituent constants (19, 160). Equation 41 also applies to this wave at E_{III}, as confirmed by a study of the dependence on concentration as well as on drop time (281).

Direct electron transfer to the anion radical, to give the dianion, in general proceeds only at very negative potentials E_{IV}. This is shifted to more positive potentials in the presence of certain cations (10, 232), obviously as the result of ion pairing facilitating the reduction process.

The rate of dimerization of the anion radical R^- shows considerable differences depending on the chemical structure of the molecule. Thus the benzaldehyde anion radical dimerizes very rapidly in aqueous solution, and even in aprotic solvents its ESR spectrum seems to be troublesome to observe (427), whereas other anion radicals are stable enough to be detected and studied even in aqueous media. With oscillopolarographic methods, allowing rapid reoxidation of short-lived intermediates, incisions have been observed in alkaline solutions with substituted benzaldehydes (19) and with formaldehyde (452), showing that the transfer of the first electron is almost reversible under these conditions. Anodic waves at more positive potentials (-0.5 to -0.7 V versus SCE), however, observed upon controlled-potential electrolysis in alkaline solutions (240), do not correspond to radical reoxidation. The origin of these waves is not yet quite clear. With benzophenone, the wave might be attributable to the oxidation of the product of dimerization, benzopinacol, with the transfer of two electrons to give benzophenone; irreversible anodic waves of benzopinacol were reported not to be preceded by a dissociation equilibrium of the dimer, as concluded from half-wave potentials that do not depend on concentration (254). Other investigators, however, could not confirm these observations and believe that benzopinacol is not oxidizable (328). Also, the corresponding product of the reduction of benzaldehyde, hydrobenzoin, does not show polarographic oxidation (240). An alternative explanation of the anodic wave observed upon the reduction of benzaldehyde, the height of which does not exceed about 10% of the original cathodic wave of benzaldehyde and decreases with time according to first-order kinetics and with a first-order rate constant of 0.0017 sec^{-1}, has been suggested (240), assuming C—O rather than C—C coupling of the anion radicals produced, according to

$$\underset{O^-}{\overset{H}{-\underset{|}{C}\cdot}} + \cdot O - \underset{-}{\overset{H}{\underset{|}{C}}} - \longrightarrow \underset{O^-}{\overset{H}{-\underset{|}{C}}} - O - \underset{-}{\overset{H}{\underset{|}{C}}} - \xrightarrow{+H^+} \underset{O^-}{\overset{H}{-\underset{|}{C}}} - O - \underset{H}{\overset{H}{\underset{|}{C}}} - \quad (68)$$

The semiacetal structure of this compound could account for its instability, whereas the hydrogen atom attached to the left carbon atom may become active as a hydride ion and account for the oxidizability, as observed with similar structures [cf. literature cited in reference (240)]:

$$\begin{array}{c} \text{H} \quad \text{H} \\ -\text{C}-\text{O}-\text{C}- \\ | \quad \quad | \\ \text{O}^- \quad \text{H} \end{array} \rightarrow \begin{array}{c} \text{H} \\ -\text{C}-\text{O}-\text{C}- \\ \| \quad \quad | \\ \text{O} \quad \text{H} \end{array} + \text{H}^+ + 2e^- \tag{69}$$

Another alternative explanation of the anodic wave is the formation of mercury compounds. This is supported by observations of the behavior of benzaldehyde in acidic media, in which the formation of dibenzylmercury was confirmed by large-scale electrolysis (8) and by studies with a hanging drop applying voltammetry and oscillopolarography, the corresponding anodic incisions at rather positive potentials being attributed to this compound (227) rather than to the reoxidation of radicals (451). Mercury compounds have also been observed upon electrolysis of ketones, except when the carbonyl group is attached to a benzene ring (7). The reduction of benzaldehyde in acidic media is further complicated by the transformation of hydrobenzoin to deoxybenzoin (227). Anodic incisions at rather positive potentials on oscillopolarographic curves observed with acetophenone, however, have been attributed to the desorption of depolarizer and dimer (pinacol) molecules (227).

In the study of these radicals, not only dimerization but also dismutation reactions must be taken into account; even the products of dimerization, the corresponding pinacols, undergo dismutation reactions, via the radical stage, especially in alkaline solutions. This may be represented as:

$$(\text{HR}-\text{R}^-) \rightleftharpoons \text{RH} + \text{R}^- \longrightarrow \text{RH}^- + \text{R} \atop {+\text{H}^+ \atop \text{RH}_2}} \tag{70}$$

The acid dissociation of the pinacol:

$$(\text{HR}-\text{RH}) \rightleftharpoons (\text{HR}-\text{R}^-) + \text{H}^+ \tag{71}$$

as well as that of the radical:

$$\text{RH} \rightleftharpoons \text{R}^- + \text{H}^+ \tag{72}$$

play an important role in the course of these reactions. A detailed study has been carried out with fluorenone and other ketones (162, 226), making use of different techniques including cyclic voltammetry, oscillopolarography, controlled-potential electrolysis, and ordinary polarography in addition to spectrophotometry. The pinacol can be determined from its polarographic oxidation wave. From this work the following conclusions can be drawn. Dimerization of the radical generated in acid solutions, that is, the forward reaction

$$2RH \rightleftharpoons (HR-RH) \qquad (73)$$

proceeds considerably more rapidly than its dismutation

$$2RH \rightarrow RH_2 + R \qquad (74)$$

Thus the dimer is the product under these conditions and dismutation is not observed since the dimerization equilibrium (Eq. 73) is shifted far to the right. The radical is a stronger acid ($pK_a = 9.5$ for the fluorenone radical) than the dimer ($pK_a = 13.7$ for the fluorenone pinacol.) In alkaline media the formation of monomer from the pinacol proceeds via the anion ($HR-R^-$), the subsequent dismutation reaction taking place with a radical and an anion radical as shown in Eq. 70, except at rather high pH values at which the reaction of two anion radicals becomes predominant:

$$2R^- \longrightarrow R + R^{-2} \searrow_{+2H^+} RH_2 \qquad (75)$$

because of the dissociation equilibrium, Eq. 72 then being shifted much in favor of R^-. A detailed study of rates and equilibria for dimerization and dismutation has also been carried out for the isatin radical generated by reversible one-electron reduction (259).

Apart from generation by controlled-potential electrolysis, which is applicable to short-lived radicals only by means of special techniques (237, 240, 242), homogeneous reductions of carbonyl compounds have been carried out by means of catalytic hydrogenation with palladium sol and by continuous or flash irradiation with ultraviolet or visible light (31, 38). Radicals have been detected and their decay kinetics have been followed by the polarographic oxidation currents attributable to reoxidation:

$$R^- \rightarrow R + e^- \qquad (76)$$

Catalytic reduction with palladium sol has been studied, apart from diketones and unsaturated ketones, with fluorenone (30, 33). Photoreduction has been carried out with benzophenone (26, 36), fluorenone, anthrone, and benzanthrone (36). The use of a flashlamp is more favorable than continuous irradiation, since in the latter case the sequence of reactions may become rather involved as a result of photoinduced reactions other than the original radical generation reaction, for example, with excited molecules which may compete with the dark reactions expected (29). Decay of the radicals (in the dark) in general follows second-order kinetics, as a result of dimerization and/or dismutation, as noted above; corresponding rate constants have been determined.

Apart from photolytic reactions in the bulk of the solution producing electroactive radicals, light may also directly affect electrode processes, especially by the excitation of molecules adsorbed at the electrode. Thus adsorbed carbonyl compounds with an electronegative group attached to the CO group give rise to an anodic current under illumination; oxalate ions forming the anion radical $C_2O_4^-$ are readily oxidized by the electrode to give carbon dioxide (182). Other carbonyl compounds (acetone, carbon dioxide) may serve as scavengers during light-induced cathodic formation of hydrated electrons and, at potentials more positive than their reduction potentials, suppress the corresponding cathodic photocurrents resulting from the reoxidation of the radicals ($CH_3COCH_3^-$, CO_2^-) formed (183).

Some other work is mentioned here, the essential features of which are in accordance with the mechanism described above and schematically shown in Eq. 67, namely, the reduction of: 2-acetylthiophene (70); di-2-furoylmethane (71); 2-acetylfuran (72); substituted benzaldehydes (92), nitroacetophenone and nitrobenzaldehyde (290), in which reduction of the nitro group precedes that of the carbonyl group [cf. also reference (192)]; benzanthrone (262); a tropoid ketone (333); and terephthalaldehydic acid (339). With the last of these compounds, and also with hydroxybenzaldehydes, complications arise from the dissociation of the acid hydrogen (192).

Some unusual mechanisms are reported in the following discussion.

A new wave in addition to the usual benzaldehyde waves has been observed with p-benzaldehyde trimethylammonium ions in neutral and slightly alkaline solutions at rather negative potentials (-1.7 V versus SCE) (20); the ratio of the wave height to the concentration increased with dilution, and the wave was attributed to two-electron reduction with the elimination of trimethylamine as follows (R denoting the —C_6H_4—CHO portion of the molecule):

$$R-N(CH_3)_3^+ + e^- \rightarrow R\cdot + N(CH_3)_3$$
$$R\cdot + e^- + H^+ \rightarrow RH \tag{77}$$

or, according to the concentration,

$$2R\cdot \rightarrow R_2$$

Bond ruptures between carbon and hetero atoms upon reduction have also been found for carbonyl compounds containing an onium group in the α-position (391):

$$-\underset{\underset{O}{\|}}{C}-\underset{\underset{Z^+}{|}}{C}\diagdown \qquad (Z = NR_3, PR_3, SR_2)$$

With phosphonium compounds, for instance, whether bond rupture takes place or not depends on the nature of R; when R is aromatic, the odd electron of the radical generated by the transfer of the first electron is essentially located at the phosphonium group and cleavage of C—P takes place. However, when R is aliphatic, the odd electron is essentially located at the carbonyl group which is reduced. Even with phenacyl compounds, ϕ—CO—CH$_2$OR, although not constituting cations, cleavage of the C—O bond takes place with the elimination of RO$^-$ and transfer of a second electron to the radical ϕ—CO—CH$_2 \cdot$ (299).

No one-electron wave is observed with p-diacetylbenzene (231, 366) which, because of its structure, resembles an α,β-diketone rather than an ordinary bifunctional compound. The reversible wave in acidic solutions was explained by the transfer of two electrons to the doubly protonated compound (231, 497)

$$CH_3-\underset{\underset{H}{+O}}{C}-\phi-\underset{\underset{H}{+O}}{C}-CH_3 \quad + \quad 2e^- \quad \rightleftharpoons \tag{78}$$

the quinoid structure of the product presumably being more probable than the biradical. This state of affairs resembles the reversible two-electron reduction of p-dinitrobenzene in alkaline solutions (190, 193):

$$NO_2-\phi-NO_2 \quad + \quad 2e^- \quad \rightleftharpoons \quad \tag{79}$$

as well as the corresponding process for 2-phenylindandione-1, 3, (428, 433):

(80)

although in the latter case an ESR signal has been claimed to indicate the biradical structure (428); in alkaline media reduction proceeds with the anion formed by ready dissociation of the proton in the 2-position, giving the corresponding trianion (433). The quinoid (or biradical) product of diacetylbenzene undergoes an acid-base-catalyzed transformation to give the product that would occur on reduction of only one carbonyl group (231, 497):

$$CH_3-\underset{\underset{H}{\overset{|}{O}}}{C}=\underset{}{\bigcirc}=\underset{\underset{H}{\overset{|}{O}}}{C}-CH_3 \longrightarrow CH_3-\underset{\underset{H}{\overset{|}{O}}}{\overset{\overset{H}{|}}{C}}-\emptyset-\underset{\overset{||}{O}}{C}-CH_3 \quad (81)$$

This has been confirmed by chronopotentiometric investigations (124).

In contrast with phenylindandione, which shows two-electron reduction even in alkaline media, phthalimide (386) and its derivatives formed by substitution at the nitrogen atom (444), although similar in structure, are reduced with the transfer of two electrons only in acidic media, whereas one-electron waves giving rise to the corresponding radicals are observed in alkaline solutions.

While the radicals of carbonyl compounds are in general rather unstable toward dimerization and/or dismutation in aqueous solutions, their stabilities in non-aqueous solvents not only depend on the conditions (especially proton availability) but also vary rather markedly from one depolarizer to another. Thus benzophenone in dimethylformamide gives a stable anion radical, whereas the radicals of benzaldehyde and acetophenone dimerize readily to give the corresponding pinacolate ions (464). Small additions of water, however, also cause the benzophenone anion radical to disappear at a considerable rate which has been found to obey simultaneously first- and second-order kinetics (449) by making use of controlled-potential electrolysis and following the decay of the ESR signal as described previously (240). In pyridine, benzophenone shows only one one-electron wave with simple supporting electrolytes (lithium chloride) corresponding to the generation of the salt from the pinacolate and the supporting electrolyte cation (80); on addition of water hydrolysis to give the pinacol takes place. The two-electron product benzhydrol, however, is also observed in controlled-potential electrolysis; this is ascribed to the attack of water on the anion radical:

$$\overset{\diagdown}{\underset{\diagup}{C}}-O^- \xrightarrow{H_2O} \overset{\diagdown}{\underset{H}{C}}-O\cdot \xrightarrow{H_2O} \overset{\diagdown}{\underset{H}{C}}-OH + \cdot OH \quad (82)$$

or on the dimeric dianion resulting in its dismutation:

$$\underset{O^-}{\overset{\diagdown}{\underset{|}{C}}}\diagup\!\!\!-\!\!\!\underset{O^-}{\overset{\diagdown}{\underset{|}{C}}}\diagup \xrightarrow{H_2O} \underset{O}{\overset{\diagdown}{\underset{\|}{C}}}\diagup + \underset{O^-}{\overset{\diagdown}{\underset{|}{HC}}}\diagup \xrightarrow{H_2O} \underset{}{\overset{\diagdown}{CHOH}}\diagup \qquad (83)$$

In the presence of tetraalkylammonium perchlorate as a supporting electrolyte, the second one-electron wave is also observed (328); upon addition of a proton donor, the first wave increases at the expense of the second, resulting in one two-electron wave in the presence of two equivalents of the proton donor. Cyclic voltammetry shows three anodic peaks which are ascribed, in order, to the oxidation of the anion radical (almost reversible), the dianion (or, more probably, the pyridyl alcohol formed by attack of the carbanion on the solvent), and the protonated carbanion.

The behavior of acetophenone and benzaldehyde in pH-buffered ethanol resembles that in aqueous solutions (284); the wave, however, corresponding to the preprotonated species disappears at about pH 8 owing to the limited kinetics of preprotonation, in favor of the direct reduction of the depolarizer molecules.

The influence of proton availability has also been studied with anthanthrone in pH-buffered tetrahydrofuran (360); one $2e^- + 2H^+$ wave occurs up to pH 17, whereas separation into two one-electron waves is observed at higher pH values. The first of these waves corresponds to the formation of the anion radical, which accepts a second electron and a second proton during the reduction that corresponds to the second wave.

From the increase in the first one-electron wave of N-benzoyllactams in dimethylformamide upon addition of a proton donor, during which the slope of the wave does not change, it was concluded (116) that in the exponential factor $\alpha n = 1$ the number of electrons in the rate-determining step is $n = 1$ even in aqueous solutions, and that the transfer coefficient is $\alpha = 1$, indicating a barrierless charge transfer.

The anion radicals of fluorenone and its derivatives, showing a rather high stability even in aqueous media, can easily be generated in dimethylformamide, and the ESR spectra can be observed (83, 94). In this solvent p-diacetylbenzene, which does not show one-electron waves in aqueous media as mentioned above, is also reduced in two separate one-electron waves, the corresponding products, anion radical and quinoid dianion, being reoxidizable as shown by the application of the Kalousek technique (366). The anion radical of hexafluoroacetophenone was observed by the ESR technique in acetonitrile, whereas only a weak single line was observed upon reduction with potassium in tetrahydrofuran, which was ascribed to preponderating dimerization under these conditions (224).

Aldehydes substituted with halogen in the α-position have been investigated in dimethylformamide (331) with and without phenol as a proton donor. Different reaction paths are followed, depending on the halogen atom, branching at the α-carbon atom, and the amount of proton donor present. The possible paths are as follows (X = halogen atom) in the scheme on page 241.

Phosphorylated ketones of the type

$$\begin{array}{c} R' \\ \diagup \\ R-C-P \\ \| \| \diagdown \\ O O R'' \end{array}$$

show one-electron reduction in dimethylformamide and acetonitrile (393), which is reversible when R is an aromatic group; the ESR spectra have been observed. Similar investigations were made with compounds in which R' and/or R'' were alkoxy groups (394).

Some further work on carbonyl compounds, including acids, esters, anhydrides, and peroxides, and involving radical mechanisms, should be noted (9, 218, 312, 352, 398); the reduction of phthalimides (413) has already been dealt with in Section III-C.

As a final example the reduction of carbon dioxide in unbuffered aqueous solution should be mentioned (225, 415); the height of the wave corresponds to the transfer of one electron, the product, however, being formate (not oxalate or other recombination products). This is attributed to the formation of a hydroxyl ion during the reduction which inactivates one molecule of carbon dioxide (to give HCO_3^-). The rate-determining step, however, is suggested to be a one-electron transfer with the anion radical CO_2^- as intermediate, which is immediately reduced:

$$CO_2^- + e^- + H_2O \rightarrow HCOO^- + OH^- \tag{85}$$

Correlations of electrochemical data with other physical properties or theoretical calculations were in part mentioned in Section III-D, for instance, with the energies of optical excitation (32). Frequently, half-wave potentials have been correlated with the energies of lowest vacant molecular orbitals or corresponding quantum mechanical characteristics (76, 87, 88, 89, 132, 432, 473). Reversible potentials have not always been used for such comparisons; however, even processes determined by kinetics can be correlated with quantum energy states or with corresponding Hammett substituent constants, and comparisons will be meaningful if made for systems involving corresponding mechanisms.

$$\underset{X}{\overset{}{\text{>C-CHO}}} \xrightarrow{+e^-} \underset{X}{\overset{}{\text{>C-}\overset{H}{\underset{}{\overset{\cdot}{C}}}\text{-O}^-}} \xrightarrow{+H^+} \underset{X}{\overset{}{\text{>C-}\overset{H}{\underset{}{\overset{\cdot}{C}}}\text{-OH}}} \xrightarrow{+e^-} \underset{X}{\overset{}{\text{>C-}\overset{H}{\underset{}{\overset{-}{C}}}\text{-OH}}} \xrightarrow{-X^-} \overset{}{\text{>C-CHO}} \xrightarrow[\text{the aldehyde}]{\text{Reduction of}} \overset{}{\text{>C-CH}_2\text{OH}}$$

$$\underset{}{\overset{+e^-}{\underset{-X^-}{\searrow}}} \overset{}{\text{>C-CHO}} \xrightarrow{+H^+}$$

(84)

F. Unsaturated Carbonyl Compounds

Unsaturated carbonyl compounds with the double bond next to the carbonyl group, —C=C—CO—, show electrochemical behaviors that differ both from those of simple unsaturated compounds without carbonyl groups and from those of simple saturated carbonyl compounds (including those with the double bond and the carbonyl group more separated from one another) inasmuch as they are generally reduced at more positive potentials than either of the simple compounds.

The principles of the reduction mechanism were established, similarly to those of saturated carbonyl compounds, in early studies (123, 142, 176, 272, 346) and were confirmed by several workers [cf. references (18), (318), (320), and (492)]. Without including all the aspects attributable to differences in the participation of protons at different pH values, which resemble in some ways those involving saturated carbonyl compounds as discussed in the preceding section, the basic mechanism is:

$$\begin{array}{c}\diagup\\\diagdown\end{array}\!\!C\!=\!\overset{|}{C}\!-\!\overset{|}{C}\!=\!O \quad\xrightarrow[+H^+]{+e^-\ +H^+}\quad \begin{array}{c}\diagup\\\diagdown\end{array}\!\!\overset{\cdot}{C}\!-\!\overset{|}{C}\!=\!\overset{|}{C}\!-\!OH \quad\xrightarrow[+H^+]{+e^-\ +H^+}\quad \begin{array}{c}\diagup\\\diagdown\end{array}\!\!\overset{|}{\underset{H}{C}}\!-\!\overset{|}{\underset{H}{C}}\!-\!C\!=\!O$$

$$\Big\downarrow\qquad\qquad\qquad\qquad\xrightarrow{\text{Reduction of the carbonyl group}}$$

$$\tfrac{1}{2}\left\{\begin{array}{c}\diagup\\\diagdown\end{array}\!\!\overset{|}{\underset{H}{C}}\!-\!\overset{|}{C}\!-\!\overset{|}{C}\!=\!O\\ \begin{array}{c}\diagup\\\diagdown\end{array}\!\!\overset{|}{\underset{H}{C}}\!-\!\overset{|}{C}\!-\!\overset{|}{C}\!=\!O\right.\quad\xrightarrow{\text{Reduction of the carbonyl groups}}$$

(86)

The intermediate radicals are formulated with the odd electron in the β-position, since this is where dimerization takes place. The products of dimerization and those of subsequent reduction of the radicals are saturated carbonyl compounds, the C=O group being reduced only at more negative potentials. Actually, reduction of the C=C double bond is facilitated by the presence of the attached carbonyl group.

The effect of preprotonation of the C=O group, as noted by several investigators (18, 208, 485, 486), is similar to the effects that occur with saturated carbonyl compounds. A limited rate of preprotonation may be responsible for kinetic currents at the respective positive potentials, whereas the rest of the depolarizer is reduced without preprotonation at more negative potentials (373, 374).

Apart from spontaneous polymerization of the depolarizer, which can be followed by the polarographic technique as well (208), polymerization induced by reduction may occur beyond the dimeric stage in aprotic (411, 467) and in aqueous solutions (103, 261) owing to the activity of the intermediate radicals; this is not included in the mechanism shown above. A secondary polymerization has been observed with allyl acetate, the radicals of which were not formed by direct electrochemical reaction but by the action of SO_4^- radicals generated by polarographic reduction of a solution containing Cu^{2+} and $S_2O_8^{2-}$ ions (270).

The stability of the radicals toward subsequent charge transfer allowing dimerization, and even a relative stability toward dimerization resulting in a certain lifetime of the radicals, is related to the particular structure. Aromatic rings in resonance with the double bond and allowing the odd electron to be dispersed across the molecule contribute, as in other systems, to increased stability (142, 411, 484). Steric effects may also be responsible for enhanced stability, such as is observed in radicals having a *tert*-butyl group attached to the carbonyl group (411). In general, dimerization is fast enough to allow the application of the appropriate theoretical relations discussed in Section II-F for current versus potential (285) as well as potential versus concentration (373). Hydroxychalcones give radicals that dimerize fairly slowly and are therefore reoxidizable (327). The decay kinetics of the radicals have also been studied by recording the corresponding oxidation wave after flash photolytic reduction of perinaphthenone (402).

Deviations from the mechanism shown above occur with special groups of compounds or under special conditions. Thus controlled-potential electrolysis of methyl vinyl ketone in acidic media at potentials on the one-electron wave resulted in a mercury compound, $Hg(CH_2CH_2COCH_3)_2$, instead of the simple dimer, suggesting a reaction of the radicals with mercury to give a mercury radical which in turn reacts with another radical (201). Peculiarities are also shown by diphenylcyclopropenone (485) the radical of which, generated at low pH values by electron transfer to the propenylium ion after preprotonation, enters two competitive reactions, dimerization or hydrogen evolution, acting as a catalyst in the latter case:

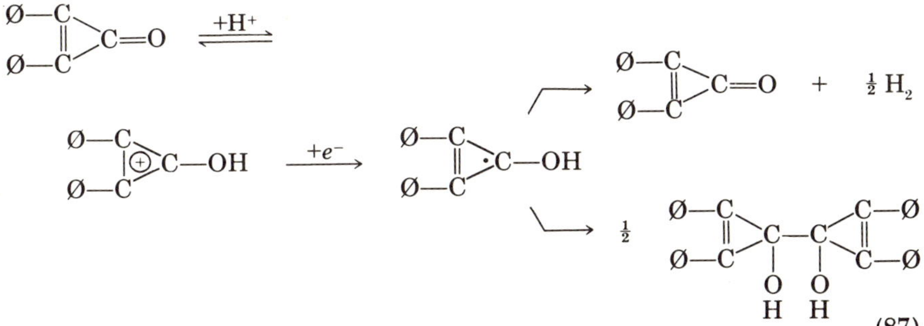

(87)

A carbonyl compound with C=S instead of C=C, ϕ—CO—CH=SR$_2$, is reduced with the transfer of two electrons and the elimination of R$_2$S to give acetophenone through an intermediate radical ϕ—CO—CH$_2$ · (440, 490). Compounds in which the carbon atom of the C=O group is a member of an unsaturated ring system, presumably show reduction of the carbonyl group instead of C=C bonds; this may apply to tropone (373, 374, 486) and tropolone (220, 337). Perinaphthenone, which also belongs to this group of compounds, was also thought to give the unsaturated alcohol (a) during the second one-electron wave (23):

$$\text{structure} \xrightarrow{+e^-} \text{structure} \xrightarrow{+e^-, +2H^+} \text{(a)} \; ; \; \text{(b)}$$

(88)

It was shown, however, that this alcohol would be unstable toward dismutation and that, at least at high pH values, reduction in the unsaturated ring system (b) is more probable (68).

The basic principles of the reduction mechanism in aqueous solutions, namely, reduction of the double bond to give saturated monomeric or dimeric carbonyl compounds, also apply to non-aqueous solvents. An unusual ease of reduction was observed with tetraphenylcyclopentadienone (484) and has been attributed to the formation of a nonbenzenoid aromatic system with six electrons in the five-membered ring of the radical. Most investigations were carried out in dimethylformamide (17, 312). The kinetics of dimerization have been followed by linear sweep polarography at different sweep rates (411); in these studies it was also shown that an increased stability of anion radicals, brought about by resonance structures or steric effects, is paralleled by an increase in dimeric products upon controlled-potential electrolysis as compared with polymeric products. Addition of water to the solvent causes, as in other systems, the first one-electron wave to increase owing to a facilitated subsequent reduction of the radical after proton transfer (310). Carboxylation may take place in the presence of carbon dioxide (467), according to the scheme on page 245.

G. Halogen-Containing and Related Compounds

Unlike the radical mechanisms dealt with so far, in which reduction is essentially a hydrogenation of double bonds (C=C, C=O), or incorporation of electrons into the molecular π-electron system, frequently with the

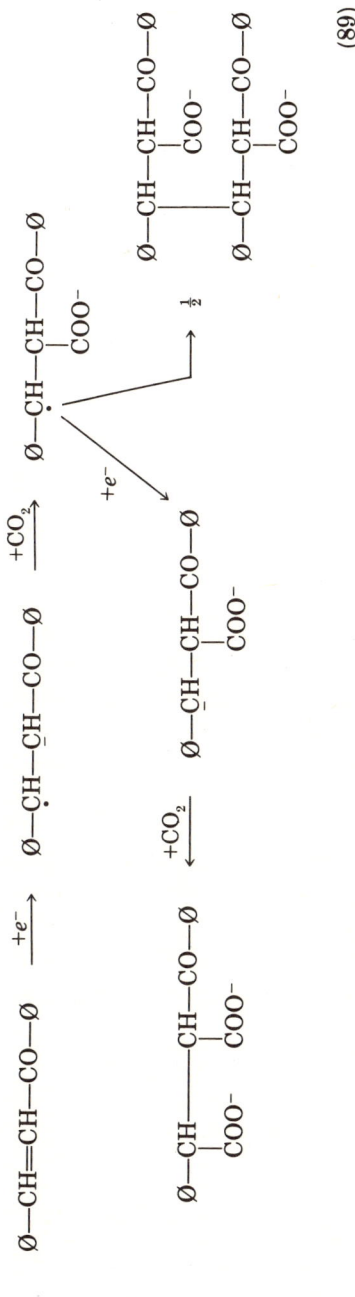

(89)

formation of new chemical bonds (C—H, C—C), the reductions of halogen compounds and of some related electroactive groups are, first, characterized by the cleavage of bonds (C—X).

The radical mechanism of these processes was established in early investigations (85, 110, 390, 421); essentially it is:

$$\begin{array}{c}\text{>}C-X + e^- \longrightarrow \text{>}C\cdot + X^- \\ \downarrow \qquad \searrow \\ \tfrac{1}{2}\ \text{>}C-C\text{<} \qquad \xrightarrow[+H^+]{+e^-}\ \text{>}C-H \end{array} \qquad (90)$$

Especially with simple alkyl halides, reduction of the radical takes place at the potential at which it is formed, not allowing dimerization; it may be doubted whether or not these radicals are at all stable enough to escape from the metal interface prior to the transfer of the second electron. Instead of the two-electron waves observed under these conditions, however, one-electron waves giving rise to the corresponding dimer have often been recorded in aqueous as well as in non-aqueous media. Another aspect of these compounds is the frequently observed formation of mercury compounds or, with other electrode materials, the corresponding organometallics.

The detailed mechanism of the first electron transfer involving the elimination of X^- has repeatedly been a subject of consideration, frequently based upon experimental observations with dihalogen compounds and steric implications [(111, 382, 404); cf. these references for further literature]. The previously suggested transfer of the electron to the C—X bond "through" the carbon atom was rejected by such considerations in favor of a transfer via the halogen atom (404):

$$(\text{Me})e^- + \text{Br}-\text{C}\text{<} \rightarrow (\text{Me})\text{Br}_{ad}^- + \cdot\text{C}\text{<} \qquad (91)$$

However, from a study (121) of the behavior of compounds $X-CH_2CH_2-I$ ($X = H$, OCH_3, OH, $COOH$, Cl, CN) with different metals, it was concluded that an anion radical with covalent bonding to the metal is formed first, which releases I^-:

$$(\text{Me})e^- + \text{RI} \rightarrow (\text{Me})-\text{RI}^- \rightarrow (\text{Me})-\text{R}\cdot + \text{I}^- \qquad (92)$$

The radical is desorbed if covalent bonding is weak, as it is with copper:

$$(\text{Me})-\text{R}\cdot \rightarrow (\text{Me}) + \text{R}\cdot \qquad (93)$$

and R· dimerizes; however, dismutation may take place according to

$$2\text{MeR}\cdot \rightarrow (\text{Me}) + \text{MeR}_2 \tag{94}$$

Additional elimination of X, however, takes place if X = Cl or OH:

$$(\text{Me})-(\text{X}-\text{CH}_2\text{CH}_2\text{I}^-) + e^- \rightarrow (\text{Me}) + \text{X}^- + \text{I}^- + \text{CH}_2=\text{CH}_2 \tag{95}$$

The electron distribution about the halogen atom is of course rather important for the rate of the reaction; correlations have therefore been established for various substituted iodobenzenes as well as for aliphatic iodo compounds between half-wave potentials and nuclear quadrupole resonance attributable to the quadrupole nature of the iodine nucleus (65). Further correlations have been established between half-wave potentials and Hammett substituent constants for benzyl chlorides and related compounds, and deviations have been interpreted in terms of substituent effects on the stability of the intermediate radicals (436).

Dimerization in aqueous solutions has been found, for instance, with bromomaleic acid (114), which gives butadienetetracarboxylic acid in addition to maleic and fumaric acids. Also, benzyl chloride was found to give bibenzyl upon controlled-potential electrolysis (317). Iodo- and bromopropionitriles show two one-electron waves, the first of which corresponds to the formation of the dimer (120); the second wave is shifted to more negative potentials with increasing depolarizer concentrations, which is not only a result of the enhanced rate of dimerization, reflected in Eq. 43, but also of an inhibition effect brought about by the adsorption of the dimer. The benzyl radical formed upon reduction of benzyl bromides in methanol (164) may dimerize to give bibenzyl or react with mercury resulting in dibenzylmercury, depending on the substituents as well as the supporting electrolytes. The same is true for the reduction of perhalogenated benzene, such as C_6F_5X or C_6Cl_5X (X = halogen), in dimethylformamide (364); it was thought that the primary product of reduction is the anion radical, for example, $C_6F_5I^-$, which upon elimination of I^- dimerizes or reacts with mercury or, upon electron transfer in the second polarographic wave, gives $C_6F_5^- + I^-$. The prominent role of adsorption during the formation of mercury compounds through radical mechanisms has been established for the reduction of bromoacetate and related compounds (115). Iodo compounds may react with mercury without net charge transfer and the product, RHgI, undergo stepwise electron transfer through organomercury radicals, RHg· (216). For further discussion see Section III-K.

During the reduction of α,β-dihalogen compounds, stabilization of the intermediate radicals usually takes place through formation of double bonds rather than through dimerization; thus fumaric and maleic acids result from the reduction of dibromosuccinate (111). Under proper conditions cyclization may also occur, for instance, with 1,3-dihalocyclobutane:

$$X-\diamondsuit-X \xrightarrow{+e^-} X-\diamondsuit\cdot + X^- \qquad (96)$$
$$\downarrow{+e^-}$$
$$X-\diamondsuit| \longrightarrow \bowtie + X^-$$

forming bicyclobutanes (382); spiro compounds may also be formed.

Relatively stable radicals are generated by the reduction of iodomethyl-trialkyl(aryl)silanes in aqueous ethanol (306).

Compounds bearing a nitro group and a halogen atom at the same carbon atom give rise to the corresponding aci-nitro compound upon reduction (417):

$$R_2C\begin{smallmatrix}X\\ \\NO_2\end{smallmatrix} \xrightarrow{+e^-} R_2\overset{\cdot}{C}-NO_2 + X^- \qquad (97)$$
$$\xrightarrow{+e^-} R_2C=NO_2^-$$

The reduction of 6-chloroquinoline in dimethylformamide which, under aprotic conditions, gives quinoline which is reduced in two further one-electron waves, proceeds without the elimination of chloride ions in the presence of water or other proton donors in which the quinoline ring is attacked first (136). Although not a radical as judged from the number of electrons, CCl_2 was suggested as an intermediate in the reduction of carbon tetrachloride in acetonitrile (465).

Whereas elimination of halide ions is usually observed on reduction of halogen compounds, the corresponding cyano compounds (nitriles) show elimination only under certain conditions, probably because of the lower polarization of the C—CN bond. Thus isonicotinonitrile (401) generates a radical by transfer of one electron with slow liberation of CN^- and subsequent dimerization:

$$R-\overset{+}{N}\bigcirc-CN \xrightarrow{+e^-} R-\overset{+}{N}\bigcirc-CN \longrightarrow R-\overset{+}{N}\bigcirc + CN^- \qquad (98)$$
$$\searrow \tfrac{1}{2}\ R-\overset{+}{N}\bigcirc-\bigcirc\overset{+}{N}-R$$

Elimination takes place with phthalonitrile in dimethylformamide, however, only subsequent to the transfer of a second electron at more negative potentials, which is immediately followed by the transfer of a third electron (381):

(99)

Different ESR spectra, corresponding to phthalo- and benzonitrile anion radicals have therefore been observed on reduction at potentials of the first and second polarographic waves. The benzonitrile anion radical, although found to be stable in dimethylformamide (381), was thought to stabilize by dismutation, giving the dihydro compound (406):

$$2\emptyset\text{—CN}^- \longrightarrow \emptyset\text{—CN} + \text{(cyclohexadiene)—CN} \quad (100)$$

Stable radicals are also formed in the reductions of polycyano compounds (1, 325, 378) such as:

These and similar compounds were found to form strong complexes with electron-donating molecules (hexamethylbenzene, pyrene) in non-aqueous solvents (353); consequently, half-wave potentials of anion radical formation are shifted to less positive values, and from the shifts stability constants can be calculated, whereas those of the second wave are not affected, showing that complex formation does not take place with the anion radicals.

Acrylonitrile, $CH_2\text{=}CH\text{—}CN$, as mentioned in Section III-A, shows a polarographic two-electron wave in aqueous solutions, whereas large-scale electrolysis gives rise to both propionitrile and adiponitrile in proportions depending upon the conditions; the latter product is of commercial interest as an intermediate in the manufacture of nylon (6) (see reference (6) for further references).

Although not directly attributable to the presence of the nitrile group, it may be mentioned here that the oxidation of the solvent acetonitrile is induced by the electrooxidation of the supporting electrolyte anion ClO_4^- (396):

$$ClO_4^- \rightarrow ClO_4 + e^-; \quad ClO_4 + CH_3CN \rightarrow HClO_4 + \cdot CH_2CN$$
$$2 \cdot CH_2CN \rightarrow NC-CH_2-CH_2-CN \qquad (101)$$

H. Compounds Containing the Groups NO, PO, SO, and SO_2

A variety of special mechanisms seems to be involved with compounds containing the groups NO, PO, SO, and SO_2; each of these could appropriately be discussed in other sections of this chapter.

The radical mechanism of the reduction of azopyridine dioxide (387) and azomethine N-oxide (494) has been mentioned in Section III-B; simple N-oxides in dimethylformamide behave similarly, giving anion radicals (5), the corresponding half-wave potentials of this reaction being correlated to calculations of lowest vacant molecular orbital energies (275). Further reference is made to the investigation of N-oxides of bispyridyl disulfides and related compounds (106).

The polarographic behavior of the radical $(CH_3)_2\dot{N}O$, prepared by controlled-potential electrolysis of trimethylnitromethane, has been studied in acetonitrile (184); both reversible one-electron oxidation and reduction were observed. Similarly, $(CF_3)_2\dot{N}O$ was prepared by anodic oxidation of the corresponding hydroxylamine, $(CF_3)_2NOH$, at platinum electrodes and isolated as a solid substance (445). The polarographic behavior of this radical has been investigated in acetonitrile; the diffusion coefficient of this radical, given in reference (445) as 2.73×10^{-6} cm^2 sec^{-1}, is so low as to give rise to suspicion that the wrong number of electrons may have been assumed for its reduction, but unfortunately the original literature was not available.

Organophosphorous compounds of the types

$$\phi-P\begin{smallmatrix}OR\\ \\R'\end{smallmatrix} \quad \text{and} \quad \phi-CH=CH-P\begin{smallmatrix}OR\\ \\R'\end{smallmatrix}$$

were shown to be reduced reversibly in dimethylformamide with the transfer of one electron, whereas a second electron is irreversibly transferred (294).

For a discussion of the reduction of phosphorylated ketones (393, 394), see Section III-E.

Anion radicals stable enough for studying their ESR spectra have also been generated by the reduction of diaryl sulfoxides R_2SO in dimethylformamide (144). However, the anion radical formed upon reduction of phenylmethyl sulfoxide in different solvents (78) is reduced with the elimination of the oxygen atom during the second stage of the process without separation into two polarographic waves:

$$\phi\text{—S(=O)—CH}_3 + e^- \longrightarrow \phi\text{—}\dot{\text{S}}(\text{O}^-)\text{—CH}_3 \xrightarrow[+2H^+]{+e^-} \phi\text{—S—CH}_3 + H_2O \tag{102}$$

Whether the electron or a proton is transferred first to the anion radical depends upon the proton availability.

The mechanism proposed for the polarographic reduction of 4-methylimidazol-2-yl-thiosulfuric acid in aqueous solutions (60), involves the elimination of the radical SO_3^- which is immediately reduced to give HSO_3^-, the latter ion giving rise to a second reduction wave (to give sulfoxylic acid):

$$\text{>C—S—SO}_3\text{H} + e^- \rightleftharpoons \text{>C—SH} + SO_3^- \xrightarrow[+H^+]{+e^-} HSO_3^- \xrightarrow{\text{Second wave}} \tag{103}$$

Formaldehyde sulfoxylate decomposes in acidic media according to

$$\text{HO—CH}_2\text{—SO}_2^- + H^+ \rightarrow HSO_2 + CH_2OH \tag{104}$$

and the radical HSO_2 is thought to be responsible for the polarographic reduction wave observed (122):

$$HSO_2 + e^- \rightarrow HSO_2^- \tag{105}$$

The reduction of naphthalenesulfonic acid (56), corresponding to the transfer of one electron, was suggested to proceed according to

$$\text{R—SO}_3\text{H} + e^- + H^+ \rightarrow R\cdot + H_2SO_3 \tag{106}$$

It is not known how the radical $R\cdot$ stabilizes, since only naphthalene but no dimer was found as the product of electrolysis. Since reduction has been carried out in unbuffered solutions with potassium chloride as supporting electrolyte, it can be inferred that a mechanism analogous to Eqs. 4 and 5 takes place with only half of the depolarizer being reduced, the other half being deactivated by serving as proton donors ($R\text{—}SO_3^-$ not undergoing reduction):

$$R\text{—}SO_3H + 2e^- \rightarrow RH + SO_3^{-2} \qquad (107)$$

$$R\text{—}SO_3H + SO_3^{-2} \rightarrow R\text{—}SO_3^- + HSO_3^- \qquad (108)$$

I. Onium Compounds

Carbonium, ammonium, phosphonium, sulfonium, and iodonium ions form another group of compounds that are analogous in chemical structure and that undergo reductions involving radicals, but that show only restricted similarities to one another. Some oxonium cations have been dealt with in Section III-C; further reference is made to the reduction of oxonium cations formed by preprotonation of quinones and carbonyl compounds.

One-electron reduction to give triphenylmethyl was observed with triphenylcarbonium ions in an aprotic solvent (CH_2Cl_2), whereas a two-electron wave appears in sulfuric acid since the radical readily reacts with the solvent to give $\phi_3\text{—}CH$ (220). Carbonium ions of triphenylmethane dyes are oxidized in liquid sulfur dioxide at platinum electrodes to give the corresponding radicals (169). Stabilization of the radical $\phi\text{—}N_2\cdot$ at the mercury interface was suggested to take place during the reduction of phenyldiazonium ions (385).

A particular characteristic of ammonium ions of the type R_3NH^+, which contain at least one hydrogen atom and in which R may represent widely different groups, is their ability to catalyze the evolution of hydrogen and to lower the corresponding high overvoltage by providing an alternative to the direct discharge of hydrogen ion at mercury electrodes (131). The mechanism of this catalysis has been investigated and quantitatively accounted for by several investigators (130, 302, 340, 423, 424) and has been reviewed in articles (303), monographs (309), and textbooks on polarography (181). It is beyond the scope of this chapter to present details, and only the general principle is discussed. The base R_3N is in equilibrium with the ammonium ion which is adsorbed at the interphase. The radical formed on the transfer of a single electron decomposes to regenerate the base and evolve hydrogen:

$$R_3N \underset{}{\overset{+H^+}{\rightleftharpoons}} R_3NH^+_{ad} \xrightarrow{+e^-} R_3NH_{ad} \longrightarrow R_3N + \tfrac{1}{2}H_2 \qquad (109)$$

Depending on the structure of the base, and on the conditions under which the reaction takes place, different factors may play a role; the equilibrium constant, the rate of the preceding protonation (which may also take place with the base R_3N being adsorbed), and the rate of charge transfer are affected by the double-layer structure and therefore depend on the composition of the solution (solvent, supporting electrolyte), as well as on the extent of adsorption of the involved species; this in turn depends on the electrode potential, thus occasionally giving rise to potential-dependent limiting currents (catalytic maxima).

Even gelatin, commonly used as a maximum suppressor and frequently exhibiting inhibition effects, can act as a catalyst (172, 207). The catalytic action of bases is not restricted to aqueous solutions but was also observed in anhydrous methanol with p-dimethylaminoaniline (222), even when the latter compound was generated only by a preceding electrode process as found in the polarographic study of p-dimethylaminonitrobenzene (223) and p-dimethylaminoazobenzene (206). Also, intermediate products formed during the reduction of p-dinitrobenzene in aqueous solution give rise to catalytic hydrogen evolution (205). The effect is not restricted to amino bases but is also observed with other species capable of forming cations, as mentioned in previous sections, for example, with diphenylcyclopropenylium cations (485).

Quaternary ammonium ions, which, owing to the absence of hydrogen (424), usually do not show any catalytic effect, were nevertheless found to give rise to the evolution of hydrogen by the attack of their corresponding amalgams on water at rather negative potentials (418):

$$R_4N^+ + e^- \longrightarrow R_4N(Hg) \xrightarrow{H_2O} R_4N^+ + OH^- + \tfrac{1}{2}H_2 \qquad (110)$$

Elimination of one R-group (e.g., benzyl), however, may also take place which, depending on concentrations as well as on the temperature, may be further reduced or dimerize (321); a corresponding process occurring during the reduction of p-benzaldehydetrimethylammonium ions (20) has been mentioned in Section III-E.

Such amalgams have been observed in aprotic solvents (dimethylformamide, acetonitrile) on reduction of R_4N^+ and were shown by x-ray powder photography to be crystalline solids appearing as many-faceted dendrites to the naked eye (295). They are capable of transferring electrons to reducible organic compounds, the generation of the corresponding anion radicals of which were shown by ESR spectra; a sequence of reduction potentials according to different R-groups has been established (335). The dependence of the potential of final current rise resulting from amalgam formation of NR_4^+ ions on the methanol content of water–methanol mixtures has been explained in terms of inhibition by adsorbed methanol molecules at low (negative shift) and progressive desolvation of the ions at high contents of methanol (positive shift) (475). Amalgam formation and behaviors analogous to that of R_4N^+ have been found for R_4P^+ and R_3S^+ ions (86) in non-aqueous media.

Phosphonium ions, R_4P^+, undergo reversible one-electron reductions in aqueous solutions (319):

$$R_4P^+ + e^- \rightleftharpoons R_3P + RHg \cdot \qquad (111)$$

with subsequent dimerization of the liberated R· and possibly catalytic hydrogen evolution by formation of R_3PH^+. In dimethylformamide the cation ϕ_4P^+ liberates a phenyl radical in the first polarographic wave, whereas the second wave corresponds to the formation of an anion (468):

First wave: $\qquad \phi_4P^+ + e^- \rightarrow \phi_3P + \phi \cdot \qquad (112)$

Second wave: $\qquad \phi_4P^+ + 2e^- \rightarrow \phi_3P + \phi^- \qquad (113)$

Triphenylphosphine is reduced during the third wave, again liberating a phenyl radical which abstracts a hydrogen atom from the solvent to give benzene:

$$\emptyset_3P + e^- \longrightarrow \emptyset_2P^- + \emptyset \cdot \xrightarrow{+H} \emptyset-H \qquad (114)$$

A mercury radical is formed on reduction of triphenylsulfonium ions in aqueous solutions, which may decompose or be reduced in a second wave, according to (322)

$$\emptyset_3S^+ + e^- \longrightarrow \emptyset_3SHg \cdot \begin{array}{l} \xrightarrow{+e^-,\ +H^+} \emptyset_2S + \emptyset-H \\ \searrow \emptyset_2S + \tfrac{1}{2}\emptyset_2Hg \end{array} \qquad (115)$$

The limiting currents corresponding to the transfer of one electron during the reduction of cyanomethyldimethylsulfonium cations in dimethylformamide (457) were presumed to be attributable to a two-electron reduction, half of the depolarizer, however, being deactivated by reaction with the liberated CH_2CN^-:

$$\overset{+}{Me_2S}-CH_2CN + 2e^- \rightarrow Me_2S + CH_2CN^- \qquad (116)$$

$$\overset{+}{Me_2S}-CH_2CN + CH_2CN^- \rightarrow \overset{+}{Me_2S}-\overset{-}{CHCN} + CH_3CN \qquad (117)$$

which is in accord with Eqs. 4 and 5; on addition of proton donors, the wave height increases and reaches a value corresponding to a two-electron reduction. The corresponding cyanoethyl and cyanopropyl compounds, however, undergo genuine one-electron reductions:

$$\text{Me}_2\overset{+}{\text{S}}\text{—(CH}_2)_n\text{CN} + e^- \rightarrow \text{Me}_2\text{S} + \cdot(\text{CH}_2)_n\text{CN} \tag{118}$$

with subsequent dimerization of the radicals, whereas their further reduction in a second wave again gives rise to deactivation of part of the depolarizer molecules, which can only be prevented by addition of a proton donor; otherwise only an unusual maximum is observed instead of a second wave.

The reduction of iodonium ions R_2I^+ resembles that of sulfonium ions in aqueous solutions and may schematically be shown as (15, 82, 459):

$$R_2I^+ \xrightleftharpoons[\text{First wave}]{+e^-} R_2I\cdot \xrightarrow[\text{Subsequent waves}]{e^-} RI, RH, I^- $$
$$\downarrow$$
$$RI + R\cdot$$
$$\downarrow$$
$$R_2 \text{ or } RH \tag{119}$$

The first wave is reversible at low concentrations; mercury compounds may also form.

The reduction of carbonyl compounds containing ammonium, sulfonium, and phosphonium groups in the α-position have been dealt with in Section III-E (391).

J. Sulfur Compounds

Whereas heterocyclics with sulfur atoms and sulfonium compounds have been discussed in Section III-C and -I, respectively, some other types of sulfur compounds are considered here.

Much attention has been given to disulfides such as cystine and sulfhydryl compounds such as cysteine, since in the presence of cobalt ions they give rise to the well-known catalysis of hydrogen evolution discovered by Brdička (46). It is well established that radicals such as $RS\cdot$ are involved in these reactions (437), and in general the adsorption at mercury and the formation of mercury compounds including those of a radical type play an important role in reduction and oxidation of these compounds.

The reduction of cystine (RSSR) was found to be catalyzed by iron(II) ions (268), either by direct attack with the formation of radicals $RS\cdot$

$$RSSR + Fe(II) \longrightarrow RS\cdot + RS^- + Fe(III) \quad (120)$$
$$ \xrightarrow{+e^-} RS^-$$

or by a reaction with radicals, $RS\cdot$, formed during the preceding reaction with mercury:

$$RSSR + Hg \longrightarrow RSHg\cdot + RS\cdot \rightleftharpoons (RS)_2Hg \quad (121)$$
$$ \xrightarrow{+Fe(II)} RS^- + Fe(III)$$

Diphenyl disulfide and the corresponding diselenide were found, subsequent to dissociation to the radicals, ϕ—$S\cdot$ and ϕ—$Se\cdot$, to form the mercury radicals, ϕ—$SHg\cdot$ and ϕ—$SeHg\cdot$, which are reduced to give thiophenol (selenophenol) (341):

$$\phi\text{---SHg}\cdot + e^- + H^+ \rightarrow \phi\text{---SH} + Hg \quad (122)$$

A mechanism proposed previously (228):

$$\phi\text{---SS---}\phi + e^- \longrightarrow \phi\text{---SS---}\phi^- \xrightarrow{+e^-} 2\phi\text{---S}^- \quad (123)$$

is somewhat doubtful; the application of polarographic theory in this work is also questionable since relations applying to reversible and irreversible reactions have been confused. Reduction in two one-electron waves to give the radical and the mercaptan has been observed with diethyldisulfide in dimethyl sulfoxide (151). Mercury radicals, $RSHg\cdot$, form upon oxidation of mercaptans (11, 229, 276).

The oxidation of ethylxanthate, $C_2H_5O\cdot CSS^-$, has been followed by measuring the differential double-layer capacity (372); five peaks were observed, the first and second of which were attributed to adsorption and reorientation of the anion, while the others correspond, in order, to the oxidation reactions:

$$Hg + EtOCSS^-_{ad} \rightarrow EtOCSSHg\cdot + e^- \quad (124)$$

$$EtOCSSHg\cdot + EtOCSS^- \rightarrow (EtOCSS)_2Hg + e^- \quad (125)$$

$$Hg + 2EtOCSS^- \rightarrow (EtOCSS)_2Hg + 2e^- \quad (126)$$

Similarly, alkyldithiocarbamates, $RNH\cdot CSS^-$, are oxidized in two successive one-electron waves to give Hg(I) and Hg(II) compounds, the second reaction being preceded by the dissociation of a proton ($\overline{RN}\cdot CSSHg\cdot$); accordingly, the second wave is not observed with dialkyl compounds $R_2N \cdot CSS^-$ (171). Esters of monothiobenzoic acid $\phi \cdot COSR$ (R = phenyl or benzyl) reduce in aqueous solutions in two one-electron waves with a radical intermediate; this differs from the two-electron reduction observed when R = alkyl (277).

K. Organometallic Compounds

The formation of mercury radicals during the reduction of several kinds of organic compounds has been mentioned repeatedly in preceding sections, especially Sections III-E, -G, -I, and -J. Adsorption of products always plays an important role in such reactions.

This is also true when mercury compounds are used as depolarizers. Most research has involved the study of compounds of the type RHgX in which X is a halogen atom, OH group, and so on. The formation of mercury radicals was observed in early investigations (25)

$$\varnothing\text{—HgOH} \xrightarrow[+H^+]{+e^-} \varnothing\text{—Hg}\cdot + H_2O$$
$$\hookrightarrow \tfrac{1}{2}\varnothing\text{—HgHg—}\varnothing \longrightarrow \tfrac{1}{2}\varnothing_2\text{Hg} + \text{Hg} \tag{127}$$

Alternatively to other mechanisms, an analogous process has been discussed for the reduction of benzylmercury iodide during the first polarographic wave in acetonitrile (466):

$$\phi\text{—CH}_2\text{HgI} + e^- \rightarrow \phi\text{—CH}_2\text{Hg}\cdot + I^- \tag{128}$$

In general, a dissociation is suggested to take place prior to the reduction, the electron transfer taking place with the cation RHg^+. Thus the half-wave potentials of the reduction of RHg^+ have been found to depend upon the concentration of Br^- ions present, being shifted to more negative potentials with increasing concentration of the anion as a result of complex formation ($RHgBr$, and even $RHgBr_2^-$) (62). A corresponding influence was observed with Cl^- (447), an excess of Cl^- rendering the reaction reversible because of an inhibition of dimerization. In the absence, as well as with an excess of Br^- (62), the current–voltage curve can be described by Eq. 38 since rapid irreversible dimerization follows a reversible electron transfer:

$$RHg^+ + e^- \rightleftharpoons RHg\cdot \rightarrow \tfrac{1}{2}R_2Hg + \tfrac{1}{2}Hg \tag{129}$$

However, if RHgBr is used in supporting electrolytes without complexing anions, the change in concentration of Br^- during consumption of RHg^+ and the corresponding change in the extent of complexation during the reaction must be taken into account, leading to a relation different from Eq. 38, namely (62)

$$\frac{1 - (i/i_d)}{(i/i_d)^{5/3}} = \text{constant} \cdot \lambda \qquad (130)$$

A correlation can be established between half-wave potentials of RHg^+ and the pK_a values referring to acid dissociation of the corresponding hydrocarbons, RH, since the affinity of R for mercury parallels that for hydrogen (61, 62). Reference is made to further work on such compounds (24, 55, 287).

Similar results have been obtained with compounds of metals other than mercury. Thus, in a solvent composed of ethanol and benzene, allyl compounds of palladium, C_3H_5PdX, are reduced to give two one-electron waves, the radical $C_3H_5Pd\cdot$ being formed in the first and propene, C_3H_6, in the second wave (165). The primary product of the reduction of corresponding magnesium compounds is $RMg\cdot$ which decomposes to give $R\cdot$ (377). Thallium compounds, ϕ_2Tl^+, are reduced via adsorbed intermediates, $Hg\phi_2Tl$ ($\to \phi_2Hg$), and $Hg\phi Tl$ ($\to \phi_2Hg$, or with further reduction to give ϕH) (102). During the reduction of organotin compounds, $(CH_3)_3SnX$, the radical formed in the first wave, $(CH_3)_3Sn\cdot$, may either dimerize or be reduced during the second wave to give $(CH_3)_3SnH$ (98), whereas the corresponding intermediate of the reduction of $(C_2H_5)_2SnX_2$, which is $(C_2H_5)_2Sn$, undergoes polymerization but is strictly speaking not a radical (97). Monoalkyl compounds, $RSnX_3$, undergo hydrolysis first to give $RSnOO^-$; transfer of three electrons giving $RSn\cdot$ is followed by polymerization (99, 100). When $R = \phi$, $\phi Hg\cdot$ is an intermediate (100).

Some further work can be mentioned here in which the metal is a more accidental component of the compound since the odd electron in the corresponding radical is delocalized throughout the organic part of the molecule. This is true for biphenyl complexes of germanium and silicon in dimethylformamide, the anion radicals of which allow correlation of half-wave potentials and ESR parameters in terms of the lowest vacant molecular orbital energies (90). Anion radicals are also formed in this solvent upon reduction of metal complexes of dithiodiketones, $Me(R—CS—CS—R)_n$ (343), of metal and metal-free phthalocyanines (383), of ferrocene and similar metalloorganic compounds (476), and of numerous different metalloorganic compounds investigated in comprehensive work with different techniques including polarography in dimethoxyethane (96). A radical is formed upon reduction of chlorophyll in dimethyl sulfoxide, the lifetime of which (0.1 sec), as determined by application of the Kalousek technique (21, 37), is smaller than that of the radical obtained by oxidation (425).

L. Nitro and Nitroso Compounds

For the electrochemical reduction of organic nitro compounds, the well-known Haber mechanism has in general been adopted,

$$R-NO_2 \xrightarrow{2e^-} R-NO \xrightarrow{2e^-} R-NHOH \xrightarrow{2e^-} R-NH_2 \qquad (131)$$

$$\searrow R-\underset{\underset{O}{\downarrow}}{N}=N-R \xrightarrow{2e^-} R-N=N-R \xrightarrow{2e^-} R-NH-NH-R$$

the pathway followed depending on the acidity of the solution. This scheme, however, by no means represents what is usually described as a mechanism of cathodic reduction. The nitroso stage is never present at an appreciable concentration since it is immediately reduced at the potential of its formation. Hence the hydroxylamine stage is the first stable product of reduction, except for aprotic solvents in which the anion radical, $R-NO_2^-$, resulting from one-electron transfer is stable (see below) and may, under certain conditions, reduce in the second wave to the dianion, $R-NO_2^{-2}$. In solutions of low pH values, as well as under the influence of certain substituents, the reduction proceeds immediately to the amine stage. The azoxy compounds therefore are not products of simple reduction processes. They can only be formed by the intermediate oxidation of the hydroxylamine stage to the nitroso compounds, for example, in the presence of oxygen, or if the anode is immersed in the same solution giving rise to direct oxidation of the hydroxylamine, or through its production of oxygen. Similarly, if the electrode potential is changed during the experiment, for example, in cyclic voltammetry (252, 258), the hydroxylamine formed may be oxidized to give the nitroso compound at more positive potentials, and this may then react with the remaining hydroxylamine. These facts should be emphasized prior to dealing with details of the reduction mechanisms of nitro compounds since they have not been given enough recognition even in more recent work.

A first conclusion as to the actual mechanism was derived from an analysis of the wave shape during the reduction of nitrobenzene, suggesting that less than one electron is involved in the formation of the transition state (351); this was confirmed later (39). Only a more detailed study of the splitting of this wave in presence of surfactants, which had been observed before without interpretation (179), led to the convincing explanation that the potential-determining step is a reversible one-electron transfer with the formation of an anion radical (186, 187, 189):

$$\phi\text{—}NO_2 + e^- \rightleftharpoons \phi\text{—}NO_2^- \qquad (132)$$

The appearance of this reversible one-electron wave in aqueous alkaline solutions was attributed to the inhibiting action of the surfactants on subsequent reactions, which allowed the transfer of further electrons. (A less specific explanation ascribed the action of inhibitors to a decrease in the product αn (434), α being the transfer coefficient and n the number of electrons in the rate-determining step.) The increase in the wave height with decreasing pH values was explained by the transfer of a proton to the anion radical to give the uncharged radical which readily undergoes subsequent reduction (188, 232):

$$\phi\text{—}NO_2^- + H^+ \rightarrow \phi\text{—}NO_2H \rightarrow \text{Immediate reduction} \qquad (133)$$

In alkaline solutions [as well as in anhydrous methanol (236)], the transfer of the first electron is immediately followed, in the absence of inhibitors, by the transfer of a second electron coupled with rapid transformation of the dianion (involving proton transfer) to give phenylhydroxylamine in a single four-electron wave (202). In the presence of inhibitors, the rate of transfer of a second electron decreases sufficiently to produce a separate wave at more negative potentials for the entire reduction exceeding one electron (195), whereas the one-electron transfer remains reversible even in the presence of strong inhibitors, as shown by a study of the faradaic impedance (238). A quantitative theory that relates the action of inhibitors to an adsorption displacement was based on these and similar observations (196, 199, 234). Cations of the supporting electrolyte were shown to exert a rather strong accelerating effect on the reduction of the anion radical, which was explained in terms of the formation of ion pairs facilitating access to the negatively charged electrode (239, 241).

The anion radical was shown to be relatively stable in aqueous alkaline solutions (several minutes or more depending on the concentration) by studying the decay of the corresponding oxidation current or light absorption (237) subsequent to its electrochemical generation. The decay is attributable to the dismutation reaction induced by

$$2\phi\text{—}NO_2^- \rightarrow \phi\text{—}NO_2 + \phi\text{—}NO_2^{-2} \qquad (134)$$

with the dianion immediately reacting with further anion radicals (involving proton transfer) to realize the net reaction

$$4\phi\text{—}NO_2^- + 2H_2O \rightarrow 3\phi\text{—}NO_2 + \phi\text{—}NHOH + 3OH^- \qquad (135)$$

A corresponding dismutation reaction had already been suggested previously (232). The dismutation, rather slow in alkaline solutions, is accelerated at lower pH values by the preceding proton transfer equilibrium

$$\phi-NO_2^- + H^+ \rightleftharpoons \phi-NO_2H \tag{136}$$

with very fast subsequent dismutation according to

$$\phi-NO_2^- + \phi-NO_2H \rightarrow \phi-NO_2 + \phi-NO_2H^- \tag{137}$$

and a net reaction similar to Eq. 135. These subsequent chemical reactions cause the first wave, appearing in the presence of inhibitors, to increase as a result of the regeneration of the depolarizer (237). The kinetics of homogeneous reactions have also been followed by the decay of ESR signals subsequent to electrochemical generation of the anion radicals (84, 232, 240, 242, 243, 271).

At even lower pH values, or under the influence of less active inhibitors and in the absence of inhibitors, proton transfer to the anion radical proceeds as a heterogeneous reaction with immediate further reduction of the $\phi-NO_2H$ radical (242, 244, 271).

In acidic solutions containing inhibitors, the dependence of currents on potential and hydrogen ion concentration suggest a preprotonation of the nitro group with subsequent rate-determining electron transfer:

$$\phi-NO_2 + H^+ \rightleftharpoons \phi-NO_2H^+ \xrightarrow{+e^-} \phi-NO_2H \longrightarrow \tag{138}$$

with the possible intermediate formation of the $\phi-NO_2H$ radical which is immediately reduced (202). Preprotonation also takes place in the absence of surfactants, but analysis of the dependence of currents on potential and pH values shows that the reactions are more involved and suggests heterogeneous protonation with the depolarizer and intermediates being adsorbed (125, 202).

Hence the following is a general reduction scheme for nitro groups (202, 244), neglecting reduction to the amine stage at low pH values or under the influence of certain substituents:

$$\begin{array}{c}
\text{Dismutation IV} \\
\hline
\end{array}$$

$$R-NO_2 \underset{I}{\overset{e^-}{\rightleftharpoons}} R-NO_2^- \xrightarrow[II]{e^-} R-NO_2^{-2} \xrightarrow[III]{2e^- + 2H_2O} R-NHO^-$$

$$\Big\updownarrow H^+ \qquad VI \Big| H^+ \text{(Heterogeneous reaction)} \qquad \Big\updownarrow H^+$$

$$R-NO_2H^+ \xrightarrow[VII]{} R-NO_2H \xrightarrow[VIII]{3e^- + 3H^+} R-NHOH$$

$$\tag{139}$$

The pathways followed in the absence of inhibitors are probably:

pH > 9: I, II, III ($E \sim -0.7$ V versus SCE)

0 < pH < 9: I, VI, VIII, or V, VII, VIII ($E \sim -0.1$ to -0.7 V versus SCE)

whereas under the influence of inhibitors the reduction proceeds:

pH > 10: First wave: I ($E \sim -0.7$ V versus SCE)
Second wave: II, III ($E < -1$ V versus SCE)

6 < pH < 10: First wave: I, IV, or I, VI, VIII (limiting current controlled by homogeneous, IV, or heterogeneous, VI, kinetics; $E \sim -0.7$ V versus SCE)
Second wave: II, III ($E < -1$ V versus SCE)

0 < pH < 6: V, VII, VIII ($E \sim -0.2$ to -0.7 V versus SCE)

This scheme, virtually established for nitrobenzene and corresponding aromatic nitro compounds, in principle also applies to other nitro compounds, although different kinetic parameters may cause deviations, for instance, in the respective pH regions and potentials.

The mechanism shown, especially the participation of the radical stage, has been confirmed in parts by investigations of several authors with different compounds, techniques, and electrodes. Thus a two-wave reduction with one and three electrons has been found in alkaline solutions with carbon paste electrodes (2), as well as with rotating disk electrodes made of different noble metals (203). Nitrophenols have been investigated with ac polarography (50, 77); from measurements including phase angle determinations, it was concluded that, different from the scheme shown above, the anion radical is reduced with the direct transfer of two electrons in the second step (50). The formation of radicals was also confirmed by oscillopolarography (180) and by the use of cyclic voltammetry at a hanging mercury drop (252, 253, 258) and at stationary platinum electrodes (369); with these techniques, as mentioned above, the oxidation of the hydroxylamine stage at rather positive potentials gives rise to the products of the Haber mechanism. The ESR technique has been applied in some work to follow the decay kinetics of the anion radical generated by electrochemical reduction which is a result of dismutation as shown above. Whereas the half-lives of the negative ions of nitrobenzene (237, 242) and of m-nitrophenol (84) are several minutes or more (second-order rate constant $k_2 \sim 1\ M^{-1}\ \text{sec}^{-1}$), the decay proceeds faster with the p-hydroxy and considerably faster with the o-hydroxy isomers (84). A fast decay has also been observed with nitrofuran (141)

(0.1–1 sec) and with aliphatic nitro compounds (369) (2–5 sec). [Anion radicals from aliphatic nitro compounds generated by homogeneous electron transfer from an electron donor rather than by electrochemical reduction may undergo chemical reactions under the prevailing conditions and give rise to different radical species (163).]

Whereas the investigations mentioned so far refer to strongly alkaline solutions (about pH 12–13), the kinetics for the nitrobenzene anion radical decay can be followed without special techniques only down to about pH 9, at which the half-life is only about 1 sec, owing to the reactions shown in Eqs. 136 and 137 (240, 271). A special technique, generating the anion radical in alkaline solutions and, after rapid mixing with the solution of a proper buffer acid, following the decay in a flow capillary inserted in the ESR cavity, allowed the determination of rate constants at even lower pH values (242, 243). ESR spectra have also been observed during the reduction of nitrofuran and some of its derivatives (431); splitting of the four-electron reduction into two waves is observed with this compound at a dropping electrode of short drop time (0.15 sec) in solutions containing 10% or more ethanol without addition of a particular inhibitor, indicating kinetic control of currents by homogeneous proton transfer with subsequent dismutation and/or heterogeneous proton transfer with immediate subsequent reduction, as shown in the scheme above. The decrease in the first wave from four- to one-electron reduction appeared at pH values that decreased with increasing alcohol content.

Further work confirming the radical mechanism was done with nitrofurazone (258), chloramphenicol (301), aromatic and heterocyclic nitrocarboxylic acids (363), ω-nitrostyrene (197), p-nitroazobenzene (204), and nitrofuran (430). Unusual behavior occurs with some nitro compounds. Thus the half-wave potentials of N-nitropyrazoles remain constant over nearly the entire accessible pH range, which was interpreted in terms of a rate-determining transfer of the first electron without participation of protons (283). p-Dinitrobenzene does not show one-electron reduction in aqueous solutions because of the favored formation of the quinoid dianion (190, 193) (see Eq. 79); this is in contrast with the behavior in aprotic solvents in which anion radicals are formed (149, 314), obviously because of the unfavorable conditions for dianions. Uncharged radicals are formed upon reduction of gem-halonitro compounds with the elimination of halide ions (417):

$$R_2C\begin{smallmatrix}NO_2\\X\end{smallmatrix} + e^- \longrightarrow R_2\dot{C}\begin{smallmatrix}NO_2\\\\\end{smallmatrix} + X^- \qquad (140)$$
$$\xrightarrow{+e^-} R_2C=NO_2^-$$

and subsequent reduction to give aci-nitro compounds; an analogous mechanism applies to the corresponding nitroso compounds. Whereas the anion radical of trimethylnitromethane is stable enough to allow further reduction in acetonitrile giving the corresponding hydroxylamine, those derived from $(CH_3)_2CX(NO_2)$ (with X = CN, $CONH_2$, $COOC_2H_5$) decompose with the elimination of nitrite ions:

$$R_2C{\overset{NO_2}{\underset{X}{\diagup\!\!\!\diagdown}}} + e^- \rightarrow R_2CX\cdot + NO_2^- \qquad (141)$$

with subsequent stabilization of the radical by reaction with the solvent or by dimerization (395); cleavage of the C—N bond and formation of NO_2^-, however, is decreased in the presence of proton donors owing to immediate reduction of the protonated radical $R_2CX(NO_2H)$ prior to its decomposition.

Because of the reversibility of the first electron transfer in the alkaline reduction of nitrobenzene, this compound may act as a catalyst for kinetically hindered reduction processes; thus catalytic currents have been observed with periodate, peroxodisulfate, and complexes of copper(II) in the presence of small amounts of nitrobenzene (235, 237). The kinetics of the reaction with peroxodisulfate:

$$\phi\text{—}NO_2^- + S_2O_8^{-2} \rightarrow \phi\text{—}NO_2 + SO_4^{-2} + SO_4^- \qquad (142)$$

with subsequent consumption of another anion radical, presumably by

$$\phi\text{—}NO_2^- + SO_4^- \rightarrow \phi\text{—}NO_2 + SO_4^{-2} \qquad (143)$$

have been followed independently by electrochemical generation of the anion radical, rapid admixing of a solution containing $S_2O_8^{-2}$, and following the decay of the ESR signal observed in a flow system as described above for the dismutation reaction of the anion radical at lower pH values (242, 243).

Numerous studies have been made with nitro compounds in non-aqueous solvents, especially in aprotic acetonitrile and dimethylformamide, frequently with the joint application of ESR techniques initiated by the application of the electrochemical generation of radicals inside the microwave cavity (146, 147, 315). A review of polarography and ESR of nitrobenzene derivatives was published several years ago in Japanese (266); further details including the influence of solvents (370), cations, and so on, are considered elsewhere (247).

Similarly to numerous other electroactive compounds, the nitro group exhibits a reversible one-electron wave in aprotic solvents resulting from the formation of the anion radical. Half-wave potentials are more negative than those for the corresponding process in aqueous solutions and depend considerably upon the cation of the supporting electrolyte, which has been explained by ion pair formation (200, 202). The number of electrons transferred in the second wave, as well as the potential of the latter, differ greatly depending on the solvent and on the special structure of the compound, namely, the substituents. For instance, nitrobenzenes with an electron-withdrawing substituent in a para position may give rise to a stabilization of the dianion by permitting the formation of a quinoid structure:

$$\underset{RO}{\overset{\bar{O}}{\underset{\|}{C}}}-\underset{}{\bigcirc}-NO_2^- + e^- \longrightarrow \underset{RO}{\overset{-O}{\underset{\|}{C}}}=\underset{}{\bigcirc}=\overset{+}{N}\underset{O^-}{\overset{O^-}{\diagup}} \qquad (144)$$

whereas other substituents give rise to the transfer of three electrons, with the participation of protons from the solvent, and the formation of the phenylhydroxylamine derivatives (200).

In general, anion radicals are very stable in aprotic solvents, allowing the observation of light absorption spectra when generated at mercury pool electrodes (255, 256), although such spectra have also been observed in aqueous solution (237). This stability has been attributed (237) to their inability to undergo dismutation, since the initiating equilibrium

$$2R-NO_2^- \rightleftharpoons R-NO_2 + R-NO_2^{-2} \qquad (145)$$

is largely shifted to the left and dismutation can only proceed to a considerable extent if the dianion is consumed by reaction with a proton donor such as water. This is in accordance with the effect of adding water as a proton donor to dimethylformamide (221, 292) or to acetonitrile (200), by which the second polarographic wave of nitro compounds is shifted to more positive potentials. This behavior must be explained either by participation of protons in the rate-determining step of charge transfer to the anion radical or, more probably, by proton transfer to the dianion generated by reversible transfer of the second electron. Different from this effect is that of the stronger phthalic acid in acetonitrile (64), which produces a wave at more positive potentials than that of the first reversible electron transfer; this suggests protonation of the nitro group prior to the transfer of the first electron. This explanation, however, has been modified somewhat, accounting for the fact that the potential (in cyclic voltammetry at a hanging mercury drop) is shifted somewhat to more negative potentials with increasing acid concentrations (at low acid/depolarizer ratios). The formation of a hydrogen-bonded complex acid–nitro group, being more strongly bonded in the electrical double layer than in the bulk of the solution, has been proposed to account for the observations.

Correlations of half-wave potentials have been established with Hammett's constants (443, 495) and with ESR data (93, 297), as well as with the energy of light absorbed during charge transfer in complexes of nitrobenzenes with electron donors (354).

Further investigations in dimethylformamide and/or acetonitrile were made with tetraisopropylnitrobenzene (323), ortho-substituted nitrobenzenes (324), pentafluoronitrobenzene (57), trinitrobenzene (156), and 9-nitrotriptycene (178).

Some halonitrobenzenes show special behavior in eliminating halide ions (see Section III-G) prior to the formation of the nitro group anion radicals. In voltammetry with a hanging mercury drop electrode, three peaks have been observed with iodonitrobenzenes (265) and attributed to, in order, halide elimination (presumably followed by proton transfer from the solvent), formation of the nitrobenzene negative ion, and further reduction of the latter:

$$I\text{-}C_6H_4\text{-}NO_2 + 2e^- \longrightarrow {}^{\cdot}C_6H_4\text{-}NO_2 + I^- \quad (146)$$

$$\xrightarrow{+H^+} \varnothing\text{-}NO_2 \xrightarrow{+e^-} \varnothing\text{-}NO_2^{\cdot-} \xrightarrow{\text{Further reduct}}$$

Hence no ESR signal was observed when reduction was performed at the potential of the first peak, and only the spectrum of the nitrobenzene negative ion was exhibited upon reduction at the potential of the second peak. The same is true, to some extent, for bromo, but not for chloro derivatives, the latter retaining the halogen during formation of the anion radicals. No elimination was observed with 2,6-dichloronitrobenzene in dimethylformamide and with 2,6-dibromonitrobenzene in a 1:1 mixture of this solvent with water (267); rather large nitrogen coupling constants in the ESR spectra of the anion radicals suggest decoupling of the π-electron system owing to steric effects of the halogen atoms. The same is true for pentachloronitrobenzene negative ion in methanol showing, apart from ^{15}N and ^{13}C effects, only three lines separated by 20.5 G, which approaches the corresponding values of aliphatic nitro radicals and indicates that the odd electron is localized largely at the nitro group (245). Dehalogenation has also been found for chloronitrobenzenes during prolonged electrolysis or reduction at more negative potentials (133, 134). Partial dehalogenation with replacement by the methoxy group has also been confirmed to be responsible for the change with time of the ESR spectrum observed during reduction of pentachloronitrobenzene in methanol (245); this replacement reaction proceeds even with the nitro compound prior to reduction, although at a considerably lower rate than with the anion radical.

Nitro anion radicals have also been observed with methanol (236) and liquid ammonia (175) as solvents.

Nitrosobenzene, showing a reversible two-electron wave in aqueous solutions, exhibits two one-electron waves in dimethylformamide (255, 257) and acetonitrile (329), corresponding to the formation of the anion radical and dianion:

$$\phi-NO + e^- \longrightarrow \phi-NO^- \xrightarrow{+e^-} \phi-NO^{-2} \tag{147}$$

The anion radical is rather unstable toward dimerization with elimination of water, presumably through participation of protons from the solvent:

$$2\phi-NO^- + 2H^+ \rightarrow \phi-N\!\!=\!\!N-\phi + H_2O \tag{148}$$
$$\qquad\qquad\qquad\qquad\quad \downarrow$$
$$\qquad\qquad\qquad\qquad\quad O$$

as well as toward small amounts of oxygen resulting in the oxidation to nitrobenzene. In solutions carefully purified of oxygen and water, the anion radical is stable enough to permit observation of its ESR spectrum (329). This radical has also been obtained in aqueous solution as a product of oxidation, by peroxodisulfate, of phenylhydroxylamine resulting from dismutation of electrolytically generated nitrobenzene anion radicals (242); the anion radical $\phi-NO^-$ exhibited a half-life of about 30 sec in these alkaline aqueous solutions containing $S_2O_8^{-2}$.

Acknowledgment: This work was carried out in part at the Chemisches Institut der Hochschule Bamberg; support by the Bundesministerium für Wirtschaft is gratefully acknowledged.

References

1. Acker, D. S., and W. R. Hertler, *J. Am. Chem. Soc.*, **84**, 3370 (1962).
2. Adams, R. N., *Rev. Polarog. (Kyoto)*, **11**, 71 (1963).
3. Adams, R. N., *J. Electroanal. Chem.*, **8**, 151 (1964).
4. Allendoerfer, R. D., and P. H. Rieger, *J. Am. Chem. Soc.*, **87**, 2336 (1965).
5. Anthoine, G., J. Nasielski, E. van der Donckt, and N. Vanlautem, *Bull. Soc. Chim. Belges*, **76**, 230 (1967).

6. Arad, Y., M. Levy, I. R. Miller, and D. Vofsi, *J. Electrochem. Soc.*, **114**, 899 (1967).
7. Arai, T., *Bull. Chem. Soc. Japan*, **32**, 184 (1959).
8. Arai, T., and T. Oguri, *Bull. Chem. Soc. Japan*, **33**, 1018 (1960).
9. Arai, T., *Nippon Kagaku Zasshi*, **89**, 188 (1968); *Chem. Abstr.*, **68**, 110840 (1968).
10. Ashworth, M., *Collection Czech. Chem. Commun.*, **13**, 229 (1948).
11. Asthana, M., R. C. Kapoor, and H. L. Nigam, *Electrochim. Acta*, **11**, 1587 (1966).
12. Atherton, N. M., J. N. Ockwell, and R. Dietz, *J. Chem. Soc., A*, 771 (1967).
13. Austen, D. E. G., P. H. Given, D. I. E. Ingram, and M. E. Peover, *Nature*, **182**, 1784 (1958).
14. Aylward, G. H., J. L. Garnett, and J. H. Sharp, *Chem. Commun.* 137 (1966); *Anal. Chem.*, **39**, 457 (1967).
15. Bachofner, H. E., F. M. Beringer, and L. Meites, *J. Am. Chem. Soc.*, **80**, 4269 (1958).
16. Bannerjee, N. R., and S. K. Vig, *J. Chem. Soc., B*, 484 (1967).
17. Bargain, M., *Compt. Rend.*, **254**, 130; **255**, 1948 (1962); **256**, 1990 (1963).
18. Barnes, D., and P. Zuman, *J. Electroanal. Chem.*, **16**, 575 (1968).
19. Bartel, E. T., and Z. R. Grabowski, *Roczniki Chem.* **31**, 323 (1957).
20. Bartel, E. T., Z. R. Grabowski, and W. Kemula, *Roczniki Chem.*, **31**, 13 (1957).
21. Bauer, E., and H. Berg, *Studia Biophys. (Berlin)*, **1**, 143 (1966).
22. Bauer, E., *J. Electroanal. Chem.*, **14**, 351 (1967).
23. Beckmann, P., *Australian J. Chem.*, **14**, 229 (1961).
24. Beletskaya, I. P., K. P. Butin, and O. A. Reutov, *Zh. Org. Khim.*, **3**, 231 (1967); *Chem. Abstr.*, **66**, 110999 (1967).
25. Benesch, R. E., and R. Benesch, *J. Phys. Chem.*, **56**, 648 (1952).
26. Berg, H., and H. Schweiss, *Nature*, **191**, 1270 (1961); H. Berg, *Naturwissenschaften*, **49**, 11 (1962).
27. Berg, H., *Naturwissenschaften*, **48**, 100 (1961).
28. Berg, H., *Naturwissenschaften*, **48**, 714 (1961).
29. Berg, H., *Rev. Polarog. (Kyoto)*, **11**, 29 (1963).
30. Berg, H., and K. Weller, 6th International Symposium on Free Radicals, Cambridge 1963, paper G.
31. Berg, H., *Polarographische Analyse* (T. A. Krjukowa, S. I. Sinjakowa, and T. W. Arefjewa, Eds.), VEB Deutscher Verlag für Grundstoffindustrie, Leipzig, 1964, p. 614; *Abhandl. Deut. Akad. Wiss. Berlin, Kl. Chem. Geol. Biol.*, 128 (1964).

32. Berg, H., and K. Kramarczyk, *Ber. Bunsenges. Physik. Chem.,* **68,** 296 (1964).
33. Berg, H., *Z. Phys. Chem. (Leipzig),* **229,** 138 (1965).
34. Berg, H., and K. Kramarczyk, *Talanta,* **12,** 1127 (1965).
35. Berg, H., and F. A. Gollmick, *Collection Czech. Chem. Commun.,* **30,** 4192 (1965).
36. Berg, H., *Z. Anal. Chem.,* **216,** 165 (1966).
37. Berg, H., and K. Kramarczyk, *Biochim. Biophys. Acta,* **131,** 141 (1967).
38. Berg, H., H. Schweiss, E. Stutter, and K. Weller, *J. Electroanal. Chem.,* **15,** 415 (1967).
39. Bergman, I., and J. C. James, *Trans. Faraday Soc.,* **48,** 956 (1952); **50,** 60 (1954).
40. Bergman, I., *Polarography 1964* (G. J. Hills, Ed.), Macmillan, London, 1966, p. 985.
41. Bezuglyi, V. D., L. A. Mel'nik, and V. N. Dimitrieva, *Zh. Obshch. Khim.,* **34,** 1048 (1964); *Chem. Abstr.,* **61,** 1512f (1964).
42. Bezuglyi, V. D., L. Ya. Kheifets, N. A. Sobina, N. S. Dokunikhin, and N. B. Kolokolov, *Zh. Obshch. Khim.,* **37,** 778 (1967); *Chem. Abstr.,* **67,** 96270 (1967).
43. Billon, J. P., *Bull. Soc. Chim. France,* 1784 (1960); J. P. Billon, G. Cauquis, J. Combrisson, and A. M. Li, ibid., 2062 (1960).
44. Billon, J. P., G. Cauquis, and J. Combrisson, *J. Chim. Phys.,* 374 (1964).
45. Böckel, W., and H. Berg, *Abhandl. Deut. Akad. Wiss. Berlin, Kl. Chem. Geol. Biol.,* 465 (1964).
46. Brdička, R., *Collection Czech. Chem. Commun.,* **5,** 112 (1933).
47. Brdička, R., and E. Knobloch, *Z. Elektrochem.,* **47,** 721 (1941); R. Brdička, ibid., **48,** 686 (1942).
48. Brdička, R., *Z. Elektrochem.,* **59,** 787 (1955).
49. Breslow, R., W. Bahary, and W. Reinmuth, *J. Am. Chem. Soc.,* **83,** 1763 (1961).
50. Britz, D., and H. H. Bauer, *Electrochim. Acta,* **13,** 347 (1968).
51. Broadbent, A. D., and Hch. Zollinger, *Helv. Chim. Acta,* **47,** 2140 (1964).
52. Brodskii, A. I., and L. L. Gordienko, *Teor. Eksperim. Khim., Akad. Nauk Ukr. SSR,* **1,**, 452 (1965); *Chem. Abstr.,* **64,** 3054f (1966).
53. Brodskii, A. I., L. L. Gordienko, and L. S. Degtyarev, *Zh. Vses. Khim. Obshch. Mendeleeva,* **11,** 196 (1966), *Chem Abstr.,* **65,** 8730d (1966).
54. Brodskii, A. I., L. L. Gordienko, and Yu. A. Kruglyak, *Teor. Eksperim. Khim.,* **3,** 98 (1967), *Chem. Abstr.,* **67,** 69337 (1967).
55. Broman, R. F., and R. W. Murray, *Anal. Chem.,* **37,** 1408 (1965).
56. Brook, P. A., and J. A. Crossley, *Electrochim. Acta,* **11,** 1189 (1966).

57. Brown, J. K., and W. G. Williams, *Trans. Faraday Soc.*, **64**, 298 (1968).
58. Broze, M., Z. Luz, and B. L. Silver, *J. Chem. Phys.*, **46**, 4891 (1967).
59. Brück, D., and G. Scheibe, *Z. Elektrochem.*, **61**, 901 (1957).
60. Bullerwell, R. A. F., *Polarography 1964* (G. J. Hills, Ed.), Macmillan, London, 1966, p. 863.
61. Butin, K. P., I. P. Beletskaya, A. N. Kashin, and O. A. Reutov, *J. Organometal. Chem.*, **10**, 197 (1967).
62. Butin, K. P., I. P. Beletskaya, A. N. Ryabtsev, and O. A. Reutov, *Elektrokhimiya*, **3**, 1318 (1967).
63. Butler, J. N., *J. Electroanal. Chem.*, **14**, 89 (1967).
64. Cadle, S. H., P. R. Tice, and J. Q. Chambers, *J. Phys. Chem.*, **71**, 3517 (1967).
65. Caldwell, R. A., and S. Hacobian, *Australian J. Chem.*, **21**, 1 (1968).
66. Čapka, O., *Collection Czech. Chem. Commun.*, **15**, 965 (1950).
67. Cardinali, M., L. Rampazzo, and A. Trazza, *Ric. Sci. Rend., Sez. A*, **8**, 1361 (1965).
68. Cardinali, M., I. Carelli, and A. Trazza, *Ric. Sci.*, **37**, 956 (1967).
69. Castellan, A., F. Masetti, U. Mazzucato, and E. Vianello, *J. Phot. Sci.*, **14**, 164 (1966).
70. Caullet, C., J. M. Bessin, and J. C. Bodard, *Compt. Rend.*, **261**, 1848 (1965).
71. Caullet, C., G. Laur, and A. Nonat, *Compt. Rend.*, **261**, 1974 (1965).
72. Caullet, C., M. Salaün, and M. Hébert, *Compt. Rend.*, **264**, 2006 (1967).
73. Cauquis, G., J. P. Billon, J. Raison, and Y. Thibaud, *Compt. Rend.*, **257**, 2128 (1963).
74. Cauquis, G., and G. Fauvelot, *Bull. Soc. Chim. France*, 2014 (1964).
75. Cauquis, G., and G. Fauvelot, *Polarography 1964* (G. J. Hills, Ed.), Macmillan, London, 1966, p. 847.
76. Chaudhuri, J. N., and S. Basu, *Nature*, **182**, 179 (1958).
77. Cheah, E. P. T., S. Hacobian, and A. J. Harle, *Australian J. Chem.*, **19**, 1117, 1609 (1966).
78. Chiorboli, P., G. Davolio, G. Gavioli, and M. Salvaterra, *Electrochim. Acta*, **12**, 767 (1967).
79. Chopart-dit-Jean, L. H., and E. Heilbronner, *Helv. Chim. Acta*, **36**, 144 (1953).
80. Cisak, A., and P. J. Elving, *Rev. Polarography (Kyoto)*, **11**, 21 (1963).
81. Cisak, A., and P. J. Elving, *Electrochim. Acta*, **10**, 935 (1965).
82. Colichman, E. L., and H. P. Maffei, *J. Am. Chem. Soc.*, **74**, 2744 (1952); E. L. Colichman and J. T. Matschiner, *J. Org. Chem.*, **18**, 1124 (1953).
83. Corvaja, C., P. L. Nordio, M. V. Pavan, and G. Rigatti, *Ric. Sci. Rend. Sez. A*, **4**, 297 (1964).

84. Corvaja, C., G. Farnia, and E. Vianello, *Electrochim. Acta,* **11,** 919 (1966).
85. Costa, G., *Ric. Sci.,* **22,** Suppl. 191 (1952).
86. Cottrell, W. R. T., and R. A. N. Morris, *Chem. Commun.,* 409 (1968).
87. Coulson, D. M., and W. R. Crowell, *J. Am. Chem. Soc.,* **74,** 1290 (1952).
88. Coulson, D. M., and W. R. Crowell, *J. Am. Chem. Soc.,* **74,** 1294 (1952).
89. Coulson, D. M., W. R. Crowell, and S. K. Tendick, *J. Am. Chem. Soc.,* **79,** 1354 (1957).
90. Curtis, M. D., and A. L. Allred, *J. Am. Chem. Soc.,* **87,** 2554 (1965).
91. Damaskin, B. B., I. P. Mishutushkina, V. M. Gerovich, and R. I. Kaganovich, *Zh. Fiz. Khim.,* **38,** 1797 (1964).
92. Davydovskaya, Yu. A., and Yu. I. Vainshtein, *Tr. Vses. Nauch. Issled. Inst. Khim. Reaktivov Osobo Chist. Khim. Veshchestv* No. 28, 238 (1966); *Chem. Abstr.,* **67,** 28734 (1967).
93. Degtyarev, L. S., L. N. Garnyuk, A. M. Golubenkova, and A. I. Brodskii, *Dokl. Akad. Nauk SSSR,* **157,** 1409 (1964).
94. Dehl, R., and G. K. Fraenkel, *J. Chem. Phys.,* **39,** 1793 (1963).
95. Demange-Guérin, G., *Compt. Rend., Ser. C,* **266,** 784 (1968).
96. Dessy, R. E. et al., *J. Am. Chem. Soc.,* **88,** 5112, 5117, 5121, 5124, 5129 (1966); **90,** 1995, 2001, 2005 (1968).
97. Devaud, M., *Compt. Rend., Ser. C,* **263,** 1269 (1966); *J. Chim. Phys.,* **64,** 791 (1967).
98. Devaud, M., *J. Chim. Phys.,* **63,** 1335 (1966).
99. Devaud, M., P. Souchay, and M. Person, *J. Chim. Phys.,* **64,** 646 (1967).
100. Devaud, M. and P. Souchay, *J. Chim. Phys.,* **64,** 1778 (1967).
101. Dietz, R., and M. E. Peover, *Trans. Faraday Soc.,* **62,** 3535 (1966).
102. DiGregorio, J. S., and M. D. Morris, *Anal. Chem.,* **40,** 1286 (1968).
103. Dineen, E., T. C. Schwan, and C. L. Wilson, *Trans. Electrochem. Soc.,* **96,** 226 (1949).
104. Dvořák, V., I. Němec, and J. Zýka, *Microchem. J.,* **12,** 99, 324, 350 (1967).
105. Edsberg, R. L., D. Eichlin, and J. J. Garis, *Anal. Chem.,* **25,** 798 (1953).
106. El-Khiami, I., and R. M. Johnson, *Talanta,* **14,** 745 (1967).
107. Elofson, R. M., and R. L. Edsberg, *Can. J. Chem.,* **35,** 646 (1957).
108. Elving, P. J., O. H. Müller, S. Wawzonek, M. J. Astle, and L. Meites, *Anal. Chem.,* **22,** 482 (1950).
109. Elving, P. J., and C. Teitelbaum, *J. Am. Chem. Soc.,* **71,** 3916 (1949).
110. Elving, P. J., and C. S. Tang, *J. Am. Chem. Soc.,* **72,** 3244 (1950).
111. Elving, P. J., I. Rosenthal, and A. J. Martin, *J. Am. Chem. Soc.,* **77,** 5218 (1955).
112. Elving, P. J., and J. T. Leone, *J. Am. Chem. Soc.,* **80,** 1021 (1958).

113. Elving, P. J., and J. M. Markowitz, *J. Phys. Chem.*, **65**, 686 (1961).
114. Elving, P. J., I. Rosenthal, J. R. Hayes, and A. J. Martin, *Anal. Chem.*, **33**, 330 (1961).
115. Ershler, A. B., G. A. Tedoradze, M. Fakhmi, and K. P. Butin, *Elektrokhimiya*, **2**, 319 (1966).
116. Ershler, A. B., G. A. Tsagareli, and G. A. Tedoradze, *Elektrokhimiya*, **4**, 116 (1968).
117. Evans, D. H., *J. Electroanal. Chem.*, **6**, 419 (1963).
118. Feldman, M., and S. Winstein, *Tetrahedron Letters*, 853 (1962).
119. Felton, R. H., and H. Linschitz, *J. Am. Chem. Soc.*, **88**, 1113 (1966).
120. Feoktistov, L. G., and S. I. Zhdanov, *Electrochim. Acta*, **10**, 657 (1965).
121. Feoktistov, L. G., A. P. Tomilov, Yu. D. Smirnov, and M. M. Gol'din, Elektrokhimiya, **1**, 887 (1965).
122. Fernandez-Martin, R., R. G. Rinker, and W. H. Corcoran, *Anal. Chem.*, **38**, 930 (1966).
123. Fields, M., and E. R. Blout, *J. Am. Chem. Soc.*, **70**, 930 (1948).
124. Fischer, O., L. Kisová, and J. Štěpánek, *J. Electroanal. Chem.*, **17**, 233 (1968).
125. Fleischmann, M., I. N. Petrov, and W. F. K. Wynne-Jones, *Proceedings of the First Australian Conference on Electrochemistry 1963*, (J. A. Friend, F. Gutmann, Eds.), Pergamon Press, Oxford, 1965, p. 500.
126. Floch, L., M. S. Spritzer, and P. J. Elving, *Anal. Chem.*, **38**, 1074 (1966).
127. Foffani, A., and M. Fragiacomo, *Ric. Sci.*, **22**, Suppl. 139 (1952).
128. Fritsch, J. M., Sh. V. Tatwawadi, and R. N. Adams, *J. Phys. Chem.*, **71**, 338 (1967).
129. Frost, J. G., and J. H. Saylor, *Rec. Trav. Chim.*, **82**, 828 (1963).
130. Frumkin, A. N., and E. P. Andreeva, *Dokl. Akad. Nauk SSSR*, **40**, 417 (1953).
131. Frumkin, A. N., *Advances in Electrochemistry and Electrochemical Engineering* (P. Delahay, Ed.), Vol. I, Interscience, New York, 1961, p. 65.
132. Fueno, T., K. Morokuma, and J. Furukawa, *Bull. Inst. Chem. Res., Kyoto Univ.*, **36**, No. 4, 87, 96 (1958).
133. Fujinaga, T., Y. Deguchi, and K. Umemoto, *Bull. Chem. Soc. Japan*, **37**, 822 (1964).
134. Fujinaga, T., K. Umemoto, and T. Arai, *Denki Kagaku*, **34**, 135 (1966); *Chem. Abstr.*, **66**, 81868 (1967).
135. Fujinaga, T., K. Izutsu, and K. Takaoka, *J. Electroanal. Chem.*, **16**, 89 (1968).
136. Fujinaga, T., and Takaoka, *J. Electroanal. Chem.*, **16**, 99 (1968).

137. Fujinaga, T., K. Takaoka, T. Nomura, and K. Yoshikawa, *Nippon Kagaku Zasshi*, **89**, 185 (1968); *Chem. Abstr.*, **68**, 110807 (1968).
138. Funt, B. L., and D. G. Gray, *Can. J. Chem.*, **46**, 1337 (1968).
139. Galus, Z., and R. N. Adams, *J. Am. Chem. Soc.*, **84**, 2061 (1962); Z. Galus, R. M. White, F. S. Rowland, and R. N. Adams, ibid,. **84**, 2065 (1962).
140. Galus, Z., *Bull. Acad. Polon. Sci., Ser. Sci. Chim.*, **13**, 63 (1965).
141. Gavars, R., J. Stradins, and S. Hillers, *Dokl. Akad. Nauk SSSR*, **157**, 1424 (1964); *Zavodsk. Lab.*, **31**, 41 (1965).
142. Geissmann, T. A., and S. L. Friess, *J. Am. Chem. Soc.*, **71**, 3893 (1949).
143. Gelb, R. I., and L. Meites, *J. Phys. Chem.*, **68**, 2599 (1964).
144. Gerdil, R., and E. A. C. Lucken, *Mol. Phys.*, **9**, 529 (1965).
145. Gerdil, R., and E. A. C. Lucken, *J. Am. Chem. Soc.*, **88**, 733 (1966).
146. Geske, D. H., and A. H. Maki, *J. Am. Chem. Soc.*, **82**, 2671 (1960).
147. Geske, D. H., and J. L. Ragle, *J. Am. Chem. Soc.*, **83**, 3532 (1961).
148. Geske, D. H., and A. L. Balch, *J. Phys. Chem.*, **68**, 3423 (1964).
149. Geske, D. H., J. L. Ragle, M. A. Bambenek, and A. L. Balch., *J. Am. Chem. Soc.*, **86**, 987 (1964).
150. Geske, D. H., and G. R. Padmanabhan, *J. Am. Chem. Soc.*, **87**, 1651 (1965).
151. Giang, B. Y., G. D. Christian, and W. C. Purdy, *J. Polarograph. Soc.*, **13**, 17 (1967).
152. Gilbert, B. C., P. Hanson, R. O. C. Norman, and B. T. Sutcliffe, *Chem. Commun.*, 161 (1966).
153. Given, P. H., and M. E. Peover, *J. Chem. Soc.*, 2674 (1958).
154. Given, P. H., and M. E. Peover, *Collection Czech. Chem. Commun.*, **25**, 3195 (1960); *J. Chem. Soc.*, 385 (1960).
155. Given, P. H., *Abhandl. Deut. Akad. Wiss. Berlin, Kl. Chem. Geol. Biol.*, 481 (1964).
156. Glarum, S. H., and J. H. Marshall, *J. Chem. Phys.*, **41**, 2182 (1964).
157. Gollmick, F. H., and H. Berg, *Ber. Bunsenges. Physik. Chem.*, **69**, 196 (1965).
158. Gordienko, L. L., *Elektrokhimiya*, **1**, 1497 (1965).
159. Gough, T. A., and M. E. Peover, *Polarography 1964* (G. J. Hills, Ed.), Macmillan, London, 1966, p. 1017.
160. Grabowski, Z. R., *Roczniki Chem.*, **28**, 513 (1954).
161. Grabowski, Z. R., *Prace Konferencji Polarograficznej 1956*, Panstw. Wydawn. Naukowe, Warsaw, 1957, p. 91.
162. Grabowski, Z. R., and M. K. Kalinowski, 6th International Symposium on Free Radicals, Cambridge 1963, paper F.
163. Griffiths, W. E., G. F. Longster, J. Myatt, and P. F. Todd, *J. Chem. Soc., B*, 533 (1967).

164. Grimshaw, J., and J. S. Ramsey, *J. Chem. Soc., B*, 60 (1968).
165. Gubin, S. P., and L. I. Denisovich, *Izv. Akad. Nauk SSSR, Ser. Khim.*, 149 (1966); *Chem. Abstr.*, **64**, 12194d (1966).
166. Gulick, W. M., and D. H. Geske, *J. Am. Chem. Soc.*, **88**, 4119 (1966).
167. Gutmann, V., and G. Schöber, *Angew. Chem.*, **70**, 98 (1958).
168. Hale, J. M., *J. Electroanal. Chem.*, **8**, 181 (1964).
169. Hall, D. A., M. Sakuma, and P. J. Elving, *Electrochim. Acta*, **11**, 337 (1966).
170. Hall, D. A., and P. J. Elving, *Electrochim. Acta*, **12**, 1363 (1967); *Anal. Chim. Acta*, **39**, 141 (1967).
171. Halls, D. J., A. Townshend, and P. Zuman, *Anal. Chim. Acta*, **40**, 459; **41**, 51, 63 (1968).
172. Hans, W., and W. Jensch, *Z. Elektrochem.*, **56**, 648 (1952).
173. Hanuš, V., *Chem. Zvesti*, **8**, 702 (1954).
174. Harle, A. J., and L. E. Lyons, *J. Chem. Soc.*, 1575 (1950).
175. Harris, W. S., *U.S. Naval Ordnance Lab. Corona, 6th Symposium on Ammonia Batteries, 1964;* NAVWEPS Rept. No. 8193 (1964).
176. Hartnell, E. D., and C. E. Bricker, *J. Am. Chem. Soc.*, **70**, 3385 (1948).
177. Hausser, K. H., A. Häbich, and V. Franzen, *Z. Naturforsch.*, **16a**, 836 (1961).
178. Heller, P. H., and D. H. Geske, *J. Org. Chem.*, **31**, 4249 (1966).
179. Heyrovský, J., F. Šorm, and J. Forejt, *Collection Czech. Chem. Commun.*, **12**, 11 (1947).
180. Heyrovský, J., *Advances in Polarography* (I. S. Langmuir, Ed.), Vol. I, Pergamon Press, Oxford, 1960, p. 1.
181. Heyrovský, J., and J. Kůta, *Principles of Polarography*, Academic Press, New York, 1966.
182. Heyrovský, M., *Abhandl. Deut. Akad. Wiss. Berlin, Klass. Medizin*, 409 (1966); *Nature*, **209**, 708 (1966).
183. Heyrovský, M., *Proc. Roy. Soc., Ser. A*, **301**, 411 (1967).
184. Hoffmann, A. K., and A. T. Henderson, *J. Am. Chem. Soc.*, **83**, 4671 (1961).
185. Hoijtink, G. J., et al., *Rec. Trav. Chim.*, **73**, 355 (1954); **74**, 277 (1955); **75**, 487 (1956); **76**, 885 (1957).
186. Holleck, L., and H. J. Exner, *Proceedings of the 1st International Congress on Polarography Prague 1951*, Prirodoved. vydavatelstvi, Prague, 1952, Vol. 1, p. 97.
187. Holleck, L., and H. J. Exner, *Z. Naturforsch.*, **6a**, 763 (1951).
188. Holleck, L., *Z. Naturforsch.*, **7a**, 282 (1952).
189. Holleck, L., and H. J. Exner, *Z. Elektrochem.*, **56**, 46 (1952).
190. Holleck, L., and H. J. Exner, *Z. Elektrochem.*, **56**, 677 (1952).
191. Holleck, L., and H. Marsen, *Z. Elektrochem.*, **57**, 301 (1953).

192. Holleck, L., and H. Marsen, Z. *Elektrochem.*, **57**, 944 (1953).
193. Holleck, L., and H. Schmidt, *Naturwissenschaften*, **41**, 87 (1954).
194. Holleck, L., and B. Kastening, Z. *Elektrochem.*, **60**, 127 (1956).
195. Holleck, L., and B. Kastening, Z. *Elektrochem.*, **63**, 177 (1959).
196. Holleck, L., and B. Kastening, *Vorträge des III. Internationalen Kongresses für Grenzflächenaktive Stoffe Köln 1960*, Verlag der Universitätsdruckerei Mainz, 1960, Vol. II, p. 288; *Korrosion* 16 (E. Rabald and W. Fritsche, Eds.), Verlag Chemie, Weinheim, 1963, p. 29.
197. Holleck, L., and D. Jannakoudakis, Z. *Naturforsch.*, **16b**, 396 (1961).
198. Holleck, L., and O. Lehmann, *Monatsh. Chem.*, **92**, 499 (1961); *Ber. Bunsenges. Physik. Chem.*, **67**, 609 (1963).
199. Holleck, L., B. Kastening, and R. D. Williams, Z. *Elektrochem.*, **66**, 396 (1962).
200. Holleck, L., and D. Becher, *J. Electroanal. Chem.*, **4**, 321 (1962).
201. Holleck, L., and D. Marquarding, *Naturwissenschaften*, **49**, 468 (1962).
202. Holleck, L., and B. Kastening, *Rev. Polarog. (Kyoto)*, **11**, 129 (1963).
203. Holleck, L., B. Kastening, and H. Vogt, *Electrochim. Acta*, **8**, 255 (1963).
204. Holleck, L., and G. Holleck, *Monatsh. Chem.*, **95**, 990 (1964).
205. Holleck, L., Z. *Anal. Chem.*, **224**, 236 (1967); L. Holleck, S. Vavřička, and M. Heyrovský, Z. *Naturforsch.*, **22b**, 1226 (1967).
206. Holleck, L., D. Jannakoudakis, and A. Wildenau, *Electrochim. Acta*, **12**, 1523 (1967).
207. Holleck, L., J. M. Abd-El-Kader, and A. M. Shams-El-Din, *J. Electroanal. Chem.*, **17**, 401 (1968).
208. Holleck, L., and S. Mahapatra, *Monatsh. Chem.*, **100**, 1928 (1969).
209. Holý, A., J. Krupička, and Z. Arnold, *Collection Czech. Chem. Commun.*, **30**, 4127 (1965).
210. Hopin, A. M., and S. I. Zhdanov, *J. Polarograph. Soc.*, **13**, 37 (1967).
211. Hume, D. N., *Anal. Chem.*, **38**, 261R (1966); **40**, 174R (1968).
212. Hünig, S., H. Balli, H. Conrad, and A. Schott, *Liebigs Ann. Chem.*, **676**, 52 (1964).
213. Hünig, S., *Pure Appl. Chem.*, **15**, 109 (1967).
214. Hünig, S., and J. Gross, *Tetrahedron Letters*, 2599 (1968).
215. Hünig, S., et al., *Angew. Chem.*, **80**, 343 (1968).
216. Hush, N. S., and K. B. Oldham, *J. Electroanal. Chem.*, **6**, 34 (1963).
217. Iida, Y., and H. Akamatu, *Bull. Chem. Soc. Japan*, **40**, 231 (1967).
218. Il'yasov, A. V., Yu. M. Kargin, Ya. A. Levin, and V. Kh. Ivanova, *Izv. Akad. Nauk SSSR, Ser. Khim.*, 583 (1966).
219. James, J. C., and J. C. Speakman, *Trans. Faraday Soc.*, **48**, 474 (1952).

220. James, M. I., and P. H. Plesch, *Chem. Commun.*, 508 (1967).
221. Jannakoudakis, D., and G. Stalidis, *Chim. Chronika (Athens)*, **29**, 248 (1964).
222. Jannakoudakis, D., A. Wildenau, and L. Holleck, *J. Electroanal. Chem.*, **15**, 83 (1967).
223. Jannakoudakis, D., and A. Wildenau, *Z. Naturforsch.*, **22b**, 603 (1967).
224. Janzen, E. G., and J. L. Gerlock, *J. Phys. Chem.*, **71**, 4577 (1967).
225. Jordan, J., and P. T. Smith, *Proc. Chem. Soc.*, 246 (1960).
226. Kalinowski, M. K., and Z. R. Grabowski, *Trans. Faraday Soc.*, **62**, 918, 926 (1966).
227. Kalvoda, R., and G. Budnikov, *Abhandl. Deut. Akad. Wiss. Berlin Klass. Chem. Geol. Biol.*, 459 (1964).
228. Kapoor, R. C., M. Asthana, and H. L. Nigam, *J. Polarograph. Soc.*, **10**, 41 (1964).
229. Kapoor, R. C., M. Asthana, and H. L. Nigam, *Rev. Polarog. (Kyoto)*, **14**, 399 (1967).
230. Kardos, A. M., P. Valenta, and J. Volke, *J. Electroanal. Chem.*, **12**, 84 (1966).
231. Kargin, Yu., O. Manoušek, and P. Zuman, *J. Electroanal. Chem.*, **12**, 443 (1966).
232. Kastening, B., and L. Holleck, *Z. Elektrochem.*, **63**, 166 (1959).
233. Kastening, B., *Z. Elektrochem.*, **64**, 82 (1960).
234. Kastening, B., and L. Holleck, *Z. Elektrochem.*, **64**, 823 (1960); B. Kastening, *Ber. Bunsenges. Physik. Chem.*, **68**, 979 (1964); *Polarography 1964* (G. J. Hills, Ed.), Macmillan, London, 1966, p. 359.
235. Kastening, B., *Naturwissenschaften*, **47**, 443 (1960).
236. Kastening, B., *Naturwissenschaften*, **49**, 130 (1962).
237. Kastening, B., *Electrochim. Acta*, **9**, 241 (1964).
238. Kastening, B., H. Gartmann, and L. Holleck, *Electrochim. Acta*, **9**, 741 (1964).
239. Kastening, B., and L. Holleck, *Talanta*, **12**, 1259 (1965).
240. Kastening, B., *Collection Czech. Chem. Commun.*, **30**, 4033 (1965).
241. Kastening, B., *Proceedings of 4th International Congress of Polarography, Prague 1966*, J. Heyrovský Institute of Polarography, Prague, 1966, p. 42.
242. Kastening, B., *Z. Anal. Chem.*, **224**, 196 (1967).
243. Kastening, B., and S. Vavřička, *Ber. Bunsenges. Physik. Chem.*, **72**, 27 (1968).
244. Kastening, B. and L. Holleck, *J. Electroanal. Chem.*, **27**, 355 (1970).
245. Kastening, B. and S. Vavřička, *J. Electroanal. Chem.*, **29**, 195 (1971).
246. Kastening, B., *Anal. Chem.*, **41**, 1142 (1969).

247. Kastening, B., *Advances in Analytical Chemistry and Instrumentation: Electroanalytical Techniques and Applications* (C. N. Reilley and H. W. Nürnberg, Eds.), Interscience, in press.
247a. Kastening, B., *Chem. Ing. Tech.*, in press; *J. Electroanal. Chem.*, in preparation.
247b. Kastening, B., *Angew. Chem. Internat. Ed.*, in press.
248. Katz, T. J., *J. Am. Chem. Soc.*, **82**, 3784 (1960); T. J. Katz, W. H. Reinmuth, and D. E. Smith, ibid., **84**, 802 (1962).
249. Katz, T. J., M. Yoshida and L. C. Siew, *J. Am. Chem. Soc.*, **87**, 4516 (1965).
250. Kaye, R. C., and H. I. Stonehill, *J. Chem. Soc.*, 27 (1951).
251. Kaye, R. C., and H. I. Stonehill, *J. Chem. Soc.*, 3240 (1952).
252. Kemula, W., and Z. Kublik, *Roczniki Chem.*, **32**, 941 (1958).
253. Kemula, W., and Z. Kublik, *Bull. Acad. Polon. Sci., Ser. Sci. Chim. Geol. Geogr.*, **6**, 653 (1958); W. Kemula, *Advances in Polarography* (I. S. Longmuir, Ed.), Vol. I, Pergamon Press, Oxford, 1960, p. 105.
254. Kemula, W., Z. R. Grabowski, and M. K. Kalinowski, *Collection Czech. Chem. Commun.*, **25**, 3306 (1960).
255. Kemula, W., and R. Sioda, *Bull. Acad. Polon. Sci., Ser. Sci. Chim.*, **10**, 507 (1962).
256. Kemula, W., and R. Sioda, *Bull. Acad. Polon. Sci., Ser. Sci. Chim.*, **10**, 513 (1962); *Nature*, **197**, 588 (1963); *J. Electroanal. Chem.*, **7**, 233 (1964).
257. Kemula, W., and R. Sioda, *J. Electroanal. Chem.*, **6**, 183 (1963).
258. Kemula, W., and A. Chodkowska, *Roczniki Chem.*, **41**, 1373 (1967).
259. Kemula, W., M. K. Kalinowski, and A. Girdwoyn, *Roczniki Chem.*, **41**, 1975 (1967).
260. Kemula, W., and M. K. Kalinowski, *Z. Anal. Chem.*, **224**, 383 (1967).
261. Kern, W., and H. Quast, *Makromol. Chem.*, **10**, 202 (1953).
262. Kheifets, L. Ya., E. A. Preobrazhenskaya, and V. D. Bezuglyi, *Zh. Obshch. Khim.*, **35**, 1703 (1965); *Chem. Abstr.*, **64**, 3049c (1966).
263. Kheifets, L. Ya., and V. D. Bezuglyi, *Elektrokhimiya*, **2**, 800 (1966).
264. Khopin, A. M., and S. I. Zhdanov, *Elektrokhimiya*, **4**, 228 (1968).
265. Kitagawa, T., T. P. Layloff, and R. N. Adams, *Anal. Chem.*, **35**, 1086 (1963).
266. Kitagawa, T., *Rev. Polarog. (Kyoto)*, **12**, 11 (1964).
267. Kitagawa, T., and R. Nakashima, *Rev. Polarog. (Kyoto)*, **13**, 115 (1966).
268. Kolthoff, I. M., W. Stricks, and N. Tanaka, *J. Am. Chem. Soc.*, **77**, 5215 (1955).
269. Kolthoff, I. M., *J. Polarograph. Soc.*, **10**, 22 (1964).
270. Kolthoff, I. M., and R. Woods, *J. Am. Chem. Soc.*, **88**, 1371 (1966).
271. Koopmann, R., and H. Gerischer, *Ber. Bunsenges. Physik. Chem.*, **70**, 127 (1966).

272. Korshunov, I. A., and Yu. V. Vodzinskii, *Zh. Fiz. Khim.*, **27**, 1152 (1952), *Chem. Abstr.*, **48**, 5674c (1954).
273. Koutecký, J., and J. Koryta, *Collection Czech. Chem. Commun.*, **19**, 845 (1954).
274. Koutecký, J., and V. Hanuš, *Collection Czech. Chem. Commun.*, **20**, 124 (1955).
275. Kubota, T., and H. Miyazaki, *Bull. Chem. Soc. Japan*, **39**, 2057 (1966).
276. Kumar, A. N., H. L. Nigam, and T. D. Seth, *J. Polarograph. Soc.*, **12**, 93 (1966).
277. Kunz, D., S. Scheithauer, and R. Mayer, *Z. Chem.*, **7**, 194 (1967).
278. Kuwana, T., *Electroanalytical Chemistry* (A. J. Bard, Ed.), Vol. I, Marcel Dekker, New York, 1966, p. 197.
279. Kuwata, K., and D. H. Geske, *J. Am. Chem. Soc.*, **86**, 2101 (1964).
280. Laitinen, H. A., and S. Wawzonek, *J. Am. Chem. Soc.*, **64**, 1765 (1942).
281. Laviron, E., *Collection Czech. Chem. Commun.*, **30**, 4219 (1965).
282. Laviron, E., and Ch. Degrand, *Bull. Soc. Chim. France*, 2194 (1966).
283. Laviron, E., and P. Fournari, *Bull. Soc. Chim. France*, 518 (1966).
284. Laviron, E., and J. C. Lucy, *Bull. Soc. Chim. France*, 2202 (1966).
285. Lavrushin, V. F., V. D. Bezuglyi, and G. G. Belous, *Zh. Obshch. Khim.*, **33**, 1711 (1963).
286. Leach, S. J., J. H. Baxendale, and M. G. Evans, *Australian J. Chem.*, **6**, 395 (1953).
287. Leach, S. J., *Australian J. Chem.*, **13**, 520 (1960).
288. Lee, H. Y., and R. N. Adams, *Anal. Chem.*, **34**, 1587 (1962).
289. Le Guillanton, G., *Bull. Soc. Chim. France*, 2359 (1963).
290. Le Guyader, M., *Compt. Rend., Ser. C*, **262**, 1383 (1966).
291. Levich, V. G., and V. Yu. Filinovskii, *Bull. Acad. Polon. Sci., Ser. Sci. Chim.*, **11**, 705 (1963).
292. Levin, E. S., and Z. I. Fodiman, *Zh. Obshch. Khim.*, **34**, 1055 (1964); *Chem. Abstr.*, **61**, 1511f (1964).
293. Levin, E. S., Z. I. Fodiman, and Z. V. Todres, *Elektrokhimiya*, **2**, 175 (1966).
294. Levin, Ya. A., Yu. M. Kargin, V. S. Galeev, and V. I. Sannikova, *Izv. Akad. Nauk SSSR, Ser. Khim.*, 411 (1968); *Chem. Abstr.*, **69**, 7802 (1968).
295. Littlehailes, J. D., and B. J. Woodhall, *Chem. Commun.*, 665 (1967).
296. Lorenz, W., and U. Gaunitz, *Collection Czech. Chem. Commun.*, **31**, 1389 (1966); W. Lorenz, *Z. Phys. Chem.*, **244**, 65 (1970).
297. Lucken, E. A. C., *J. Chem. Soc., A*, 991 (1966).
298. Lund, H., *Acta Chem. Scand.*, **13**, 249 (1959).
299. Lund, H., *Acta Chem. Scand.*, **14**, 1927 (1960).

300. Lund, H., *Österr. Chem. Ztg.*, **68**, 152 (1967).
301. Macris, C. G., P. P. Georgakopoulos, and D. Jannakoudakis, *Chim. Chronika (Athens)*, **A32**, 104 (1967).
302. Mairanovskii, S. G., *Dokl. Akad. Nauk SSSR*, **110**, 593 (1956); **114**, 1272 (1957); **120**, 1294 (1958); **132**, 1352 (1960); **133**, 162 (1960); **142**, 1327 (1962); *Proc. Acad. Sci. USSR, Sect. Phys. Chem.*, **114**, 437 (1957), *J. Electroanal. Chem.*, **4**, 166 (1962); *Abhandl. Deut. Akad. Wiss. Berlin, Kl. Chem. Geol. Biol.*, 369 (1964).
303. Mairanovskii, S. G., *Russ. Chem. Rev.*, **33**, 38 (1964).
304. Mairanovskii, S. G., and V. N. Pavlov, *Zh. Fiz. Khim.*, **38**, 1804 (1964).
305. Mairanovskii, S. G., *Polarography 1964* (G. J. Hills, Ed.), Macmillan, London, 1966, p. 719.
306. Mairanovskii, S. G., V. A. Ponamarenko, and N. V. Barashkova, *Izv. Akad. Nauk SSSR, Ser. Khim.*, 1951 (1964).
307. Mairanovskii, S. G., N. V. Barashkova, and Yu. B. Vol'kenshtein, *Izv. Akad. Nauk SSSR, Ser. Khim.*, 1539 (1965).
308. Mairanovskii, S. G., and V. N. Pavlov, *Izv. Akad. Nauk SSSR, Ser. Khim.*, 1669 (1967); *Elektrokhimiya*, **1**, 226 (1965).
309. Mairanovskii, S. G., *Catalytic and Kinetic Waves in Polarography*, Plenum Press, New York, 1968.
310. Mairanovskii, V. G., and G. I. Samokhvalov, *Elektrokhimiya*, **2**, 62 (1966).
311. Mairanovskii, V. G., L. A. Vakulova, and G. I. Samokhvalov, *Elektrokhimiya*, **3**, 23 (1967).
312. Mairanovskii, V. G., I. E. Valashek, and G. I. Samokhvalov, *Elektrokhimiya*, **3**, 611 (1967).
313. Maki, A. H., and D. H. Geske, *J. Chem. Phys.*, **30**, 1356 (1959).
314. Maki, A. H., and D. H. Geske, *J. Chem. Phys.*, **33**, 825 (1960).
315. Maki, A. H., and D. H. Geske, *J. Am. Chem. Soc.*, **83**, 1852 (1961).
316. Mamedzade, R. Yu., et al., *Izv. Akad. Nauk SSSR, Ser. Khim.*, **306**, 1009 (1967); *Chem. Abstr.*, **66**, 121539 (1967); **67**, 60307 (1967).
317. Marple, L. W., L. E. I. Hummelstedt, and L. B. Rogers, *J. Electrochem. Soc.*, **107**, 436 (1960).
318. Martinet, P., and J. Simonet, *Bull. Soc. Chim. France*, 3533 (1967).
319. Matschiner, H., and K. Issleib, *Z. Anorg. Allgem. Chem.*, **354**, 60 (1967).
320. Maturová, M., A. Němečková, and F. Šantavý, *Collection Czech. Chem. Commun.*, **27**, 1021 (1962).
321. Mayell, J. S., and A. J. Bard, *J. Am. Chem. Soc.*, **85**, 421 (1963).
322. McKinney, P. S., and S. Rosenthal, *J. Electroanal. Chem.*, **16**, 261 (1968).
323. McKinney, T. M., and D. H. Geske, *J. Chem. Phys.*, **44**, 2277 (1966).

324. McKinney, T. M., and D. H. Geske, *J. Am. Chem. Soc.*, **89**, 2806 (1967).
325. Melby, L. R., et al., *J. Am. Chem. Soc.*, **84**, 3374 (1962).
326. Melchior, M. T., and A. H. Maki, *J. Chem. Phys.*, **34**, 471 (1961).
327. Meunier, J. M., M. Person, and P. Fornari, *Bull. Soc. Chim. France*, 2872 (1967).
328. Michielli, R. F., and P. J. Elving, *J. Am. Chem. Soc.*, **90**, 1989 (1968).
329. Möbius, K., *Z. Angew. Physik*, **17**, 534 (1964).
330. Modiano, J., *Ann. Chim. (Paris)*, **10**, 541 (1955).
331. Moe, N. S., *Polarography 1964* (G. J. Hills, Ed.), Macmillan, London, 1966, p. 1077.
332. Mooney, B., and H. I. Stonehill, *J. Chem. Soc., A*, 1 (1967).
333. Morotomi, Y., and A. Sekine, *Sankyo Kenkyusho Nempo*, **17**, 93 (1965); *Chem. Abstr.*, **67**, 39568 (1967).
334. Müller, O. H., *Ann. N.Y. Acad. Sci.*, **40**, 91 (1940).
335. Myatt, J., and P. F. Todd, *Chem. Commun.*, 1033 (1967).
336. Nakaya, J., and H. Kinoshita, *Bull. Univ. Osaka Prefect. Ser. A*, **14**, 83 (1965).
337. Neish, W. J. P., and O. H. Müller, *Rec. Trav. Chim.*, **72**, 301 (1953).
338. Nicholson, R. S., J. M. Wilson, and M. L. Olmstead, *Anal. Chem.*, **38**, 542 (1966).
339. Nishiyama, M., M. Mayurama, and H. Hamaguchi, *Bull. Chem. Soc. Japan*, **37**, 616 (1964).
340. Nürnberg, H. W., *Advances in Polarography* (I. S. Longmuir, Ed.), Vol. II, Pergamon Press, Oxford, 1960, p. 694; H. W. Nürnberg, G. van Riesenbeck, and M. von Stackelberg, *Z. Elektrochem.*, **64**, 130 (1960).
341. Nygård, B., *Acta Chem. Scand.*, **20**, 1710 (1966).
342. O'Donnell, J. F., and C. K. Mann, *J. Electroanal. Chem.*, **13**, 157 (1967).
343. Olson, D. C., V. P. Mayweg, and G. N. Schrauzer, *J. Am. Chem. Soc.*, **88**, 4876 (1966).
344. Parkanyi, C., and R. Zahradník, *Abhandl. Deut. Akad. Wiss. Berlin, Kl. Chem. Geol. Biol.*, 363 (1964).
345. Parravano, G., *J. Am. Chem. Soc.*, **73**, 628 (1951).
346. Pasternak, R., *Helv. Chim. Acta*, **31**, 753 (1948).
347. Patzak, R., and L. Neugebauer, *Monatsh. Chem.*, **82**, 662 (1951); **83**, 776 (1952).
348. Paul, D. E., D. Lipkin, and S. I. Weissman, *J. Am. Chem. Soc.*, **78**, 116 (1956).
349. Pavlov, V., *Z. Chem.*, **4**, 392 (1964).
350. Pavlov, V. N., Ya. M. Zolotovitskii, and S. G. Mairanovskii, *Elektrokhimiya*, **1**, 427 (1965).

351. Pearson, J., *Trans. Faraday Soc.*, **44**, 683 (1948).
352. Peover, M. E., *Trans. Faraday Soc.*, **58**, 2370 (1962).
353. Peover, M. E., *Trans. Faraday Soc.*, **60**, 417 (1964).
354. Peover, M. E., *Trans. Faraday Soc.*, **60**, 479 (1964).
355. Peover, M. E., and J. D. Davies, *Trans. Faraday Soc.*, **60**, 476 (1964).
356. Peover, M. E., and J. D. Davies, *Polarography 1964* (G. J. Hills, Ed.), Macmillan, London, 1966, p. 1003.
357. Peover, M. E., and B. S. White, *J. Electroanal. Chem.*, **13**, 93 (1967).
358. Peover, M. E., *Electroanalytical Chemistry* (A. J. Bard, Ed.), Vol. 2, Marcel Dekker, New York, 1967, p. 1.
359. Perichon, J., and R. Buvet, *Electrochim. Acta*, **9**, 587 (1964).
360. Perichon, J., and R. Buvet, *Bull. Soc. Chim. France*, 1282 (1968).
361. Perone, S. P., Dissertation, Univ. Wisconsin, *Dissertation Abstr.*, **24**, 2245 (1963).
362. Perrin, C. L., *Progress in Physical Organic Chemistry* (S. G. Cohen, A. Streitwieser, R. W. Taft, Eds.), Vol. 3, Interscience, New York, 1965, p. 165.
363. Person, M., *Bull. Soc. Chim. France*, 1832 (1966).
364. Petrov, V. P., *Izv. Sib. Otd. Akad. Nauk SSSR, Ser. Khim. Nauk*, 74 (1966); *Chem. Abstr.*, **67**, 17163 (1967).
365. Petrovich, J. P., and M. M. Baizer, *Electrochim. Acta*, **12**, 1249 (1967).
366. Philp, R. H., R. L. Flurry, and R. A. Day, *J. Electrochem. Soc.*, **111**, 328 (1964).
367. Philp, R. H., T. Layloff, and R. N. Adams, *J. Electrochem. Soc.*, **111**, 1189 (1964).
368. Pietrzyk, D. J., *Anal. Chem.*, **38**, 278R (1966); **40**, 194R (1968).
369. Piette, L. H., P. Ludwig, and R. N. Adams, *Anal. Chem.*, **34**, 916 (1962); *J. Am. Chem. Soc.*, **83**, 2671 (1960).
370. Piette, L. H., P. Ludwig, and R. N. Adams, *J. Am. Chem. Soc.*, **84**, 4212 (1962).
371. Plieth, W. J., and K. J. Vetter, *Ber. Bunsenges. Physik. Chem.*, **72**, 1052 (1968); **73**, 79 (1969); *Z. Phys. Chem. (Frankfurt)*, **61**, 282 (1968); *Collection Czech. Chem. Commun.*, **36**, 816 (1971).
372. Pomianowski, A., *Roczniki Chem.*, **41**, 775 (1967); *Chem. Abstr.*, **67**, 60288 (1967).
373. Pozdeeva, A. A., and S. I. Zhdanov, *Izv. Akad. Nauk SSSR, Ser. Khim.*, 2156 (1964).
374. Pozdeeva, A. A., and S. I. Zhdanov, *Polarography 1964* (G. J. Hills, Ed.), Macmillan, London, 1966, p. 781.
375. Prelog, V., and O. Häflinger, *Helv. Chim. Acta*, **32**, 2088 (1949).

376. Prevost, C. A., P. Souchay, and J. Chauvelier, *Bull. Soc. Chim. France,* 714 (1951).
377. Psarras, T., and R. E. Dessy, *J. Am. Chem. Soc.,* **88**, 5132 (1966).
378. Rieger, P. H., I. Bernal, and G. K. Fraenkel, *J. Am. Chem. Soc.,* **83**, 3918 (1961).
379. Rieger, P. H., Thesis, Columbia Univ., New York, 1961 [according to reference (413)].
380. Rieger, P. H., and G. K. Fraenkel, *J. Chem. Phys.,* **37**, 2811 (1962).
381. Rieger, P. H., I. Bernal, W. H. Reinmuth, and G. K. Fraenkel, *J. Am. Chem. Soc.,* **85**, 683 (1963).
382. Rifi, M. R., *J. Am. Chem. Soc.,* **89**, 4442 (1967).
383. Rollmann, L. D., and R. T. Iwamoto, *J. Am. Chem. Soc.,* **90**, 1455 (1968).
384. Rosenthal, I., J. R. Hayes, A. J. Martin, and P. J. Elving, *J. Am. Chem. Soc.,* **80**, 3050 (1958).
385. Rüetschi, P., and G. Trümpler, *Helv. Chim. Acta,* **36**, 1649 (1953).
386. Ryvolová, A., *Collection Czech. Chem. Commun.,* **25**, 420 (1960).
387. Sadler, J. L., and A. J. Bard, *J. Electrochem. Soc.,* **115**, 343 (1968).
388. Sadler, J. L., and A. J. Bard, *J. Am. Chem. Soc.,* **90**, 1979 (1968).
389. Sartori, G., and C. Cattaneo, *Gazz. Chim.,* **71**, 713 (1941).
390. Sartori, G., and G. Costa, *Ric. Sci.,* **22**, Suppl. 175 (1952).
391. Savéant, J. M., *Bull. Soc. Chim. France,* 481, 486, 493 (1967). J. M. Savéant and H. Veillard-Royer, *Bull. Soc. Chim. France,* 2415 (1967).
392. Savéant, J. M., and E. Vianello, *Electrochim. Acta,* **12**, 1545 (1967).
393. Savicheva, G. A., M. B. Gazizov, A. V. Il'yasov, and A. I. Razumov, *Zh. Obshch. Khim.,* **37**, 2785 (1967); *Chem. Abstr.,* **68**, 110808 (1968).
394. Savicheva, G. A., M. B. Gazizov, and A. I. Razumov, *Zh. Obshch. Khim.,* 622 (1968); *Chem. Abstr.,* **69**, 7834 (1968).
395. Sayo, H., Y. Tsukitani, and M. Masui, *Tetrahedron,* **24**, 1717 (1968).
396. Schmidt, H., and J. Noack, *Z. Anorg. Allgem. Chem.,* **296**, 262 (1958).
397. Schneider, F., *Z. Instrumentenk.,* **72**, 11 (1964).
398. Schulz, M., and K. H. Schwarz, *Z. Chem.,* **7**, 176 (1967).
399. Schwabe, K., *Z. Phys. Chem. (Leipzig),* Sonderheft, 289 (1958).
400. Schwabe, K., and H. J. Baer, *Rev. Polarog. (Kyoto),* **11**, 117 (1963).
401. Schwarz, W. M., E. M. Kosower, and I. Shain, *J. Am. Chem. Soc.,* **83**, 3164 (1961).
402. Schweiss, H., and H. Berg, *Abhandl. Deut. Akad. Wiss. Berlin, Kl. Medizin,* 391 (1966).
403. Scott, J. M. W., and W. H. Jura, *Can. J. Chem.,* **45**, 2375 (1967).
404. Sease, J. W., P. Chang, and J. L. Groth, *J. Am. Chem. Soc.,* **86**, 3154 (1964).

405. Sevast'yanova, I. G., and A. P. Tomilov, *Zh. Obshch. Khim.*, **33**, 2815 (1963).
406. Sevast'yanova, I. G., and A. P. Tomilov, *Elektrokhimiya*, **3**, 563 (1967).
407. Shapoval, G. S., and V. I. Shapoval, *Ukr. Khim. Zh.*, **31**, 1080 (1965); *Chem. Abstr.*, **64**, 10749e (1966).
408. Shapoval, G. S., E. M. Skobets, and N. P. Markova, *Dopov. Akad. Nauk Ukr. RSR, Ser. B*, **30**, 141 (1968); *Chem. Abstr.*, **68**, 101181 (1968).
409. Shimanskaya, N. P., L. A. Pavolova, and V. D. Bezuglyi, *Zh. Obshch. Khim.*, **37**, 974 (1967); *Chem. Abstr.*, **67**, 96265 (1967).
410. Shriver, D. F., D. E. Smith, and P. Smith, *J. Am. Chem. Soc.*, **86**, 5153 (1964).
411. Simonet, J., *Compt. Rend., Ser. C*, **263**, 685, 1546 (1966).
412. Sioda, R. E., and W. S. Koski, *J. Am. Chem. Soc.*, **87**, 5573 (1965).
413. Sioda, R. E., and W. S. Koski, *J. Am. Chem. Soc.*, **89**, 475 (1967).
414. Smith, D. L., and P. J. Elving, *J. Am. Chem. Soc.*, **84**, 1412, 2741 (1962).
415. Smith, P. T., and J. Jordan, *Polarography 1964* (G. J. Hills, Ed.), Macmillan, London, 1966, p. 407.
416. Solon, E., and A. J. Bard, *J. Am. Chem. Soc.*, **86**, 1926 (1964).
417. Souchay, P., J. Armand, and S. Deswarte, *Polarography 1964* (G. J. Hills, Ed.), Macmillan, London, 1966, p. 811.
418. Southworth, B. C., R. Osteryoung, K. D. Fleischer, and F. C. Nachod, *Anal. Chem.*, **33**, 208 (1961).
419. Spritzer, M., and L. Meites, *Anal. Chim. Acta*, **26**, 58 (1962).
420. Spritzer, M. S., J. M. Costa, and P. J. Elving, *Anal. Chem.*, **37**, 211 (1965).
421. Stackelberg, M.v., and W. Stracke, *Z. Elektrochem.*, **53**, 118 (1949).
422. Stackelberg, M.v., and P. Weber, *Z. Elektrochem.*, **56**, 806 (1952).
423. Stackelberg, M.v., and H. Fassbender, *Z. Elektrochem.*, **62**, 834 (1958).
424. Stackelberg, M.v., W. Hans, and W. Jensch, *Z. Elektrochem.*, **62**, 839 (1958).
425. Stanienda, A., *Z. Phys. Chem. (Leipzig)*, **229**, 257 (1965).
426. Stapelfeldt, H. E., and S. P. Perone, *Anal. Chem.*, **40**, 815 (1968).
427. Steinberger, N., and G. K. Fraenkel, *J. Chem. Phys.*, **40**, 723 (1964).
428. Štradins, J., *Electrochim. Acta*, **9**, 711 (1964).
429. Štradins, J., and V. Terauds, *Latvijas PSR Zinatnu Akad. Vestis, Kim. Ser.*, 169 (1964); 43 (1965); *Chem. Abstr.*, **61**, 12953c (1964); **64**, 1639h (1966).
430. Štradins, J., and S. Hillers, *Nitro Compounds—Proceedings of the International Symposium, Warsaw, 1963*, Pergamon Press, Oxford, 1964, p. 409; J. P. Štradins and G. O. Reikhmanis, *Elektrokhimiya*, **3**, 178 (1967).
431. Štradins, J., G. Reikhmanis, and R. Gavars, *Elektrokhimiya*, **1**, 955 (1965); J. Štradins, R. Gavars, G. Reikhmanis, and S. Hillers, *Abhandl. Deut. Akad. Wiss. Berlin, Kl. Medizin*, 601 (1966).

432. Štradins, J., E. Grens, V. Kampars, and G. Vanags, *Zh. Obshch. Khim.*, **35**, 222 (1965); *Chem. Abstr.*, **63**, 1478g (1965).
433. Štradins, J. P., I. K. Tutane, and G. J. Vanag, *Zh. Analit. Khim.*, **20**, 1239 (1965).
434. Strassner, J. E., and P. Delahay, *J. Am. Chem. Soc.*, **74**, 6232 (1952).
435. Strauss, H. L., T. J. Katz, and G. K. Fraenkel, *J. Am. Chem. Soc.*, **85**, 2360 (1963).
436. Streitwieser, A., and C. Perrin, *J. Am. Chem. Soc.*, **86**, 4938 (1964).
437. Sunahara, H., *Rev. Polarog. (Kyoto)*, **9**, 222 (1961).
438. Suzuki, M., and P. J. Elving, *J. Phys. Chem.*, **65**, 391 (1961).
439. Takahashi, R., and P. J. Elving, *Electrochim. Acta*, **12**, 231 (1967).
440. Tang, S., and P. Zuman, *Collection Czech. Chem. Commun.*, **28**, 829, 1524 (1963).
441. Thiec, J., and J. Wiemann, *Bull. Soc. Chim. France*, 177 (1956).
442. Thomas, C. W., L. L. Leveson, and M. Bailes, *J. Polarograph. Soc.*, **13**, 43 (1967).
443. Tirouflet, J., *Bull. Soc. Chim. France*, 274 (1956).
444. Tirouflet, J., R. Robin, and M. Guyard, *Bull. Soc. Chim. France*, 571 (1956).
445. Tomilov, A. P., Yu. D. Smirnov, S. S. Dubov, and S. P. Makarov, *Zh. Vses. Khim. Obshch. Mendeleeva*, **11**, 473 (1966); *Chem. Abstr.*, **65**, 16501f (1966).
446. Tompkins, P. C., and C. L. A. Schmidt, *Univ. Calif. Publ. Physiol.*, **8**, 237, 247 (1944) (according to I. M. Kolthoff and J. J. Lingane, *Polarography*, Interscience, New York, London, 1952).
447. Toropova, V. P., M. K. Saikina, and M. G. Khakimov, *Zh. Obshch. Khim.*, **37**, 47 (1967); *Chem. Abstr.*, **66**, 81865 (1967).
448. Uehara, M., *Nippon Kagaku Zasshi*, **86**, 901 (1965); *Chem. Abstr.*, **64**, 10760g (1966).
449. Umemoto, K., *Bull. Chem. Soc. Japan*, **40**, 1058 (1967).
450. Vajda, M., *Advances in Polarography* (I. S. Longmuir, Ed.), Vol. II, Pergamon Press, Oxford, 1960, p. 786.
451. Valenta, P., *Advances in Polarography* (I. S. Longmuir, Ed.), Vol. III, Pergamon Press, Oxford, 1960, p. 1004.
452. Valenta, P., *Collection Czech. Chem. Commun.*, **25**, 853 (1960).
453. Vincenz-Chodkowska, A., and Z. R. Grabowski, *Electrochim. Acta*, **9**, 789 (1964).
454. Volke, J., and J. Holubek, *Collection Czech. Chem. Commun.*, **27**, 1777 (1962).
455. Volke, J., and M. M. Amer, *Collection Czech. Chem. Commun.*, **29**, 2134 (1964).
456. Voorhies, J. D., and E. J. Schurdak, *Anal. Chem.*, **34**, 939 (1962).

457. Wagenknecht, J. H., and M. M. Baizer, *J. Electrochem. Soc.*, **114**, 1095 (1967).
458. Wawzonek, S., and J. W. Fan, *J. Am. Chem. Soc.*, **68**, 2541 (1946).
459. Wawzonek, S., *Anal. Chem.*, **26**, 65 (1954).
460. Wawzonek, S., E. W. Blaha, R. Berkey, and M. E. Runner, *J. Electrochem. Soc.*, **102**, 235 (1955).
461. Wawzonek, S., R. Berkey, E. W. Blaha, and M. E. Runner, *J. Electrochem. Soc.*, **103**, 456 (1956).
462. Wawzonek, S., R. Berkey, and D. Thomson, *J. Electrochem. Soc.*, **103**, 513 (1956).
463. Wawzonek, S., and D. Wearring, *J. Am. Chem. Soc.*, **81**, 2067 (1959).
464. Wawzonek, S., and A. Gundersen, *J. Electrochem. Soc.*, **107**, 537 (1960).
465. Wawzonek, S., and R. C. Duty, *J. Electrochem. Soc.*, **108**, 1135 (1961).
466. Wawzonek, S., R. C. Duty, and J. H. Wagenknecht, *J. Electrochem. Soc.*, **111**, 74 (1964).
467. Wawzonek, S., and A. Gundersen, *J. Electrochem. Soc.*, **111**, 324 (1964).
468. Wawzonek, S., and J. H. Wagenknecht, *Polarography 1964* (G. J. Hills, Ed.), Macmillan, London, 1966, p. 1035.
469. Wawzonek, S., *Talanta*, **12**, 1229 (1965).
470. Wawzonek, S., and T. W. McIntyre, *J. Electrochem. Soc.*, **114**, 1025 (1967).
471. Wettig, K., *Z. Phys. Chem. (Leipzig)*, **233**, 423 (1966).
472. Wettig, K., *Elektrokhimiya*, **3**, 269 (1967).
473. Wettig, K., *Z. Chem.*, **7**, 107 (1967).
474. Whitman, W. E., and L. A. Wiles, *J. Chem. Soc.*, 3016 (1956).
475. Wiesner, W., and K. Schwabe, *J. Electroanal. Chem.*, **15**, 73 (1967).
476. Wilkinson, G., et al., *J. Am. Chem. Soc.*, **74**, 6149 (1952); **75**, 3586 (1953); **76**, 1970, 4281 (1954).
477. Yasukouchi, K., Y. Ono, et al., *Nippon Kagaku Zasshi*, **88**, 428, 538 (1967); *Chem. Abstr.*, **67**, 17277, 39581 (1967).
478. Zahlan, A. B., and R. H. Linnell, *J. Am. Chem. Soc.*, **77**, 6207 (1955).
479. Zahradník, R., and C. Parkanyi, *Talanta*, **12**, 1289 (1965).
480. Zhdanov, S. I., *Z. Phys. Chem. (Leipzig)*, Sonderheft, 235 (1958).
481. Zhdanov, S. I., and A. N. Frumkin, *Dokl. Akad. Nauk SSSR*, **122**, 412 (1958).
482. Zhdanov, S. I., and L. S. Mirkin, *Collection Czech. Chem. Commun.*, **26**, 370 (1961).
483. Zhdanov, S. I., and L. G. Feoktistov, *Izv. Akad. Nauk SSSR, Otd. Khim. Nauk*, 53 (1963); *Chem. Abstr.*, **65**, 10123e (1966).
484. Zhdanov, S. I., and A. A. Pozdeeva, *Collection Czech. Chem. Commun.*, **30**, 4143 (1965).

485. Zhdanov, S. I., and A. A. Pozdeeva, *Elektrokhimiya,* **2,** 1047 (1966).
486. Zhdanov, S. I., *Results of the Electrochemistry of Organic Compounds* (in Russian) (Inst. Elektrokhim. Akad. Nauk, Ed.), Izdatelstvo "Nauka," Moscow, 1966, p. 144.
487. Zhdanov, S. I., V. Sh. Tsveniashvili, and Z. V. Todres, *J. Polarograph. Soc.,* **13,** 100 (1967).
488. Zuman, P., *Collection Czech. Chem. Commun.,* **15,** 839 (1950).
489. Zuman, P., *Z. Phys. Chem. (Leipzig),* Sonderheft, 243 (1958).
490. Zuman, P., and V. Horák, *Advances in Polarography* (I. S. Longmuir, Ed.), Vol. III, Pergamon Press, Oxford, 1960, p. 804; *Collection Czech. Chem. Commun.,* **26,** 176 (1961).
491. Zuman, P., J. Chodkowski, and F. Šantavý, *Collection Czech. Chem. Commun.,* **26,** 380 (1961).
492. Zuman, P., and J. Michl, *Nature,* **192,** 655 (1961).
493. Zuman, P., *Organic Polarographic Analysis,* Pergamon Press, Oxford, 1964.
494. Zuman, P., *Polarography 1964* (G. J. Hills, Ed.), Macmillan, London, 1966, p. 687.
495. Zuman, P., *Progress in Physical Organic Chemistry* (A. Streitwieser, R. W. Taft, Eds.), Vol. 5, Interscience, 1967, p. 81.
496. Zuman, P., *J. Polarograph. Soc.,* **13,** 53 (1967).
497. Zuman, P., *Z. Anal. Chem.,* **224,** 374 (1967).
498. Zychiewicz-Zajdel, Z., *Ann. Univ. Mariae Curie-Sklodowska Lublin (Poland),* **17,** 83 (1962).

CHAPTER V

DOUBLE LAYER AND RELATED EFFECTS IN CLASSICAL POLAROGRAPHY

S. G. Mairanovskii

N. D. Zelinskii Institute of Organic Chemistry, Academy of Sciences of U.S.S.R., Moscow, U.S.S.R.*

Contents

I. Introduction	288
II. Fundamental Ideas of the Double-Layer Structure	288
III. The Evaluation of ψ_1-Potentials	295
IV. The Double-Layer Structure and Electrode Kinetics	300
A. Uncharged depolarizers	300
B. Electrode processes involving ions	301
C. Electrode processes with chemical reactions. Volume kinetic and catalytic waves	309
D. Surface kinetic and quasidiffusion waves	315
E. Some specific features of anion reduction	322
V. Depolarizer Adsorption and Characteristics of Polarographic Waves	325
A. The adsorption effect on the shape and slope of the waves	325
B. The reduction of the quasiunadsorbed particles	331
C. The effect of the distances between the electrode, the plane of closest approach of the ions, and the depolarizer reaction center	334
D. The salting-out effect	340
E. The heredity effect	342
VI. The Effect of the Supporting Electrolyte Nature on the Electrode Process Kinetics	344
A. The cation nature effect	344
B. The specific anion effect	348
C. The influence of the organic depolarizer structure on the specific effect value upon changing the nature of the supporting electrolyte cation	354
VII. Conclusion	359
References	360

* Translated into English by Vera Reingold.

I. Introduction

At the interface between a metal and a surrounding solution containing ions or dipolar molecules, there arises an electrical double layer at which charges equal in value but opposite in sign are present in the boundary regions of the two phases.

The electrical double layer at the electrode surface has significant effects on the rates of electrochemical processes; in the case of a dropping mercury electrode (DME), it also affects the capacity current (charging current) because of the periodic renewal of the dropping electrode surface.

Frumkin was the first who took into account the effects of the structure of the electrical double layer on the rates of electrochemical reactions when he developed the theory of slow hydrogen discharge in 1933 (1–3). Nowadays, Frumkin's ideas are commonly accepted; they are widely used by electrochemists in different countries in the study of electrode kinetics. Double-layer structure and its effects on the electrode processes have been treated by Delahay (4) in his monograph (which, even though published only recently, is already well known) and also in numerous reviews [see, for example, references (5) through (12)]. Therefore this work treats briefly only the general ideas that are indispensable for further understanding of the text. Emphasis is placed mainly on the principal features of the effects of the double layer on the characteristics of polarographic waves. Some related double-layer effects, the effects that become evident in studying double-layer phenomena, are considered as well.

The objective of this chapter is to acquaint investigators who use the polarographic technique without having had sufficient training in electrochemistry with the essential ideas of the double layer and related effects so that they can consciously take into consideration—to the extent possible at the present level of knowledge—the influence of these effects upon the data obtained in polarographic experiments. Although adsorption phenomena, being closely connected with those of the double layer (4, 8, 9), also significantly affect electrode reactions, their influences are considered only briefly, when it is necessary to consider them at all, because of the limited scope of the article.

II. Fundamental Ideas of the Double-Layer Structure

According to Stern (13), who made use of the Gouy (14), Chapman (15), and Helmholtz (15) concepts, the structure of the electrical double layer at the electrode–solution interface may be represented by the following scheme (Fig. 1). The charged electrode surface (in Fig. 1a the electrode charge is

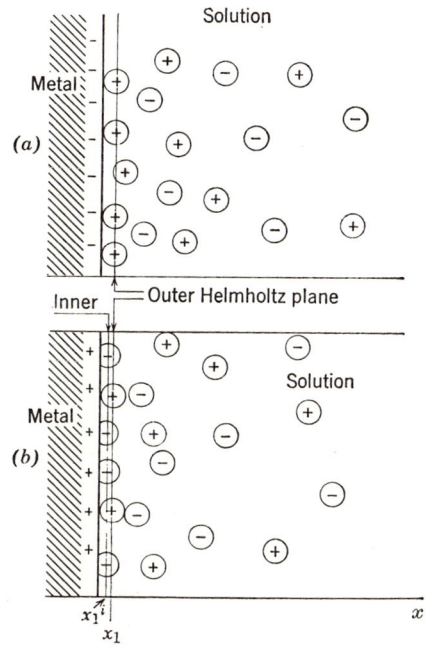

FIG. 1. Scheme illustrating the ion distribution close to a negatively charged electrode surface in the absence of specific adsorption (a) and with anion adsorption on a positively charged surface (b).

shown as being negative) electrostatically attracts ions having the opposite sign. Some of the ions immediately adjacent to the electrode surface form the compact (or Helmholtz) layer; because of the thermal Brownian movement which counteracts the electrostatic electrode attraction other ions form the blurred or diffuse layer.

In the compact layer the ions are at the closest approach to the electrode surface, and their centers form the so-called Helmholtz plane. Grahame (5) took into account the possibility of specific ion adsorption; in this case ions are bound to the electrode surface by forces that resemble chemical ones in nature, and their centers are closer to the electrode surface than those of ions that are also in the compact layer but are attracted only by electrostatic forces (see Fig. 1b). Therefore, according to Grahame, one should consider two Helmholtz planes: the inner plane formed by the centers of the specifically adsorbed ions with distorted solvation shells (x_1^i in Fig. 1b) and the outer plane formed by the centers of the ions closest to the electrode surface, usually solvated, and not specifically adsorbed.

Among inorganic ions many anions (iodide and bromide in particular) undergo specific adsorption, especially at positive and small negative charges of the electrode surface; at negatively charged surfaces, however, some cations, including Cs^+ (17), Tl^+ (18), Pb^{2+}, and others, also exhibit slight specific adsorption.

If the difference between the electrode potential and the potential of zero charge is denoted as ϕ_a, that is, $\phi_a = E - E_z$, and the potential in the bulk of the solution is taken to be zero, the change in the potential difference with distance from the electrode surface is shown schematically in Figure 2. In the absence of specific adsorption, a monotonic change in the potential from ϕ_a to 0 with distance from the electrode surface is observed. Within the compact part of the double layer, there are no charges, and if we take its dielectric constant to be independent of the distance from the electrode surface, the potential changes linearly from ϕ_a to ψ_1, which corresponds to the potential of the outer Helmholtz plane (curves 1 and 2 in Fig. 2). When specific adsorption of ions occurs, a break in the potential–distance curve is observed at the Helmholtz inner plane. If the charge of the adsorbed ion and that of the electrode are of opposite signs and the potential drop in the compact part is high enough, the potential drop ψ_1^i in the diffuse layer proves to be opposite in sign to the overall change in potential ϕ_a between the electrode surface and the bulk of the solution (curve 3 in Fig. 2). Such is

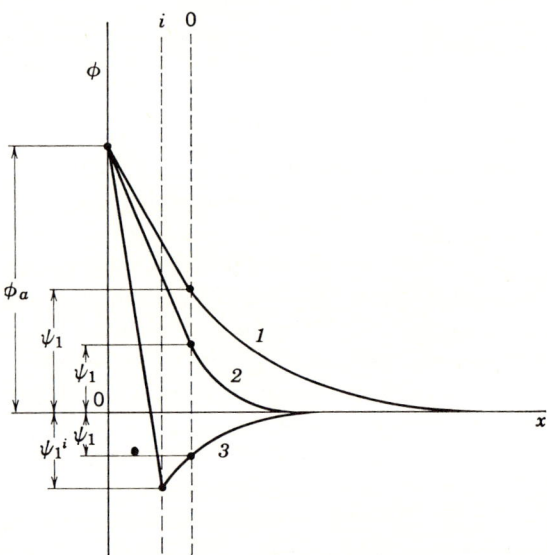

FIG. 2. Variation of potential in the layer at the electrode: (1 and 2)—in the absence of specific ion adsorption (the ion concentration increasing from 1 to 2) and (3) with adsorption of the ions with the charge opposite to that of the electrode.

the case when there is only a small positive charge on an electrode in a solution containing specifically adsorbed anions. This phenomenon is called surface charge inversion (3).

The potential difference across the diffuse part of the double layer is determined by the distribution of ions with respect to distance from the electrode surface or, more precisely, from the Helmholtz plane. This distribution obeys the Boltzmann law, and therefore the concentration c_s of the ion just at the electrode surface (at the Helmholtz plane) can be expressed by

$$c_s = c_0 \exp\left(-\frac{z\psi_1 F}{RT}\right) \quad (1)$$

where c_0 is the concentration of the ion in the bulk of the solution, z the ionic charge, F the faraday (96,500 C/g-equiv), R the universal gas constant, and T the absolute temperature. From Eq. 1 it follows that if the sign of z, the charge of the ions under consideration, and the sign of the potential drop ψ_1 in the diffuse part of the double layer are the same, the concentration c_s of the ion at the Helmholtz plane will be much lower than its concentration c_0 in the volume of the solution (electrostatic repulsion). If the signs of z and ψ_1 are opposite, the concentration c_s of the ion at the electrode surface will be much higher than in the bulk of the solution because of electrostatic attraction. Hence within the diffuse layer, the number of ions having the same sign as that of the potential ψ_1 is much smaller than the number of ions having the opposite sign. Therefore the diffuse layer as a whole has a certain charge, its value q_s in the absence of specific adsorption being equal, but opposite in sign, to the value of the electrode charge q_m (electrical neutrality of the electrode–solution system). With specifically adsorbed ions, whose charge q_i is localized in the Helmholtz inner plane, the charge of the diffuse part of the double layer is equal in value but opposite in sign to the algebraic sum of the charges of the electrode and the specifically adsorbed ions:

$$q_s = -(q_m + q_i) \quad (2)$$

The charges are usually expressed per unit area (square centimeter) of the electrode surface so that, to be precise, the values q_m and q_i are the densities of the charges and q_s is the overall charge of a column of solution perpendicular to the electrode surface and having a cross-sectional area of 1 cm². To simplify the matter we further refer to "the value of charge density" as to "the value of electrode charge."

On increasing the concentration of a salt whose ions contribute to the diffuse part of the double layer, the thickness of this layer decreases and with it the absolute value of the potential ψ_1 (which corresponds to the transition from curve *1* to curve *2* in Fig. 2). As the electrostatic forces of the electrode in the electrolyte solution operate only close to the electrode surface and decrease quickly with distance, the thickness of the diffuse layer is rather small. To be exact, the thickness of the diffuse part of the double layer has an uncertain value, since as the distance from the electrode surface increases, the ionic concentration approaches only asymptotically the value characteristic of the bulk of the solution. Likewise, the potential varies with distance, tending to zero in the bulk of the solution. Nevertheless, if we conventionally assume the distance at which the potential drop attains, for example, 99% of the overall fall in the potential ψ_1, to be the thickness of the diffuse layer $\delta_{99\%}$ then, according to the theory based on the Debye–Hückel concepts of ion atmosphere, it is possible to express the value $\delta_{99\%}$ which for an aqueous solution of a z, z-charged electrolyte at 25°C as equal to (8):

$$\delta_{99\%} = \frac{1.4 \times 10^{-7}}{z c_0^{1/2}} \text{ cm} \tag{3}$$

where the electrolyte concentration c_0 is given in moles per liter.

If we take the distance from the electrode that is 90% of the overall drop in ψ_1 to be the boundary of the diffuse part of the double layer, then $\delta_{90\%}$ is half as large as $\delta_{99\%}$ and 2.3 times as large as the thickness δ of the ionic atmosphere of an ion according to Debye–Hückel. Thus, in an 0.1 M solution of a uni-univalent electrolyte at 25°C, $\delta_{99\%}$ is 4.5×10^{-7} cm, but it is 10 times as large if the electrolyte concentration is only 0.001 M. Detailed accounts of the structure of the double layer and its properties can be found in the publications already cited and, in particular, in the monograph by Delahay (4) and in the review by Mohilner (8).

The properties of the double layer are characterized by its electrical capacity. One should distinguish between the integral capacity $\bar{C}$, which is a proportionality factor between the electrode charge (q_m) and its potential referred to the point of zero charge ($\phi_a = E - E_z$):

$$\bar{C} = \frac{q_m}{\phi_a} \tag{4}$$

and the differential capacity determined by the derivative of charge with respect to potential:

$$C = \frac{\partial q_m}{\partial \phi} = -\frac{\partial q_m}{\partial E} \tag{5}$$

The overall potential drop ϕ_a between the electrode and the bulk of the solution in the absence of specific adsorption is the sum of the potential drop in the compact part of the double layer $(\phi_a - \psi_1)$ and that in the diffuse part (ψ_1) (Fig. 2):

$$\phi_a = (\phi_a - \psi_1) + \psi_1 \qquad (6)$$

Differentiating ϕ_a with respect to the charge, we obtain

$$\frac{\partial \phi_a}{\partial q} = \frac{\partial(\phi_a - \psi_1)}{\partial q} + \frac{\partial \psi_1}{\partial q} \qquad (7)$$

or, with Eq. 5 taken into account,

$$\frac{1}{C} = \frac{1}{C_{\text{comp}}} + \frac{1}{C_{\text{dif}}} \qquad (8)$$

In accordance with the laws of electrostatics, it follows from Eq. 8 that the total differential capacity of the double layer C is nominally equal to that of two condensers connected in series, the capacity C_{comp} of the compact layer being represented by one condenser and that of the diffuse layer C_{dif} by the other. Only the differential capacity C_{dif} of the diffuse part of the double layer, which is directly related to the value of ψ_1, can be calculated from the Gouy–Chapman–Stern theory:

$$C_{\text{dif}} = \frac{2|z|F}{RT}\left(\frac{RT\varepsilon c}{2\pi}\right)^{1/2} \cosh\left(\frac{|z|F}{2RT}\psi_1\right) \qquad (9)$$

For aqueous solutions at 25°C, this equation assumes the form (5):

$$C_{\text{dif}} = 228.5|z|c^{1/2} \cosh(19.5\psi_1|z|) \qquad (10)$$

According to Grahame (5), the capacity of the compact layer does not depend on electrolyte concentration but is a complex function of the electrode potential alone. Therefore in moderately dilute solutions and at potentials not very close to the point of zero charge the value of C_{dif} is much greater than that of C_{comp}, so that according to Eq. 8 the total differential capacity is determined mainly by the value of C_{comp}. C_{dif} exerts a greater influence on the value of the total capacity as the electrolyte concentration decreases. A plot of C_{dif} against the potential is V-shaped, the minimum being at the potential of zero charge. Therefore the minima on differential capacity curves become more pronounced as the supporting electrolyte concentration decreases (see Fig. 3).

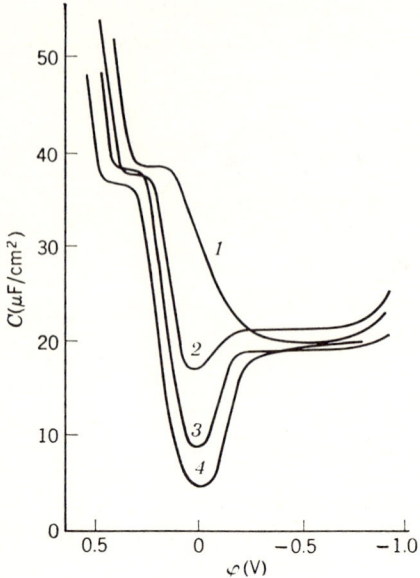

Fig. 3. The differential capacity of the double layer at the mercury electrode as a function of its potential in solutions with varying electrolyte concentration: (1) 0.1 N KCl; (2) 0.01 N KCl; (3) 0.001 N KCl; (4) 0.0001 N HCl. [From reference (7).]

With the ion adsorption at the electrode surface, the capacity of the double layer increases as the distance between "the condenser plates," that is, between the electrode surface and the centers of the adsorbed ions, decreases. On the contrary, when uncharged organic molecules are adsorbed from aqueous or aqueous-organic solutions, the capacity of the double layer decreases because of the separation of the condenser plates caused by the compound adsorbed and because of the decrease in the dielectric constant between them. The adsorption of organic compounds is strongly dependent upon the electrode potential (see Section V-A). Therefore in the region of potentials where the adsorption of organic compounds takes place the double-layer capacity is markedly diminished (Fig. 4). The boundaries of the range of potentials in which adsorption occurs exhibit more-or-less pronounced "pseudocapacity" peaks resulting from periodic changes in double-layer capacity attributable to the adsorption–desorption process when a small sinusoidal current is superimposed on a direct current flowing through the electrode. The peak height increases, and so does the width of the range of potentials over which adsorption is observed, as the concentration of the adsorbed compound is increased (see Fig. 4). A more detailed account of adsorption phenomena and their present theory can be found in the publications already mentioned (4–8), as well as in a monograph (19) especially concerned with adsorption on electrodes.

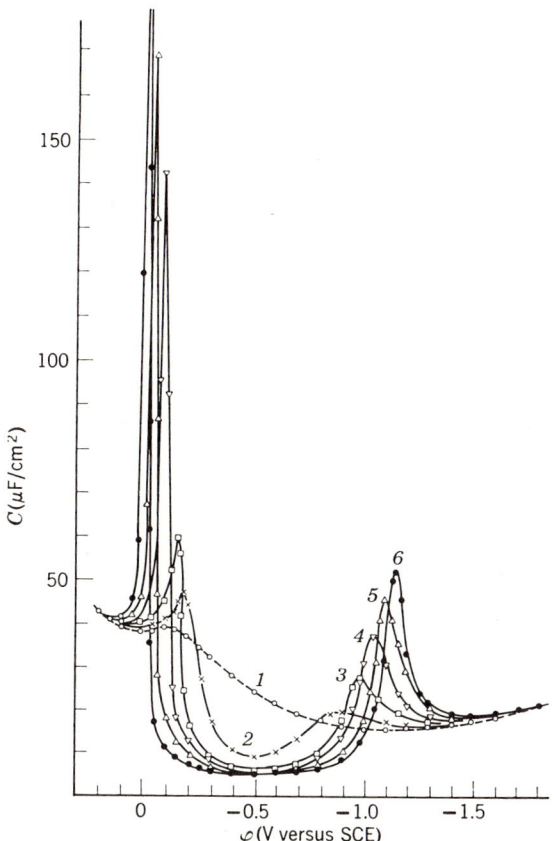

FIG. 4. Differential capacity curves at the mercury electrode in 1.0 N Na_2SO_4 in the presence of varying quantities of $n\text{-}C_5H_{11}OH$. (1) 0; (2) 5×10^{-3}; (3) 1.5×10^{-2}; (4) 2.5×10^{-2}; (5) 3.5×10^{-2}; (6) 5×10^{-2} M. [From reference (20).]

III. The Evaluation of ψ_1-Potentials

The active center of a particle taking part in an electrochemical reaction is likely to be close to the outer Helmholtz plane. Therefore the potential corresponding to this plane (ψ_1) is an important factor affecting the kinetics of electrode processes (see Section IV).

In the absence of specific adsorption, when the charge of the diffuse layer q_s is equal in value (but opposite in sign) to that of the electrode surface q_m, the value of the potential ψ_1 in a solution containing only one electrolyte, of which the anion and cation have the same charge ($z_a = z_c = z$), is connected with q_m (or $-q_s$), according to the Gouy–Chapman theory, by the relationship

$$q_m = (8RT\varepsilon)^{1/2} c_0^{1/2} \sinh\left(\frac{zF\psi_1}{2RT}\right) \tag{11}$$

in which c_0 is the concentration of electrolyte in the bulk of the solution and ε is the dielectric permittivity equal to the product of the dimensionless dielectric constant K_D (for water at 25°C $K_D = 78.5$) and the vacuum permittivity ε_0 ($\varepsilon_0 = $ constant $= 8.85 \times 10^{-14}$ F/cm). It follows from Eq. 11 that

$$\psi_1 = \left(\frac{2RT}{zF}\right) \text{Arsinh}\left[\frac{q_m}{(8RT\varepsilon c_0)^{1/2}}\right] \tag{12}$$

In calculating ψ_1 in the case of specific adsorption, one should substitute the value of $q_m + q_i = -q_s$ for that of q_m (see Eq. 2). Thus the charge of electrode surface (or that of the diffuse part of the double layer) at a given potential should be known in order to find the value of ψ_1.

The value of the charge q_m for an ideally polarized electrode can be calculated from the relationship between the surface tension and the electrode potential (i.e., from electrocapillary curves) by using the well-known Lippmann equation:

$$q_m = -\frac{\partial \gamma}{\partial E} \tag{13}$$

or from differential capacity curves by integrating them from the electrocapillary zero E_z to a given potential E:

$$q_m = \int_{E_z}^{E} C\, dE \tag{14}$$

One can also calculate the value of q_m by such an integration and then calculate ψ_1 from Eq. 12 in the case in which there are uncharged organic compounds adsorbed on the electrode (6). However, if the whole electrode surface is covered with the adsorbed compound, the capacity of the double layer frequently proves to be constant within a fairly wide range of potentials including E_z. In this case the charge can be evaluated by multiplying this constant value of the capacity by the difference $(E - E_z)$ between the potential E, at which ψ_1 is being evaluated, and E_z.

The charge density at the surface of a DME (in microcoulombs per square centimeter) can be determined from polarographic charging currents (21, 22):

$$i_c = 8.5 \times 10^{-3} m^{2/3} t^{-1/3} q_m \tag{15}$$

where i_c is the charging current averaged over the drop life (in microamperes) and m and t are characteristics of the dropping electrode (in milligrams per second and in seconds, respectively). The charging currents, however, are somewhat smaller than the whole residual current observed, since in the most thoroughly purified solutions there still remain some traces of reducible substances (especially oxygen). The faradaic reduction diffusion current decreases as the drop time decreases, while the capacity current rises. Therefore, to increase the accuracy of polarographic determination of the surface charge density, one should work with dropping electrodes yielding short drop times with simultaneous use of some device for enforced detachment of drops (23–25).

The dropping electrode charge can be found with some precision by using a different relationship between the charging and faradaic components of the residual polarographic current i_0 and the drop time t (23). For this purpose a plot of $i_0/t^{1/6}$ against $t^{-1/2}$ is constructed. It is linear, and q_m is calculated from its slope which is equal to $8.5 \times 10^{-3} \, m^{2/3} q_m$ for currents averaged over the drop time.

The value of q_m can also be found by integrating the oscillographically observed dependence of current on time as the electrode potential is varied stepwise from the point of zero charge to the potential under study. One can make similar use of the current–time curves obtained by superimposing small (up to 10 mV) rectangular voltage impulses on the electrode at a constant potential [see, for example, reference (26)].

In the literature some data are reported on the values of ψ_1 at various electrode potentials in the absence of specific adsorption. Russell (27), for instance, presents some detailed tables of values of ψ_1 calculated from the data obtained experimentally by Grahame (28). The curves presented in Figure 5 show how ψ_1 depends on the electrode potential (measured versus NCE) at different sodium fluoride concentrations. These curves were constructed by using the data of reference (27) (as indicated by McCoy and Mark (29), it is erroneously stated in Russell's work (27) that the potentials are referred to the SCE). Parsons (6) has also calculated the dependence of ψ_1 on ϕ_a from the data supplied by Grahame for sodium fluoride.

If the value of q_m is assumed to be almost independent of the potential, which is justifiable if the latter is sufficiently negative, an approximate equation can be derived from Eq. 12. However, in the absence of specific adsorption, this equation expresses the actual dependence of ψ_1 on the concentration c_0 of the supporting electrolyte (3):

$$\psi_1 \simeq \text{constant} + \frac{RT}{zF} \ln(c_0) - \frac{2RT}{zF} \ln|-\phi_a| \qquad (16)$$

where constant $\simeq -0.06$ V.

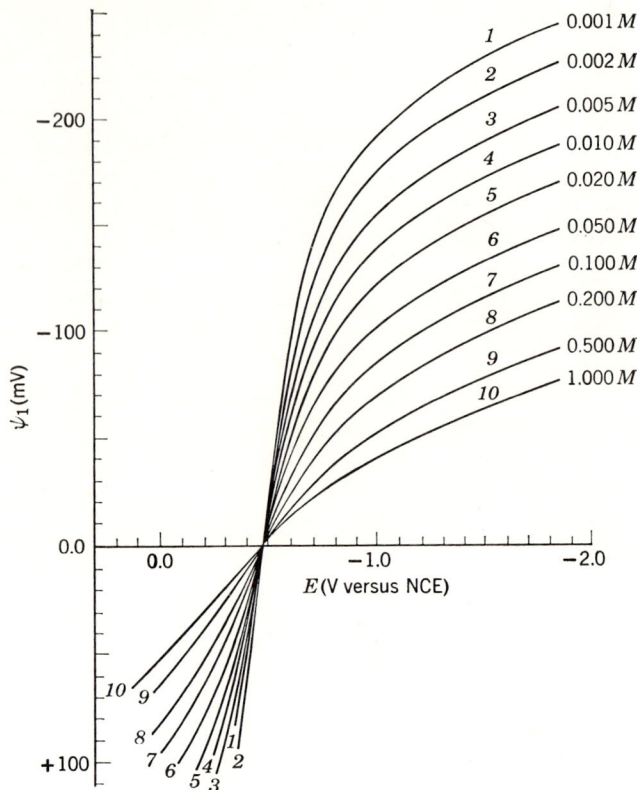

FIG. 5. The dependence of ψ_1 on the electrode potential in solutions with varying NaF concentration [constructed from the data of reference (27)].

To evaluate ψ_1 from Eq. 12 if one ion is specifically adsorbed, the value of q_i should be known (see Eq. 2). The latter can be calculated either from electrocapillary curves or differential capacity curves [see, for example, references (4), (5), (6), (8), and (30)]. The latter publications also present curves showing how ψ_1 varies with the electrode potential for solutions containing different surface-active anions (see, for instance, Fig. 6). Grahame (5, 28) presumed that only fluoride ions would not exhibit any appreciable specific adsorption at positively charged mercury electrode surfaces. However, Palanker, Skundin, and Bagotskii (31) have established the occurrence of some fluoride ion adsorption on a mercury electrode by studying the double-layer capacity in potassium fluoride solutions. Independently,

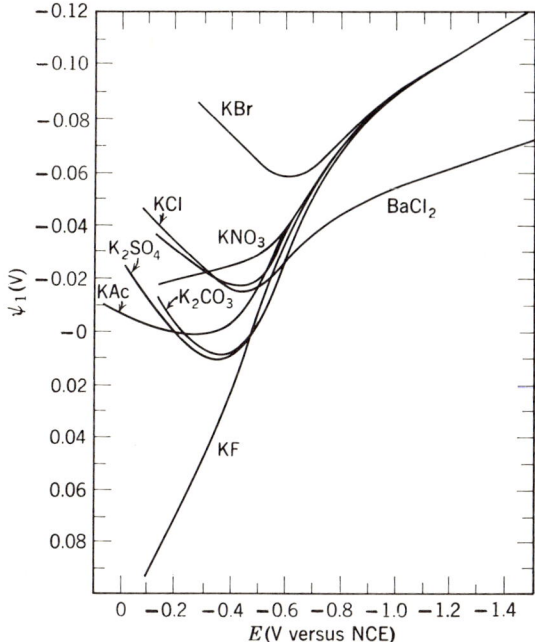

FIG. 6. The dependence of ψ_1 on the mercury electrode potential in 0.1 N solutions of different salts. [From reference (30).]

McCoy and Mark (29), having compared a sufficient number of experimental data on q_i, the charge densities with various adsorbed anions, have also concluded that some appreciable adsorption of fluoride ion occurs in aqueous solutions at potentials more positive than −0.4 V (versus NCE). These investigators constructed a hypothetical plot of electrode charge against potential in the complete absence of ion adsorption (Fig. 7). The difference between the value of the surface charge found experimentally and that corresponding to the same potential according to this hypothetical curve is assumed to be equal to the charge of the adsorbed ions q_i. It is seen from Figures 6 and 7 that no anion adsorption occurs in solutions of potassium salts at potentials more negative than −1.0 V, so that ψ_1 can be evaluated either from the tables (27) or from the plot in Figure 5.

The vaule of $\psi_1{}^i$ depends on the ratio of the distance between the inner and outer Helmholtz planes $(x_1 - x_1{}^i)$ to that between the outer Helmholtz plane and the electrode surface (x_1), as well as on the charge values q_s and

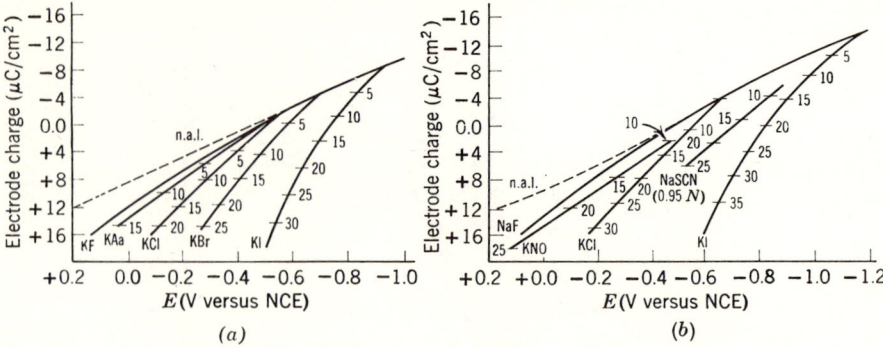

FIG. 7. The charge density of the electrode surface as a function of its potential in 0.1 N (a) and 1.0 N (b) solutions of different salts. The dotted curve corresponds to the hypothetical values of the charge in the absence of specific adsorption. The figures beside the curves are the values of q_i for the adsorbed anions. [From reference (29).]

q_i (32). At a small q_i the potential distribution in the compact part of the double layer can be considered linear, so that

$$\psi_1^i \simeq \psi_1 \frac{x_1^i - x_1}{x_1}$$

In conclusion it should be noted that no strictly quantitative theory concerning the compact part of the double layer has been developed so far, whereas the diffuse-layer theory has been sufficiently elaborated to allow a good comparison with experimental data.

IV. The Double-Layer Structure and Electrode Kinetics

A. Uncharged Depolarizers

The active center of a particle taking part in an electrochemical reaction is usually at some distance from the electrode surface. This distance is likely to correspond approximately to the outer Helmholtz plane. Thus the effective potential drop that causes the electron transfer is equal to $E - \psi_1$ (see Fig. 2). In the simplest case of the irreversible electrochemical reduction of an uncharged compound when no chemical reactions at the electrode are involved, the half-wave potential value $E_{1/2}$ is the measure of the electron transfer rate constant (32). Therefore, in order to find the true half-wave potential ($E_{1/2}^{\text{true}}$) of the wave of such a process without any double-layer structure effects, the value of $E_{1/2}$ found experimentally should be corrected for the potential drop in the diffuse layer, that is,

$$E_{1/2}^{\text{true}} = E_{1/2} - \psi_1 \tag{17}$$

On changing the double-layer structure, hence ψ_1, the experimentally accessible value of the half-wave potential changes as well. If $E'_{1/2}$ and $E''_{1/2}$ denote the values determined before and after changing the double-layer structure, with ψ'_1 and ψ''_1 standing for the corresponding values of the potential drop in the diffuse part of the double layer, then it follows from Eq. 17 that

$$E_{1/2}^{\text{true}} = E'_{1/2} - \psi' = E''_{1/2} - \psi''_1$$

or

$$E'_{1/2} - E''_{1/2} = \psi'_1 - \psi''_1, \text{ that is, } \Delta E_{1/2} = \Delta \psi_1 \tag{18}$$

Thus it follows that in the case of the irreversible electrode process, with an uncharged compound being involved, but without any antecedent chemical reactions, the change in the half-wave potential value on changing the double-layer structure, which is observed during the experiment, is equal both in value and in sign to the change in ψ_1.

Levin and Fodiman (33) were the first to take into consideration the effects of changing the double-layer structure on half-wave potentials and were followed by others (34, 35). Nowadays, the validity of Eq. 18 has been supported by a great many studies. For instance, the half-wave potential shift of the second reduction waves of aromatic aldehydes and ketones at different ionic strength was reported (9) to obey this equation quite well. A fairly good agreement is attained for the half-wave potential shift of the reduction waves of different organic iodo derivatives with increasing potassium chloride concentration in the solution (36).

B. Electrode Processes Involving Ions

The rate of the irreversible electrochemical reactions involving charged particles is strongly affected by changing the concentration of these particles at the electrode caused by the electrode field; the higher the potential drop in the diffuse part of the double layer ψ_1, the greater the difference between the concentration at the electrode surface and that in the bulk of the solution (see Eq. 1). Therefore both the ion concentration at the electrode and the effective potential drop between the electrode surface and the discharging particle change with changing ψ_1. Consideration of the two factors results in the following relationship for the change in half-wave potential of the reduction waves of charged species and that in the ψ_1-potential value:

$$\Delta E_{1/2} = \Delta\psi_1 \frac{\alpha n_a - z}{\alpha n_a} \qquad (19)$$

where z is the charge of the particle entering an electrochemical reaction, α the transfer coefficient ($0 < \alpha < 1$), and n_a is the number of electrons transferred in the potential-determining step (usually $n_a = 1$). At $z = 0$ Eq. 19 reduces to Eq. 18.

In the reduction of anions ($z < 0$), which as a rule proceed at a negatively charged electrode surface, the electrode field effects, as seen from Eq. 19, cause the value of $\Delta E_{1/2}$ to correspond in sign to the value of $\Delta\psi_1$. However, the former is much greater than the latter in its absolute value. In other words, the effect of changing the double-layer structure is most strongly pronounced if the anions are reduced at a negatively charged electrode surface. In this case some specific phenomena can be observed, such as a wave decrease, which is discussed in particular in Section IV-E.

During the reduction of cations ($z > 0$) at a negatively charged surface, the electrostatic action of the field facilitates their approach to the electrode surface and as seen from Eq. 19 the signs of $\Delta E_{1/2}$ and $\Delta\psi_1$ are opposite, that is, the higher the potential drop in the diffuse part of the double layer, the more the cation reduction is facilitated.

One can easily see that the half-wave potential shift of a wave at different ψ_1-potentials increases sharply as the charge of the particle entering an electrode reaction increases, this shift being very significant for polycharged particles. For example, during the reduction of the diprotonated form of diaziridine, which bears a double positive charge in strongly acidic medium, the 10-fold increase in the ionic strength of the solution caused by adding potassium chloride results in a nearly 200-mV shift in the half-wave potential to negative potentials (37). Fedorovich and Frumkin (38, 39) have shown that a much more considerable shift in half-wave potential can be observed when anions bearing a multiple charge are reduced.

The effect of the double-layer structure on the electroreduction of charged particles accounts for the considerable (up to several hundred millivolts) deviation in the half-wave potentials observed as compared to those calculated theoretically from certain correlation ratios. So, the half-wave potential of the wave of p-iodophenolate anion reduction proves to be approximately 300 mV (40) more negative than that calculated by using the molecular orbital method, with the negative charge of phenolate oxygen atom being taken into account. When p-iodoanilinium cation is reduced, however, the half-wave potential observed is found to be greater than 300 mV more positive than that calculated (40).

Similarly, the electrode field effect appears to cause strong shift in the half-wave potential of the reduction wave on changing the depolarizer charges. Protonated particles are reduced more readily than unprotonated ones and, on the contrary, the anions of acids are reduced with much greater difficulty than their undissociated molecules. Changing the extent (and rate) of ionization or dissociation of similar compounds at different pH of the medium causes the dependence of half-wave potentials of their waves on pH.

The change in the charge of a depolarizer resulting from the loss or attachment of a proton is accompanied by the electron density rearrangement. Therefore if an atom (or a group of atoms) bearing a charge is close to the reaction center of the depolarizer, the effect of such rearrangement is added to the action because of the electrode field effect. Thus the resulting shift in half-wave potential of the wave may attain a value greater than 1 V. However, it should be remembered that in this case the half-wave potential shift is frequently affected by a change in adsorption of the original compounds and products as well as by other factors.

Equation 19 allows estimation of the half-wave potential shift fraction contributed by the electrode field effect at different depolarizer charges (provided, of course, the reaction mechanism remains unchanged) as is the case, for instance, in the reduction of the halogeno derivatives of carboxylic acids and their anions. If the half-wave potential of the wave is in the -1.2 to -1.8 V potential range and the ionic strength of the buffer solutions composed of K^+ and Na^+ salts is of the order of $0.1\ N$ then, as seen from Figure 5, ψ_1 would be predicted to be approximately equal to -100 mV, and the change in half-wave potential from only the electrode field effect at $\alpha n_a \approx 0.5$ would be about 200 mV, the half-wave potential becoming more positive when the molecule assumes the positive charge and more negative with the transition from the neutral molecule to the anion. At smaller values of αn_a, the half-wave potential shift of the reduction of ions resulting from a change in their charge by a unit would be considerably greater, as may be seen from Eq. 19: while at $\alpha n_a = 0.4$, $\Delta E_{1/2} \approx 250$ mV, at $\alpha n_a = 0.3$, $\Delta E_{1/2} \approx 330$ mV.

As it is possible to vary the character and value of the effect of ψ_1 (for instance, by increasing the supporting electrolyte concentration) on the half-wave potentials of the waves of particles having different charges, the value and the sign of the charge of a particle entering an electrochemical reaction can be determined with the aid of Eq. 19 (41, 42). Also, one can deduce the change in ψ_1 from the observed shift in half-wave potentials or from the change in the electron transfer rate on increasing the supporting electrolyte concentration. For example, the change in ψ_1 attributable to the adsorption of a small amount of bromide ions in the presence of 0.01, 0.10, and 1.00 g-equiv/l of ClO_4^- has been estimated (43) from the potential shift in the reduction of cobalt(III) to cobalt(II).

When the effect of the double layer upon ion reduction is considered, it is often possible to improve the conditions significantly, and sometimes even to carry out a polarographic analysis that would be otherwise impossible. For instance, in ac (44) and square-wave (45) polarography, the peak heights of Zn^{2+} and Ni^{2+} diminish with increasing concentration of the supporting electrolyte, which becomes more pronounced when Na^+ is replaced by Ba^{2+} in the electrolyte. With I^- anions introduced into the solution, the Zn^{2+} peak increases (44). In these cases the peak height increases because of an increase in the discharging ion concentration at the electrode resulting from an increase in the negative ψ_1-potential. Therefore in order to determine the concentration of the cations discharging on the negatively charged mercury surface by employing the ac polarographic method, one should use (46) a supporting electrolyte solution of very low concentration. On the contrary, in the polarographic determination of anions, the wave of which under usual conditions is concealed by the background discharge current as is the case with the reduction of terephthalic acid anions (47), one can observe the wave of such anions before the background discharge current by increasing the ionic strength of the solution and in particular by using electrolytes with polyvalent cations.

The double-layer structure affects the wave slope as well. Therefore the apparent value of $\alpha n_{a(app)}$ found directly from the semilogarithmic plot of the wave can be markedly different from its true value (6). The relationship between the value observed and the true value of αn_a is expressed by (4, 48):

$$\alpha n_a = \frac{\alpha n_{a(app)} - z(\partial \psi_1/\partial \phi)_{E_{1/2}}}{1 - (\partial \psi_1/\partial \phi)_{E_{1/2}}} \quad (20a)$$

or

$$\alpha n_{a(app)} = \alpha n_a + (z - \alpha n_a)\left(\frac{\partial \psi_1}{\partial \phi}\right)_{E_{1/2}} \quad (20b)$$

From these expressions it follows that the difference between αn_a and $\alpha n_{a(app)}$ is at a maximum for waves having half-wave potentials close to the potential of zero charge E_z since close to it the value of the derivative $\partial \psi_1/\partial \phi$ is the largest (see Fig. 5). The $\partial \psi_1/\partial \phi$ value at E_z changes appreciably with the potential even if the latter changes in a very narrow range. If this range is in the rising part of the wave, it causes the distortion of the semilogarithmic plot. In this case $(\partial \psi_1/\partial \phi)_{E_{1/2}}$ should be replaced by $(\partial \psi_1/\partial \phi)_E$ corresponding to the various potentials of the rising part of the wave. As the distance between $E_{1/2}$ and E_z increases, the value of $\partial \psi_1 / \partial \phi$ becomes practically the same in the whole potential range in which the wave is observed, and its value becomes smaller so that $\alpha n_{a(app)}$ approximates αn_a. For example, at $E_{1/2}$ of the order of -1.9 V, $\partial \psi_1 / \partial \phi \approx 0.003$ and $\alpha n_{a(app)}$ is almost the same as αn_a. From Eq. 20 it also follows that for anions ($z < 0$) and uncharged depolarizers, αn_a is larger than $\alpha n_{a(app)}$, while for cations αn_a is somewhat smaller than $\alpha n_{a(app)}$. Thus, according to Eq. 20a, for uncharged substances the largest slope of the wave (i.e., the smallest $\alpha n_{a(app)}$ value) would be observed close to E_z. It should be especially emphasized that Eq. 20a is valid only in the absence of the depolarizer adsorption at the electrode surface, which can considerably change the slope of the wave (see Section V), this change for the uncharged depolarizers being opposite to that predicted from Eq. 20a. It is the influence of adsorption, rather than the ψ_1-effect as was suggested in reference (49), that accounts for the sharp increase in the wave slope (i.e., the $\alpha n_{a(app)}$ value) close to E_z, which is observed for the reduction waves of some organic halogeno derivatives.

It is common knowledge that hydrogen ions are not adsorbed on mercury. Therefore their reduction can supply at least a qualitative illustration of the effect on the slope of the polarizing curve of changing the ψ_1-potential. Figure 8 reproduces in terms of the coordinates adopted in polarography the

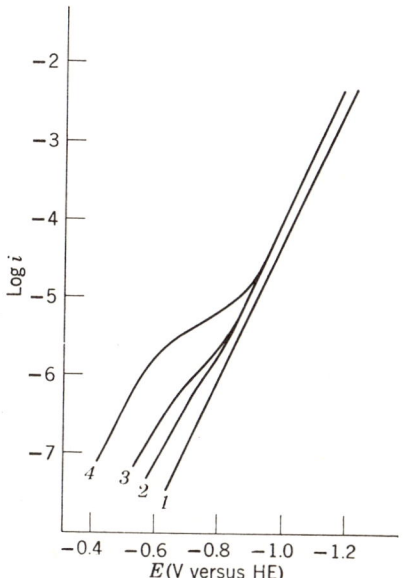

Fig. 8. The dependence of the current density at a large mercury cathode on the potential (versus the hydrogen electrode in the same solution) for the solutions: (1) 1 N Na_2SO_4 + 0.1 N H_2SO_4; (2) 1 N KCl + 0.1 N HCl; (3) 1 N KBr + 0.1 N HCl; (4) 1 N KI + 0.1 N HCl. [From reference (50).]

results of rather accurate measurements by Iofa et al. (50) of polarizing curves (in the absence of the concentration polarization) of hydrogen discharge on a large mercury cathode in solutions containing surface-active anions that strongly change the ψ_1-potential. At sufficiently negative potentials, $\partial\psi_1/\partial\phi$ is positive, almost constant, and small in its absolute value (the extreme right parts of the curves for potassium bromide and potassium chloride in Fig. 6), which according to Eq. 20a corresponds to some value of $\alpha n_{a(\text{app})} > \alpha n_a$ (the extreme right-hand portions of the curves in Fig. 8). On decreasing the cathode potential and approaching the point of zero charge, $\partial\psi_1/\partial\phi$, as follows from Figure 6, increases, which corresponds to the sharp increases of slope in the curves in Figure 8. With further decrease in the cathode potential, $\partial\psi_1/\partial\phi$ decreases, passes through zero (minimum in the curves in Fig. 6), and then becomes more negative and attains some constant value. Such character of changing $\partial\psi_1/\partial\phi$ accounts for the inflection in each curve in Figure 8, after which the curve straightens and its steepness, hence $\alpha n_{a(\text{app})}$, is smaller than for the linear portion at large cathode potentials. On the left-hand portions of the curves in Figure 6, $\partial\psi_1/\partial\phi < 0$, therefore according to Eq. 20a on the left-hand portions of the curves in Figure 8 $\alpha n_{a(\text{app})} < \alpha n_a$.

As the anion adsorptivity increases in the series $Cl^- < Br^- < I^-$, ψ_1 at small surface charges becomes more negative, which results in an acceleration of the electrochemical hydrogen evolution, and the inflection in the polarizing curves becomes more pronounced (see Fig. 8). In the absence of surface-active anions, when no acceleration in the hydrogen ion discharge at small charges of the electrode surface takes place, the polarizing curves do not show any inflection. However, it is necessary to note that Eq. 20b and the data on ψ_1 in Figure 6 cannot supply a quantitative description of the shape of the curves in Figure 8. This is likely a result of the fact (62) that the hydronium ion centers are closer to the electrode surface than the outer Helmholtz plane. Thus the ψ_1-potential and its changes are different from those in Figure 6.

Strictly speaking, the true value of α is not constant. As shown by Krishtalik (51), for very large changes in the potential α should vary from 0 to 1. However, the conditions under which the polarograms are recorded allow the α-value to be considered constant with great accuracy.

In order to construct true plots of the current intensity as a function of the potential, which are not distorted by the double-layer structure effect, Delahay et al. (52) suggested that advantage be taken of the so-called corrected Tafel plot (CTP) which is constructed with the coordinates $\log\{[i/(i_d - i)] + (zF\psi_1/2.3RT)\}$ versus $E - \psi_1$ (instead of $\log[i/(i_d - i)]$, use is often made of a more accurate value, the Koutecký function $\log \chi$, calculated (53) for different ratio values of i/i_d instant or $\bar{i}/\bar{i}_d$ averaged polarographic currents, which represents a value proportional to the rate constant of the electron transfer). Lately, many investigators have expressed the polarizing curves found experimentally as CTP (54–61), which apparently should not depend on the nature and concentration of the supporting electrolyte. Figure 9 shows as an example some standard semilogarithmic plots of the waves (a) and CTP (b) for the waves of the trichloroacetate anion reduction in the presence of potassium salts with different anions (60). As follows from Figure 9, standard CTP plots for various anions fit well into one line, the position of which is independent of the salt concentration (59). However, on changing the cation nature, there arises a series of lines instead of one CTP (Fig. 10).

The CTP for hydrogen ion discharge consists of straight lines independent of the supporting electrolyte concentration (55). These are certain to shift parallel to one another on changing nature of the background cation. The distortion of the CTP is sometimes observed as in the case of $S_2O_8^{2-}$ reduction (55), although even in this case on increasing the supporting electrolyte concentration neither position nor CTP shape is changed. However, such changes may reappear if the background cation is changed.

All these, as well as some other deviations of experimental data from the

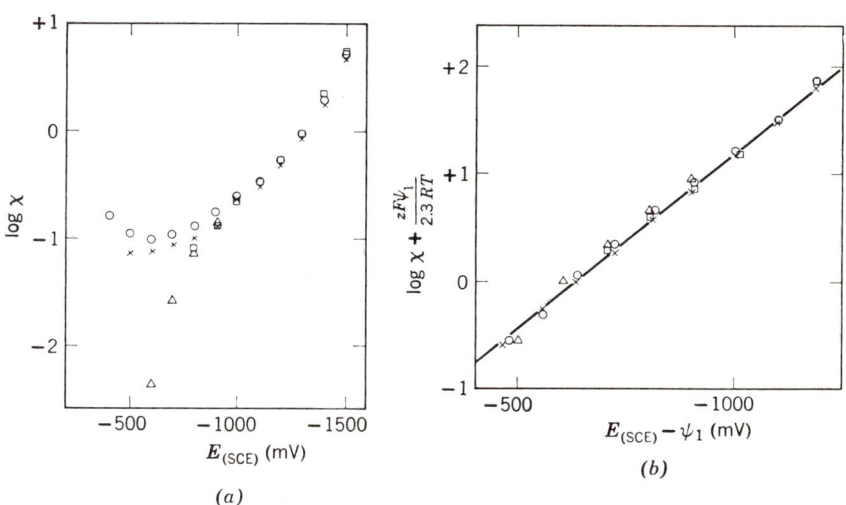

Fig. 9. The effect of the nature of the anions of the supporting electrolyte (1.0 N) on the semilogarithmic plots log χ–E (a) and CTP (b) for the reduction wave of trichloroacetate anions (1×10^{-4} M): (○) KF; (□) KI; (×) KCl; and 3×10^{-3} M: (△) KI. [From reference (60).]

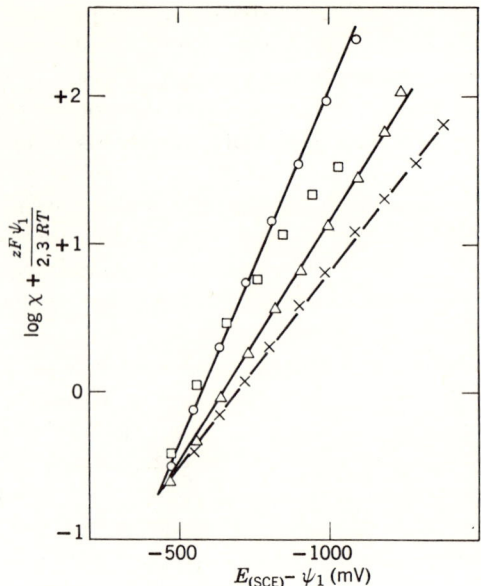

FIG. 10. CTP of the reduction wave in 1×10^{-4} M solutions of trichloroacetate anions containing different supporting electrolyte cations (0.1 N). (×) LiCl; (△) KCl; (○) CsCl; (□) BaCl$_2$. [From reference (60).]

theory, result from the fact that the above considerations of the double-layer structure represent only a rough approximation to the actual state of the matter. A number of factors such as the effect of the discrete nature of ions in the plane of closest approach (62), or which is the same thing, the possibility of forming ion pairs (63) or bridges (64) at the electrode, the combined adsorption of anions and cations as well as the depolarizer adsorption and specific influence of the supporting electrolyte nature upon it are not taken into account in the simple theory. Then, the position of the active center of a reacting particle in the transition state is related to the outer Helmholtz plane somewhat at random. As shown by Petrii and Frumkin (55), the CTP shape and position depend to a great extent upon selecting the distance between the active center of a particle and the electrode surface, for it is upon this distance that ψ_1 depends. Thus CTP calculated for the $S_2O_8^{2-}$ reduction in a supporting electrolyte containing cesium chloride consists of straight lines if the active center is assumed to be either in the Helmholtz plane or within the diffuse layer at some distance from it, which does not exceed 0.6Å. If the active center is assumed to be at a greater distance from the electrode, the CTP changes slope and becomes distorted (55). In the same publication (55), it is shown that in order to straighten the CTP of reduced $Fe(CN)_6^{3-}$ it is necessary to assume that the particle center is at a distance of approximately 10 Å from the outer Helmholtz plane. This, however, results in the unreasonably small value of $\alpha \approx 0.05$. These facts have not been adequately explained so far.

To find a more accurate value of ψ_1, one should also take into account (65) the structural changes in the solvent in the close neighborhood of the electrode surface, in particular the dielectric saturation of the solvent which results in decreasing the ion concentration at the electrode, the sign of the ions being opposite to that of the electrode charge so that the absolute value of ψ_1 proves to be larger than that in some effective plane of the maximum approach of the unadsorbed ions (65).

To conclude this section it should be emphasized that the theory in question describes the double-layer structure effect on electrode kinetics correctly (and in some cases quantitatively) although it omits some factors. The effect of some of these factors on electrode processes are considered in the following sections.

C. *Electrode Processes with Chemical Reactions. Volume Kinetic and Catalytic Waves*

In the early 1940s, Brdička and Wiesner (66–68) discovered the first electrode processes complicated by chemical reactions. It has since become known that most electrode processes, in particular those involving organic substances, are accompanied by chemical reactions at the electrode. In some cases such reactions precede the electron transfer, and it is because of the reactions that the depolarizer is formed. If the antecedent reaction rate ρ is not very high, it may limit the rate of the entire electrode process. The currents depending on such reactions are called kinetic currents.

There are several types of antecedent electrode chemical reactions resulting in kinetic currents or waves. These are mainly reactions of decomposition or formation of complexes, oxidation of ions of metals with varying valency, protonation, or dehydration. The two last-mentioned reactions are especially common in polarography of organic compounds. Let us consider electrode chemical reactions in the case of the most important and common antecedent protonation reaction when an electrochemically inactive form of a depolarizer (B is uncharged or A^- is negatively charged) accepts a proton of a proton donor DH^+ (which may be either a cation, such as NH_4^+, or an uncharged molecule as CH_3COOH, or an anion as $H_2PO_4^-$:

$$B + DH^+ \underset{\rho\sigma}{\overset{\rho}{\rightleftarrows}} BH^+ + D \quad \text{or} \quad A^- + DH^+ \underset{\rho\sigma}{\overset{\rho}{\rightleftarrows}} AH + D \qquad (21)$$

where D is the base conjugated with the proton donor.

The value of ρ is an overall reaction rate constant (of the first order) determined by contributions from all the proton donors present in the solution:

$$\rho = k_1[\mathrm{D_1H^+}] + k_2[\mathrm{D_2H^+}] + \cdots \qquad (22)$$

while the equilibrium constant σ depends only upon the value of the acidic constant of $\mathrm{BH^+}$ or AH dissociation K_a and the pH of the solution:

$$\sigma = \frac{K_a}{[\mathrm{H^+}]} = \frac{[\mathrm{B}]}{[\mathrm{BH^+}]} \quad \text{or} \quad \sigma = \frac{K_a}{[\mathrm{H^+}]} = \frac{[\mathrm{A^-}]}{[\mathrm{AH}]} \qquad (23)$$

Other reactions preceding the electron transfer can be represented similarly. Therefore what is presented here regarding kinetic currents limited by a protonation reaction can be applied in principle for processes limited by other reactions.

Electrode reactions occur both in the solution close to the electrode surface and on the surface itself, involving the substances adsorbed at it. Accordingly, volume and surface kinetic currents and waves can be distinguished (69, 70).

In the case of volume kinetic waves, chemical reactions proceed in a layer close to the electrode surface, its thickness μ corresponding (71) to the distance to which the electrochemically active particle $\mathrm{BH^+}$ or AH (see Eq. 21) can be transported by diffusion until it is converted by the reverse reaction of Eq. 21 with the rate constant $\rho\sigma$ into the electrochemically inactive form B or $\mathrm{A^-}$. The value of μ is equal (72, 73) to:

$$\mu = \sqrt{D/\rho\sigma} \qquad (24)$$

where D is the diffusion coefficient of the electrochemically active particle.

Other conditions being equal, the kinetic current depends on the chemical reaction component concentration at the electrode. If one or several of these components are ions, their concentration at the electrode is strongly affected by the double-layer structure (see Eq. 1). It can be easily shown that the thinner the reaction layer, that is, the closer to the electrode surface the chemical reaction proceeds, the greater the double-layer effect upon the kinetic current (34, 74–76). On the contrary, if the reaction space μ is sufficiently large and greatly exceeds the thickness of the diffuse part of the double-layer δ (see Eq. 3), then the double layer has practically no effect on the concentration of the charged components of the chemical reaction in the reaction space, hence does not affect the process rate. In general, because of nonlinear concentration distribution in the reaction and diffuse layers, the double-layer effect does not lend itself to simple quantitative estimation. In several studies (74–76), rather complicated equations have been developed to take into account the double-layer effect, the ratio of the reaction layer thickness to that of the diffuse part of the double layer μ/δ being an important factor in determining this effect.

The effect of the ratio μ/δ is clearly manifested in the case of volume kinetic waves in unbuffered aqueous dipotassium maleate solutions, in which water is the only proton donor DH^+, with hydroxyl ion as its conjugated base D (77). The electrochemical reduction of the monohydrogen maleate anion gives rise to hydroxyl ions (78), enhancing the reverse reaction of Eq. 21 which annihilates the electrochemically active form of the depolarizer. Therefore the reaction layer thickness μ is a function of the current flowing (77, 78); the greater the current, the thinner the reaction layer. The value of δ does not depend on the current intensity. Hence the ratio μ/δ decreases as the current on the wave rise increases. Figure 11 shows the polarograms of a dipotassium maleate solution in a supporting electrolyte that contains various concentrations of potassium chloride (79). As the potassium chloride concentration increases, the wave becomes more positive. In the upper part of the wave where, as noted earlier, the ratio μ/δ is smaller, the protonation rate increase is greater than that in the lower part, which results in a greater comparative increase in the current in the upper part of the wave. This effect accounts for the increase in the steepness of the wave and greater shift of its upper part to positive potentials compared to its lower part. So, the value of $E_{3/4}$ (i.e., the potential at which the current is $\frac{3}{4}$ of its limiting value) with a 10-fold increase of potassium chloride concentration shifts by 165 mV while $E_{1/4}$ shifts only by 75 mV.

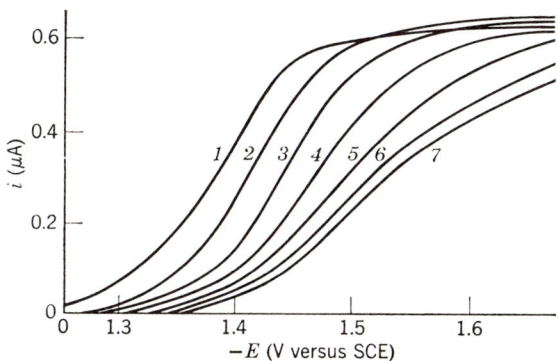

FIG. 11. Polarograms of dipotassium maleate (15 mM) in neutral KCl solutions of varying concentration. (1) 2.0, (2) 1.0, (3) 0.5, (4) 0.25, (5) 0.1, (6) 0.075, and (7) 0.05 M. The electrode characteristics: $m = 0.95$ mg/sec, $t = 0.34$ sec. [From reference (79).]

In the reduction of the dianions of maleic acid in an unbuffered solution considered above, the increase in the ionic strength of the solution results mainly in an increase in the electrode concentration of the depolarizer, that is, maleic acid dianions, as well as that of hydroxyl ions. As the concentration of the latter increases, the effect of enhancing the entire electrode process diminishes to some extent. In buffer solutions the increase in ionic strength (with pH retained constant) is responsible for the increase in pH close to the negatively charged electrode (9, 80), which causes a decrease in the protonation rate. In any buffer acid–base system (DH^+ and D), the charge of the acid form DH^+ is larger than that of the base from D by a unit. Therefore, as a result of the electrostatic forces, the concentration of DH^+ at the cathode is much higher than in the bulk of the solution, that is, the acidity at the electrode is much higher than that in the bulk of the solution. As the ionic strength increases (with pH in the bulk of the solution being constant), on account of a decrease in the absolute value of ψ_1, the difference between the pH value at the electrode and in the bulk of the solution becomes smaller as a result of the pH increase in the layer close to the electrode. The quantitative estimation of the double-layer effect at various ionic strengths of the buffer solution on the rate of the electrode process limited by the volume antecedent protonation reaction has been carried out (81) in the case of the limiting catalytic hydrogen current caused by pyridine. Figure 12 presents the dependence of the limiting current function $\log [i_l''/(i_l - i_l'')^2]$ on the overall concentration of uni-univalent electrolytes (i_l and i_l'' are limiting catalytic currents with and without the double-layer effect correction, respectively; the value of i_l is observed at a very high supporting electrolyte concentration). As follows from the figure, the theoretical line describes well the observed current change (81).

A sharp change in the apparent (i.e., uncorrected for the double-layer effect) protonation rate constant ρ' in the case of volume catalytic current in a pyridine solution is observed as the reaction layer thickness decreases with an increase in the pH of the solution. This effect is greatly pronounced in ammonia buffer solutions with ammonium ions forming the outer double-layer plate and serving at the same time as proton donors. It is of interest that the graphs of ρ' as a function of the ammonium ion concentration $C_{NH_4^+}$ are no longer linear at the high concentration (Fig. 13); as $C_{NH_4^+}$ increases, the ρ' increase slows down (82) and at a very high $C_{NH_4^+}$ the value of ρ' asymptotically approaches a limit that corresponds to the neutralization of electrode charge by the ammonium ions that are in a thin layer near the electrode surface (83). As expected with decreasing potassium chloride concentration, when the NH_4^+ ion contribution to the outer double-layer plate becomes greater, distortion in the ρ'–$C_{NH_4^+}$ function graph can be

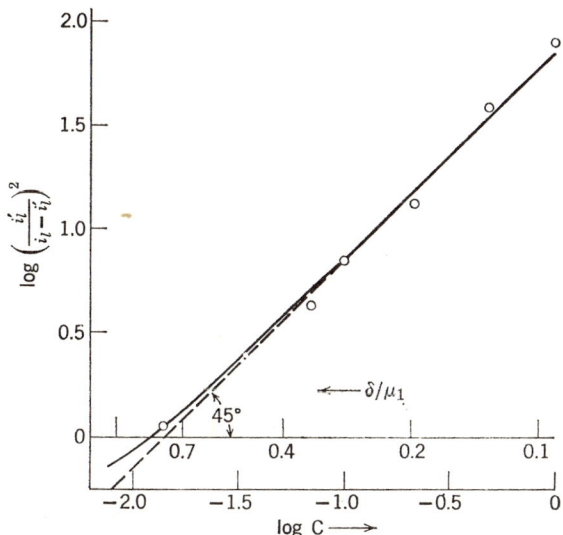

FIG. 12. The dependence on ionic strength (KCl added) of $\log [i'_l/(i_l - i'_l)^2]$ for the catalytic reduction of hydrogen in borate buffers containing pyridine and having constant pH and buffer capacity. The dots are experimental values; the straight line is the dependence theoretically predicted. [From reference (81).]

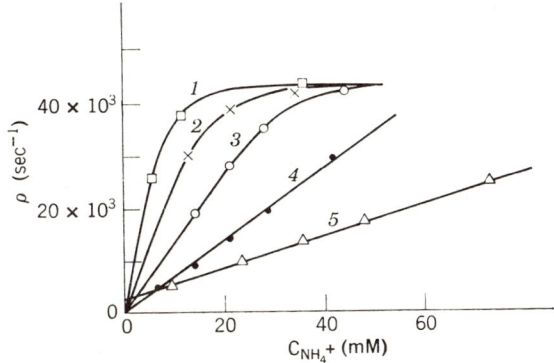

FIG. 13. The apparent rate constant for the protonation of pyridine as a function of the concentration of ammonium ion in ammoniacal buffer solutions with pH 9.20 (curves 1–4) and 7.90 (curve 5) at varying ionic strength (KCl added). (1 and 5) 0.05; (2) 0.1; (3) 0.25; (4) 0.5 F. [From reference (83).]

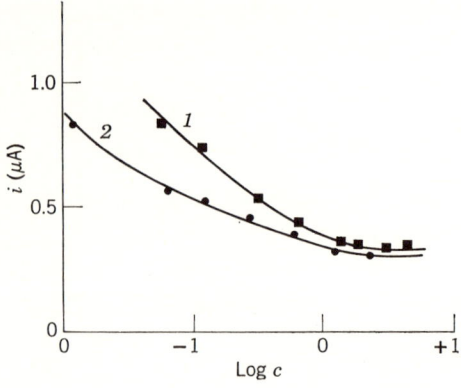

Fig. 14. The limiting current of pyridine in ammoniacal buffer solutions of pH 9.20 as a function of the logarithm of the molar concentration of (1) KCl, (2) CaCl$_2$. [From reference (83).]

observed at smaller NH$_4^+$ concentrations. However, the limit to which ρ' tends is the same (at a constant pH value in the bulk of the solution).

On adding the supporting electrolytes to the buffer ammonia solution, the limiting catalytic current decreases appreciably, tending toward some constant value (Fig. 14). In the case of ammonia solutions, this value is not a true rate constant, as it is significantly affected by the double-layer structure (83).

To determine the true rate constants undistorted by the double-layer effect, the apparent values of particular protonation constants k' (equal to the slope of the lines showing ρ' as a function of the concentration of a given proton donor of the solution) determined directly from an experiment should be extrapolated to the infinitely large reaction layer thickness at which no electrode field effect is exercised. With this in view, the k'-values found for several pH values should be plotted as a function of a reciprocal hydrogen ion concentration value $[H^+]^{-1}$. This function for volume catalytic waves is rectilinear (82) and k' extrapolated to $[H^+]^{-1} = 0$ corresponds to the true rate constant of proton transfer from a donor to a catalyst molecule.

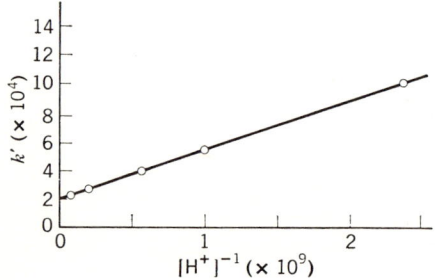

Fig. 15. The extrapolation to infinite reaction-layer thickness of the apparent rate constant for proton transfer from the amino group of the glycine zwitterion to pyridine. [From reference (84).]

Figure 15 illustrates such extrapolation for the rate constant of the proton transfer from the protonated amino group of glycine zwitterion to pyridine acting as a catalyst (84).

D. Surface Kinetic and Quasidiffusion Waves

Only a few instances of volume kinetic and catalytic waves are known. The volume waves caused by organic depolarizers, which have been most thoroughly studied, were discussed in the previous section. As a rule, because of the high surface activity of many compounds, organic ones in particular, electrode chemical reactions proceed at the electrode surface involving the adsorbed reagents (9, 12). As shown by a simple calculation (85), coverage of a few tenths of a per cent of the electrode surface by the adsorbed reagent is sufficient for the limiting current to be caused mainly by a chemical reaction which proceeds involving the adsorbed compound, the contribution of the same reaction proceeding in the reaction volume becoming insignificant. In this case the concentration of the charged particles at the electrode is determined by a simple ratio (Eq. 1), and the double-layer effect can be described by comparatively simple equations.

Thus when the difference in charges of the acidic and basic forms of the components in the buffer solutions are taken into account, the expression for the concentration ratio of both these forms at the electrode and hydrogen ions to those in the volume has been obtained (9, 85):

$$\left(\frac{[DH^+]}{[D]}\right)_s \bigg/ \left(\frac{[DH^+]}{[D]}\right)_0 = \frac{[H^+]_s}{[H^+]_0} = \exp\left(-\frac{\psi_1 F}{RT}\right) \qquad (25)$$

As the charge of the acidic forms of the buffer components is always greater than that of the corresponding basic forms by a unit, the acid–base equilibrium at the negatively charged electrode surface is shifted (compared to the bulk of the solution) toward the acidic form. This shift can be quite prominent Thus, in a 0.1 M solution of uni-univalent electrolyte at the electrode potential about -1.5 V (versus SCE), the pH at the cathode surface is two units lower than that in the solution volume.

The difference between the proton donor concentration at the electrode surface and that in the solution volume, as follows from Eq. 25, depends on ψ_1. Therefore, in the case of processes with antecedent protonation [when the half-wave potential depends on pH (9, 12)], the pH change at the electrode is an additional factor affecting the half-wave potential of the waves when the double-layer structure changes. In the most often occurring case of electrode processes with the preceding reaction—surface quasidiffusion (12, 86) waves (i.e., the waves the height of which owing to the sufficiently high preceding reaction rate practically attains the level of the limiting diffusion current of the depolarizer in the solution, with the half-wave potential still being a function of the preceding chemical reaction rate)—the dependence of the half-wave potential on ψ_1 taking into account the effective potential drop change as well as the change in the concentration of charged particles at the electrode is expressed by (9, 12):

$$\Delta E_{1/2} = \Delta \psi_1 \left(\frac{\alpha n_a - z}{\alpha n_a} + \frac{\partial E_{1/2}}{\partial \mathrm{pH}} \frac{F}{2.3RT} \right) \qquad (26)$$

In this equation z is the charge of a particle in the solution, that is, before its surface protonation. It is evident that with no antecedent protonation, when $\partial E_{1/2}/\partial \mathrm{pH} = 0$, Eq. 26 becomes Eq. 19.

The change in the half-wave potential if pH is increased by one unit is often equal to the slope of the wave semilogarithmic plot (12), that is,

$$\frac{\partial E_{1/2}}{\partial \mathrm{pH}} = - \frac{2.3RT}{\alpha n_a F} = -b \qquad (27)$$

In this case Eq. 26 takes the form

$$\Delta E_{1/2} = \Delta \psi_1 \frac{\alpha n_a - z - 1}{\alpha n_a} \qquad (28)$$

It follows from comparing Eq. 28 with Eq. 19 that the antecedent protonation effect, provided the ratio of Eq. 27 is valid, is nominally equal to increasing the charge of a particle diffusing from the solution to the electrode by a unit.

One can infer from Eqs. 26 and 28 that the antecedent protonation may result in an apparent decrease in the effect of changing the double-layer structure on the half-wave potential because of the mutual compensation of various effects. This compensation is significant in the case of the quasi-diffusion waves corresponding to the reduction of the depolarizers uncharged in the solution ($z = 0$) with antecedent surface protonation, when $\Delta E_{1/2}$ proves to be much smaller than $\Delta \psi_1$. The signs of $\Delta E_{1/2}$ and $\Delta \psi_1$ may be either equal or opposite depending on $\partial E_{1/2}/\partial \mathrm{pH}$. They are opposite if $|\partial E_{1/2}/\partial \mathrm{pH}| > 2.3RT/F$ ($= 59$ mV at 25°C); then the absolute value of the second term in parentheses of Eq. 26 is more than one unit and as the sign of this term is always negative (i.e., on increasing the pH of the solution the half-wave potential of the waves with antecedent protonation becomes more negative), then the signs of $\Delta E_{1/2}$ and $\Delta \psi_1$ are opposite. On the contrary, if $|\partial E_{1/2}/\partial \mathrm{pH}|/ < 2.3RT/F$ the signs of $\Delta E_{1/2}$ and $\Delta \psi_1$ are the same. Provided that $|\partial E_{1/2}/\partial \mathrm{pH}| = 2.3RT/F$, the half-wave potential is independent of changes in the double-layer structure, as is also true for reversible waves.

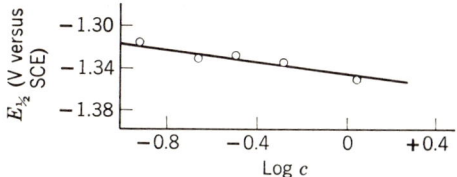

FIG. 16. The dependence of $E_{1/2}$ of the reduction wave of 2-chloropyridine on the logarithm of the ionic strength (KCl added) in acetate buffers of pH 2.8. [From reference (87).]

Figure 16 illustrates the dependence of $E_{1/2}$ of the wave of 2-chloropyridine reduction (87) on the ionic strength (adding potassium chloride) of the acetate buffer solution with pH = constant = 2.8. The reduction, that is, the electrochemical fission of the C—Cl bond in a 2-chloropyridine molecule is facilitated by its protonation (on its nitrogen atom), which accounts for the half-wave potential shift observed on changing the pH (in the acidic medium $\partial E_{1/2}/\partial \mathrm{pH} = -86$ mV). The value of $\Delta E_{1/2}$ calculated from Eq. 26 with the ionic strength of the solution increased 10-fold (at $z = 0$), -26 mV, practically coincides with that found experimentally (Fig. 16), which is equal to -28 mV.

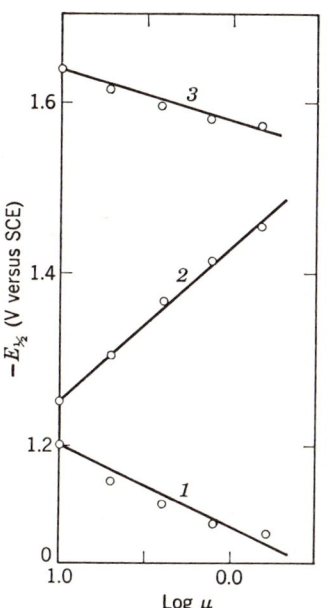

FIG. 17. The dependences of $E_{1/2}$ and $E_{p/2}$ on the ionic strength (KCl added) in borate buffer solutions at pH 8.4 for: (1) The surface kinetic wave for the reduction of Co(II) complex with cysteine; (2) the discharge of Co^{2+} from its aquo complex; (3) the catalytic hydrogen wave caused by the Co(II) complex with cysteine. [From reference (88).]

Equation 26 allows conclusions to be made concerning the charge of a particle moving to the electrode from the value of the shift in half-wave potential of the wave corresponding to the process with antecedent protonation at various values of ψ_1. Figure 17 shows as an example the plots of the half-wave potential of the Co^{2+} ion discharge diffusion wave as a function of ionic strength, the surface kinetic wave of reduction Co^{2+} complexes with cysteine (the height of this wave is limited by the rate of interaction of Co^{2+} with cysteine adsorbed on the electrode surface), as well as the half-peak potential $E_{p/2}$ (the potential at which the current attains one-half its value at the maximum of the hump-shaped wave) of the surface catalytic hydrogen wave, caused by cobalt–cysteine complexes (88). The half-wave potential of the wave of Co^{2+} discharge becomes more negative as the ionic strength increases, the value of $\Delta E_{1/2}/\Delta \log c_{salt} = -180$ mV being in good agreement with that calculated from Eq. 19 (at $z = +2$ and $\alpha n_a = 0.5$). For the kinetic wave the value of $\Delta E_{1/2}/\Delta \log c_{salt} = 90$ mV corresponds formally (at $\partial E_{1/2}/\partial pH \approx 0$ and $\alpha n_a = 0.5$) to the complex charge $-\frac{1}{4}$. The complex in question is likely to be entirely electroneutral, and the additional wave shift to the positive potentials on increasing the ionic strength results from increasing the adsorptivity of the cysteine anion. This shift might also be attributable to the negative terminal (cysteine anion) of the cobalt(II)–cysteine complex being turned to the electrode surface.

For the catalytic wave of hydrogen, $\Delta E_{1/2}/\Delta \log c_{salt} = 60$ mV, which according to Eq. 26 is indicative of the negative charge of the catalytically active complex in the solution. After surface protonation this complex becomes neutral and as such enters the electrochemical reaction responsible for catalytic hydrogen evolution.

The formation of the kinetic wave of nickel in the presence of o-phenylenediamine and the double-layer structure effect on it have been thoroughly studied by Mark and McCoy (89). They took into consideration the surface nature of the chemical reaction responsible for the prewave height, which significantly simplified the calculation of the supporting electrolyte concentration effect on the nickel ion concentration at the electrode. They also found the dependence of o-phenylenediamine adsorption on its volume concentration and the electrode potential from the differential capacity curves. Consideration of the double-layer effect gave (89) a semiquantitative agreement between the value calculated from the prewave height and that found experimentally, showing the dependence of changing reagent surface concentrations on different factors. It is of interest that with a change in the nature of the supporting electrolyte anion, in particular with the transition from Cl^- to Br^- to I^-, an appreciable increase in the prewave height and its shift to less negative potentials can be observed, obviously on account of increasing the negative ψ_1-potential as a result of increasing the specific adsorption in the anion series in question ($E_{1/2}$ of the nickel prewave in an o-phenylenediamine solution is about -0.7 V versus SCE).

Under conditions in which specific anion adsorption takes place, the difference between $\psi_1{}^i$ and ψ_1 should manifest itself. This difference is likely to be the reason (4, 9) for the unusual shift in half-wave potentials of the reduction wave of nitromethane in acidic medium, which is observed at potentials on the positive side of the electrocapillary curve close to the point of zero charge at which strong adsorption of iodide ions takes place. An increase in potassium iodide concentration in the solution results (34) in a half-wave potential shift to negative potentials. The absolute value of this shift ($\Delta E_{1/2}$) is approximately twice as large as the theoretical change in $\Delta\psi_1$, which moreover has an opposite sign (i.e., toward increasing the positive ψ_1-potential). Simultaneously, because of iodide ion adsorption a negative shift in the potential within the inner Helmholtz plane $\Delta\psi_1{}^i$ takes place, and the changes in $\Delta\psi_1$ and $\Delta\psi_1{}^i$, which are opposite in sign, have approximately the same absolute values. The observed $-\Delta E_{1/2}$ shift of the reduction wave of nitromethane has been accounted for (4, 9) by the addition of two effects, that is, the effect of decreasing the actual potential drop by $\Delta\psi_1{}^i$ (the reaction center of nitromethane is presumably at the inner Helmholtz plane) and that of increasing the pH of the solution at the electrode as a result of increasing the ψ_1-potential on the outer Helmholtz plane (the centers of the hydrated hydrogen ions do not seem to approach the electrode surface closer than the outer plane).

The ψ_1-potential change strongly affects the heights of the surface kinetic waves limited by the rates of reactions involving ions. As a result of changing the adsorptivity of a substance at the electrode at various potentials (see Section V-A), such waves are uusally hump-shaped, the limiting current of such waves being not a constant value but depending on the electrode potential (86, 90). The limiting surface current is attained on the descending part of the wave, at potentials somewhat more negative than the wave maximum. However, for simplicity, with the precision sufficient for many purposes, the height of the maximum is assumed to be the limiting current of the hump-shaped wave (12).

In the above case concerning the surface kinetic wave of the cobalt(II) reduction from its complex with cysteine, the height of this wave appreciably decreases as the ionic strength of the solution increases. Under conditions in which the wave was studied (88), the cysteine concentration at the electrode attributable to its adsorption was considerably higher than that of cobalt ions. The decrease in the latter with increasing ionic strength of the solution is likely to be the main cause of decreases in the kinetic wave height. However, by comparing the decrease in the limiting current of this wave corrected for the concentration polarization of cobalt ions (with this in view, i_l was replaced by $i_l/(i_d - i_l)$, where i_d is the diffusion current corresponding to the analytical concentration of cobalt ions) with the decrease in the $[Co^{2+}]_s$ concentration at the electrode surface (see Eq. 1), the wave height has been shown to change less than the cobalt concentration at the electrode (see Fig. 18). Provided the salt effect is taken into account, however, a greater change in the wave height should be expected. Apparently, in the case under consideration, the effect of increasing the concentration of the cysteine anions at the electrode surface and their adsorptivity with increasing ionic strength of the solution does manifest itself.

A considerable decrease in the surface catalytic current on introducing sodium chloride into the borate buffer solution is observed in the case of the second wave caused by quinine (Fig. 19). However, in this case as well, the catalytic wave height drop does not correspond precisely to the decrease in hydrogen ion concentration at the electrode surface; the wave height drop is somewhat stronger because of the additional effect of quinine adsorptivity decrease with increasing ionic strength of the solution (80).

As shown above, the decrease in the absolute value of the negative ψ_1-potential results in a decrease in the protonation rate of the depolarizer uncharged molecules [see also the pioneering work of Grabowski and Bartel (91), which considers the effect of changing the double-layer structure on electrode processes with antecedent protonation]. However, in rare cases it is possible to observe an increase in the rate of the electrode process. For example, on introducing small amounts of tetra-substituted ammonium salt into the phosphate buffer solution, the increase in the second hydrogen

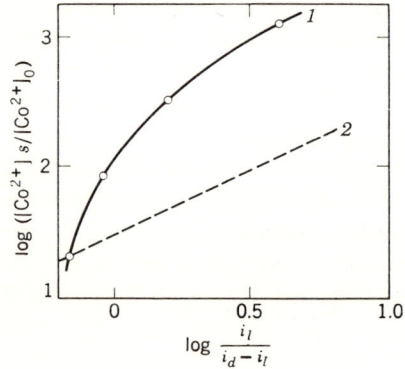

Fig. 18. The relation between the corrected height of the surface kinetic wave for the reduction of cobalt–cysteine complex and the concentration of the free ions $[Co^{2+}]_s$ at the electrode surface. (1) Observed dependence; (2) calculated dependence without taking into account the change in adsorptivity and salt effect. [From reference (88).]

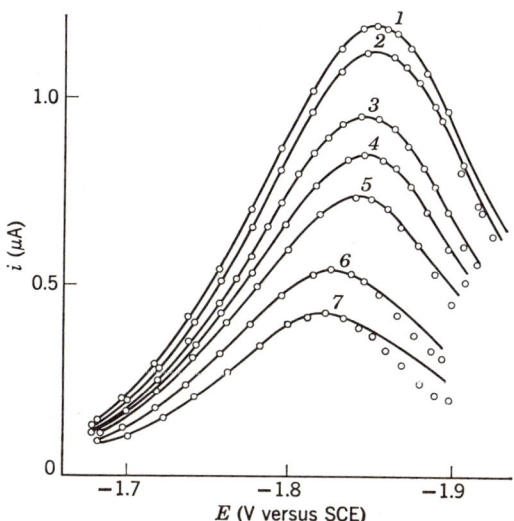

FIG. 19. Catalytic hydrogen waves in 3×10^{-6} M solutions of quinine in borate buffers at pH 9.5 with varying Na^+ ion concentration (NaCl added). (1) 0.04; (2) 0.045; (3) 0.50; (4) 0.055; (5) 0.060; (6) 0.070; (7) 0.080 M. [From reference (80).]

catalytic wave caused by quinine can be observed (92). This phenomenon is accounted for (93) by the fact that in weak alkaline phosphate solution the HPO_4^{2-} and $H_2PO_4^-$ anions are the main proton donors; their approach to the electrode surface is facilitated by decreasing the absolute value of the negative ψ_1-potential.

Mention should be made of the temperature dependence of the effect of the double-layer structure on electrode reaction kinetics. The difference between the ion concentration in the volume and that at the electrode surface, including hydrogen ions, is determined by Eqs. 1 and 25. At the same time, one can consider, as an approximation, that the absolute value of ψ_1 changes linearly with T (see Eq. 16). Thus temperature does not appreciably affect the relationship between the ion concentration in the volume and that at the electrode surface. It is for this reason that temperature does not actually affect the relative changes in the apparent pyridine protonation rate constants on changing the reaction layer thickness, that is, it does not affect the slope of a plot similar to that in Figure 15 (94).

To conclude this section it should be noted that because the charges of protonated and unprotonated depolarizer forms are different, these forms exhibit different adsorptivity at the electrode surface in consequence of which the dissociation or ionization constant values for such compounds when in the adsorbed state are different from those for the same compounds in the bulk of the solution. It has been shown by a simple calculation that the ratio of the acid dissociation constant for the adsorbed compound to that for the unadsorbed one is equal to their adsorptivity ratio in the basic and acidic form (95). Just as the adsorptivity is significantly affected by the electrode potential, so the above ratio between the dissociation constants is a function of the potential.

The dissociation constants for the acids just at the electrode surface are also considerably affected by the electrode field.

The above considerations should be borne in mind in comparing the surface protolytic reaction rate constants with the surface constants (equilibrium) of ionization and dissociation.

E. Some Specific Features of Anion Reduction

As already noted, the effect of the double-layer structure is strongly pronounced in the case of anion electroreduction at a negatively charged electrode surface (38, 39, 96, 97). As the absolute value of ψ_1 decreases, the effect of increasing the potential drop is added in this case to that of increasing the depolarizer–anion concentration at the electrode. This, in particular, results in a strong shift in half-wave potential of the wave with increasing ionic strength toward less negative potentials. This shift is considerably larger than that of the uncharged particle reduction and also much greater (in its absolute value) than that of cation reduction (see Eq. 19).

If anions are reduced at potentials close to the electrocapillary zero, then on increasing the negative potential of the electrode, especially in dilute solutions of a supporting electrolyte, a sharp increase in the negative ψ_1-potential takes place (see Fig. 5). This may prevent the approach of the anions to the electrode surface to such an extent that the current descent appears in the polarograms.

Kryukova was the first to observe such a descent (or "minimum") in the polarograms of anion reduction (in the case of persulfate reduction waves) (98). This was accounted for by a decrease in the electrode process rate because of the sharp drop in the reducing anion concentration at the electrode. On increasing the supporting electrolyte concentration as a result of decreasing the anion repulsion from the cathode surface, the minimum in the polarograms becomes less deep and at a sufficiently high concentration of the indifferent salt disappears altogether.

The first quantitative theory concerning the descent on the anion reduction wave was suggested by Frumkin and Florianovich (96, 97), who took into consideration the decrease in the anion concentration at the electrode on increasing the negative ψ_1-potential. As the potential becomes more negative, the increase in the electron transfer rate becomes so strong that it starts to counterbalance the decrease in the electrode process rate caused by the anions repulsion. The polarograms begin to show a rise in the current which at sufficiently negative potentials may attain the level of the limiting diffusion current.

The high sensitivity of the anion reduction electrode process to rather small changes in the double-layer structure allows applications of these processes to investigation of less marked effects, such as the very weak specific adsorption of Cs^+ ions (99, 100). Especially high sensitivity is observed in the reduction of $Fe(CN)_6^{3-}$ (101, 102), in which the reaction rate is affected even by traces of cations that are eluted into the solution because of the chemical instability of glass (103). Nikolaeva-Fedorovich (39, 104) developed an elegant method for establishing the changes in the double-layer effect in the course of the discharge of the cations of some metals. This method is based on changing the character of the hindered reduction of $S_2O_8^{2-}$ anions. In Figure 20, curve 1 represents a portion of the polarographic curve corre-

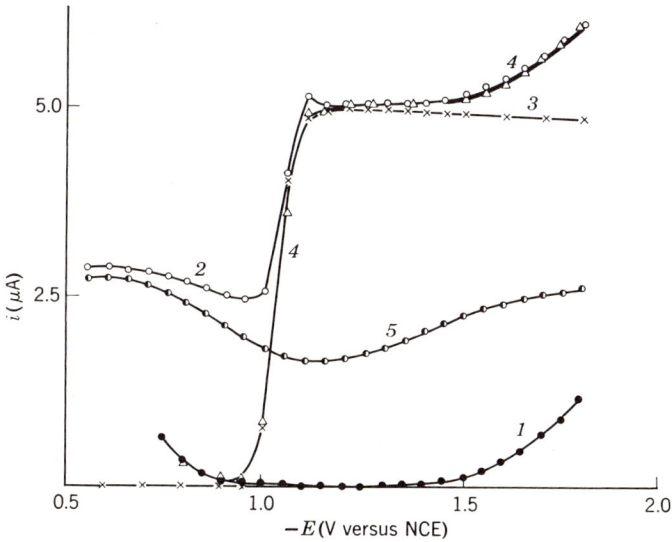

FIG. 20. Parts of the polarograms obtained with solutions containing (1) $10^{-3} N$ $K_2S_2O_8$; (2) $10^{-3} N$ $K_2S_2O_8$ + $10^{-3} N$ $ZnSO_4$; (3) $10^{-3} N$ $ZnSO_4$ + $10^{-3} N$ K_2SO_4; (4) sum of curves 1 and 3; (5) $10^{-3} N$ $K_2S_2O_8$ + $10^{-3} N$ $MgSO_4$. [From reference (104).]

sponding to the hindered reduction of peroxydisulfate anions. With zinc ions added to the solution, the current of $S_2O_8^{2-}$ reduction before the Zn^{2+} reduction potentials is about to attain the level of the diffusion current (the left part of curve 2). At a potential of approximately -1.0 V, the current rises because of the reduction of zinc ions approximately up to the level of the diffusion current of Zn^{2+} (curve 3). The current value at the potentials more negative than -1.0 V in the solution containing $K_2S_2O_8$ and $ZnSO_4$ (curve 2) is precisely equal to the sum of the Zn^{2+} diffusion current and that of the hindered reduction of $S_2O_8^{2-}$ (curve 4). This indicates the complete absence of the effect of reduced Zn^{2+} ions on the electrical double-layer structure (103). Figure 20 also shows a portion of a polarogram of $S_2O_8^{2-}$ in the presence of Mg^{2+} ions (curve 5). These, as seen from the figure, eliminate almost completely the hindrance of $S_2O_8^{2-}$ reduction.

The method suggested by Nikolaeva-Fedorovich was applied (105) to prove the adsorption of Ni^{2+}-pyridine complexes at the electrode surface, nickel being reduced more readily from these complexes than from aquo complexes (106–108). The effect of the nickel ions on diminishing the hindrance of $S_2O_8^{2-}$ anion reduction becomes considerably stronger in the presence of pyridine (Fig. 21). This indicates a strong adsorption of nickel–pyridine complexes (in comparing the curves the shift in the nickel wave

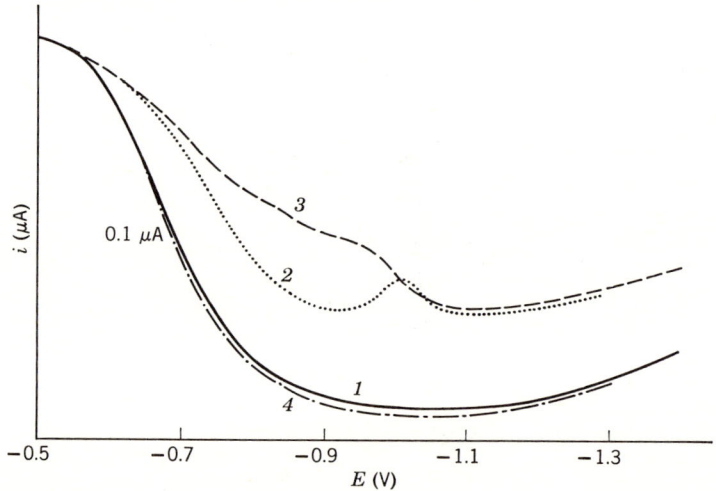

FIG. 21. Polarograms in an 0.5 mM solution of $K_2S_2O_8$ containing (1) 1 mM KCl; (2) with 0.2 mM $NiCl_2$ added; (3) with 0.2 mM $NiCl_2$ + 0.2 mM pyridine added; (4) with 0.2 mM pyridine added to solution (1) in the absence of $NiCl_2$. [From reference (105).]

toward positive potentials in the presence of pyridine has been taken into consideration).

If the anionic form of the depolarizer is in mobile equilibrium with the uncharged one (or with the anion bearing a smaller negative charge) then, as noted, the preceding chemical reaction resulting in decreasing the negative charge of the depolarizer particle eliminates (or diminishes) the effect of the electrostatic repulsion, and hence facilitates the electrode process. Therefore while considering the anion electroreduction the competition of antecedent chemical reactions should be considered. Antecedent reactions are most frequently protonations (already discussed) and decompositions or formations of complexes (109–111). Anion protonations are of particular importance in media of low proton donor activity such as alkaline and unbuffered aqueous solutions or solutions in non-aqueous organic solvents such as dimethylformamide, acetonitrile, and others. The anion radical resulting from a one-electron transfer to the neutral molecule requires considerably more negative potentials for its further reduction. However, with proton donors present in the solution, the protonation of the anion radical takes place, resulting in a loss of its negative charge. Radicals formed in such a reaction easily undergo a subsequent electrochemical reaction (112–117).

V. Depolarizer Adsorption and Characteristics of Polarographic Waves

A. The Adsorption Effect on the Shape and Slope of the Waves

Depolarizer adsorption results in considerable increase in its concentration at the electrode surface. This as a rule causes an increase in the electrochemical and preceding chemical reaction rates.

Adsorption of the substance taking part in the electrode process is known in many cases to accelerate it (12). For instance, an increase in the size of the alkoxy group in esters of α-oxy-β,β,β-trichloroethylphosphorus acid appreciably increases their adsorptivity at the mercury electrode and at the same time considerably facilitates their reduction. This accounts for the shift in half-wave potential of the wave to less negative potentials (118).

An increase in the amount of substance adsorbed has a greater effect on electrode processes accompanied by antecedent chemical reactions, especially in the case of hydrogen catalytic waves, in which the wave height is independent of the diffusion rate of the substances. Bullerwell (119) found that on increasing the size of the alkyl substituent in position 4 of imidazol-2-thione the height of the catalytic wave in the acetate buffer solution markedly increases. Figure 22, constructed from the data of reference (119),

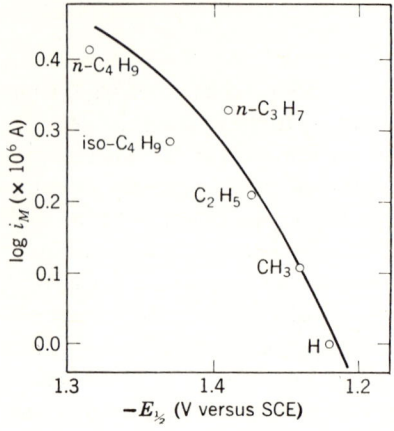

FIG. 22. The dependences of the maximum height and the half-peak potential of the catalytic hydrogen wave on the nature of the alkyl substituent at the 4-position of imidazolthione-2 catalyst. [Constructed from the data of reference (119).]

shows that the increase in the wave height is accompanied by a shift to positive potentials, although from the change in the inductive effect of the substituents which results in increasing pK_a of the catalyst oppositely a shift to more negative potentials would be expected (120). The increase in the height of the catalytic wave and its shift to less negative potentials are caused by increasing the catalyst adsorptivity on increasing size of the alkyl substituent.

The adsorption of an organic substance at the electrode depends to a considerable extent on its potential. According to Frumkin (121), adsorption reaches a maximum at a potential of "maximum adsorption" E_m, which is close to the electrocapillary zero E_z. However, as the potential shifts from E_m, either to the positive or negative side, the adsorption decreases:

$$\beta = \beta_0 \exp\left(-\frac{\bar{C} - \bar{C}'}{2RT\Gamma_\infty}\phi^2\right) = \beta_0 \exp(-a\phi^2) \tag{29}$$

In this equation β expresses the adsorption of a given substance at the electrode at a given potential; at a small coverage of the electrode surface by the adsorbed substance θ (i.e., when Henry's law is valid) β is a proportionality factor between the coverage value θ and the concentration of the substance adsorbed in the solution c. At the potential of maximum adsorption E_m, β attains its highest value β_0. The values $\bar{C}$ and $\bar{C}'$ are integral capacities of the double layer in the absence of the surface-active substance and at the complete coverage of the electrode surface by it, respectively, ϕ is the potential measured from the point of maximum adsorption in a given solution, that is, $\phi = E - E_m$; and Γ_∞ is the number of moles of the substance adsorbed per square centimeter of the electrode surface at its complete monolayer coverage.

As the potential becomes increasingly more negative than E_m, the surface concentration of the adsorbed substance decreases considerably. If the rate of the electrode process depends not only on the potential but also on the depolarizer surface concentration, the polarograms may show a decrease in current that formally resembles current–voltage curves of anion reduction as discussed in the previous section.

The effect of decreasing the amount of the adsorbed substance with increasingly negative potential is most prominent in the case of the electrode processes accompanied by antecedent surface chemical reactions. Therefore the surface kinetic and catalytic waves frequently exhibit the typical humplike shape (86, 122–124). The formation of the current decrease in the catalytic hydrogen waves caused by the catalyst desorption with increasingly negative electrode potential is shown schematically in Fig. 23 (123). The shape of the surface kinetic waves can be described quantitatively (12, 86).

Quadratic dependence of the substance adsorptivity on the electrode potential (Eq. 29) also causes the changes in the slope of waves of electrode

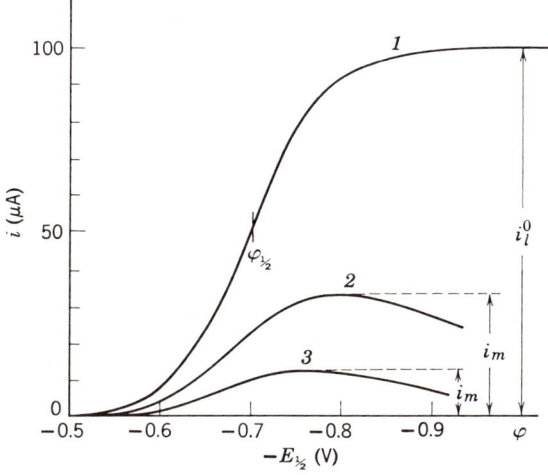

FIG. 23. Scheme illustrating the formation of the humplike surface catalytic wave. (*1*) Hypothetical curve which would be observed if the quantity of adsorbed catalyst were independent of the electrode potential; (*2*) and (*3*) waves calculated by taking the variation of adsorption with potential, according to Eq. 29, into account, the assumed values of a being 1.63 and $a = 3.27 \text{ V}^{-2}$, respectively. [From reference (123).]

processes accompanied by depolarizer adsorption, resulting from an increase in the distance between the potential region where these waves appear and the potential of maximum adsorption (86). This effect is much greater than that of the change in ψ_1-potential on the wave slope considered in Section IV-B. Therefore in the case of the electrode processes complicated by adsorption phenomena (i.e., in the case of almost all electrode processes involving organic substances), the transfer coefficient value α cannot be determined directly from the wave slope.

The polarographic wave shape for processes with depolarizer adsorption is described by Eq. 30 (12, 86):

$$\phi = \phi_{1/2} - \frac{bb'}{b' - b} \log \frac{i}{i_l - i} \qquad (30)$$

in which b corresponds to the true wave slope that would have taken place in the absence of the depolarizer adsorption and b' is a factor that takes into account the effect of adsorption and its changes with the potential:

$$b' = -\frac{2.3}{2a\phi_{1/2}} \qquad (31)$$

Equation 30 is valid for surface kinetic and catalytic waves including quasi-diffusion ones, as well as for diffusion-limited waves without antecedent chemical reactions but with depolarizer adsorption. In the two last-mentioned cases, i_l in Eq. 30 corresponds to the limiting diffusion current i_d.

As follows from Eq. 30, the wave slope, that is, the slope of the $\log[i/(i_l - i)]$–ϕ plot, is equal in value to

$$\frac{bb'}{b' - b} = b/(1 + 0.87ab\phi_{1/2}) \qquad (32)$$

which is not constant but is dependent on the difference between the measured value of the half-wave potential and the value of E_m of the depolarizer. As seen from Eq. 32, the wave exhibits maximal steepness at $\phi_{1/2} = 0$, that is, at $E_{1/2} = E_m$. In this case the value of $bb'/(b' - b)$ found from the experiment is equal to the true value of b. As the difference between the value of $E_{1/2}$ and that of E_m increases, the wave slope, that is, $bb'/(b' - b)$ decreases. Equation 32 corresponds to a linear dependence of the reciprocal value of the wave slope, that is, $[bb'/(b' - b)]^{-1}$ on the half-wave potential. Figure 24 shows the reciprocal value of the experimentally measured slope of the wave of the reduction (which corresponds to an electrochemical cleavage of the C—Br bond) of α-bromocaproic acid in acidic medium as a function of the half-wave potential (125). To change this half-wave potential, various

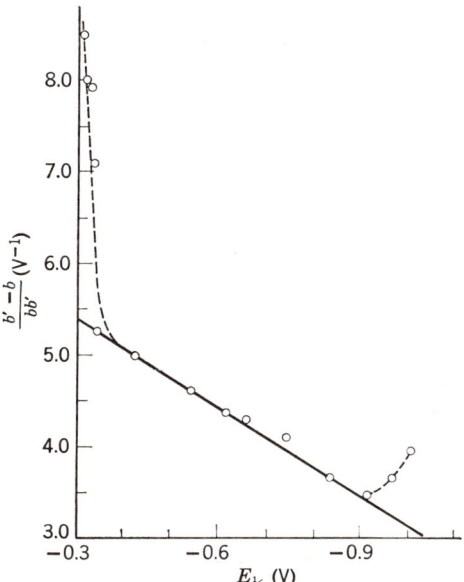

FIG. 24. The effect of the electrode potential on the reciprocal value of the logarithmic plot slope, that is, $[bb'/(b' - b)]^{-1}$, for the reduction waves of 0.116 mM α-bromocaproic acid in acidic solutions containing various concentrations of ethyl alcohol. [From reference (125).]

quantities of ethyl alcohol were added to the solution. Although ethanol affects E_m of the depolarizer and also causes some other effects that have not been taken into consideration in constructing the plot (Fig. 24) (for example, no account has been taken of the liquid junction potential changes at the boundary between the aqueous-alcoholic solution under study and aqueous potassium chloride solution of the calomel reference electrode) nevertheless a very large shift in half-wave potential observed allows the effect of these relatively small additional effects to be neglected. The linear dependence of the reciprocal value of the wave slope on the half-wave potential followed (Fig. 24) over a wide range of $E_{1/2}$ (nearly 0.6 V) and the slope of this dependence gives 3.7 V^{-2} for the value of a. This corresponds to the area occupied by one adsorbed molecule of α-bromocaproic acid, approximately 23 Å^2. At very low ethanol concentrations (extreme left, Fig. 24) the orientation of the bromocaproic acid molecules adsorbed changes (126), while at high alcohol contents complete desorption of bromocaproic acid from the electrode surface takes place (125) (see previous section).

These two factors are likely to account for the deviations from the linear dependence predicted by Eq. 32, shown in the extremes of the plot (Fig. 24).

The depolarizer adsorption and its changes with the electrode potential affect not only the wave slope but also the slope of the dependence of the half-wave potential on the pH for electrode processes with antecedent surface protonation. For the half-wave potentials of such waves (12):

$$\phi_{1/2} = \frac{bb'}{b'-b}\left(\log\frac{0.81 k'_{el} t^{1/2}}{D^{1/2} K_{as}} + \log[\text{H}^+]_s + \log\beta_0\Gamma_\infty - 0.43 a\phi_{1/2}^2\right) \quad (33)$$

where K_{as} is the acid dissociation constant of the depolarizer in the adsorbed state. If the wave is located close to the potential of maximum adsorption of the depolarizer, then the absolute value $\phi_{1/2}$ is not large, and therefore, neglecting the changes of the last term of Eq. 33, one may consider in the first approximation that

$$\frac{\Delta\phi_{1/2}}{\Delta\text{pH}} \simeq -\frac{b'b}{b'-b} \quad (34)$$

Inasmuch as $bb'/(b'-b)$ itself is a function of the potential, $\Delta\phi_{1/2}/\Delta\text{pH}$ should also depend on $\phi_{1/2}$. Figure 25 shows the dependence of $\Delta E_{1/2}/\Delta\text{pH}$ on $E^0_{1/2}$ (i.e., the value of $E_{1/2}$ at pH = 0) for the reduction waves of pyridine

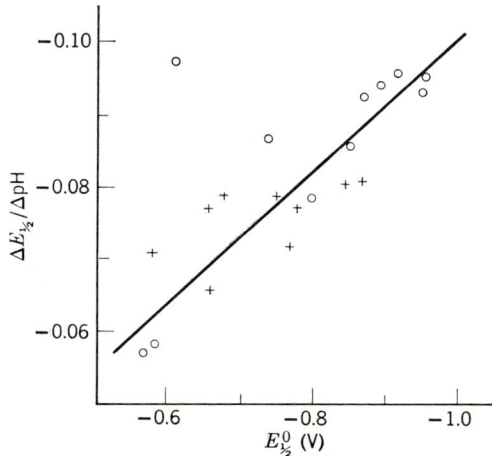

FIG. 25. The relation between $\Delta E_{1/2}/\Delta\text{pH}$ and $E^0_{1/2}$ for the reduction waves of pyridine N-oxide derivatives having substituents at position 2 (+) and 3 (○). [Constructed from the data of reference (127).]

N-oxide derivatives that have different substituents at the 3- or 4-position of the pyridine ring [constructed from the data given in reference (127)]. Data for the compounds that have NH_2 and NHOH substituents and can be protonated at sites other than the $N \to O$ group were not taken into account. Electrochemical reduction of an $N \to O$ group of pyridine N-oxide is preceded by its surface protonation (128). Therefore the waves observed in the acidic medium are quasidiffusion ones and the $\Delta E_{1/2}/\Delta pH$ tends to change with $E_{1/2}^0$ (Fig. 25). This change is apparently caused by a decrease in the adsorptivity of the unprotonated form of the pyridine N-oxide derivatives as $E_{1/2}^0$ of the wave shifts away from the region of the potential of maximum adsorption. As $E_{1/2}^0$ approaches the potential of maximum adsorption E_m, when $bb'/(b'-b)$ decreases tending to b, the value of $\Delta E_{1/2}/\Delta pH$ also decreases.

B. The Reduction of the Quasiunadsorbed Particles

In discussing the nature of changes resulting from varying ethanol concentrations (Fig. 24), it has been noted that in solutions containing a preponderance of an organic solvent, in which the adsorptivity of the organic substance sharply decreases, a deviation from the dependence described by Eq. 32 is observed. This deviation has been attributed (125) to a displacement of the depolarizer from the electrode surface where it was adsorbed. In such solutions the depolarizer enters the electrochemical reaction close to the electrode surface as if in an unadsorbed (quasiunadsorbed) state. Various factors are likely to affect differently the reduction of the same depolarizer when it is in the adsorbed or quasiunadsorbed state. In the former case the depolarizer concentration at the electrode apparently depends on the electrode potential according to the Frumkin equation (Eq. 29), while in the latter case it is practically independent of it. Therefore the factors causing the shift in the half-wave potential, such as various contents of the organic solvent in an aqueous organic mixture or various pH values of the solution, should affect the slope of only those waves that correspond to processes with depolarizer adsorption. The slope of the reduction waves of the quasiunadsorbed substance should not change, however, provided, of course, the reduction mechanism does not change. The waves of methyl iodide and butyl bromide reduction may illustrate this point. Methyl iodide is reduced at relatively positive potentials (about -1.6 V versus SCE). At such potentials its adsorption at the electrode surface is still appreciable. Therefore increasing the amount of the organic solvent in the solution causes both a shift in the half-wave potential of the wave and a change in its slope (129). Butyl bromide is reduced at much more negative potentials (ca. -2.25 V). At these potentials it cannot be actually adsorbed on mercury, and although by increasing the ethanol concentration in the solution it is possible to cause a shift in the half-wave potential to more negative potentials, the wave slope is not affected (129).

The half-wave potentials of the reduction wave of γ-hexachlorocyclohexane (130) when water is mixed with different organic solvents vary considerably (by ~0.3 V). The higher the surface activity of the organic solvent, the more negative the half-wave potential. With a decrease in the water content in the mixture the half-wave potentials become more negative, and with almost nonaqueous media they tend to practically the same value in different solvents. This half-wave potential probably corresponds to the reduction of the quasiunadsorbed depolarizer; the molecules of the solvent displace the depolarizer from the electrode surface [see, for example, references (131) and (132)]. Similarly, it is possible to explain why half-wave potentials of depolarizers with similar structure tend to the same value on increasing organic solvent content. Thus the half-wave potentials of reduction waves of α-bromosubstituted carboxylic acids in acidic solutions containing low concentrations of ethanol vary considerably with the bromoacid structure (up to 0.5 V). With increasing ethanol concentration the half-wave potentials of these acids approach approximately the same value (Fig. 26), which

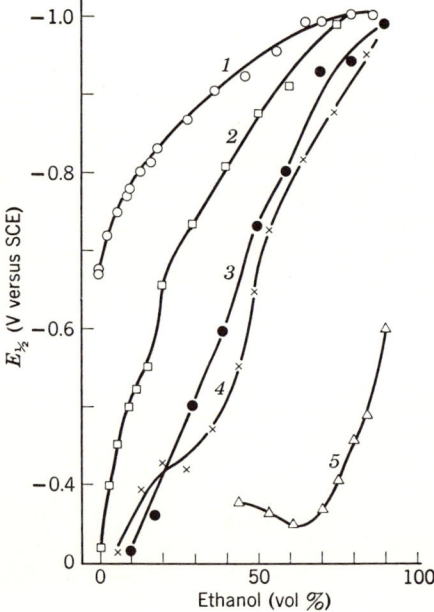

FIG. 26. The effect of ethyl alcohol concentration in acidic alcoholic-aqueous solutions on the half-wave potentials of the reduction waves of α-bromocarboxylic acids (at a concentration of approximately 1 mM). (1) Acetic; (2) butyric; (3) caproic; (4) enanthic; (5) palmitic. [From reference (125).]

corresponds either to the reduction of the acid molecules with similar orientation (perpendicular to the electrode surface, with the hydrocarbon chain toward the solution), or to the reduction of the quasiunadsorbed molecules (125).

In a recently published review, while considering the relationship between the time needed to attain the adsorbed state at the electrode and the kinetics of the electro reduction of adsorbed molecules, Ershler (133) has shown that at very negative potential values and for not very small a-values of the Frumkin equations (Eq. 29) (not less than $2\ V^{-2}$) the number of particles that can be adsorbed in unit time cannot provide the current density values observed experimentally. From this Ershler concludes that in the extremely negative potential range either the depolarizer adsorbed at the electrode is in the state corresponding to very large values of Γ_∞, that is, to very low a-values (Eq. 29), or the reduction of the quasiunadsorbed particles takes place.

Grabowski (134), Elving (135), Laviron (136), and others have mentioned the possibility of the reduction of organic molecules unadsorbed at the electrode. For more detailed information the reader is referred to the Ershler review (133).

As the potential region of the depolarizer reduction becomes more negative and farther from the potential of the maximum adsorption (for example, with an increase in the pH of the solution), first, the slope of the wave increases according to Eq. 32 while the depolarizer is in the adsorbed state. When reduction occurs in the potential range in which depolarizer adsorption does not take place, the steepness of the wave increases somewhat (the slope decreases) up to the value corresponding to the process with the quasiunadsorbed depolarizer (86). Such changes in the wave slope are observed, for example, in the case of the reduction waves of 2-acetyl-5-bromothiophene [see Fig. 43 in reference (12)] and aromatic nitrosohydroxylamines (137) (see Fig. 27).

It is possible to assume (129) that during the reduction of the quasiunadsorbed depolarizer the reaction center of the depolarizer in the transition state is at some distance from the electrode surface. It is possible that between the depolarizer and the electrode surface there are one or several water molecules (or some other solvent), which carry out the electron transfer according to the relay mechanism of the Grotthuss type (138), or contribute to the composition of the activated complex of the electrochemical reaction (139). In some solvents such as a hexamethylphosphoramide and ethanol mixture the electron may pass into the solution to form a solvated electron which may then transfer to the reduced substance (140). The lifetime of the solvated electron in the hexamethylphosphoramide solution is approximately 12 min (140).

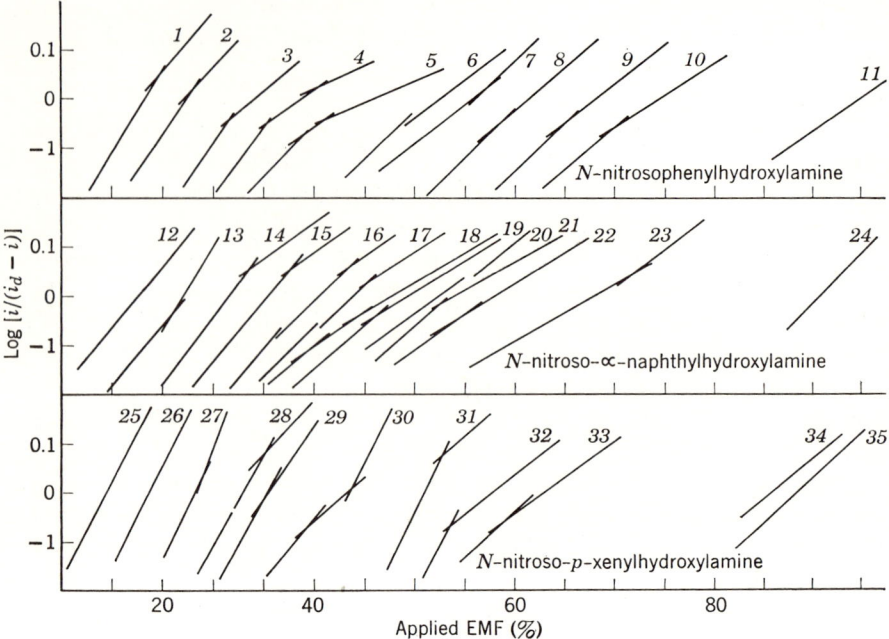

FIG. 27. Semilogarithmic plots of the reduction waves of (*1–11*) N-nitrosophenylhydroxylamine, (*12–24*) N-nitroso-α-naphthylhydroxylamine, and (*25–35*) N-nitroso-p-xenylhydroxylamine in buffer solutions at various pH values, increasing from left to right. [From reference (137).]

C. *The Effect of the Distances between the Electrode, the Plane of Closest Approach of the Ions, and the Depolarizer Reaction Center*

The rate of electron transfer to a quasiunadsorbed depolarizer resulting from its center's being at some distance from the electrode surface is likely to be lower than that to the same depolarizer when in the adsorbed state, with its reaction center (especially if the adsorbed molecule orientation is favorable) close to the electrode surface and forming a transition complex with it.

Apparently, the increased distance between the reaction center and the electrode surface may also account for rendering the reduction of tetraalkyl-substituted ammonium cation more difficult with an increase in size of the alkyl substituents, at least for the lower members of the homologous series— see Figure 28 constructed from the data reported by Wiesener and Schwabe (141).

In solutions containing not more than 60% methanol, the reduction of the

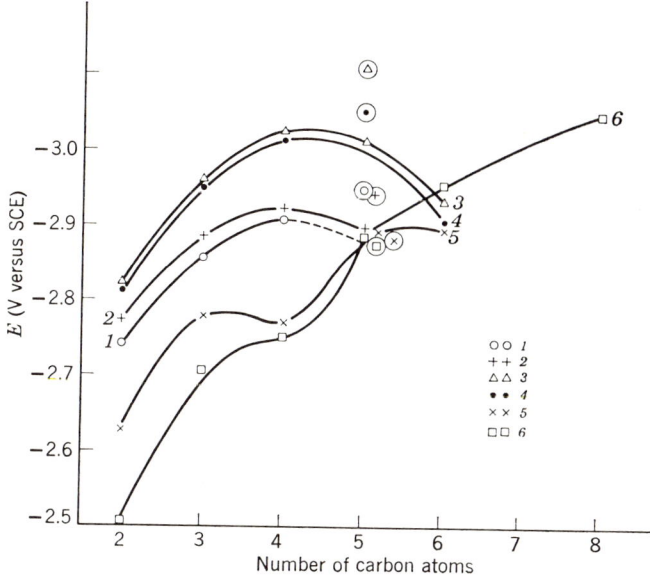

FIG. 28. The dependence of the potential of the final current rise on polarograms of a 0.1 M solution of the iodides (4 and 6) and hydroxides (1–3 and 5) of tetra-substituted ammonium cations on the number of carbon atoms in normal (and, circled, iso-) alkyl chains of substituents in (1) water and in aqueous methanol solutions containing (2) 20; (3 and 4) 60; (5 and 6) 90 volume percent methanol. [Constructed from the data of reference (141).]

tetraalkylammonium cations with the number of carbon atoms in one alkyl group exceeding four is facilitated. This is possibly a result of the long alkyl chains being deformed (142), which results in the reaction center (a positively charged nitrogen atom) approaching the electrode surface. When normal alkyl groups are replaced by their isomeric branched isoalkyls, the reduction of the tetraalkylammonium ions is inhibited. This is likely attributable to lower deforming ability of the branched alkyl groups. However, in 90% methanol there is almost no difference in the reduction potentials of tetraalkylammonium ions with normal or branched alkyl groups (141), nor is the facilitation of the reduction observed with increasing chain length, even for the tetra-n-octyl derivative. Therefore it is possible to assume that the deformation of alkyls in tetra-n-alkylammonium ions on the electrode surface in a medium with a high organic solvent content is inhibited.

Tetraalkylammonium salts are usually used in polarography as supporting electrolytes. The size of tetraalkylammonium cations, which determines the distance to the outer Helmholtz plane, may affect significantly electrode process kinetics. For instance, Nikolaeva-Fedorovich, Fokina, and Petrii (142) found that tetrabutylammonium salts accelerate the electron transfer to the $PtCl_4^{2-}$ anion only at potentials more negative than -1.2 V. At less negative potentials tetrabutylammonium cations inhibit the process. These phenomena are associated with the planar structure of $PtCl_4^{2-}$ anion which enables it to be very close to the electrode surface during the electrochemical reaction—closer than the centers of large organic cations—so that the field of these cations affects the reduction kinetics of $PtCl_4^{2-}$ only slightly. However, at potentials more negative than -1.2 V, tetrabutylammonium cations adsorbed at the electrode are deformed [a small peak of pseudo-capacity in the differential capacity curves corresponds to this deformation (143)]. As a result of this deformation, the positive center of the cation approaches the electrode surface and begins to affect the rate of the reduction of $PtCl_4^{2-}$. The inhibiting effect of tetrabutylammonium cations at the potentials more positive than -1.2 V (142) results from a displacement of the cations of alkali metals from the compact part of the double layer by adsorbed tetrabutylammonium cations. The alkali metal cations facilitate the electron transfer to PCl_4^{2-} because of their small size.

Similar phenomena are also observed for the reduction of other planar anions, for example, of $PtBr_4^{2-}$ (144). However, in the case of tetrahedral ions, such as $PtCl_6^{2-}$, the organic cations accelerate reduction over the entire potential range in which cations are adsorbed (144). It is of interest that planar cations such as N-methylpyridinium also accelerate the reduction of $PtCl_4^{2-}$ anion in the whole potential range of the methylpyridinium cation adsorption (39).

The change in half-wave potential of the reduction wave of n-heptanone observed by Wiesener and Schwabe (141) can also be explained by a change in the position of the plane of the cation's center of closest approach to the electrode surface on increasing the size of the alkyl groups in tetraalkyl-ammonium cations. In 0.1 M solution of hydroxides of the cations $(C_2H_5)_4N^+$, $(n\text{-}C_3H_7)_4N^+$, $(n\text{-}C_4H_9)_4N^+$, $(n\text{-}C_5H_{11})_4N^+$, and $(\text{iso-}C_5H_{11})_4N^+$, the half-wave potentials of heptanone in 20% aqueous methanol are equal to -2.540, -2.645, -2.526, -2.282, and -2.497 V (versus aqueous SCE) respectively. The shift in the half-wave potential to negative values with variation from tetraethyl- to tetrapropylammonium is apparently caused by increasing the distance from the electrode surface to the cation centers in the compact layer. This results in a decrease in the absolute value of the effective potential drop between the electrode and the plane in which the reaction center of the heptanone being reduced at the electrode is located. Figure 29 schematically illustrates the potential distribution close to the electrode surface, with the outer plate of the double layer formed by the cations with smaller (curve *1*)

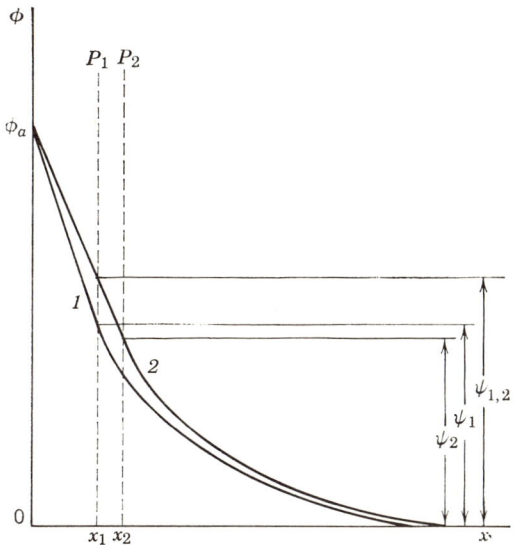

FIG. 29. Scheme illustrating the change in the effective potential drop in the plane of closest approach of depolarizer reaction centers (P_1) with transition from tetraethylammonium cations (values are marked with subscript 1) to tetra-n-propylammonium (2) in the supporting electrolyte.

and larger (curve 2) radii. For simplicity, the reaction center of the substance entering the electrochemical reaction is assumed to be in the plane of closest approach of the cations with smaller radii (the subsequent considerations remaining valid as well for some other position of the depolarizer reaction center). At the same electrode potential, its charge is somewhat smaller when cations with larger radii are present in the solution (the condenser capacity increases as the distance between its plates decreases). Thus the potential drop in the diffuse layer between the P_2 plane and the bulk of the solution is somewhat smaller than that between P_1 and the bulk of the solution, that is, $|\psi_1| > |\psi_2|$. Despite this potential drop outside the plane P_1, where the reaction centers of the depolarizer molecules are located in the solution containing cations of larger radii, $|\phi_{1,2}|$ (Fig. 29) is greater than $|\psi_1|$ in the case of smaller cations forming the double layer. Therefore, other conditions being equal, the effective potential drop determining the electrochemical reaction rate is greater for solutions with smaller cations, that is, $|\phi_a - \psi_1| > |\phi_a - \psi_{1,2}|$. It is for this reason that the electrochemical reduction of heptanone in the solution containing $(C_2H_5)_4N^+$ proceeds more readily than that with $(C_3H_7)N_4^+$

With variation from $(C_3H_7)_4N^+$ to $(C_4H_9)_4N^+$ and then to $(n\text{-}C_5H_{11})_4N^+$, the heptanone reduction is facilitated, probably owing to the increase in the effective potential drop caused by the approach of positively charged cation centers to the electrode surface because of the deformation of the long alkyl chains. However, with the change from $(n\text{-}C_5H_{11})_4N^+$ to $(\text{iso-}C_5H_{11})_4N^+$, the heptanone reduction is hindered because of a smaller ability of the branched chain to be deformed.

As noted, in solutions with high organic solvent content no appreciable deformation of the alkyl group of tetra-substituted ammonium cations is likely to take place at the electrode surface. This is in agreement with the observation that in non-aqueous dimethylformamide solution the half-waves of depolarizers that are reduced at potentials more negative than -1.3 V (versus the mercury electrode in 0.1 M solution of tetraalkylammonium iodide in dimethylformamide) become more negative with an increase in the length of an n-alkyl in tetraalkylammonium ion (145) (Fig. 30). Data for tetramethylammonium iodide were not included in constructing Figure 30 because of the lower solubility of this salt in dimethylformamide (not

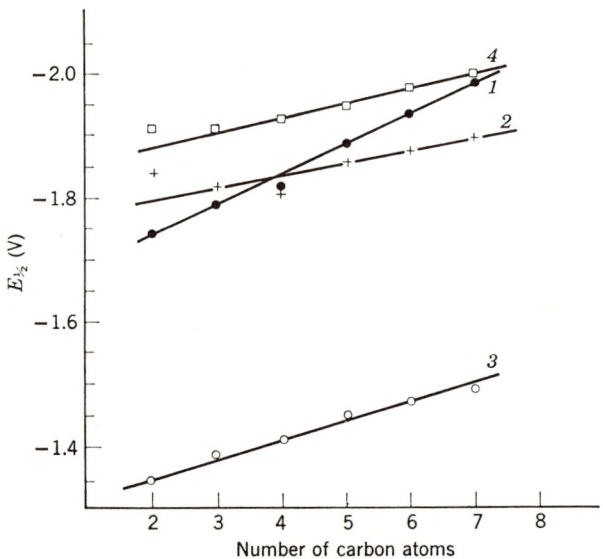

FIG. 30. The effect of the length of the alkyl radical in tetra-n-alkylammonium cations on the half-wave potentials of the reduction waves of different depolarizers in anhydrous dimethylformamide. (*1*) The second wave of ethyl cinnamate; (*2*) the wave of ethyl crotonate; (*3* and *4*) the first and second waves of cyclo-octatetraene, respectively. [Constructed from the data of reference (145).]

exceeding 0.02 M) when compared with other 0.1 M salts. For depolarizers reduced at relatively positive potentials which are apparently subjected to electrode reaction in the adsorbed state, the size of the tetraalkylammonium cation practically does not affect half-wave potentials (145).

From comparison of slopes of the straight lines in Figure 30, it is apparent that the larger the size of the depolarizer molecule the stronger the effect of the size of the supporting electrolyte cation. Only for small depolarizers are the deviations from linear plots (Fig. 30, lines 2 and 4) observed, which can be accounted for by a greater adsorptivity of their molecules at the negative potentials (from the dimethylformamide solution), as well as by their greater tendency to form complexes with the cations of smaller radii (145).

The effect of supporting electrolytes is usually studied at their various concentrations. On increasing the concentration of tetraalkylammonium salts a tendency can be observed to reach a limiting value corresponding to a complete neutralization of the electrode charge by the cations in the compact layer. At higher concentration of the salt, the concentration of the cations at the electrode does not follow that in the bulk of the solution. This has been already considered in the discussion of ammonia buffer solutions (see Section IV-C). The plot of dependence of half-wave potentials on the concentration of tetraalkylammonium salts (Fig. 31) therefore shows the regions of "saturation" in which the ψ_1-potential decreases to zero. The

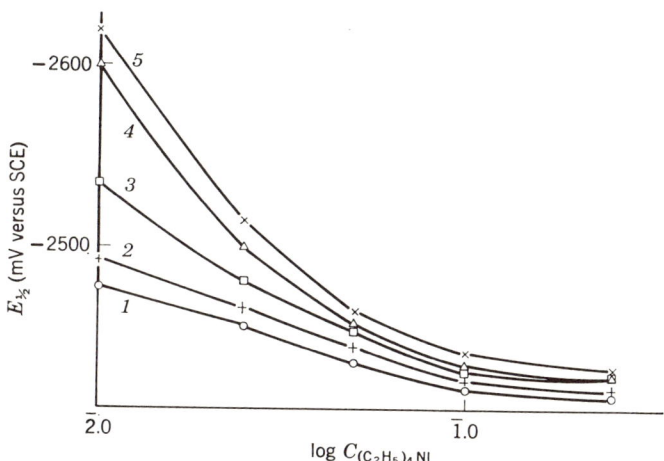

FIG. 31. The effect of the concentration of tetraethylammonium iodide in 50% aqueous dimethylformamide on the half-wave potential for the reduction of α, α'—dithienyl sulfide at concentrations of (1) 0.5, (2) 1.0, (3) 2.0, (4) 4.0, and (5) 8.0 mM. [From reference (146).]

shift in the curves (Fig. 31) to more negative potentials on increasing the concentration of the depolarizer—dithienyl sulfide—is caused by its strong surface activity that results in the competition between adsorption of the depolarizer and that of the tetra-substituted ammonium cation at the electrode surface (146).

The effects of salt concentration and the organic solvent content on the kinetics of the electrochemical reaction when the distance of its transition state from the electrode varies have been considered by Ershler (133).

D. The Salting-Out Effect

With the change in the supporting electrolyte concentration, some other effects are observed in addition to the variation in ψ_1-potential. These effects influence the kinetics of the electrode process. In particular, the "salting-out" or "salting-in" effects are observed. These are caused by changing the activity coefficient of the uncharged particles present in the solution by increasing the concentration of dissolved salts. Then a decrease in the solubility of the uncharged substance is usually observed and, consequently, an increase in their adsorptivity at the electrode surface (147), resulting from their salting-out of the solution. It is considerably less frequent that an increase in solubility of the neutral substance occurs on increasing the salt concentration in the solution. This effect is called salting-in. Upon salting-in the substance adsorptivity on the electrode decreases.

The relationship between the solubility of the uncharged substance S and the salt concentration in the solution c_{salt} is given by the Sechenoff equation (148, 149):

$$\log \frac{S_0}{S_i} = \log \frac{f_i}{f_0} = K_{\text{salt}} c_{\text{salt}} \tag{35}$$

where S_0 and f_0 are the solubility and the activity coefficient of the uncharged substance in a pure solvent, and S_i and f_i are those in the salt solution; K_{salt} is the salting-out constant, positive for the salting-out effect and negative for salting-in. The value of K_{salt} depends to some extent on both the nature of the uncharged substance dissolved and that of the salt.

The half-wave potential shift of the uncharged substance reduced in adsorbed state at the electrode, which results from changing the solubility (adsorptivity) on increasing the salt concentration is described (150) by the expression:

$$E_{1/2} - \psi_1 = \frac{2.3RT}{\alpha n_a F} (\text{constant} + K_{\text{salt}} c_{\text{salt}} - 0.5 \log D) \tag{36}$$

where D is the diffusion coefficient of the substance dissolved (depolarizer).

To show the order of the half-wave potentials shifts predicted by Eq. 36, use can be made of K_{salt}; it is almost the same for various benzene derivatives (aniline, nitrobenzene, phenol, benzoic acid, salicylic acid) and in aqueous solutions of potassium chloride it is close to 0.15 liter/mole (149). Thus the shift in half-wave potentials on increasing the potassium chloride concentration from 0.1 to 1.0 M at $\alpha n_a = 0.5$ caused solely by the salting-out effect should be approximately $+10$ mV. The shift in half-wave potentials observed for the reduction wave of iodobenzene (150) is of a similar order of magnitude and can be attributed to the salting-out effect only.

As noted, the solubility of organic substances is in direct relation to their adsorptivity. Therefore the salting-out effect can be studied by measuring the change in the adsorptivity of the substance on increasing the concentration and changing the nature of the supporting electrolyte. Thus, for example, at high concentrations of the supporting electrolyte the pseudocapacity peak in the differential capacity curves, which is caused by *tert*-amyl alcohol desorption, has been shown to shift to negative potentials owing to the salting-out of the amyl alcohol from the solution (151). Increasing n-butyl alcohol adsorption on the electrode is observed on increasing the supporting electrolyte concentration in the solution (152). The effect of the nature of the supporting electrolyte cations and anions on the salting-out of pyridine from aqueous solutions has been studied quantitatively on the basis of the shift in the pyridine desorption potential in studies with a DME (153, 154). On increasing the $MgSO_4$ concentration, the salting-out of tetrabutyl-ammonium cations and the increase in their adsorption on the electrode take place. This considerably affects the kinetics of the electrode process (155).

The salting-in effect is known to play a role in several cases. Thus on increasing the sodium sulfate concentration in an aqueous solution, the solubility of coumarin in it increases and, simultaneously, coumarin adsorption at the electrode decreases. The ionic strength affects not only the adsorption potential but also the potential at which orientation of the coumarin molecules becomes perpendicular to the electrode surface rather than parallel to it. On increasing the sulfate concentration, both these potentials become less negative (156).

Many investigators studied the effect of salts on the solubility of organic compounds in aqueous solutions (157–159). Baizer (158, 159) used the specific effect of some tetra-substituted ammonium salts on acrylonitrile solubility for the electrosynthesis of adipodinitrile. Baizer's works were followed by a number of investigations of the effect of tetra-substituted ammonium cations on acrylonitrile adsorption at the electrode (160) and its state in the solution (161).

The shift of half-wave potential caused by the salting-out effect is relatively small, as seen from Eq. 36. However, if the electrode process is limited by an antecedent surface reaction rate, very small changes in the depolarizer adsorption may considerably affect the limiting kinetic (or catalytic) current. The salting-out effect on quinine (80) and cysteine adsorption (see Section IV-D) is probably responsible for the difference in experimentally found and theoretically predicted dependence of a limiting current corresponding to a surface wave on ψ_1-potential changes on increasing the ionic strength of the solution. At a high supporting electrolyte concentration, the salting-out of o-phenylenediamine takes place, resulting in a change in the surface kinetic wave of Ni^{2+} in solutions containing o-phenylenediamine different from that described by the theory which does not take into account the salting-out effect (89).

It should be taken into consideration in such calculations that increasing the ionic strength of the solution alters acid dissociation constants and complex stability constants because of the secondary salt effect [see, for example, reference (111)].

E. The Heredity Effect

In some cases the kinetics of the subsequent steps of multistage electrode processes is strongly influenced by the so-called "heredity effect" (12). The main assumption made in interpreting this effect is that the depolarizer generated at the electrode as a result of the first stages of the electrode process within a given (very short) time retains or "inherits" some properties of the original substance. Therefore its properties are somewhat different from those of the same depolarizer introduced into the solution in "ready-made" form. This effect is responsible, for instance, for the differences in the values of half-wave potentials of the reduction waves of some bromine derivatives of thiophene and those of the same compounds formed on the electrode as a result of electrochemical elimination of bromine from a bromo-substituted thiophene with a larger number of bromine atoms in a molecule (162).

The heredity effect is also caused by a depolarizer's retaining the solvation shell of its predecessor, the original substance. For instance, acetaldehyde formed in the reduction of α-halogenoacetaldehyde retains the solvation shell of the latter; its polarographic behavior is considerably different from that of the acetaldehyde introduced directly into the solution (163). An analogous observation of the difference between the depolarizing properties of a complex formed at the electrode surface and of the same complex introduced directly into the solution was interpreted (164) as attributable to the fact that the complex compound formed on the electrode retains the configuration of the parent complex. The reason for the heredity effect in the polarography of complexes is that the equilibrium between the complex inner sphere composition corresponding to the central metal ion of high valency and that of a complex with the metal of low valency formed on the electrode establishes relatively slowly (164). For example, the slow substitution of water for ammonia in inner sphere of hexamminocobalt(II) ion formed on the electrode during the reduction of hexamminocobalt(III) ion results in the half-wave potential of the second wave of $[Co(NH_3)_6]^{3+}$ being different by 30–50 mV from the half-wave potential of the cobalt(II) wave in the same solution. In the first case the $[Co(NH_3)_6]^{2+}$ complex retains the inner sphere of the trivalent cobalt before reduction, while in the added reduced species $[Co(NH_3)_5H_2O]^{2+}$ predominates as corresponds to the equilibrium composition of the solution under the experimental conditions (164).

The heredity effect can be used in explanation of the differences between the effect of the cobalt(III) ion, forming cobalt(II) at an electrode, and that of cobalt(II) introduced into the solution observed for catalytic hydrogen evolution in solutions of proteins and some sulfur-containing amino acids.

The most important consequence of the heredity effect, especially in organic polarography, is that the new depolarizer resulting from the electrochemical reaction retains the adsorption properties of the original compound, that is, its orientation at the electrode surface and its distance from it (12). For instance, the half-wave potential of the second wave in phenylvinyl sulfone is about 15 mV more positive than that of the wave of phenylethyl sulfone solution (166) [at the first stage of the phenylvinyl sulfone reduction, hydrogenation of the double bond takes place to yield phenylethyl sulfone (166)]. Even greater differences are observed between the half-wave potential of the second reduction wave of α,α'-dithienyl sulfoxide and that of the wave of dithienyl sulfide (167). The third waves in the polarograms of various nitrochalcones corresponding to the reduction of the aminochalcones formed on the electrode are 70–90 mV more negative than those for solutions of the corresponding aminochalcones (168). The second two-electron waves in the polarograms of the substituted hydrazones of acetone and cyclohexanone correspond in acidic media to the reductions of the protonated imines of acetone and cyclohexanone, respectively, resulting from the first stage of the process. However, these waves attributable to the heredity effect are considerably lower than those observed in solutions of the imines. And the half-wave potentials of the second waves of the above hydrazones are mostly (depending on the solution composition) more negative than the half-wave potentials of the waves of the imines introduced directly into the solution (169). In the case of di- and trichloroacetanilides, the half-wave potential of the last wave of their solutions (corresponding to the monochloroacetanilide reduction) in 0.05 M $(CH_3)_4NI$ as supporting electrolyte in 50% ethyl alcohol is equal to -1.52 V, while the half-wave potential of the wave in the monochloroacetanilide solution is -1.45 V (170). Some similar examples of the heredity effect are given in reference (12).

In our opinion the systematic study of various factors affecting the half-wave potential shifts for waves of organic substances formed at the electrode and those introduced directly into the solution may provide valuable evidence concerning the dependence of electrode processes upon the orientation of particles reacting at the electrode and upon the distance of their reaction centers from the electrode surface.

VI. The Effect of the Supporting Electrolyte Nature on Electrode Process Kinetics

It has already been shown (Sections III, VI-B, and IV-C) that anion adsorption at the electrode surface radically changes the ψ_1-potential value, and hence considerably affects electrode processes. For instance, the anion adsorption increasing in the series $SO_4^{2-} < Cl^- < ClO_4^- < NO_3^- < Br^- < SCN^-$ observed in 0.9 N solutions of their sodium salts results in a decrease in the kinetic wave of chromate anion reduction in an unbuffered (neutral) medium (171). By introduction of surface-active anions (halides, SCN^-) into an alkaline solution the chromate wave is completely supressed (172). This is explained by a sharp decrease in CrO_4^{2-} anion concentration in the reaction layer at the electrode caused by an increase in the negative value of the ψ_1-potential caused by specific adsorption of anion. There are many similar examples of the anion adsorption effect (12). The effect of the adsorbing anions is usually great.

Inorganic cations are very little adsorbed at the mercury electrode. Nevertheless, even insignificant differences in the surface activities of various cations and their size differences sometimes considerably affect electrode process kinetics.

A. The Cation Nature Effect

The supporting electrolyte cation nature effect on electrode kinetics seems to have first been discovered by Herasymenko and Slendyk (173) while studying the electrochemical hydrogen evolution in the presence of various alkali and alkali earth metals salts. However, it was Frumkin and his co-workers (17, 174, 175) who correctly explained the cation nature effect observed when some increase in the double-layer differential capacity with the variations from Li^+ to Cs^+ and from Ca^{2+} to Ba^{2+} had been found (176). This was assigned to specific adsorption displayed by cations with large radii (17).

From many published papers (12, 38, 39, 103, 123, 171, 175, 177, 178, 179), it follows that the absolute value of the negative ψ_1-potential at the negatively charged electrode surface decreases in the following sequence when the cation of the supporting electrolyte is changed: $Li^+ > Na^+ > K^+ > Rb^+ > Cs^+ > Ca^{2+} > Sr^{2+} > Ba^{2+} > La^{3+}$.

Accordingly, a marked effect of the nature of the supporting electrolyte on half-wave potentials and other characteristics of the waves depending on the double-layer structure has been observed. For instance, the half-wave potentials of the reduction of acetaldehyde and 3-phenylpropionaldehyde have been shown to tend to decrease (i.e., the reduction is facilitated) on increasing the size of the alkali metal cations added as chlorides to a borate buffer (180).

The effect of the nature of the supporting electrolyte cation is especially prominent in anion reductions (39). For example, Ashworth (181) observed the second waves of benzophenone and fluorenone to shift sharply to less negative potentials (in alkaline medium) on increasing the alkali metal radius. The decrease in the absolute value of the negative ψ_1-potential as the supporting electrolyte cation size grows makes approach of the anion to the electrode surface easier, and hence eliminates the current decreases observed on the plateaus of anion waves (38, 39, 175), whereas in the case of the kinetic waves, with the anion being an electrochemically inactive form of the depolarizer, a rise in the kinetic waves is observed. Such an increase in the kinetic wave heights occurs, for instance, in the reduction of chromate (171) and chlorite (179), as well as of the last waves in polarograms of aromatic aldehydes and ketones in alkaline media (182, 183). In the case of the second wave in an alkaline solution of cinnamaldehyde—attributable to the reduction of its anion radical—the increase in the limiting current in the presence of cesium salts is so significant that Barnes and Zuman (183) suggested that this effect could be used in analytical determinations of cesium concentration.

With cation reduction as well as with electrode processes determined by antecedent protonation of uncharged molecules, an increase in the supporting electrolyte cation radius results in inhibition of the process. For instance, with the variation from lithium to cesium salts added to a borate buffer solution, the catalytic hydrogen wave caused by quinine sharply decreases (80). Likewise does the hydrogen catalytic wave caused by the reduction products of p-dinitrobenzene decrease with the variation from Li^+ to Cs^+ in a supporting electrolyte (184). Zhdanov and Zuman (185) observed a decrease in the hydrogen catalytic wave in solutions of N-troponylamino acids on increasing the cation radius of the supporting electrolyte.

While studying the effect of the cation nature on the electrode processes with antecedent limiting protonation, it is necessary to consider the change in "acid strength" depending on the cation nature. For instance, Schwabe and co-workers (186) showed the pH of dilute hydrochloric acid solutions (0.01 N) to decrease markedly upon the addition of alkali metal halides, this effect being greatest for lithium chloride ($\Delta \mathrm{pH}/\Delta c = -0.25$) and gradually decreasing as the atomic number of the cation increases, so that for RbCl it drops almost to zero. Cesium chloride solutions have an opposite effect and the pH of the solution somewhat increases. The "acidity" of different alkali metal cations in acetic acid medium has been reported (187, 188).

The specific role of lithium ions in electrode processes with antecedent protonation is considered next. These ions can act as protons, increasing the kinetic current. Thus the increase in the height of the catalytic wave caused by pyridine with increasing lithium chloride concentration in unbuffered aqueous solutions lead to a suggestion (189) that Li^+ ions, similarly to H^+ ions, could combine with pyridine nitrogen to form lithium pyridinium cations entering the electrochemical reaction as well as the usual pyridinium ions do. Zhdanov and co-workers noted (185, 190) the specific role of lithium ions in the formation of hydrogen catalytic waves. It is possible that it is these properties of lithium ions that cause shifts in half-wave potentials of acetaldehyde and phenylpropionaldehyde in lithium chloride solution to positive potentials (180) compared with other alkali metal salts. By increasing the lithium chloride concentration the reversibility of the indium ion electrochemical reduction in aqueous solution increases (191).

In returning to the comparison of the surface activity of individual cations, it can be noted that tetraalkylammonium cations exhibit even greater specific adsorptivity on the mercury electrode than cesium ions do. Therefore the absolute value of the negative ψ_1-potential decreases in the presence of these cations very considerably. However, tetraalkylammonium cations, similarly to uncharged organic molecules adsorb specifically only within a certain potential range (192) and their effect is most strongly expressed within this potential region (175). The adsorptivity of these cations increases on increasing the alkyl length (193, 194). Consequently, their action (within the adsorption potential range) also increases with their size [see reference (195) as well as some examples in reference (12)].

Levin and Fodiman (33) were the first to associate the effect of tetrasubstituted ammonium cations with changes in the double-layer structure. Nowadays, the effect of these salts is considered in almost all investigations concerned with double-layer effects on electrode kinetics.

Because of the strong surface activity of tetraalkylammonium ions, inhibition of the electrode process can be observed even at relatively low concentrations (of the order of several millimoles per liter) of these ions, where complete coverage of the electrode surface by these adsorbed ions is achieved (196–199).

At sufficiently negative potentials tetraalkylammonium cations do not display any appreciable specific adsorption and their effect is determined by the ratio of the ionic center distance from the electrode surface in the compact layer to the depolarizer size (see Section V-C).

For instance, hexafluorobenzene in a 75% aqueous dioxane solution with 0.05 M tetra-n-butylammonium iodide as supporting electrolyte gives rise to two waves (with $E_{1/2} = -2.22$ and -2.79 V). In the presence of tetramethyl- and tetraethylammonium salts, however, only one wave, with $E_{1/2} = -2.29$ V, is observed (200). The possibility of the tetrabutylammonium cation's being deformed in the aqueous dioxane solution is probably responsible for facilitating hexafluorobenzene reduction in the presence of the tetrabutylammonium salt.

Chlorobenzene in anhydrous dimethylformamide gives rise to a distinct reduction wave in a 0.05 M tetraethylammonium salt solution, whereas in the presence of tetrabutylammonium cations, which are not deformed in non-aqueous media, the wave is almost concealed by the base discharge current (201). Upon adding tetraethylammonium salt to a tetrabutylammonium salt solution, a distinct chlorobenzene reduction wave appears (201). Consequently, the tetraethylammonium cations smaller in size contribute less to the potential drop in the solution outside the plane where the depolarizer reaction centers are located, hence facilitate the reduction.

When studying the effect of the nature of supporting electrolyte cations in non-aqueous media, it is necessary to take into account the differences between cation adsorptivities in these media and in water or aqueous organic mixtures. In anhydrous dimethylformamide lithium ions exhibit greater surface activity than tetraethylammonium cations (202). Therefore, for instance, the half-wave potential of the second waves in the polarograms of anthraquinone and its 1-hydroxy derivative (corresponding to the reduction of anions resulting from the first step of the process) is nearly 300 mV more positive in a lithium chloride solution than in a tetraethylammonium salt solution (202). The half-wave potentials of aromatic nitro compounds in anhydrous dimethylformamide and acetonitrile become more negative on changing the supporting electrolyte cation nature in the sequence $Li^+ < Na^+ < Cs^+ < (C_2H_5)_4N^+$, that is, on increasing the radius of the non-solvated cation (203). Similarly, the first one-electron waves of the reduction of dinitrobenzene isomers in dimethylformamide are noticeably more positive in a sodium perchlorate solution than in one of tetraethylammonium perchlorate (204). This difference is especially pronounced in the wave of *o*-dinitrobenzene (see Section VI-C). The above waves are of quasireversible character, therefore the half-wave potential shift observed probably corresponds to the formation of a complex between the resulting anion radicals and the supporting electrolyte cations, which increases with cation size (205).

Murray and Hiller (206) have established that the one-electron wave of the acetylacetonate iron(III) reduction in acetonitrile in the presence of tetraethylammonium salt shifts to less negative potentials upon the addition of lithium perchlorate. This shift results from the rapid interaction of Li^+ cations with the primary products of the electroreduction of the above complex. The effect of the formation of ion pairs with the product of the reversible electrode reaction or the dissociation of the complex resulting from this reaction on the character of the wave observed has also been considered (206). It is shown how to decide between the dissociation of the electrode product and the formation of ion pairs with the supporting electrolyte components from experimental data.

Kheifets and Bezuglyi (207) provide numerous data on the half-wave potentials of both reduction waves of various anthraquinone derivatives in anhydrous dimethylformamide with lithium chloride and tetraethylammonium iodide as supporting electrolytes. In the lithium salt solution the half-wave potentials of the waves are as a rule less negative. This shift is particularly large for the half-wave potentials of the second waves (apparently, of the anion radical reduction), so that the difference in half-wave potentials of both waves in the lithium chloride solution proves to be considerably smaller than that in the presence of tetraethylammonium iodide.

In anhydrous dimethylformamide benzil in lithium chloride as a base electrolyte gives rise to one two-electron reversible wave, whereas in the presence of tetrabutylammonium iodide it produces two one-electron waves (208). In the authors' opinion (208), a small Li^+ cation stabilizes the resulting dianion to a greater extent than does the tetrabutylammonium ion.

B. *The Specific Anion Effect*

The specific anion effect was discovered and thoroughly studied by Gierst and co-workers (177, 195, 209). In the presence of the adsorbing cations, the electrochemical reaction kinetics is also affected by the nature of the anion, but there is usually no correlation between the sequence of the anion effect intensity and the changes in their adsorptivity. For example, ClO_4^- anions considerably diminish the accelerating effect of tetrabutylammonium cations on reduction of chromate anions, and the ClO_4^- ion effect is considerably stronger than that of anions I^- and Br^- which exhibit strong adsorptivity (195).

The specific anion effect depends to a considerable extent on the nature of the adsorbed cations; for instance, in the case of tetramethylammonium, it is almost absent (195). The specific anion effect results from the ion pair formation involving adsorbed cations and anions. Its intensity therefore depends both on the nature of the anions and of the cations (177, 195).

The detailed work of Gierst and co-workers (177) analyzes thoroughly the effect of the differing nature of supporting electrolyte anions and cations on the electrochemical reduction processes of depolarizers with charges varying in value and in sign. Comparison of the specific anion effect in a series of anions (coupled with the same cations) with the sequence of changing their different properties in a solution (adsorption at the water–air boundary, coefficient characterizing the ionic exchange on ion exchange resins, distribution coefficient between water and organic solvents, viscosity of the aqueous solutions) shows a certain parallelism (177). The more intense the effect of the ions with opposite charges on the water-ordering in the water layer in the vicinity of the ions, the stronger the anion and cation interaction resulting in the ion pair formation (177). Gierst ordered the ions in a sequence showing increasing water-ordering ability around the ions (which increases from Cs^+ to Li^+ and from Ba^{2+} to Mg^{2+} as well as from I^- to F^-). Gierst also proposed the opposite series, indicating an increase in the ability to destroy the water structure. Tetra-substituted ammonium cations and monovalent anions of large size, such as ClO_4^-, MnO_4^-, and ReO_4^- are the most effective in this respect. According to Gierst, the formation of ion pairs and the combined adsorption of the ions is the reason for the appearance of the specific anion effect.

Damaskin and Nikolaeva-Fedorovich (143) established that tetraalkylammonium cations adsorbed on the electrode introduce anions (halides, for example) into the compact double layer. These anions in turn increase cation adsorptivity. The above anion effect increases with increasing surface activity. Tza Chuan-sin and Iofa (210) have shown that adsorption of tetrabutylammonium cations increases in the presence of iodide ions. This combined ion adsorption significantly affects the kinetics of hydrogen ion evolution.

Tedoradze and Asatiani (211, 212) found that the nature of the supporting electrolyte anion affects the shape and position of the polarizing curves for the reduction of hydrogen ion at mercury electrodes from acidic solutions in the presence of p-toluidine and tribenzylamine. They explained this effect by the combined action of the anions and organic cations adsorbed on the electrode, which changes the double-layer structure and thereby either facilitates or inhibits, depending on the conditions of hydrogen ion reduction. However, in our opinion (95, 213) the results can be explained by a specific anion effect on catalytic hydrogen evolution. In their studies Tedoradze and Asatiani (211, 212) used amine cations that contain hydrogen atoms and that consequently (12) can give rise to catalytic hydrogen evolution. In the publications by Gierst and by Iofa (210), the tetra-substituted ammonium derivatives were used as adsorbed cations.

We have thoroughly studied (213) a similar specific anion effect in the case of rather distinct catalytic hydrogen waves caused by quinine in acidic media and found the anion nature to affect considerably the catalytic wave height as well as its half-wave potential (to be precise, the half-peak potential, $E_{p/2}$). Figures 32 and 33 show the effect of varying the nature of the anion on the dependence of wave parameters on concentration of different supporting electrolytes added to a 3×10^{-5} M solution of quinine in the 0.05 N H_2SO_4. The sequence of the anion effect changes on changing the concentration of the added salt which results in some curves crossing (Figs. 32 and 33). By comparing the effect of the anion nature separately on i_l and $E_{p/2}$, one can follow a certain tendency; the higher the catalytic wave in the presence of the given anions, the more negative $E_{p/2}$. This dependence is known to be

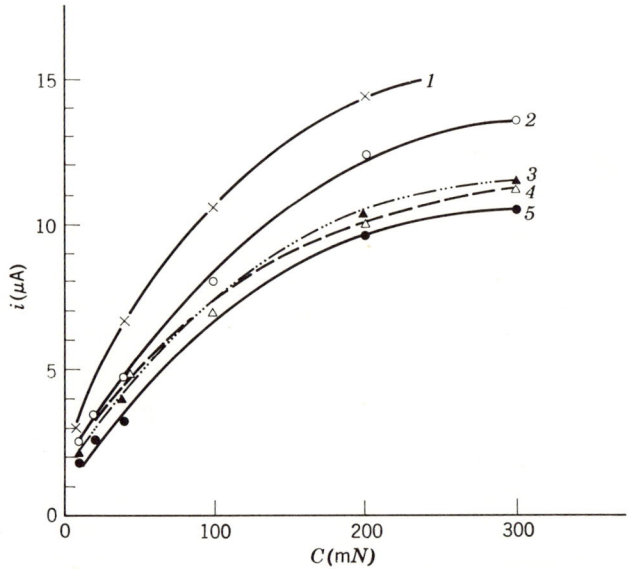

FIG. 32. The dependence of the maximum current of the catalytic hydrogen wave in 3×10^{-5} M quinine solutions containing 0.05 N H_2SO_4 on the concentrations of different supporting electrolytes. (1) K_2SO_4; (2) KNO_3; (3) KI; (4) KBr; (5) KCl. [From reference (213).]

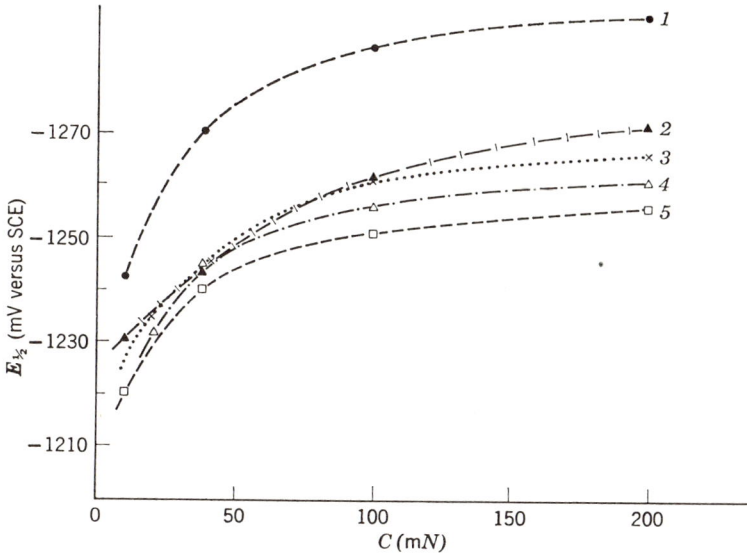

FIG. 33. The effect of supporting electrolyte concentration on the half-peak potential of the catalytic hydrogen wave in quinine solutions containing 0.05 N H_2SO_4. (1) K_2SO_4; (2) KI; (3) KBr; (4) KNO_3; (5) KCl. [From reference (213).]

determined in the case of the volume catalytic hydrogen waves by changing the acid–base properties of a catalyst (12, 214). In the case of surface catalytic waves, this dependence is partly distorted by adsorption effects (12, 215). To confirm that the changes in the catalytic activity of quinine with various supporting electrolyte anions are associated with changing its acid-base properties the quinine pK'_a values (of the first ionization step, that is, in the more acid solutions) in 0.2 N potassium salt solutions with various anions have been determined. From the dependence of log i_l and $E_{p/2}$ catalytic waves in 0.2 N solutions of varying salts and 0.05 N H_2SO_4 on the pK'_a of quinine in the same salt solutions (Fig. 34), it follows that the main reason for changing the quinine reactivity under the influence of varying salts is the change in its pK'_a attributable to the formation of ion pairs or complexes (at one or both amino groups of quinine) with anions. Experiments (213) involving inhibition of the reduction of Cd^{2+} by quinine adsorbed at the electrode in the presence of salts with various anions have shown quinine adsorption to increase on substituting the anions in the series: $Br^- < Cl^- < SO_4^{2-} \leq NO_3^-$, which does not coincide with the sequence of the anions that increase the catalytic activity of quinine. However, the increase in the

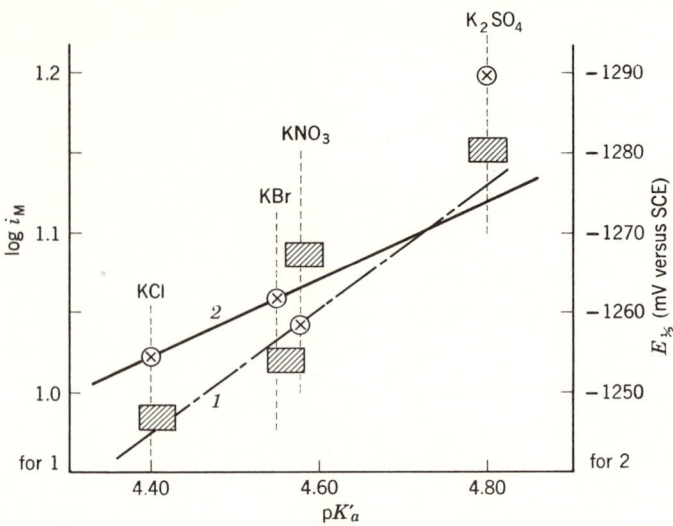

FIG. 34. The dependence of i_l (1) and $E_{p/2}$ (2) of the catalytic hydrogen wave in solutions containing 3×10^{-5} M quinine, 0.05 N H_2SO_4, and different salts at a concentration of 0.2 equivalent/liter on pK'_a of quinine in the same salt solutions (without H_2SO_4). [From reference (213).]

limiting current in the presence of nitrate and, to a lesser extent, of sulfate, as well as the upward deviation of the corresponding dots from the straight line 1 in Figure 34 may account for the increase in the quinine adsorptivity. The considerable shift in $E_{p/2}$ to negative potentials in the sulfate solution has been explained (213) by the hindered discharge of the quinine complex with the bivalent SO_4^{2-} anion that seems to have a less positive charge than similar complexes with univalent anions.

Still more clearly does the dependence of the catalytic activity in hydrogen evolution on the nature of the anions of the supporting electrolyte manifest itself in the case of the volume catalytic waves in the pyridine solutions (see Fig. 35) when the catalytic waves are not influenced by adsorption effects.

The above considerations support Gierst's (177) suggestion that the specific anion effect results from their interaction with the supporting electrolyte cations or with those of a depolarizer. Hence the change is attributable to either the ψ_1-potential value changes (in the case of the adsorbed cations) or the reactivity of the particles taking part in the electrode process. The change in the reactivity of the particles in the solution upon the formation of the ion pairs correspondingly alters the other properties of

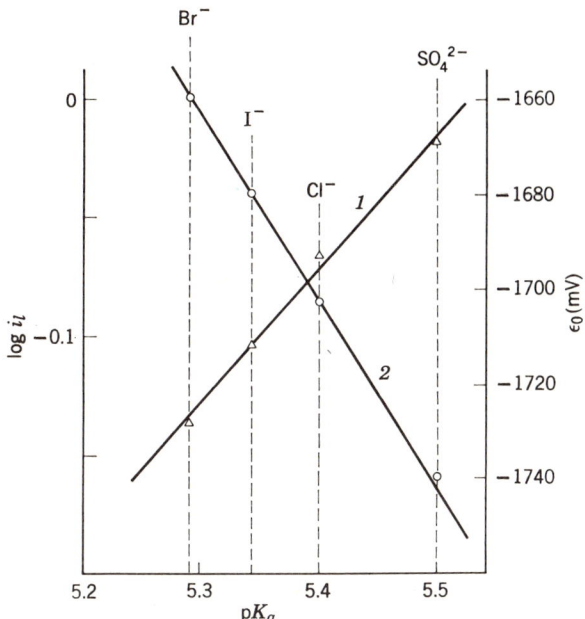

FIG. 35. The effect of pK_a for pyridinium ion in solutions containing different salts at a concentration of 0.2 equivalent/liter on the values of $\log i_l$ (1) and ϵ_0 (2) for the catalytic hydrogen waves 0.03 M pyridine in unbuffered solutions of the same salts. [From reference (216).]

these particles (177). In particular, the effect of the nature of the anions of the supporting electrolyte on the desorption potentials and other properties of pyridine have been found by Gierst and co-workers (153, 154) to be correlated with changing the pK_a of pyridine in solutions of different salts (216).

Among publications concerned with the specific effect of the anion nature on electrode kinetics, it is possible to point out the effect of addition of halide ions to a solution of acetatpentammincobalt(III) which results in formation of a kinetic prewave. Its height is governed by the rate of the penetration of the halide ion into the inner sphere of the complex (replacing the acetate ion). This reaction is accelerated (217) under the influence of the electrode field. In a different way, depending on the nature of the anion, the supporting electrolyte concentration affects the height of the gallium(III) reduction wave that shows the kinetic character (218), SCN^- anion effect being most significant. Gorodetskii and Losev (219) found that halide ions interact with the intermediates of the electrode process and accelerate the ionization of bismuth from its amalgam or the reduction of its ions from the solutions of perchloric acid. The increase in the rate constant of the quasi-reversible reduction of Mn^{2+}, which is attributable to a change in the anion, increases in the sequence: $ClO_4^- < Cl^- < Br^- < I^-$, although in this case the half-wave potential is not affected by the nature of the anion (220). The formation of a complex between tropylium cation and sulfuric acid gives rise to a shift in the reversible wave of the tropylium ion reduction to negative potentials (221).

There are many publications dealing with the effect of the nature of the anion on the quasireversible wave for the reduction of oxygen. The effect is due mainly to the adsorption of the anion and the resulting change of ψ_1-potential.

Among the latest papers in this field those of Kůta and Koryta (222) and Shams El Din and Holleck (223) are especially worthy of attention.

C. The Influence of the Organic Depolarizer Structure on the Specific Effect Value upon Changing the Nature of the Supporting Electrolyte Cation

The influence of the nature and position of substituents in iodo- and bromobenzenes on the half-wave potential of the cleavage of C—I or C—Br bonds has been studied (224), and half-wave potentials for 35 iodobenzene derivatives have been obtained both in the presence of 1 M $LiClO_4$ and 0.1 M $(CH_3)_4NCl$ (concerning the solvent and some other conditions of the experiment (224), no information was given, but it is assumed that they were the same in all the experiments, and some indirect data indicate that the solvent was an aqueous-organic solution). The half-wave potentials for bromobenzene derivatives have been determined only in 0.1 M $(CH_3)_4NCl$ solution.

The half-wave potentials of a number of substituted iodobenzenes have been found (225) to be markedly more positive in lithium perchlorate solutions than in tetramethylammonium chloride solutions. The difference between the half-wave potentials in these media is not constant, but varies from one compound to another in the range 128–243 mV.

The decrease in the absolute value of the negative ψ_1-potential is partly responsible (see Section VI-A) for facilitating the reductions of uncharged organic derivatives in solutions of lithium salts as compared to those in solutions of readily adsorbed tetra-substituted ammonium salts (192). However, the absolute value of the change in half-wave potential ($\Delta E_{1/2}$) is much higher than ψ_1 under the experimental conditions used (224); in a 1.0 M $LiClO_4$ solution at the potentials before -2.0 V, $|\psi_1|$ is less than 100 mV (see Fig. 5).

A specific interaction of the organic depolarizer bearing the polar groups with supporting electrolyte cations seems to occur in solutions of tetra-substituted ammonium salts (225). This results in the depolarizer being incorporated into the compact double layer by adsorbed cations of tetra-substituted ammonium, as is the case with the anion reduction (193, 226). Upon incorporation of the depolarizer into the double layer, along with the effects that are discussed later, the depolarizer concentration at the electrode increases and so does the rate of the electrode process (12).

Specific effects exceeding in their value the ψ_1-potential changes have been repeatedly observed in the presence of varying cations upon the reduction of different organic halogeno derivatives (227–229). These have been interpreted as being attributable to a change in the orientation of the particle reduced at the electrode (227), to a local ψ_1-potential (228), or to formation of a cation bridge or a kind of a complex between the cation and halogeno derivative (227–229). Similar complexes with tetra-substituted ammonium cations which increase the adsorptivity of organic molecules are likely to form in the case of the nitriles of unsaturated carboxylic acids (160, 230).

The effect of the nature of the cation depends both on the ability of the organic molecule to form complexes with different cations and on the readiness with which the organic molecule can be incorporated by means of the adsorbed cation into the compact part of the double layer, that is, the adsorptivity of the complex formed. The organic structure appears to affect considerably both these factors.

Increasing size of the hydrocarbon substituent in the benzene ring of iodobenzene in the sequence CH_3, C_2H_5, C_6H_5 gives rise to an increase in $\Delta E_{1/2}$, that is, it results in increasing the specific effect of tetramethylammonium cations. Growth in the size of other substituents except Cl, Br, and CF_3 groups also increases the specific effect of tetramethylammonium. Figure 36 shows the dependence of $\Delta E_{1/2}$ for several iodobenzene derivatives having a substituent in the ortho position on the geometric sizes of a substituent [the size of the substituent radius is taken from reference (224)]. The increase in the size of the hydrocarbon and other substituents results in an increase in the adsorptivity of the organic molecule and may be assumed to contribute to its incorporation into the compact double layer through the action of the adsorbed $(CH_3)_4N^+$ cations, which results in an increase in $\Delta E_{1/2}$.

Comparing $\Delta E_{1/2}$ for isomers shows that in the case of hydrocarbon substituents their effect in the ortho position is greater as a rule. Thus the $\Delta E_{1/2}$ values for o- and p-phenyliodobenzene are equal to 243 and 209 mV, respectively. $\Delta E_{1/2}$ of 2,6-dimethyliodobenzene (202 mV) is markedly higher than that of the 2,4-isomer (191 mV). This effect of the aryl and alkyl

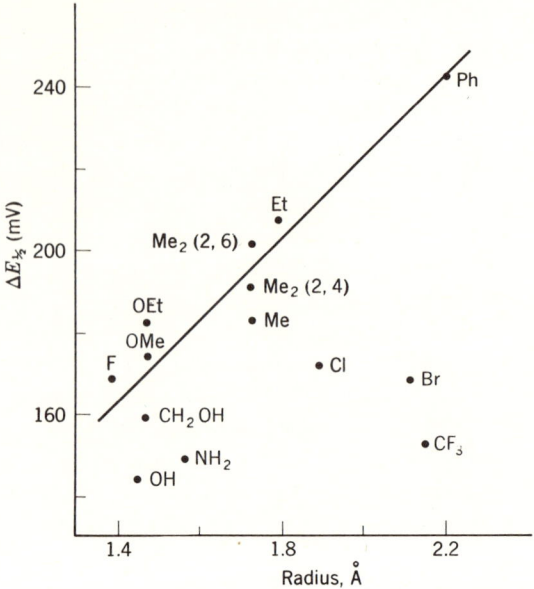

FIG. 36. The dependence of the difference between the half-wave potentials for the reduction waves of *para*-substituted derivatives of iodobenzene in 1.0 M LiClO$_4$ and in 0.1 M (CH$_3$)$_4$NCl 0.1 M on the radius of the substituent group. [Constructed from the data of reference (224).]

substituent position may be accounted for by assuming the deformation of the C—I bond as the result of these substituents being in the ortho position to contribute to the formation of a complex with (CH$_3$)$_4$N$^+$ (in the case of organic halide derivatives, the halide atoms are mainly responsible for the complex formation). It is also possible that with nonparallel orientation of the adsorbed organic molecule (which is likely to take place in the potential range under study, of the order of -1.5 V versus SCE) the occurrence of the hydrophobic substituent close to the reaction center of the depolarizer molecule draws it nearer to the electrode surface ("pushes" it out of the solution) thereby facilitating the electrochemical reaction. For instance, the rate of the electrochemical fission of the C—Br bond is known to decrease as the distance of the reaction center (C—Br bond) from the electrode surface increases as a result of steric hindrance (231, 232).

In accordance with the above considerations the values of $\Delta E_{1/2}$ for iodobenzene derivatives having hydrophilic substituents (COOH, CN, OH, NH$_2$) are minimal; for *p*-COOH $\Delta E_{1/2}$ is equal to 128 mV; for *p*-CN to 130 mV, those of ortho derivatives being especially low (the $\Delta E_{1/2}$ values of o-, m-, and p-iodophenols are equal to 144, 174, and 180 mV, respectively, and those of o-, m-, and p-iodoaniline are 149, 187, and 181 mV, respectively). Thus the occurrence of hydrated groups in the depolarizer molecule close to the reaction center draws it from the electrode surface into the solution, diminishing the specific effect of the $(CH_3)_4N^+$ ion, that is, retards the electrochemical reaction. In the case of chloro- and bromo- (and probably trifluoromethyl) iodobenzenes, the decrease in the specific effect, that is, the decrease in $\Delta E_{1/2}$ (see Fig. 36), may be attributable to the increase in the distance between the reaction center of the depolarizer (C—I bond) and the electrode surface, which result from partial formation of a complex with Br or Cl atoms (or the CF_3 group) and different orientation of the depolarizer molecule to the electrode surface.

It is possible to assume that at more negative potentials both tetra-substituted ammonium cations (193) and the hydrate shells attached to hydrophilic groups become deformed, resulting in an increase in the specific cation effect. To verify this assumption the differences in half-wave potentials of corresponding iodo- and bromobenzene derivatives under the same conditions have been compared (225) (see Table I). If the mechanism of the electrochemical fission of C—I and C—Br bonds in these compounds is assumed to be the same and a complex with tetramethylammonium cations is formed in both cases then the difference in $E_{iodo} - E_{bromo}$ values for the derivatives with various substituents should indicate the deforming effect of the electrode field. In the case of methyl derivatives $E_{iodo} - E_{bromo}$ is almost independent of the position of the methyl group (Table I). For derivatives with hydrophilic substituents the differences $E_{iodo} - E_{bromo}$ are somewhat smaller, especially for ortho isomers. In other words, the specific effect of tetramethylammonium cations, which was found to decrease as a result of the occurrence of hydrophilic groups in the neighborhood of the C—I bond of iodobenzenes, is increased for bromobenzene derivatives which are reduced at more negative potentials, probably on account of the deformation of those groups under the electrode field effect, which causes the reaction center to approach the electrode surface.

The deviations observed in establishing the relation between the structure of organic compounds and the half-wave potentials of their reduction waves in some cases may be explained by the occurrence of the specific effect whose value depends on the position and nature of the substituent in the presence of the same supporting electrolyte. Many deviations of this kind can be found in the well-known book by Zuman (233) concerned with the correlations in organic polarography. For instance, in Figure III-5 [p. 54 in reference (233)], the half-wave potentials of nitrobenzene derivatives having $N(CH_3)_2$

TABLE I. Difference in the Half-Wave Potentials between Analogous Iodo- and Bromobenzene Derivatives in Solutions Containing 0.1 M $(CH_3)_4NCl^a$

Substituent and its position	$E_{iodo} - E_{bromo}$, V
CH$_3$	
Ortho	0.62
Meta	0.63
Para	0.61
NH$_2$	
Ortho	0.59
Meta	0.61
Para	0.62
OH	
Ortho	0.58
Meta	0.60
Para	0.62
CF$_3$	
Ortho	0.48
Meta	0.55
Para	0.55

^{a} From the data of reference (224).

and OH groups in the para position are, respectively, 65 and 100 mV more negative than predicted by the correlation equation [the authors of the original work (234) account for this deviation by the possible formation of the quinoid structure]. In the case of ortho-substituted benzaldehydes, the half-wave potential shift to positive potentials (75 mV) as compared to the values corresponding to the correlation equation takes place with methyl derivatives [Fig. III-10, p. 76 in reference (233)]. Small positive deviations are observed upon introducing one or two methyl groups into the ortho position of iodobenzene [Fig. III-11, p. 77 in reference (233)]. As follows from Table III-10 [pp. 80–83 in reference (233)], the difference between the half-wave potentials of various ortho and para isomers of benzene derivatives (with CH$_3$ being one of substituents) is positive (except in nitrotoluenes); however, in the case of benzaldehyde and benzoic acid derivatives with different substituents in the ortho and para positions it is negative.

Comparing the half-wave potentials of nitrophthalic acids at pH 9.0, with both carboxylic groups dissociated, shows half-wave potentials of the wave of the NO$_2$ group reduction of the isomer with the NO$_2$ group being in the ortho and meta positions with respect to the two neighboring carboxylic groups to be 140 mV more negative than that of the other isomer [p. 90 in reference (233), from the data in reference (235)].

When the carboxylic group of isomeric nitrobenzoic and o-nitrotoluic acids (CH_3 and NO_2 being in ortho position in respect to each other) changes its position from para to ortho, it gives rise to a shift in the waves (at pH 2) to negative potentials, by 100 and 130 mV, respectively [p. 89 in reference (233), from the data of reference (235)].

It is interesting that the marked deviations from the correlation line for the half-wave potentials of reversible one-electron waves to positive potentials in the case of chlorine derivatives and to negative ones in the case of amino derivatives have been observed (236) in polarograms of various piazselenazole and piazthiazole derivatives in anhydrous dimethylformamide solutions containing tetraethylammonium salts. In this case the shift in the wave is likely to result from the complex formation similar to that discussed in the previous section.

VII. Conclusions

The very limited scope of this review prevented us from considering many other problems concerning the effect of the double-layer structure on electrode process kinetics, such as the influence of adsorption of the uncharged molecule on the double-layer structure (156, 237–239), the effect of changing the position of the zero charge potential in the case of amalgam dropping electrodes (240–242) [or, the mercury electrode with a metal having been discharged on it (243)], the effect of the electrode field on the reactivity of the particles directly at its surface (244–246), and so on. Some rather important problems concerning the double-layer effect in chronopotentiometry, voltammetry, and chronoamperometry with linear potential sweep [see, for example, references (247–250)], have been omitted.

It was not possible to refer to all the literature known to us concerning the problems discussed in this chapter. The text includes only the most typical examples.

The considerations and conclusions of this article represent mainly the viewpoint of the author and therefore are certainly disputable. Moreover, this viewpoint sometimes does not coincide with that of the authors of the original papers from which some material or experimental data have been taken to be used in this review.

The author will acknowledge all the criticism received concerning the problems presented here, for it is discussion that permits the truth to be revealed and thereby contributes to the advance of science.

References

1. Frumkin, A., *Z. Phys. Chem.*, **A164**, 121 (1933).
2. Frumkin, A. N., *Zh. Fiz. Khim.*, **24**, 244 (1950).
3. Frumkin, A. N., V. S. Bagotskii, Z. A. Iofa, and B. N. Kabanov, *Kinetika Elektrodnych Processov* (in Russian), M.G.U., Moscow, 1952.
4. Delahay, P., *Double Layer and Electrode Kinetics*, Interscience, New York, 1965.
5. Grahame, D. C., *Chem. Rev.*, **41**, 441 (1947).
6. Parsons, R., *Advances in Electrochemistry and Electrochemical Engineering* (P. Delahay, Ed.), Vol. I, Interscience, New York, 1961, Chap. 1.
7. Damaskin, B. B., *Usp. Khim.*, **30**, 230 (1961).
8. Mohilner, D. M., in *Electroanalytical Chemistry* (A. J. Bard, Ed.), Vol. I, Marcel Dekker, New York, 1966, p. 242.
9. Mairanovskii, S. G., *J. Electroanal. Chem.*, **4**, 166 (1962).
10. Nürnberg, H. W., and M. W. Stackelberg, *J. Electroanal. Chem.*, **4**, 1 (1962).
11. Parsons, R., in *Modern Aspects of Electrochemistry* (J. O'M. Bockris, Ed.), Vol. I, Butterworths, London, 1954, Chap. 3.
12. Mairanovskii, S. G., *Catalytic and Kinetic Waves in Polarography*, Plenum Press, New York, 1968.
13. Stern, O., *Z. Elektrochem.*, **30**, 508 (1924).
14. Gouy, G., *Compt. Rend.*, **149**, 654 (1910).
15. Chapman, D. L., *Phil. Mag.*, [6], **25**, 475 (1913).
16. Helmholtz, H. L. F., *Ann. Physik.*, [2], **89**, 211 (1853); [3], **7**, 337 (1879).
17. Frumkin, A. N., B. B. Damaskin, and N. V. Nikolaeva-Fedorovich, *Dokl. Akad. Nauk SSSR*, **115**, 751 (1957).
18. Frumkin, A. N., and A. S. Titievskaja, *Zh. Fiz. Khim.*, **31**, 485 (1958).
19. Damaskin, B. B., O. A. Petrii, and V. V. Batrakov, *Adsorbtsia Organicheskikh Soedinenii na Elektrodakh*, Nauka, Moscow, 1968.
20. Lerkh, R., and B. B. Damaskin, *Zh. Fiz. Khim.*, **38**, 1154 (1964).
21. Mairanovskii, S. G., *Elektrokhimiya*, **1**, 164 (1965).
22. Vavřička, S., L. Němec, and J. Koryta, *Collection Czech. Chem. Commun.*, **31**, 947 (1966).
23. Wåhlin, E., and Å. Bresle, *Acta Chem. Scand.*, **10**, 935 (1956); Å. Bresle, ibid, **10**, 943 (1956).
24. Skobetz, E., and N. Kavetskii, *Zavodsk. Lab.*, **15**, 1299 (1949).
25. Wolf, S., *Angew. Chem.*, **72**, 449 (1960).
26. Pangarov, N., I. Christova, M. Atanasov, and V. Kertov, *Electrochim. Acta*, **12**, 717 (1967).
27. Russell, C. D., *J. Electroanal. Chem.*, **6**, 486 (1963).

28. Grahame, D. C., *J. Am. Chem. Soc.*, **76**, 4819 (1954).
29. McCoy, L. R., and A. B. Mark, Jr., *Electroanal. Chem.*, **15**, 15 (1967).
30. Grahame, D. C., and B. A. Soderberg, *J. Chem. Phys.*, **22**, 449 (1954).
31. Palanker, V. Sh., A. M. Skundin, and V. S. Bagotskii, *Elektrokhimia*, **3**, 371 (1967).
32. Parry, J. M., and R. Parsons, *Trans. Faraday Soc.*, **59**, 241 (1963).
33. Levin, E. S., and Z. I. Fodiman, *Zh. Fiz. Khim.*, **28**, 601 (1954).
34. Breiter, M., M. Kleinerman, and P. Delahay, *J. Am. Chem. Soc.*, **80**, 5111 (1958).
35. Reinmuth, W. H., L. B. Rogers, and L. C. I. Hummelstedt, *J. Am. Chem. Soc.*, **81**, 2947 (1959).
36. Mairanovskii, S. G., V. A. Ponomarenko, N. V. Barashkova, and A. D. Snegova, *Dokl. Akad. Nauk SSSR*, **134**, 387 (1960).
37. Kitaev, Yu. P., and G. K. Budnikov, *Collection Czech. Chem. Commun.*, **30**, 4178 (1965).
38. Frumkin, A., and N. Nikolaeva-Fedorovich, in *Progress in Polarography*, (P. Zuman and I. M. Kolthoff, Eds.), Vol. 1, Interscience, New York, 1962, p. 223.
39. Fedorovich, N. V., Thesis, Moscow Univ., 1968.
40. Rigatti, G., in *Advances in Polarography* (I. S. Longmuir, Ed.), Vol. 3, Pergamon Press, Oxford, 1960, p. 904.
41. Frumkin, A. N., and O. A. Petrii, *Dokl. Akad. Nauk SSSR*, **147**, 418 (1962).
42. Frumkin, A. N., and O. A. Petrii, and N. V. Nikolaeva-Fedorovich, *Electrochim. Acta*, **8**, 177 (1963).
43. Anson, F. C., and Teh-Liang Chang, *Inorg. Chem.*, **5**, 2092 (1966).
44. Breyer, B., H. H. Bauer, and J. R. Beevers, *Australian J. Chem.*, **14**, 479 (1961).
45. Sturm, F., and M. Ressel, *Microchem. J.*, **5**, 53 (1961).
46. Salikhdzhanova, R. M. -F., Thesis, Moscow Institute of Chemical Engineering, 1966.
47. Bezuglyi, V. D., and E. Yu. Novik, *Zavodsk. Lab.*, **27**, 544 (1961).
48. Damaskin, B. B., *Principy Sovremennykh Metodov Izuchenija Elektrokhimicheskikh Reaktsyi* (in Russian), M.G.U., Moscow, 1965, p. 94.
49. Reinmuth, W. H., L. B. Rogers, and L. E. I. Hummelstedt, *J. Am. Chem. Soc.*, **81**, 2947 (1959).
50. Iofa, Z., B. Kabanov, E. Kuchinskii, and F. Chistiakov, *Zh. Fiz. Khim.*, **13**, 1105 (1939).
51. Krishtalik, L. I., *Usp. Khim.*, **34**, 1831 (1965).
52. Asada, K., P. Delahay, and A. Sundaram, *J. Am. Chem. Soc.*, **83**, 3396 (1961).

53. Weber, J., and J. Koutecký, *Collection Czech. Chem. Commun.*, **20**, 980 (1955).
54. Frumkin, A. N., *J. Electroanal. Chem.*, **9**, 173 (1965).
55. Petrii, O. A., and A. N. Frumkin, *Dokl. Akad. Nauk SSSR*, **146**, 1121 (1962).
56. Frumkin, A. N., O. A. Petrii, and N. V. Nikolaeva-Fedorovich, *Dokl. Akad. Nauk SSSR*, **147**, 878 (1962).
57. Aramata, A., and P. Delahay, *J. Phys. Chem.*, **68**, 880 (1964).
58. Delahay, P., and E. Solon, *J. Electroanal. Chem.*, **11**, 233 (1966).
59. Torsi, G., and P. Papoff, *Z. Anal. Chem.*, **224**, 130 (1967).
60. Torsi, G., and P. Papoff, *J. Electroanal. Chem.*, **16**, 83 (1968).
61. Žežula, I., *Collection Czech. Commun.*, **33**, 2327 (1968).
62. Frumkin, A., in *Advances in Electrochemistry and Electrochemical Engineering* (P. Delahay, Ed.), Vol. 1, Interscience, New York, 1961, p. 65.
63. Zykov, V. I., and S. I. Zhdanov, *Zh. Fiz. Khim.*, **32**, 644, 791 (1958).
64. Frumkin, A. N., *Trans. Faraday Soc.*, **55**, 156 (1959).
65. Krishtalik, L. I., *Elektrokhimia*, **2**, 1351 (1966).
66. Brdička, R., and K. Wiesner, *Naturwissenschaften*, **31**, 247 (1943).
67. Wiesner, K., *Z. Electrochem.*, **49**, 164 (1943).
68. Brdička, R., and K. Wiesner, *Collection Czech. Chem. Commun.*, **12**, 39 (1947).
69. Mairanovskii, S. G., *Dokl. Akad. Nauk SSSR*, **132**, 1352 (1960).
70. Mairanovskii, S. G., and L. I. Lishcheta, *Collection Czech. Chem. Commun.*, **25**, 3025 (1960).
71. Wiesner, K., *Chem. Listy*, **41**, 6 (1947).
72. Koutecký, J., and R. Brdička, *Collection Czech. Chem. Commun.*, **12**, 337 (1947).
73. Budevsky, E., *Izv. Bulg. Akad. Nauk, Ser. Fiz.*, **3**, 43 (1952–54).
74. Matsuda, H., *J. Phys. Chem.*, **64**, 336 (1960).
75. Gierst, L., and H. Hurwitz, *Z. Elektrochem.*, **64**, 36 (1960).
76. Hurwitz, H., *Z. Elektrochem.*, **65**, 178 (1961).
77. Mairanovskii, S. G., *Zh. Fiz. Khim.*, **33**, 691 (1959).
78. Mairanovskii, S. G., *Dokl. Akad. Nauk SSSR*, **149**, 1373 (1963).
79. Polievktov, M. K., and S. G. Mairanovskii, in *Novosti Elektrokhimii Organicheskikh Soedinenii* (A. Frumkin et al., Eds.), Nauka, Moscow, 1968, p. 36; S. G. Mairanovskii, *Progress Elektrochimii Organicheskikh Soedinenii*, (A. Frumkin and A. Ershler, Eds.), Nauka, Moscow, 1969, p. 88.
80. Mairanovskii, S. G., L. D. Kliukina, and A. N. Frumkin, *Dokl. Akad. Nauk SSSR*, **141**, 147 (1961).
81. Levich, V. G., B. I. Khaikin, and S. G. Mairanovskii, *Dokl. Akad. Nauk SSSR*, **145**, 605 (1962).

REFERENCES

82. Mairanovskii, S. G., J. Koutecký, and V. Hanus, *Zh. Fiz. Khim.*, **36**, 2621 (1962).
83. Mairanovskii, S. G., and R. G. Baisheva, *Elektrokhimia*, **5**, 893 (1969).
84. Khurgin, U. I., R. G. Baisheva, and S. G. Mairanovskii, *Izv. Akad. Nauk SSSR, Ser. Khim.*, 1679 (1969).
85. Mairanovskii, S. G., *Dokl. Akad. Nauk SSSR*, **142**, 1120 (1962).
86. Mairanovskii, S. G., *Electrochim. Acta*, **9**, 803 (1964).
87. Mairanovskii, S. G., R. G. Baisheva, *Elektrochimia*, **6**, 226 (1970).
88. Mairanovskii, S. G., and E. F. Mairanovskaya, in *Novosti Elektrokhimii Organicheskikh Soedinenii*, (A. Frumkin et al., Eds.), Nauka, Moscow, 1968, p. 33; S. G. Mairanovskii, in *Progress Elektrochimii Organicheskikh Soedinenii*, (A. Frumkin and A. Ershler, Eds.), Nauka, Moscow, 1969, p. 106.
89. Mark, H. B., Jr., and L. R. McCoy, *Rev. Polarog. (Kyoto)*, **14**, 122 (1967).
90. Mairanovskii, S. G., *Dokl. Akad. Nauk SSSR*, **133**, 162 (1960).
91. Grabowski, Z. R., and T. Bartel, *Roczniki Chem.*, **34**, 611 (1960).
92. Pungor, E., and Gy. Farsang, *J. Electroanal. Chem.*, **2**, 291 (1961).
93. Mairanovskii, S. G., *Zh. Fiz. Khim.*, **37**, 451 (1963).
94. Mairanovskii, S. G., and L. I. Lishcheta, *Izv. Akad. Nauk SSSR, Otdel. Khim. Nauk.*, 227 (1962).
95. Mairanovskii, S. G., in *Progress Elektrokhimii Organicheskikh Soedinenii*, (A. Frumkin and A. Ershler, Eds.), Nauka, Moscow, 1969, pp. 99, 102.
96. Frumkin, A. N., and G. M. Florianovich, *Dokl. Akad. Nauk SSSR*, **80**, 907 (1951).
97. Florianovich, G. M., and A. N. Frumkin, *Zh. Fiz. Khim.*, **29**, 1827 (1955).
98. Kriukova, T. A., *Dokl. Akad. Nauk SSSR*, **65**, 517 (1949).
99. Frumkin, A. N., B. B. Damaskin, and N. V. Nikolaeva-Fedorovich, *Dokl. Akad. Nauk SSSR*, **115**, 751 (1957).
100. Damaskin, B. B., N. V. Nikolaeva-Fedorovich, and A. N. Frumkin, *Dokl. Akad. Nauk SSSR*, **121**, 129 (1958).
101. Petrii, O. A., N. V. Nikolaeva-Fedorovich, *Zh. Fiz. Khim.*, **35**, 1999 (1961).
102. Frumkin, A. N., and N. V. Nikolaeva-Fedorovich, *Vestn. M.G.U., Ser. Khim.*, No. 4, 169 (1957).
103. Nikolaeva-Fedorovich, N. V., O. A. Petrii, B. B. Damaskin, and G. A. Furazhkova, *Vestn. M.G.U., Ser. II, Khim.*, No. 3, 40 (1962).
104. Iakovleva, E. V., and N. V. Nikolaeva-Fedorovich, *Elektrokhimia*, **3**, 543 (1967).
105. Mairanovskii, S. G., V. P. Gultiai, and N. K. Lisitsyna, *Elektrokhimia*, **5**, 752 (1969).
106. Turian, Ia. I., and G. F. Serova, *Zh. Fiz. Khim.*, **31**, 1976 (1957).
107. Turian, Ia. I., and G. F. Serova, *Elektrokhimia*, **2**, 1185 (1966).

108. Mark, H. B., Jr., and C. N. Reilley, *Anal. Chem.*, **35**, 195 (1963).
109. Koryta, J., *Z. Elektrochem.*, **61**, 432 (1957).
110. Kern, D. M. H., and J. Koryta, *Z. Phys. Chem. N.F.*, **52**, 170 (1967).
111. Pospíšil, L., and J. Kůta, *Collection Czech. Chem. Commun.*, **31**, 735 (1966).
112. Given, P. H., and M. E. Peover, in *Advances in Polarography* (I. S. Longmuir, Ed.), Vol. 3, Pergamon Press, Oxford, 1960, p. 948.
113. Peover, M. E., in *Electroanalytical Chemistry* (A. J. Bard, Ed.), Vol. 2, Marcel Dekker, New York, 1967.
114. Kastenning, B., and L. Holleck, *Z. Elektrochem.*, **63**, 166 (1959).
115. Stradyns, J. P., G. O. Reichmanis, and R. A. Gavar, *Elektrokhimia*, **1**, 955 (1965).
116. Bezuglyi, V. D., and Yu. P. Ponomarev, in *Novosti Elektrokhimii Organicheskikh Soedinenii*, (A. Frumkin et al., Eds.), Moscow, Nauka, 1968, p. 56.
117. Petrosyan, V. A., S. G. Mairanovskii, V. I. Slovetskii, and A. A. Fainzilberg, *Izv. Akad. Nauk SSSR, Ser. Khim.*, 928 (1968).
118. Karguin, Yu. M., *III Vsesojuznoe Soveschanie po Polarografii, Kiev,* 1965, p. 40.
119. Bullerwell, R. A. F., *J. Polarograph. Soc.*, **10**, 55 (1965).
120. Mairanovskii, S. G., *Dokl. Akad. Nauk SSSR*, **142**, 1327 (1962).
121. Frumkin, A. N., *Z. Physik. Chem.*, **35**, 792 (1926).
122. Mairanovskii, S. G., *Dokl. Akad. Nauk SSSR*, **133**, 162 (1960).
123. Mairanovskii, S. G., and A. N. Frumkin, *Rev. Polarog. (Kyoto),* **11**, 96 (1963).
124. Mairanovskii, S., *Elektrochem. Methoden und Prinzipien in der Molekular-Biologie* (III Jena Symposium), Akad. Verlag, Berlin, 1966, p. 473.
125. Mairanovskii, S. G., and A. D. Filonova, in *Electrokhimicheskie Protsessy s Uchastiem Organ. Veschestv,* (Zhurnal Elektrokhimia), Nauka, Moscow, 1969, p. 68.
126. Mairanovskii, S. G., and A. D. Filonova-Krasnova, *Izv. Akad. Nauk SSSR, Ser. Khim.*, 1673 (1967).
127. Kubota, T., and H. Miyazaki, *Bull. Chem. Soc. Japan,* **39**, 2057 (1966).
128. Mairanovskii, S. G., N. V. Barashkova, and F. D. Alashev, *Zh. Fiz. Khim.*, **36**, 562 (1962).
129. Mairanovskii, S. G., and A. D. Filonova, *Elektrokhimia,* **3**, 1397 (1967).
130. Schwabe, K., *Z. Phys. Chem. (Leipzig),* Sonderheft, 293 (1958).
131. Damaskin, B. B., *Elektrokhimia,* **1**, 63 (1965).
132. Bockris, J. O'M., E. Gileadi, and K. Müller, *Electrochim. Acta,* **12**, 1301 (1967).
133. Ershler, A. B., in *Progress Elektrokhimii Organicheskikh Soedinenii,* (A. Frumkin and A. Ershler, Eds.), Nauka, Moscow, 1969, p. 176.

134. Grabowski, Z., *Trudy 4 Soveshchania po Elektrokhimii, Moscow, 1956,* (A. Frumkin and S. Zhdanov, Eds.), Izd. Akademii Nauk, Moscow, 1959, p. 236.
135. Elving, P. J., *Rec. Chem. Progr.,* **14,** 99 (1953).
136. Laviron, E., *Bull. Soc. Chim. France,* **1962,** 418.
137. Elving, P. J., and E. C. Olson, *J. Am. Chem. Soc.,* **79,** 2697 (1957).
138. Horn, R. A., R. A. Courant, and D. S. Johnson, *Electrochim. Acta,* **11,** 987 (1966).
139. Schmidt, P. P., Jr., and H. B. Mark, Jr., *J. Chem. Phys.,* **43,** 3291 (1965).
140. Sternberg, H. W., R. E. Markby, I. Wender, and D. Mohilner, *J. Am. Chem. Soc.,* **89,** 186 (1967).
141. Wiesener, W., and K. Schwabe, *J. Electroanal. Chem.,* **15,** 73 (1967).
142. Nikolaeva-Fedorovich, N. V., L. A. Fokina, and O. A. Petrii, *Dokl. Akad. Nauk SSSR,* **122,** 639 (1958).
143. Damaskin, B. B., and N. V. Nikolaeva-Fedorovich, *Zh. Fiz. Khim.,* **35,** 1279 (1961).
144. Nikolaeva-Fedorovich, N. V., and O. A. Petrii, *Zh. Fiz. Khim.,* **35,** 1270 (1961).
145. Petrovich, J. P., *Electrochim. Acta,* **12,** 1429 (1967).
146. Gladkova, L. K., S. G. Mairanovskii, and F. M. Stojanovich, *Elektrokhimia,* **7,** 325 (1971).
147. Blomgren, E., J. O'M. Bockris, and C. Jesch, *J. Phys. Chem.,* **65,** 2000 (1961).
148. Setschenow, J., *Z. Phys. Chem.,* **4,** 117 (1889).
149. Long, F. A., and W. F. McDevit, *Chem. Rev.,* **51,** 119 (1952).
150. Ershler, A. B., E. D. Belokolos, and G. A. Tedoradze, *Elektrokhimia,* **1,** 1429 (1965).
151. Grigoriev, N. B., and B. B. Damaskin, in *Novosti Elektrokhimii Organicheskikh Soedinenii,* (A. Frumkin et al., Eds.), Nauka, Moscow, 1968, p. 66.
152. Damaskin, B. B., A. A. Survila, and L. E. Rybalka, *Elektrokhimia,* **3,** 927 (1967).
153. Gierst, L., and C. Pecasse, *Polarography* 1964 (G. J. Hills, Ed.), Macmillan, London, 1966, p. 305.
154. Gierst, L., and P. Herman, *Z. Anal. Chem.,* **216,** 238 (1966).
155. Losev, V. V., *Dokl. Akad. Nauk SSSR,* **111,** 626 (1956).
156. Partridge, L. K., A. C. Tansley, and A. S. Porter, *Electrochim. Acta,* **11,** 517 (1966).
157. McKee, R. H., et al., *Trans. Electrochem. Soc.,* **62,** 203 (1932); **65,** 301, 327 (1934) [from reference (161)].
158. Baizer, M., *Tetrahedron Letters,* 973 (1963).
159. Baizer, M., *J. Electrochem. Soc.,* **111,** 215 (1964).

160. Tomilov, A. P., L. A. Fedorova, V. A. Klimov, and G. A. Tedoradze, *Elektrokhimia*, **4**, 1264 (1968).

161. Beck, F., *Ber. Bunsenges. Physik. Chem.*, **72**, 379 (1968).

162. Mairanovskii, S. G., N. V. Barashkova, and Yu. B. Volkenshtein, *Elektrokhimia*, **1**, 72 (1965).

163. Kirrman, A., and P. Federlin, *Bull. Soc. Chim. France*, 944 (1958).

164. Vlček, A. A., in *Progress in Inorganic Chemistry* (F. Cotton, Ed.), Vol. 5, Interscience, New York, 1963, p. 208.

165. Pozdeeva, A. A., S. G. Mairanovskii, and L. K. Gladkova, *Elektrokhimia*, in press.

166. Pozdeeva, A. A., S. G. Mairanovskii, and L. K. Gladkova, *Elektrokhimia*, **3**, 1127 (1967).

167. Gladkova, L. K., S. G. Mairanovskii, and F. M. Stojanovich, *Elektrokhimia*, in press.

168. Belous, G. G., V. F. Lavrushin, and V. D. Bezuglyi, *Zh. Obsch. Khim.*, **37**, 2169 (1967).

169. Kitaev, Yu. P., G. K. Budnikov, T. V. Troepolskaia, and I. M. Skrebkova, *Zh. Obsch. Khim.*, **37**, 1437 (1967).

170. Svetkin, Yu. V., L. N. Andreeva, *Zh. Obsch. Khim.*, **35**, 839 (1965).

171. Tondeur, J. J., A. Dombret, L. Gierst, *J. Electroanal. Chem.*, **3**, 225 (1962).

172. Lingane, P. J., F. C. Anson, and R. A. Osteryoung, *J. Electroanal. Chem.*, **12**, 250 (1966).

173. Herasymenko, P., and J. Slendyk, *Z. Phys. Chem.*, **A149**, 123 (1930).

174. Frumkin, A. N., *Trudy IV Vsesojusnogo Soveshchania po Elektrokhimii*, Izd. Akademii Nauk, Moscow, 1959, p. 7.

175. Nikolaeva-Fedorovich, N. V., and B. B. Damaskin, *Trudy IV Vsesojusnogo Soveshchania po Elektrokhimii*, Izd. Akademii Nauk, Moscow, 1959, p. 150.

176. Grahame, D., *J. Electrochem. Soc.*, **98**, 343 (1951).

177. Gierst, L., L. Vandenberghen, E. Nicalas, and A. Fraboni, *J. Electrochem. Soc.*, **113**, 1025 (1966).

178. Žežula, I., *Collection Czech. Commun.*, **33**, 18 (1968).

179. Gierst, L., L. Vandenberghen, and E. Nicolas, *J. Electroanal. Chem.*, **12**, 462 (1966).

180. Zuman, P., D. Barnes, and A. Ryvolová-Keyharová, *Discussions Faraday Soc.*, **45**, 202 (1968).

181. Ashworth, M., *Collection Czech. Commun.*, **13**, 229 (1948).

182. Bezuglyi, V. D., L. A. Melnik, V. N. Dmitrieva, *Zh. Obsch. Khim.*, **34**, 1048 (1964).

183. Barnes, D., and P. Zuman, *Analyst*, **93**, 589 (1968).

184. Holleck, L., S. Vavřička, and M. Heyrovský, *Z. Naturforsch.*, **22b**, 1226 (1967).
185. Zhdanov, S. I., and P. Zuman, *Collection Czech. Chem. Commun.*, **29**, 960 (1964).
186. Beinroth, G., K. Schwabe, and H.-D. Suschke, *Z. Phys. Chem. (Leipzig)*, **235**, 133 (1967).
187. Kolthoff, I. M., and A. Willmann, *J. Am. Chem. Soc.*, **56**, 1007, 1014 (1934).
188. Pifer, C. W., and E. G. Wollish, *Anal. Chem.*, **24**, 519 (1952).
189. Polievktov, M. K., and S. G. Mairanovskii, *Elektrokhimia*, **3**, 139 (1967).
190. Vasilieva, E. G., S. I. Zhdanov, and T. A. Kriukova, *Elektrokhimia*, **4**, 24 (1968).
191. Jain, D. S., and J. N. Gaur, *Electrochim. Acta*, **12**, 413 (1967).
192. Frumkin, A. N., and B. B. Damaskin, *Dokl. Akad. Nauk SSSR*, **129**, 862 (1959).
193. Damaskin, B. B., and N. V. Nikolaeva-Fedorovich, *Zh. Fiz. Khim.*, **35**, 1279 (1961).
194. Devanathan, M. A. V., and M. J. Fernando, *Trans. Faraday Soc.*, **58**, 368 (1962).
195. Gierst, L., J. Tondeur, R. Cornelissen, and F. Lamy, Moscow CITCE Meeting, 1963.
196. Missan, S. R., E. I. Becker, and L. Meites, *J. Am. Chem. Soc.*, **83**, 58 (1961).
197. Kůta, J., and J. Weber, *Electrochem. Acta*, **9**, 541 (1964).
198. Koryta, J., *Rev. Polarog. (Kyoto)*, **13**, 13 (1965).
199. Koryta, J., and S. Vavřička, *J. Electroanal. Chem.*, **10**, 451 (1965).
200. Bartle, W. W., and B. R. Eggins, *J. Polarograph. Soc.*, **12**, No. 3, 89 (1966).
201. Lambert, F. L., and K. Kobaysahi, *J. Org. Chem.*, **23**, 773 (1958).
202. Given, P. H., M. E. Peover, and J. Schoen, *J. Chem. Soc.*, 2674 (1958).
203. Holleck, L., and D. Becher, *J. Electroanal. Chem.*, **4**, 321 (1962).
204. Todres, Z. V., *Izv. Akad. Nauk SSSR, Ser. Khim.*, 1749 (1970).
205. Ward, L., *J. Am. Chem. Soc.*, **83**, 1296 (1961).
206. Murray, R. W., and L. K. Hiller, Jr., *Anal. Chem.*, **39**, 1221 (1967).
207. Kheifets, L. I., and V. D. Bezuglyi, *Elektrokhimia*, **2**, 800 (1966).
208. Philp, R. H., R. L. Flurry, and R. A. Day, Jr., *J. Electrochem. Soc.*, **111**, 328 (1964).
209. Gierst, L., J. Tondeur, and E. Nicolas, *J. Electroanal. Chem.*, **10**, 397 (1965).
210. Tza Chuan-sin, Z. A. Iofa, *Dokl. Akad. Nauk SSSR*, **125**, 1065 (1959).
211. Tedoradze, G. A., and A. L. Asatiani, *Elektrokhimia*, **4**, 960 (1968).
212. Asatiani, A. L., and G. A. Tedoradze, in *Novosti Elektrokhimii Organicheskikh Soedinenii*, (A. Frumkin et al., Eds.), Moscow, Nauka, 1968, p. 27.

213. Mairanovskii, S. G., V. P. Gultiai, and N. K. Lisitsyna, *Elektrokhimia,* **6,** 541 (1970).
214. Mairanovskii, S. G., *Dokl. Akad. Nauk SSSR,* **142,** 1327 (1962).
215. Mairanovskii, S. G., *J. Electroanal. Chem.,* **6,** 77 (1963).
216. Mairanovskii, S. G., V. P. Gultiai, and N. K. Lisitsyna, *Elektrokhimia,* **6,** 1202 (1970).
217. Vlček, A. A., and J. Kůta, *Nature,* **185,** 95 (1960).
218. Moorhead, E. D., and G. M. Frame, *Anal. Chem.,* **40,** 280 (1968).
219. Gorodetskii, V. V., and V. V. Losev, *Elektrokhimia,* **2,** 656 (1966).
220. Gaur, J. N., and N. K. Goswami, *Electrochim. Acta,* **12,** 1483 (1967).
221. Hopin, A. M., and S. I. Zhdanov, *J. Polarograph. Soc.,* **13,** 37 (1967).
222. Kůta, J., and J. Koryta, *Collection Czech. Chem. Commun.,* **30,** 4095 (1965).
223. Shams El Din, A. M., and L. Holleck, *Ber. Bunsenges. Physik. Chem.,* **71,** 739 (1967).
224. Hussey, W. W., and A. J. Diefenderfer, *J. Am. Chem. Soc.,* **89,** 5359 (1967).
225. Mairanovskii, S. G., *Elektrokhimia,* **5,** 757 (1969).
226. Nikolaeva-Fedorovich, N. V., I. E. Barbasheva, and N. P. Berezina, *Elektrokhimia,* **3,** 836 (1967).
227. Frumkin, A. N., *Osnovnye Voprosy Sovremennoi Teorettischeskoi Elektrochimii,* Mir, Moscow, 1965, p. 311.
228. Feoktistov, L. G., and S. I. Zhdanov, *Electrochim. Acta,* **10,** 657 (1965).
229. Nikolaeva-Fedorovich, N. V., and L. A. Akimova (Diplome-work of L. A. Akimova), Moscow University, 1961.
230. Tomilov, A. P., E. V. Kriukova, V. A. Klimov, and I. N. Brago, *Elektrokhimia,* **3,** 1501 (1967).
231. Sease, J. W., P. Chang, and J. L. Groth, *J. Am. Chem. Soc.,* **86,** 3154 (1964).
232. Lambert, F. L., A. H. Albert, and J. P. Hurdy, *J. Am. Chem. Soc.,* **86,** 3155 (1964).
233. Zuman, P., *Substituent Effects in Organic Polarography,* Plenum Press, New York, 1967.
234. Holleck, L., and R. Schindler, *Z. Elektrochem.,* **60,** 1138, 1142 (1956).
235. Tirouflet, J., *Bull. Soc. Chim. France,* 274 (1956).
236. Todres, Z. V., S. I. Zhdanov, and V. Sh. Tsveniashvili, *Izv. Akad. Nauk SSSR, Ser. Khim.,* 975 (1968).
237. Bockris, J. O'M., *Electrochim. Acta,* **3,** 340 (1961).
238. Mott, N. F., and R. J. Watts-Tobin, *Electrochim. Acta,* **4,** 79 (1961).
239. Venkatesan, V. K., B. B. Damaskin, and N. V. Nikolaeva-Fedorovich, *Zh. Fiz. Khim.,* **39,** 129 (1965); M. M. Andrusev, N. V. Nikolaeva-Fedorovich, and B. B. Damaskin, *Elektrokhimia,* **3,** 1094 (1967).

240. Delahay, P., and M. Kleinerman, *J. Am. Chem. Soc.*, **82**, 4509 (1960).
241. Nikolaeva-Fedorovich, N. V., E. V. Iakovleva-Stenina, and K. V. Rybalka, *Elektrokhimia*, **3**, 1502 (1967).
242. Frumkin, A. N., O. A. Petrii, and N. V. Nikolaeva-Fedorovich, *Dokl. Akad. Nauk SSSR*, **147**, 878 (1962).
243. Mairanovskii, S. G., *Rev. Polarog. (Kyoto)*, **14**, 329 (1967).
244. Grabowskii, Z. R., and E. Bartel, *Roczniki Chem.*, **34**, 611 (1960).
245. Nürnberg, H.-W., and G. Wolff, *J. Electroanal. Chem.*, **21**, 99 (1969).
246. Mairanovskii, S. G., and A. P. Churilina, *Elektrokhimia*, **6**, 1679, 1857 (1970).
247. Rodgers, R. S., and L. Meites, *J. Electroanal. Chem.*, **16**, 1 (1968).
248. Reinmuth, W. H., *Anal. Chem.*, **33**, 485 (1961); A. J. Bard, *ibid.*, **35**, 340 (1963).
249. de Vries, W. T., *J. Electroanal. Chem.*, **17**, 31, 469 (1968).
250. Gaselli, M., G. Ottombrini, and P. Papoff, *Electrochim. Acta*, **13**, 241 (1968).

Manuscript received April, 1969.

AUTHOR INDEX

Numbers in parentheses are reference numbers and show that an author's work is referred to although his name is not mentioned in the text. Numbers in *italics* indicate the pages on which the full references appear.

Abd-El-Kader, J. M., 253(207), *275*
Acker, D. S., 228(1), 237(1), 249(1), *267*
Ackermann, H., *63, 68*
Adachi, K., *18,* 43(57), *48, 49,* 50(57), *70*
Adamovsky, M., 106(22), *154*
Adams, R. N., 197(3, 369, 370), 199(139, 288), 227(128), 229(367), 237(3), 262(2, 369), 263(369), 264(370), 266(265), *267, 272, 273, 277, 278, 281*
Akamatu, H., 227(217), *275*
Akimova, L. A., 355(229), *368*
Alashev, F. D., 330(128), *364*
Albert, A. H., 356(232), *368*
Alberts, G. S., *56, 68,* 113(39), *155*
Albright, C. H., 99(9), 100(9), *154*
Allendoefer, R. D., 221(4), *267*
Allred, A. L., 258(90), *271*
Anderegg, G., 30(76), *71*
Anderko, K., 158(1), 161(1), 162(1), *189*
Andreeva, E. P., 252(130), *272*
Andreeva, L. N., 344(170), *366*
Andrusev, M. M., 359(239), *368*
Anson, F. C., 304(43), 344(171), *361, 366*
Anthoine, G., 250(5), *267*
Antropov, L. I., 168(75), *191*
Aparina, V. I., 186(156), *194*
Arad, Y., 213(6), 214(6), 250(6), *268*
Arai, T., 234(7, 8), 240(9), 266(123), *268, 272*
Aramata, A., 307(57), *362*
Armand, J., 248(417), 263(417), *283*

Armbroster, M. H., *192*
Arnold, Z., 213(209), *275*
Asada, K., 306(52), *361*
Asatiani, A. L., *349, 350, 367*
Ashworth, M., 198(10), 230(10), 232(10), 233(10), *268, 345, 366*
Asthana, M., 256(11, 228, 229), *268, 276*
Astle, M. J., 198(108), 230(108), 232(108), *271*
Atanasov, M., 297(3), *360*
Atherton, N. M., 226(12), *268*
Austen, D. E. G., 227(13), *268*
Ayabe, Y., 8(54), *13, 18, 27,* 43(56, 57), *48, 49,* 50(55-57), *51, 70*
Aylward, G. H., *20, 60, 61, 62,* 64(3), *68,* 215(14), *268*

Babkin, G. N., 181(131), *193*
Bachofner, H. E., 255(15), *268*
Baer, H. J., 226(400), *282*
Bagotskii, V. S., 288(3), 291(3), 297(3), *298, 360, 361*
Bahary, W., 218(49), *269*
Bailes, 215(442), *284*
Baimakov, Yu. V., *180,* 181(128, 129), *193*
Baisheva, R. G., 312(83), 313(83), 314(83, 84), 315(84), 317(87), *363*
Baizer, M. M., 213(365), 254(457), *281, 285,* 341(158, 159), *365*
Balch, A. L., 227(148), 263(149), *273*
Baletskaya, L. G., 183(143), 186(143), *194*

AUTHOR INDEX

Balli, H., 217(212), *275*
Bambenek, M. A., 263(149), *273*
Bannerjee, N. R., 229(16), *268*
Barashkova, N. V., 210(307), 230(307), 248(306), *279,* 301(36), 331(158), 342-(162), *361, 364, 366*
Barbasheva, I. E., 355(226), *368*
Bard, A. J., 56(28), 57(28), *69,* 215(388), 216(287), 217(416), 250(387), *279, 282, 283,* 359(248), *369*
Bargain, M., 244(17), *268*
Barnes, D., 103(18), 104(21), 106(23), 113-(18,36), 114(18), 116(23), 118(23), 119-(23), 120(21), 122(21), 130(36), 132(18), 135(76), 139(23), 140(23), 148(21,76, 89), *154, 155, 156,* 242(18), *268, 345,* 346(1180), *366*
Bartel, E. T., 121(60), *155,* 233(19), 236-(20), 253(20), *268, 320,* 359(244), *363, 369*
Bartle, W. W., 347(200), *367*
Basil'eva, L. N., 185(151), *194*
Basu, S., 240(76), *270*
Batrakov, V. V., 294(19), *360*
Batterton, J. O., 161(34), *190*
Bauer, E., 229(22), 258(21), *268*
Bauer, H. H., 262(50), *269,* 304(44), *361*
Baumberger, J. P., 116(44), *153*
Baxendale, J. H., 197(286), 219(286), *278*
Becher, D., 265(200), *275,* 347(203), *367*
Beck, F., 340(161), *366*
Becker, E. I., 347(196), *367*
Becker, M., 116(50), *155*
Beckmann, P., 244(23), *268*
Beevers, J. R., 304(44), *361*
Begli, K., 162(41), 169(41), *190*
Beinroth, G., 346(186), *367*
Beletskaya, I. P., 257(62), 258(24,61,62), *268, 270*
Belokolos, E. D., 340(150), 341(150), *365*
Belous, G., 243(285), *278,* 343(168), *366*
Benesch, R., 198(25), 257(25), *268*
Benesch, R. E., 198(25), 257(25), *268*
Bent, H. E., 164(53), *169, 190, 192*
Berezina, N. P., 355(226), *368*
Berg, H., 39(22), 66(7), *69,* 197(38), 229-(27,28,30,35,36,45,157), 230(32,34), 235(30,31,33), 236(26,29,36), 240(32), 243(402), 258(21,37), *268, 269, 273, 282*
Bergman, I., 223(40), 259(39), *269*

Beringer, F. M., 255(15), *268*
Berkey, R., 197(460), 218(462), 219(460), 227(461), 228(461), *285*
Bernal, I., 197(381), 249(378,381), *282*
Berzins, T., *13, 69*
Bessin, J. M., 236(70), *270*
Bezuglyj, V. D., 141(82), *156,* 224(409), 227(42), 228(263), 232(41), 236(262), 243(285), *269, 277, 278, 283, 285,* 304-(47), 325(116), 343(168), 345(182), *348, 361, 364, 366, 367*
Biernat, J., 65(8), *69*
Billon, J. P., 218(73), 226(43,44), *269, 270*
Bittner, H., 161(35), *190*
Blaha, 197(460), 218(460), 227(461), 228-(461), *285*
Blomgren, E., 340(147), *365*
Blomster, M. L., 65(12), *69*
Blout, E. R., 198(123), 242(123), *272*
Böckel, W., 229(45), *269*
Bockris, J. O'M., 332(132), 348(147), 359-(237), *364, 365, 368*
Bodard, J. C., 236(70), *270*
Bolla, M., 130(74), *156*
Bonnier, E., 166(38), *190*
Brago, I. N., 355(230), *368*
Brand, P., 158(3), *189*
Brauer, G., 160(9), *189*
Brdička, R., 11(9,10,42), 20(9,10,42), *60, 69, 70,* 1113(40), 114(18), 116(45,46), 137(79), 143(79), *155, 156,* 197(47), 205(47), 231(48), *255, 269, 309,* 310-(72), *362*
Breiter, M., 301(34), 310(34), 319(34), *361*
Bresle, A., 297(23), *360*
Breslow, R., 218(49), *269*
Breyer, B., 304(44), *361*
Březina, M., 116(53), *155*
Bricker, C. E., 198(176), 242(176), *274*
Bril, K., 63(11), *69*
Bril, S., 63(11), *69*
Britske, E. V., 160(20), 161(20), 162(20), *189*
Britz, D., 262(50), *269*
Broadbent, A. D., 227(51), *269*
Brodskii, A. I., 217(53), 225(54), 227(52), 266(93), *269, 271*
Broman, R. F., 258(55), *269*
Brook, P. A., 251(56), *269*
Brown, J. K., 266(57), *270*

Broze, M., 227(58), *270*
Brück, D., 230(59), *270*
Buck, R. P., 141(85), *156*
Budevsky, E., 310(73), *362*
Budnikov, G., 329(227), 234(227), *276*, 302(37), 343(169), *361, 366*
Bukhman, S. P., 180(127), 181(127), *193*
Bullerwell, R. A. F., 251(60), *270, 325*, 326(119), *364*
Butin, K. P., 247(115), 257(62), 258(24,61, 62), *268, 270, 272*
Butler, J. N., 228(63), 230(63), *270*
Buvet, R., 218(359), 220(360), 239(360), *281*
Bydale, T. J., 65(12), *69*

Cadle, S. H., 265(64), *270*
Caldwell, R. A., 247(65), *270*
Capka, O., 198(66), *270*
Cardinali, M., 229(67), 244(68), *270*
Carelli, I., 244(68), *270*
Castellan, A., 223(69), *270*
Catteneo, C., 197(389), 229(389), *282*
Caullet, C., 236(70-72), *270*
Cauquis, G., 216(74), 218(73), 225(74,75), 226(43,44), *269, 270*
Cave, G. C., 26(30), *69*
Cazelet, P. V., 175(113), *192*
Chambers, J. Q., 265(64), *270*
Chang, P., 246(404), *282*, 356(231), *368*
Chao, F., 168(72), 173(102), *191, 192*
Chapman, D. L., *288, 360*
Chaudhuri, J. N., 240(76), *270*
Chauvelier, J., 198(376), *382*
Cheah, E. P. T., 262(77), *270*
Chiorboli, P., 251(78), *270*
Chistiakov, F., 305(50), 306(50), *361*
Chodkowska, A., 259(25), 262(258), *277*
Chodkowski, J., 100(10), *154*
Chopart-dit-Jean, L. H., 197(79), 220(79), 221(79), *270*
Christian, G. D., 256(151), *273*
Christova, I., 297(26), *360*
Churilina, A. P., 359(246), *369*
Chviruk, V. P., 168(75), *191*
Cisak, A., 222(81), 238(80), *270*
Claire, J., 161(28), 164(28), *190*
Colichman, E. L., 198(82), 255(82), *270*
Combrisson, J., 266(43,44), *269*
Conant, J. B., 169(85), *192*

Conrad, H., 217(212), *275*
Cooke, W. D., 175(110), 183(137), *192, 193*
Cooper, W. Ch., *173, 192*
Corcoran, W. H., 251(122), *272*
Coriou, H., 174(109), *192*
Cornelissen, R., 345(195), 348(195), 349-(195), *367*
Corvaisier, A., 107(25), *154*
Corvaja, C., 239(83), 262(84), *270, 271*
Costa, G., 246(85,390), *271, 282*
Costa, J. M., 219(420), 222(420), *283*
Costa, M., 168(72), 173(102), *191, 192*
Cottrell, W. R. T., 253(86), *271*
Coulson, D. M., 198(87,88), 240(87-89), *271*
Courant, R. A., 333(138), *365*
Crosseley, J. A., 251(56), *269*
Crow, D. R., *2, 69*
Crowell, W. R., 198(87,88), 240(87-89), *271*
Curtis, M. D., 258(90), *271*
Cusack, N., 169(93), *192*

Damaskin, B. B., 196(91), *271*, 288(7), 289-(17), 294(7,19), 295(20), 304(48), 323-(99,100,103), 324(103), 332(131), 336-(143), 341(151,152), 344(17,175), 345-(103,175), 346(175,192,193), *349*, 355-(193), 356(193,239), *360, 361, 363, 364, 365, 366, 367, 368*
Daniel'cheko, P. T., 161(23), *190*
Daniels, M., 169(83), *192*
Davies, J. D., 227(356), 228(355), *281*
Davolio, G., 251(78), *270*
Davydovskaya, Yu. A., 236(92), *271*
Day, R. A., Jr., 229(366), 237(366), 239-(366), *281*, 348(208), *367*
DeFord, D. D., *26, 46, 69*
Degrand, Ch., 232(282), *278*
Degtyarev, L. S., 217(53), 266(93), *269, 271*
Deguchi, Y., 217(133), 266(133), *272*
Dehl, R., 239(94), *271*
Delahay, P., *6*, 9(17), 11(17), *13*, 20(17, 58), *60, 69, 70*, 260(434), *284, 288*, 292, 294(4), 301(34), 304(4), *306*, 307(57, 58), 310(34), 319(4,34), 359(240), *360, 361, 362, 368*
Delaroff, V., 130(74), *156*

AUTHOR INDEX 373

Demange-Guérin, G., 228(95), *271*
Denisovich, L. I., 258(165), *274*
Desré, P., 166(38), *190*
Dessy, R. E., 258(86,377), *271, 282*
Deswarte, S., 248(417), 263(417), *283*
Devanathan, M. A. V., 346(194), *367*
Devaud, M., 258(97,99,100), *271*
DeVet, J. F., 166(66), *191*
Diefenderfer, A. J., 354(224), 355(224), 356(224), 258(224), *368*
Dietz, R., 226(12), *268*
DiGregorio, J. S., 258(102), *271*
Dimitrieva, V. N., 232(41), *269*
Dineen, E., 198(103), 243(103), *271*
Dmitrieva, V. N., 345(182), *366*
Dokunikhin, N. S., 227(42), *269*
Dombret, A., 344(171), 345(171), *366*
Donckt, E., v.d., 250(5), *267*
Dubov, S. S., 250(445), *284*
Duty, R. C., 248(465), 257(466), *285*
Dvořák, V., 199(104), *271*
Dzhaparidze, D. I., 115(42), *155*

Ebata, K., 54(88), *55, 56,* 64(88), *71*
Edsberg, R. L., 197(105), 223(107), 227-(105), *271*
Eggins, B. R., 347(200), *367*
Eichlin, D., 197(105), 227(105), *271*
Eigin, M., 101(14), *154*
El-Khiami, I., 250(106), *271*
Elofson, R. M., 223(107), *271*
Elving, P. J., 99(9), 100(9), *154,* 198(108-111), 213(109,439), 214(384), 217(170), 218(113), 219(414,420), 222(81,126, 420,439), 223(414), 230(108,112,438), 231(431), 323(108,438), 233(328), 238-(80), 239(328), 246(110,111), 247(114), 248(111), 252(119), *270, 271, 272, 274, 280, 282, 283, 284, 333,* 334(137), *365*
Erdey-Gruz, T., 167(70), 168(70), *191*
Ershler, A. B., 203(116), 239(116), *272, 333, 340,* 341(150), *364, 365*
Evans, D. H., 232(117), *272*
Evans, E., 160(11), 161(11), 164(11), *189*
Evans, M. G., 197(286), 219(286), *278*
Eversole, J. F., 168(79), *191*
Evseev, A. M., 158(5), *189*
Exner, H. J., 198(186,187,189), 237(190), 260(186,187,189), 263(190), *274*
Exner, O., 101(15), 116(15), 121(59), *154, 155*

Faffani, A., 198(127), 215(127), *272*
Fainzilberg, A. A., 325(117), *369*
Fakhmi, M., 247(115), *272*
Fan, J. W., 197(458), 218(45), *285*
Farnia, G., 261(84), 262(84), *271*
Farsang, Gy., 321(92), *363*
Fassbender, H., 252(423), *283*
Fauvelot, G., 216(74), 225(74,75), *270*
Federlin, P., 342(163), *366*
Fedoreňko, M., 132(75), *156*
Fedorova, L. A., 341(160), 355(160), *366*
Fedorovich, N. V., *302,* 322(39), *323,* 336-(39), 345(39), *361*
Feldman, M., 219(118), *272*
Felton, R. H., 223(119), 225(118), *272*
Feoktistov, L. G., 214(483), 246(121), 247-(120), *272, 285,* 355(228), *368*
Fernandez-Martin, R., 251(122), *272*
Fernando, M. J., 346(194), *367*
Ferro, R., 160(22), 161(22), *190*
Ficker, H. K., 177(120), 181(120), *184, 193*
Fielder, M., 169(93), *192*
Fields, M., 198(123), 242(123), *272*
Filonovam, A. D., 328(125), 329(125,126), 331(125,129), 332(125,129), 333(125,129), *364*
Filinovskii, V. Yu., 202(291), *278*
Fischer, O., 130(68), *156,* 138(124), *272*
Fleet, B., 102(16), 121(16), *154*
Fleischer, K. D., 253(418), *283*
Fleischmann, M., 261(125), *272*
Flemming, J., 39(22), *69*
Floch, L., 222(126), *272*
Florianovich, G. M., *322, 363*
Flurry, R. L., 229(366), 237(366), 239-(366), *281,* 348(208), *367*
Fodiman, Z. I., 225(293), 228(292), 265-(292), *278, 301,* 346, *361*
Fokina, L. A., 335(142), *336, 365*
Forbes, G. S., 168(76), 169(76), *191*
Foreit, J., 259(179), *274*
Fornari, P., 243(327), *280*
Fornasari, E., 112(32), *155*
Fournari, P., 263(283), *278*
Fraboni, A., 345(177), 348(177), 349(177), 352(177), 353(177), *366*
Fraenkel, G. K., 197(381), 217(380), 221-

(435), 239(94), 249(378,381), *217, 282, 284*
Fragiacomo, M., 198(127), 198(127), *272*
Frame, G. M., 353(218), *368*
Franzen, V., 225(177), *274*
Freyhold, H., v., *1, 71*
Friedman, H. L., 160(17), *189*
Friess, S. L., 198(142), 242(142), 243(142), *273*
Fritsch, J. M., 227(128), *272*
Frost, B., 169(92), *192*
Frost, J. G., 225(129), *272*
Frumkin, A. N., 101(11), *154,* 220(481), 252(130,131), *272, 285,* 288(1-3), 290-(17,18), 291(3), *302,* 303(41,42), 306-(62), 307(54-56,62), *308,* 309(55), 311-(80), 320(80), 321(80), *322,* 323(99,100, 102), *326,* 327(123), 342(80), 344(17, 174), 345(38,80,123), 346(192), 355-(227), 359(242), *360, 361, 362, 363, 366, 367, 368, 369*
Fueno, T., 214(132), 240(132), *272*
Fujinaga, T., 217(133), 224(135-137), 248-(136), 266(133,134), *272, 273*
Fujisawa, T., *62,* 64(23,24), 65(23,24), *69*
Funt, B. L., 217(138), *273*
Furazhkova, G. A., 323(103), 324(103), 345(103), *363*
Furhkawa, J., 214(132), 240(132), *272*
Furman, N. H., *173, 192*

Galeev, V. S., 250(294), *278*
Galus, Z., 183(141), *184,* 186(141,158-160), 187(159), *193, 194,* 199(139), 209-(140), *273*
Gardner, H. J., 86(6), 121(63), 139(63), *154, 156*
Garis, J. J., 197(105), 227(105), *271*
Garnett, J. L., 215(14), *268*
Garnyuk, L. N., 266(93), *271*
Garrod-Thomas, R. N., 168(78), 169(78), *191*
Gartmann, H., 260(238), *276*
Gaselli, M., 359(250), *369*
Gaunitz, U., 196(296), *278*
Gaur, J. N., 346(191), 354(22), *367, 368*
Gavars, R. A., 262(141), 263(431), *273, 283,* 325(115), *364*
Gavioli, G., 251(78), *270*
Gazizov, M. B., 240(394), 251(393,

394), *282*
Geissmann, T. A., 198(142), 242(142), 243-(142), *273*
Gelb, R. I., 214(143), *273*
Georgakopoulos, P. P., 263(301), *279*
Georgans, W. P., 121(63), 139(63), *156*
Gercke, R. H., 164(44), *190*
Gerdil, R., 225(145), 251(144), *273*
Gergely, E., 121(57,58), *155*
Gerischer, H., *20, 42,* 46(26), *51, 69,* 261-(271), 263(261), *277*
Gerlock, J. L., 240(224), *276*
Gerovich, V. M., 196(91), *271*
Geske, D. H., 197(313-315,146), 213(279), 217(137), 226(150), 227(148,166), 263-(149,314), 264(146,147,315), 266(178, 323,324), *273, 274, 278, 279, 280*
Giacometti, G., 112(32), *155*
Giang, B. Y., 256(151), *273*
Gierst, L., 310(75), 341(153,154), 344-(171), 345(171,177,179), 346(195), 348-(177,195,209), *349, 350, 352, 353, 362, 365, 366, 367*
Gilbert, B. C., 226(152), *273*
Gileadi, E., 332(132), *364*
Girdwoyn, A., 235(259), *277*
Given, P. H., 220(154,155), 221(154), 227-(13,153), 228(154), 230(155), *268, 273,* 325(113), 347(202), *364, 367*
Gladkova, L. K., 339(146), 340(146), 343-(165-167), *365, 366*
Gladyschev, V. P., 163(166), *194*
Glarum, S. H., 266(156), *273*
Gollmeck, F. A., 229(35,157), *269, 273*
Golubenkova, A. M., 266(93), *271*
Gordienko, L. L., 217(53), 225(54,158), 227(52), *269, 273*
Gorodetskii, V. V., *353, 368*
Gorodovykh, V. E., 183(142), 186(142), *194*
Goswami, N. K., 354(226), *368*
Gough, T. A., 217(159), 218(159), *273*
Gouy, G., *288, 360*
Grabowski, Z. R., 121(60), *155,* 198(160), 218(161), 229(453), 232(453), 233(19, 160,254), 234(162,226), 236(20), 253-(20), *268, 273, 276, 277, 284, 320, 333,* 359(244), *363, 365, 369*
Grahame, D. C., 288(5), *289,* 293, 294(5), *297, 298,* 294(30), 245(176), *360,*

361, 366
Gray, D. G., 217(138), *273*
Grens, J. E., 240(432), *284*
Grenschall, J., *192*
Grieb, N. W., 36(45), 41(45), *70*
Griffiths, W. E., 263(163), *273*
Grigoriev, N. B., 341(151), *365*
Grimshaw, J., 247(*164*)
Gross, J., 223(214), *275*
Groth, J. L., 246(404), *282,* 356(231), *368*
Grube, G., *160, 189*
Gubin, S. P., 258(165), *274*
Gulick, W. M., 227(166), *227*
Gultiai, V. P., 324(105), 349(213), 350-(213), 351(213), 352(213,216), *363, 368*
Gunderson, A., 238(464), 243(467), 244-(467), *285*
Gut, R., 30(75,76), *71*
Gutmann, V., 197(167), *274*
Guyard, M., 224(444), 238(444), *284*

Häbich, A., 225(174), *274*
Hacobian, S., 247(65), 262(77), *270*
Häflinger, O., 198(375), 214(375), *281*
Hale, J. M., 204(168), 223(168), *274*
Hall, D. A., 217(170), 252(169), *274*
Halls, D. J., 257(171), *274*
Hamaguchi, H., 236(33), *280*
Hans, W., 252(424), 253(172,424), *274, 283*
Hanson, M., 158(1), 161(1), 162(1), *189*
Hanson, P., 226(152), *273*
Hanuš, V., 11(9), 20(9), 60(9), *69,* 137(79), 138(79), 141(84), 143(79,84), *156,* 209-(173,274), 210(274), *274, 278,* 312(82), 314(82), *363*
Haring, M. M., 164(47), 66(47), *190*
Harle, A. J., 198(174), 262(77), *270, 274*
Harris, W. S., 267(175), *274*
Hartmann, H., 167(71), 177(121), 181-(121), *191, 193*
Hartnell, E. D., 198(176), 242(176), *274*
Hashitani, T., 39(27), *69*
Hauffe, K., 160(13), *189, 192*
Haul, R. A. W., 166(66), *191*
Hausser, K. H., 225(177), *274*
Hayes, J. R., 214(384), 247(114), *272, 282*
Hayes, J. W., *20, 60, 61, 62,* 64(3), *68*
Hébert, M., 236(72), *270*
Heibronner, E., 197(79), 220(79), 221(79), *270*

Heiskel, E., 130(72), *156*
Heller, Ph. H., 266(178), *274*
Helmholtz, H. L. F., *288, 360*
Henderson, A. T., 250(184), *274*
Herasymenko, P., *344, 366*
Herman, H. B., 56(28), 57(28), *69*
Herman, P., 341(154), 353(154), *365*
Hertter, W. R., 228(1), 237(1), 249(1), *267*
Heyrovsky, J., 171(97), *192,* 197(183), 205-(181), 236(182,183), 252(181), 253(205), 259(179), 262(180), *274, 275,* 345(174), *367*
Hickling, A., *184, 194*
Hickichi, H., 64(29), 65(29), *69*
Hildebrand, J. H., 164(51), *168, 169, 190, 191*
Hiller, L. K., *343, 367*
Hillers, S., 262(141), 263(430,431), *273, 283*
Hinnüber, J., 165(58), 166(58), 167(58), *191*
Hoffmann, A. K., 250(184), *274*
Hohn, H., 166(59), *191*
Hoijtink, G. J., 198(185), 206(185), 213-(185), 218(185), 220(185), *274*
Holden, K. B., 160(18), *189*
Holleck, G., 263(204), *275*
Holleck, L., 113(34), 121(59), 130(34), *155,* 198(191,192,186-189), 207(232, 239,246), 214(194), 230(191,232,233), 231(198), 232(191,192,193), 233(192, 232), 236(192), 237(190,193), 242(208), 243(201), 253(205-207,222), 260(186-189,195,196,199,202,232,234,238,239), 261(202,232,244), 262(203), 263(190, 193,197,204), 265(200,202), *274, 275, 276,* 325(114), 345(184), 347(203), *354, 358(234), 364, 367, 368*
Holubek, J., 116(55), 117(55), 118(55), *155,* 214(454), *284*
Holý, A., 213(209), *275*
Hopin, A. M., 222(210), *275,* 354(221), *368*
Horák, V., 106(22), 130(70), *154, 156,* 244-(490), *286*
Horn, R. A., 333(130), *365*
Hume, D. N., *26, 46, 69,* 197(211), *275*
Hume-Rothery, W., 161(34), *190*
Hummelstedt, L. E. I., 247(317), *279,* 301-(35), 305(49), *361*
Hünig, S., 216(212,213), 217(215), 223-

(214), *275*
Huré, J., 175(109), *192*
Hurdy, J. P., 356(232), *368*
Hurwitz, H., 310(75, 76), *362*
Hush, N. S., 247(216), *275*
Hussy, W. W., 354(224), 355(224), 356-(224), 352(224), *368*

Iakovleva, E. V., 323(104), *363*
Iakovleva-Steina, E. V., 249(241), *369*
Iida, Y., 227(217), *275*
Ikeuchi, H., 39(31), *69*
Il'yasov, A. V., 240(218), 251(393), *275, 287*
Ilyushchenko, V. M., 181(132), 185(149), *193, 194*
Ingram, D. I. E., 227(13), *268*
Iofa, Z. A., 186(153,154), *194*, 288(3), 291(3), 297(3), 305(50), *306, 349, 350, 360, 361, 367*
Ireadale, T., 121(57,58), *155*
Irvin, M., 175(113), *192*
Issleib, K., 254(319), *279*
Ito, T., *60, 71*
Ivanov, V. F., 186(153,154), *194*
Ivanova, V. Kh., 240(218), *275*
Iwamoto, R. T., 258(383), *282*
Izutsu, K., 224(135), *272*

Jain, 346(191), *367*
James, J. C., 198(219), 259(30), *269, 275*
James, M. J., 244(220), 252(220), *276*
Jänecke, E., 160(12), *189*
Jangg, G., 159(32,42), 161(25,26,29,32), 162(25,32), *163*, 164(42,52), 166(42), *167*, 168(42), *169, 184, 190*
Jannakoudakis, D., 253(206,222,223), 263-(197,301), 265(22), *275, 276, 279*
Janzen, E. G., 240(224), *276*
Jensch, W., 252(242), 253(172,424), *274, 283*
Jesch, C., 340(147), *365*
Johnson, D. S., 333(138), *365*
Johnson, R. M., 250(106), *271*
Jordan, J., 240(225,415), *276, 283*
Joyner, R. A., 166(60), *191*
Jura, W. H., 215(403), 217(403), *282*

Kabanov, B. N., 288(3), 291(3), 297(3), 305(50), 306(50), *360, 361*

Kaganovich, R. I., 196(91), *271*
Kahlweit, M., 160(17), *189*
Kalinowski, M. K., 223(60), 232(254), 243-(162,226), 235(259), *273, 276, 277*
Kalvoda, R., 186(155), *194*, 229(227), 243-(227), *276*
Kamada, M., 35(91), 36(91), 63(90,104, 105), *71, 72*
Kampars, V., 240(432), *240*
Kane, P. O., 141(83), *156*
Kaplan, M., 161(30), *190*
Kapoor, R. C., 256(11,228,229), *268, 276*
Kapulla, H., 66(7), *69*
Kapustinskii, A. F., 160(2), 161(20), 162-(20), *189*
Kardos, A. M., 223(230), *276*
Kargin, Yu. M., 128(67), 130(67), *156, 237(231), 238(231), 240(218), 250(294), 275, 276, 278*
Karguin, Yu. M., 325(118), *364*
Karlikov, D. M., 158(4), 164(4), *189*
Kashin, A. N., 258(61), *270*
Kastening, B., 113(34), 130(34), *155*, 197-(247), 207(232,244,239,246,247b), 212-(247a), 214(194), 229(240), 230(232, 233), 232(244), 233(232,240,247), 234-(240), 235(237,240,242), 238(240), 260-(195,196,199,202,232,234,236-239), 261(202,232,240,242-244), 262(203,237, 242), 263(240,242,243), 264(235,237, 242,243,247), 265(202,237), 266(245), 267(236,245,242), *275, 276, 277*, 325-(114), *364*
Kato, K., 32, 33, 35, 36(92,94), 63(32,93, 95,96), 65(93), *69, 71, 72*
Katz, T. J., 221(248,249,435), *277, 284*
Kavetskii, N., 296(24), *360*
Kaye, R. C., 197(250,251), 225(250,251), *277*
Kemula, W., 121(60), *155*, 183(135,141), 185(147), 186(135,141,158-160), 187-(159), *193, 194*, 226(260), 233(254), 235(259), 236(20), 253(20), 259(252, 258), 262(252,253), 265(255,256), 267-(255,257), *268, 276, 277*
Kendall, P., 169(93), *192*
Kern, D. M. H., 325(110), *364*
Kern, W., 198(261), 243(261), *277*
Kertov, V., 297(26), *360*
Khaikin, B. I., 312(81), 313(81), *362*

AUTHOR INDEX 377

Khakimov, M. G., 257(447), *284*
Kheifets, L. Ya., 227(42), 228(263), 236-(262), *269, 277, 348, 367*
Khopin, A. M., 222(264), *277*
Khurgin, U. I., 314(84), 315(84), *363*
Kimura, M., 63(97), 65(97), *72*
Kinoshita, H., 225(336), *280*
Kirchmayr, H., 159(42), *163,* 164(42), 166-(42), *167,* 168(42), *169, 190, 191*
Kirrbauer, H., 175(115), *193*
Kirrman, A., 342(163), *366*
Kišová, L., 130(68), *156,* 238(124), *272*
Kitaev, Yu. P., 302(37), 343(169), *361, 366*
Kitagawa, T., 264(266), 266(265,267), *277*
Kleinerman, M., 20(58), *70,* 301(34), 310-(34), 319(34), 359(340), *361, 369*
Kleppa, O., 161(30), *190*
Klimov, V. A., 341(160), 355(160), 365-(230), *366, 368*
Kliukina, L. D., 312(80), 320(80), 321(80), 342(80), 345(80), *362*
Knobloch, E., 197(47), 205(47), *269*
Kobayashi, K., 347(201), *367*
Kodama, M., 63(33,34), *69, 70*
Kollmann, K., 165(68), 166(68), *167, 191*
Kolokolov, N. B., 227(42), *269*
Kolthoff, I. M., 197(269), 198(268), 243-(270), 255(268), *277,* 346(187), *367*
Königstein, J., 132(75), *156*
Konràd, D., 36(35), *70*
Koopmann, R., 261(271), 263(271), *277*
Kopyl, L. D., *192*
Korshumov, I. A., 198(272), 242(272), *278*
Koryta, J., *2, 4,* 11(43), 20(43), *53, 54, 60, 61,* 64(36,38,40), 65(8), *69, 70,* 209-(273), 212(273), *278,* 325(109,110), 347-(198,199), *354, 360, 364, 367, 368*
Koseki, K., *38, 39,* 41(98), *72*
Koseki, W. S., 218(412), 224(413), 240-(413), *282*
Kosower, E. M., 248(401), *282*
Koutecky, J., 11(9,42,43), *20,* 39(44), *60, 69, 70,* 101(12), 137(80), 143(80), *154, 156,* 208(273,274), 210(274), 212(273), *278,* 306(53), 310(72), 312(82), 314(82), *362, 363*
Kovaleva, L. M., 177(118,123,126), 181-(118,123,126), 186(126), *193*
Kozin, L. F., 164(50), 166(57), 169(57), *190, 191*

Kozlovskii, M. T., 163(166), 166(64,65,67), 167(67), 169(67), 173(64,65,67), 180-(1270,130), 181(127,130-134), 184(146), 185(149), 186(146,162-164), *191, 193, 194*
Kramarczyk, K., 230(32,34), 240(32), 258-(37), *269*
Krasnova, I. E., 183(140), 184(146), 185-(148), 186(140,146), *193, 194*
Kraus, C. A., 160(21), *190*
Krishtalik, L. I., *306,* 309(65), *361, 362*
Kriukova, E. V., 355(230), *368*
Kruglyak, Yu. A., 225(54), *269*
Kruikova, T. A., 322, 346(190), *363, 367*
Krumholz, P., 63(11), *69*
Krupička, J., 213(209), *275*
Kubaschewski, O., 160(11,15), 161(11), 162(15), 164(11), *189*
Kublik, Z., 183(141), 186(141,160), *193, 194,* 259(252), 262(252,258), *277*
Kubota, T., 250(275), *278,* 330(127), 331-(127), *364*
Kucharczyk, N., 106(22), *154*
Kuchinskii, E., 305(50), 306(50), *361*
Kudra, O. K., 168(74), *191*
Kumagi, T., 54(88), *55, 56,* 64(88), *71*
Kumar, A. N., 256(276), *278*
Kunz, D., 57(277), *278*
Kurnakov, N. S., 160(14), *189*
Kurtz, H. F., 160(21), *190*
Kůta, J., 114(41), 135(78), 141(81), *156,* 171(97), 185(152), *192, 194,* 204(181), 252(181), *274,* 325(111), 342(111), 347-(197), 353(217), *354, 364, 367, 368*
Kuwana, Th., 197(278), *278*
Kuwata, K., 213(279), *278*

Lafitte, M., 161(28), 164(28), *190*
Laitinen, H. A., 36(45), 41(45), *70,* 197(280), 218(280), *278*
Lambert, F. L., 347(201), 356(232), *367, 368*
Lamy, F., 346(195), 348(195), 349(195), *367*
Laur, G., 236(71), *270*
Laviron, E., 109(31), 110(31), 116(31,54), 117(31,54), *154, 155,* 232(281,282), 233-(281), 239(284), 263(283), *278, 333, 365*
Lavrushin, V. F., 243(285), *278,*

343(168), *366*
Layloff, T. P., 266(265), *277*
Leach, S. J., 197(286), 219(286), 258(287), *278*
Lee, H. J., 199(288), *278*
Legrand, M., 130(74), *156*
Le Guillanton, G., 197(289), *278*
Le Guyader, M., 236(290), *278*
Lehmann, O., 231(198), *275*
Leone, J. T., 230(112), *271*
Leontovich, E. V., 169(74), *191*
Lerkh, R., 295(20), *360*
Levenson, L. L., 215(442), *284*
Levich, V. G., 202(291), *278,* 312(81), 313-(81), *362*
Levin, E. S., 225(293), 228(292), 265(292), *278, 301, 346, 361*
Levin, Ya. A., 240(218), 250(294), *275, 278*
Levitskaya, S. A., 167(63), 169(63), 177-(119,122,125), 181(119,122,125), *191, 193*
Levy, M., 213(6), 214(6), 250(6), *268*
Li, A. M., 226(43), *269*
Liebhafsky, H. A., 168(81), *191*
Liebl, G., 166(56), 167, 169(56), *191*
Lihl, F., 161(31), *175, 190, 193*
Linek, K., 132(75), *156*
Lingane, J. J., *1, 70,* 170(96), *192*
Lingane, P. J., 344(172), *366*
Linnell, R. H., 197(478), 223(478), *285*
Linschitz, H., 223(119), *272*
Lipkin, D., 219(348), *280*
Lishcheta, L. I., 310(70), 321(94), *362, 363*
Lishczeta, S. C., 101(13), 118(13), 138(13), 139(13), 143(13), *154*
Lisitsyna, N. K., 324(105), 349(213), 350-(213), 351(213), 352(213,216), *363, 368*
Littlehailes, J. D., 253(295), *278*
Long, F. A., 340(149), *365*
Longster, G. F., 273(163), *263*
Lorenz, W., 196(296), *278*
Los, J., 116(49), *155*
Losev, V. V., 168(74), *191,* 341(155), *353, 365, 368*
Lucken, E. A. C., 225(145), 251(144), 266-(297), *273, 278*
Ludwig, P., 197(369,370), 263(369), 264-(370), *281*

Luey, J. C., 239(284), *278*
Lund, H., 107(29), 108(29,30), 121(64), 130(73), *154, 156,* 199(300), 215(298), 237(299), *278, 279*
Luoys, A. M., 175(114), *193*
Luz, Z., 227(58), *270*
Lyons, L. E., 86(6), 139(64), *154, 156,* 198(174), *274*

McCoy, H. N., 161(24), *190*
McCoy, L. R., *297, 299,* 300(29), *318,* 342-(89), *361, 363*
McDevit, W. F., 340(149), *365*
McIntyre, T. W., 199(470), *285*
McKee, R. H., 341(157), *365*
McKinney, P. S., 254(322), 266(323,324), *279, 280*
Macris, C. G., 263(301), *279*
Maffei, H. P., 198(82), 255(82), *270*
Mahapatra, S., 240(208), *275*
Mairanovskaya, E. F., 317(88), 318(88), 319(88), 320(88), *363*
Mairanovskii, S. G., 208(308), 210(307), 213(311,312), 230(304,307), 231(304, 350), 232(305,308), 240(312), 244(310, 312), 248(306), 252(302,303,309), *279,* 288(9,12), 301(9,36), 310(69,70), 311-(77-79), 312(9,80-83), 313(81,83), 314-(82-84), 315(84-86,9,12), 316(9,12), 317-(78,88), 318(88), 319(9,12,86,88,90), 320(80,88), 321(80,93,94), 322(95), 324-(105), 325(12,117), 326(12,86,120,122-124), 328(12,86,125), 332(125,129), 333-(12,86,125,129), 339(146), 340(146), 342(12,80,162), 343(12,166,167), 344-(12), 345(12,80,123), 346(12,189), 349-(95,213), 350(12,213), 351(12,213-215), 352(213,216), 354(225), 355(12,255), 356(255), 359(243,246), *360, 361, 362, 363, 365, 367, 368, 369*
Majranoskij, S. G., 77(5), 101(13), 115(42), 118(5,13), 138(13,139), 139(5,13), 143-(13), *154, 155*
Makarov, S. P., 250(445), *284*
Maki, A. H., 197(313-315,146), 199(326), 263(314), 264(146,315), *273, 279, 280*
Mamedzade, R. Yu., 227(316), *279*
Mann, C. K., 199(342), *280*
Manoušek, O., 107(28), 108(28), 126(67), 130(67,69), 138(28), 139(28), *154, 156,*

AUTHOR INDEX 379

237(231), 238(231), *276*
Mark, H. B., Jr., *297, 299,* 300(29), *318,* 324(108), 332(139), 343(89), *361, 363, 364, 365*
Markby, R. E., 333(140), *365*
Markova, N. P., 215(408), *283*
Markowitz, J. M., 218(113), *272*
Marple, L. W., 247(317), *279*
Marquarding, D., 243(201), *275*
Marsen, H., 121(61), *155,* 198(191,192), 230(191), 232(191,192), 233(192), 236-(192), *274, 275*
Marshall, J. H., 266(156), *273*
Martin, A. J., 198(111), 214(384), 246-(111), 247(114), 248(111), *271, 282*
Martinet, P., 242(318), *279*
Massetti, F., 223(69), *270*
Masui, M., 264(395), *282*
Matschiner, J. T., 198(82), 254(319), 255-(82), *270, 279*
Matsuda, H., 8(49,54), *13,* 16(48,49), *18, 20,* 22(47), *27, 43,* 50(55-57), *51, 61, 62,* 64(59), *66, 70,* 310(74), *362*
Maturová, M., 242(32), *279*
Maxwell, J., *184, 194*
Mayer, R., 257(277), *278*
Mayurama, M., 236(339), *280*
Mayweg, V. P., 258(343), *280*
Mazzucato, U., 223(69), *270*
Meites, L., 177(12), 181(120), *184, 193,* 198(108), 214(143,419), 230(108), 232-(108), 255(15), *268, 271, 272, 283,* 347-(196), 359(247), *367, 369*
Melby, L. R., 228(325), 249(325), *280*
Melchior, M. T., 199(326), *280*
Melnik, L. A., 232(41), *269,* 345(182), *366*
Mennier, N., 175(14), *192*
Menz, W., 158(3), *189*
Mesyats, N. A., 183(143), *194*
Meunier, J. M., 243(327), *280*
Michael, F., 130(72), *156*
Michielli, R. F., 233(328), 239(328), *280*
Michl, J., 148(87), 149(87), 150(87), *156,* 242(492), *286*
Miller, I. R., 213(6), 214(6), 250(6), *268*
Milner, G. W. C., 9(60), *70*
Mirkin, L. S., 219(482), 223(482), *285*
Mishutushkina, I. P., 196(91), *271*
Missan, S. R., 347(196), *367*
Miyazaki, H., 250(275), *278,* 330(127),

331(127), *364*
Möbius, K., *280*
Modiano, J., 197(33), 229(33), 267(329), *280*
Moe, N. S., 240(331), *280*
Mohilner, D. M., 288(8), *292,* 294(8), *298,* 333(140), *360, 365*
Momoki, K., *29, 70*
Mooney, B., 229(332), *280*
Moorhead, E. D., 353(218), *368*
Morinaga, K., 64(63), *71*
Morokuma, K., 214(132), 240(132), *272*
Morotomi, Y., 236(333), *280*
Morris, M. D., 258(102), *271*
Morris, R. A. N., 253(86), *271*
Mott, N. F., 358(238), *368*
Müller, K., 162(39), *190,* 332(132), *364*
Müller, O. H., 116(44), *155,* 197(334), 198-(108,337), 230(108), 232(108), 244(337), *271, 280*
Murray, R. W., 258(55), *269, 348, 367*
Myatt, J., 253(335), 263(163), *273, 280*

Nachod, F. C., 253(418), *283*
Nakashima, R., 266(267), *277*
Nakaya, J., 225(336), *280*
Nasielski, J., 250(5), *267*
Němec, I., 199(104), *271*
Němečková, A., 242(320), *279*
Neugebauer, L., 198(347), *280*
Neish, W. J. P., 198(337), 244(337), *280*
Nicalas, E., 345(177,179), 348(177,209), 349(177), 352(177), 353(177), *366, 367*
Nicholson, R. S., 56(64-66), *71,* 113(35), 130(35), *155,* 207(338), *280*
Nigam, H. L., 256(11,228,229,278), *268, 276, 278*
Nigmatullina, A. A., 164(43), 166(61), 167-(43), 169(61), 175(111), *190, 191, 192*
Nikelly, J. G., 183(137), *193*
Nikolaeva-Fedorovich, N. V., 290(17), 302-(38), 303(42), 307(56), 322(38), *323,* 324(103), 335(142), *336,* 344(17,175), 345(38,103,175), 346(175,193), *349,* 355(193,226,229), 356(193), 358(239), 35(241,242), *360, 361, 362, 363, 365, 366, 367, 368, 369*
Nishiyama, M., 236(339), *280*
Nisli, G., 106(23), 116(23), 118(23), 119-(23), 139(23), 140(23), *154*

Noack, J., 250(396), *282*
Nomura, T., 64(63), *71*, 224(137), *273*
Nonant, A., 236(71), *270*
Nordio, P. L., 239(83), *270*
Norman, O. C., 226(152), *273*
Novik, E. Yu., 141(82), *156*, 304(47), *361*
Novotny, H., 160(19), 161(35), *189, 190*
Nürnberg, H. W., 252(340), *280*, 288(10), 359(245), *360, 369*
Nygård, B., 256(341), *280*

Ockwell, J. N., 226(12), *268*
O'Donnell, J. F., 199(342), *280*
Ogawa, H., *70*
Ogino, H., 33(67), 35(67,99,101,106,107), 36(99,106,107), 37(103), 41(103), 63(68, 69,100), 65(70,100,102), *71, 72*
Oguri, T., 234(8), *268*
Oldham, K. B., 247(216), *275*
Olmstead, M. L., 56(66), *71*, 113(35), 130(35), *155*, 207(338), *280*
Olson, D. C., 258(343), *280*
Olson, E. C., 333(137), 334(137), *365*
Ono, S., 116(47, 48), *155*
Ono, Y., 227(477), *285*
Osawa, H., 35(91), 36(19), 63(104,105), *71, 72*
Osswald, E., 161(27), *190*
Osteryoung, R. A., 253(418), *283*, 344-(172), *366*
Ottombrini, G., 359(250), *369*

Padmanabhan, G. R., 226(150), *273*
Palanker, V. Sh., *298, 361*
Palman, H., 159(32), 161(132), 162(32), *190*
Pangarov, N., 297(26), *360*
Papoff, P., 64(71), *71*, 307(59,60), 308(60), 359(250), *362, 369*
Parkanyi, C., 218(344,479), *280, 285*
Parravano, G., 198(345), *280*
Parry, J. M., 300(32), *361*
Parsons, R., 288(6,11), 294(6), 296(6), *297, 298*, 300(32), 304(6), *360, 361*
Partridge, L. K., 341(156), 359(156), *365*
Pasternak, R., 198(346), 230(236), 232-(346), 242(346), *280*
Patzak, R., 198(347), *280*
Paul, D. E., 219(348), *280*
Pavan, M. V., 239(83), *270*

Pavlov, V. N., 208(308), 230(304,349), 231-(304,349,350), 232(308), *279, 280*
Pavolova, L. A., 224(409), *283*
Pearce, J. N., 168(79), *191*
Pearson, J., 198(351), 259(351), *281*
Pecasse, C., 341(153), 353(153), *365*
Peover, M. E., 217(159), 218(159,357,358), 219(358), 220(154), 221(154), 227(13, 153,356), 228(154,355), 240(352), 249-(353), 266(354), *268, 273, 281*, 325(112, 113), 347(201), *364, 367*
Perichon, J., 218(359), 220(360), 239(360), *281*
Perone, S. P., 229(426), 232(361), *281, 283*
Perrin, Ch. L., 197(362), 247(436), *281, 284*
Person, M., 243(327), 258(99), 263(363), *271, 280, 281*
Petot-Ervas, G., 166(38), *190*
Petrii, O. A., 294(19), 303(41,42), 307(55, 56), *308*, 309(55), 323(101,103), 324-(103), 335(142), *336*, 345(103), 359 (242), *360, 361, 362, 363, 365, 369*
Petrov, I. N., 261(125), *272*
Petrov, V. P., 247(364), *281*
Petrosyan, V. A., 325(117), *364*
Petrovich, J. P., 213(365), *281*, 338(145), 339(145), *365*
Philip, R. H., 229(366,367), 237(366), 239-(366), *281*, 348(208), *367*
Pietrzyk, D. J., 197(362), *281*
Piette, L. H., 197(369,370), 262(369), 263-(369), 264(370), *281*
Pifer, C. W., 346(188), *367*
Plaksin, I. P., 161(36), *190*
Plesch, P. H., 244(220), 252(220), *276*
Plieth, W. F., 196(371), *281*
Polievktov, M. K., 311(79), 346(189), *362, 367*
Poluyan, E. S., 173(98), *192*
Ponomarev, Yu. P., 325(116), *364*
Ponamarenko, V. A., 248(306), 256(372), *279, 281*, 301(36), *361*
Porter, A. S., 341(156), 359(156), *365*
Porter, J. T., 165(110), *192*
Pospišel, L., 325(111), 342(111), *364*
Pozdeeva, A. A., 221(373), 230(373), 242-(373,374,485), 243(273,484,485), 244-(373,374,484), 253(485), *281, 285, 286,*

AUTHOR INDEX 381

343(166), *366*
Prelog, V., 198(375), 214(375), *281*
Preobrazhenskaya, E. A., 236(262), *277*
Prevost, C. A., 198(376), *282*
Prikhodchenko, V. G., 168(74), *191*
Psarras, Th., 258(377), *282*
Pugachevich, P. P., 169(95), *192*
Pungor, E., 321(92), *363*
Purdy, W. C., 256(15), *273*
Pushin, N. A., 164(45,58), 166(48,45), *176, 190, 193*

Quast, H., 198(261), 243(261), *277*

Ragle, J. L. 217(147), 263(149), 264(147), *273*
Raison, J., 218(73), *270*
Rampazzo, L., 229(67), *270*
Ramsey, J. S., 247(164), *274*
Randles, J. E. B., 9(72), *71*
Razumov, A. I., 240(394), 251(393, 394), *282*
Regnolds, J., 161(34), *190*
Reichmanis, G. O., 325(115), *364*
Reikhamis, G. O., 263(430,431), *283*
Reilly, C. N., 324(108), *364*
Reinert, K. E., *63,* 66(72), *71*
Reinmuth, W. H., 56(113,114), 57(113, 114), *72,* 113(381), *155,* 197(381), 218-(49), *269, 282,* 301(35), 305(49), 359-(248), *361, 369*
Ressel, M., 304(45), *361*
Reutov, O. A., 257(62), 258(24,61,62), *268, 270*
Richards, T. W., 164(49), *167,* 169(76-78, 83,85), *190, 191, 192*
Rieger, P. H., 197(381), 217(380), 221(4), 224(379), 249(378,381), *267, 282*
Riesenbeck, G., v., 252(340), *280*
Rifi, M. R., 246(382), 248(382), *282*
Rigatti, G., 112(32), *155,* 239(83), *270,* 302(40), 303(40), *361*
Rinker, R. G., 251(122), *272*
Robin, R., 224(444), 238(444), *284*
Rodgers, R. S., 359(247), *369*
Rogers, L. B., 247(317), *279,* 301(35), 305 (49), *361*
Roiter, V. A., 173(99), *192*
Rollman, L. D., 258(383), *282*
Rolls, M. C., 160(18), *189*

Roos, G. D., 164(54), 166(54), *190*
Rosenthal, I., 198(111), 214(384), 246 (111), 247(114), 248(111), *271, 272, 282*
Rosenthal, S., 254(322), *279*
Rosenthal, T., 99(9), 100(9), *154*
Rüetschi, P., 198(385), 252(385), *282*
Runner, M. E., 197(460), 219(460), 227-(461), 228(461), *285*
Russel, A. S., *175, 192, 193*
Russell, C. D., *297,* 298(27), 299(27), *360*
Rybalka, K. V., 359(241), *369*
Rybalka, L. E., 341(152), *365*
Ryabtsev, A. N., 257(62), 258(62), *270*
Ryvolová, A., 224(386), 238(386), *282*
Ryvolvá-Kejharova, A., 104(21), 120(21), 122(21), 125(65,66), 126(66), 141(84), 143(84), 148(21,88), 149(88), 150(88), *154, 156,* 345(180), 346(180), *366*

Sadler, J. L., 215(388), 216(387), 250-(387), *282*
Sagadieva, K. Zh., 166(64,65), 173(64,65), *181, 191, 193*
Saikina, M. K., 257(447), *284*
Saito, Y., 35(106,107), 36(106,107), *72*
Sakuma, M., 252(169), *274*
Sakuma, Y., 65(108), 66(108), *72*
Salaün, M., 236(72), *270*
Salikhdzhanova, R. M. -F., 304(46), *361*
Salvaterra, M., 251(78), *270*
Samokhvalov, G. I., 213(311,312), 240 (312), 244(310,312), *279*
Sandera, J., 30(77), *71*
Sannikova, V. I., 250(274), *278*
Santavý, F., 100(10), *154,* 222(491), 242 (320), *279, 286*
Satô, G., 35(91), 36(91), 37(103), 39(31), *103, 69, 71, 72*
Sato, H., 29(62), *70*
Satori, G., 197(389), 229(389), 246(390), *282*
Sauerwald, F., 158(2,3), 160(16), 161(16, 27), 164(16), *189, 190*
Saveant, J. M., 99(7), *118,* 130(7,71), 133-(7), 134(7), *154, 156,* 209(392), 237-(391), 255(391), *282*
Savicheva, G. A., 240(394), 251(393,394), *282*
Saylor, J. H., 225(129), *272*

Sayo, H., 264(395), *282*
Scheibe, G., 230(59), *270*
Scheithauer, S., 257(277), *278*
Schindler, R., 358(234), *368*
Schmidt, C. L. A., 197(446), 218(446), 223(446), *284*
Schmidt, H., 237(193), 250(396), 263(193), *275*, *282*
Schmidt, P. P., Jr., 333(139), *365*
Schneider, A., 160(8), *189*
Schneider, F., 197(397), *282*
Schöber, G., 197(167), *274*
Schoen, J., 347(202), *367*
Schölzel, K., 167(71), 177(121), 181(121), *191*, *193*
Schott, A., 216(212), *275*
Schrauzer, G. N., 258(343), *280*
Schuler, D., *192*
Schulz, L. G., 169(94), *192*
Schulz, M., 240(398), *282*
Schwabe, K., 226(400), 232(399), 253-(475), *282*, *285*, 332(130, 334(141), 335-(141), *336*, *346*, *364*, *365*, *367*
Schwedak, E. J., 226(456), *284*
Schwan, T. C., 198(103), 243(103), *271*
Schwarz, K. H., 240(398), *282*
Schwarz, W. M., 248(401), *282*
Schwarzenbach, G., *30*, *63*, *68*, *71*
Schweiss, H., 197(38), 235(38), 236(26), 243(402), *268*, *269*, *282*
Scott, J. M. V., 215(403), 217(403), *282*
Sease, J. W., 246(404), *282*, 356(231), *368*
Sekine, A., 236(333), *280*
Selected Values of Chemical Thermodynamics Properties, 160(10), 161(10), *189*
Senda, M., 56(83), *71*, 113(37), *155*
Ser, S., 121(62), 139(62), *156*
Serova, G. F., 324(106,107), *363*
Šestáková, I., 107(26), *154*
Seth, T. D., 256(276), *270*
Setschenow, J., *340*, *365*
Sevast'yanova, I. G., 213(405), 249(406), *283*
Shain, I., *56*, *68*, *71*, 113(39), *155*, 248-(401), *282*
Shams-El-Din, A. M., 253(207), *275*, *354*, *368*
Shapoval, G. S., 214(407), *283*
Shapoval, V. I., 214(407), 215(408), *283*
Sharp, J. H., 215(14), *268*

Shimanskaya, N. P., 224(409), *283*
Shishakov, I. A., 161(33), *190*
Shriver, D. F., 226(410), *283*
Siew, L. C., 221(249), *277*
Silver, B. L., 227(58), *270*
Simonet, J., 242(318), 245(411), 243(411), *279*, *283*
Simson, C. V., 161(37), *190*
Sinyakova, S. I., 183(138), *193*
Sioda, R. E., 218(412), 224(413), 240(413), 265(225,256), 267(255,257), *277*, *283*
Skobets, E. M., 215(408), *283*, 297(23), *360*
Skrebkova, I. M., 343(169), *366*
Skundin, A. M., *298*, *361*
Slovetskii, V. I., 325(117), *364*
Slendyk, J., *344*, *366*
Smallmann, R., 169(92), *192*
Smirnov, Yu. D., 246(121), 250(445), *272*, *284*
Smith, D. E., *20*, 23(78), *24*, *25*, *61*, *71*
Smith, D. L., 219(414), 223(414), 226-(410), *283*
Smith, P., 226(410), *283*
Smith, P. T., 240(225,415), *276*, *283*
Smoler, I., 185(152), *194*
Smyth, C. P., 164(49), *190*
Snegova, A. D., 301(36), *361*
Sobina, N. A., 227(42), *269*
Soderberg, B. A., 298(30), 299(30), *361*
Solon, E., 217(416), *283*, 307(58), *362*
Solov'eva, Z. A., 173(101), *192*
Šorm, F., 259(179), 274
Sorokin, O. I., 115(42), *155*
Souchay, P., 121(62), 139(62), *156*, 198-(376), 248(417), 258(99,100), 263(417), *271*, *282*, *283*
Southworth, B. C., 253(418), *283*
Speakman, J. C., 198(219), *275*
Speranskaya, E. F., 181(134), *193*
Spritzer, M., 214(419), 219(420), 222(126, 420), *272*, *283*
Stackleberg, M., v., *1*, 39(44), *70*, *71*, *173*, *192*, 198(421), 208(422), 246(421), 252-(340), 252(423,424), 253(424), *280*, *283*, 288(10), *360*
Stalidis, G., 265(221), *276*
Stanienda, A., 258(425), *283*
Stapefeldt, H. E., 229(429), *283*
Šťastný, M., 39(82), *71*

AUTHOR INDEX 383

Štěpánek, J., 130(68), *156,* 238(124), *272*
Stepanova, O. S., 183(139), 186(156,161), *193, 194*
Stern, O., *288, 360*
Sternberg, H. W., 33(140), *365*
Stiehl, G. L., *60, 69*
Stočesová, D., 112(33), *155*
Stojanovich, F. M., 339(146), 340(146), 343(167), *365, 366*
Stonehill, H. I., 197(250,251), 225(250, 251), 229(332), *277, 280*
Stracke, W., 198(421), 246(421), *283*
Stradins, 232(428,429), 237(428,433), 238-(428,433), 240(432), 262(141), 263(430, 431), *273, 283, 284*
Stradyns, J. P., 325(115), *364*
Stráfelda, F., 39(82), *71*
Strassner, J. E., 260(434), *284*
Strauss, H. L., 221(435), *284*
Strehlow, H., 116(50), *155*
Streitwieser, A., 247(436), *284*
Strecks, W., 198(268), 255(268), *277*
Stromberg, A. G., *173,* 183(136,142), 186-(142), *192, 194*
Stromberg, E. A., 183(136), *193*
Sturm, F., 304(45), *361*
Stutter, E., 197(38), 235(38), *269*
Sunahara, H., 255(437), *284*
Sundaram, A., 306(52), *361*
Sunden, N., 166(55), 169(55), *191*
Survila, A. A., 341(152), *365*
Suschke, H. -D., 346(186), *367*
Sutcliffe, B. T., 226(152), *273*
Suvorovskaya, K. A., 161(36), *190*
Suzuki, M., 230(438), 231(438), 232(438), *284*
Svetkin, Yu. V., 344(170), *366*

Tachi, I., 56(83), *71,* 113(37), *155*
Tachoire, H., 161(28), 164(28), *190*
Takagi, M., 116(47,48), *155*
Takahari, T., 54(88), *55, 56,* 64(88), *71*
Takahashi, R., 213(439), 222(439), *284*
Takaoka, K., 224(135-137), 248(136), *272, 273*
Tammann, G., *165,* 166(58,68), *167, 175, 191, 192*
Tamamushi, R., *13, 17, 18, 20, 60, 61, 62,* 63(96), 64(59), *68, 70, 71, 72*
Tanaka, N., *10, 13, 17, 18, 32, 33, 35, 36*
(91,92,94,99,106,107), 37(103), *38, 39, 40,* 41(98,103,110), 54(88), *55, 56,* 57-(111), *58, 59, 60, 62, 63,* 74(23,24,29, 88), 65(23,24,29,70,93,97,100,102,108, 111), 66(108), *69, 71, 72,* 198(268), 255-(268), *277*
Tang, S., 99(8), 130(8), *154,* 198(110), 244-(440), 246(110), *271, 284*
Tansley, A. C., 341(156), 359(156), *365*
Tatwawadi, Sh. V., 227(128), *272*
Tedoradze, G. A., 203(116), 239(116), 247-(115), *272,* 340(150), 341(150,160), *349, 350,* 355(160), *365, 366, 367*
Teh-Laing, Chang, 304(43), *361*
Teitlebaum, C., 198(109), 213(109), *371*
Tendick, S. K., 240(89), *271*
Tenygl, J., 116(53), *155*
Terauds, V., 232(429), *283*
Testa, A. C., 56(113,114), 57(113,114), 72, 113(38), *153*
Thibaud, Y., 218(73), *270*
Thiec, J., 219(441), *284*
Thomas, C. W., 215(442), *284*
Thompson, H. E., 166(62), *191*
Thomson, D., 218(462), *285*
Tice, P. R., 265(64), *270*
Timofeicheva, O. A., 169(95), *192*
Tirouflet, J., 107(25), 109(31), 110(31), 116(31), 117(31), *154,* 224(444), 238-(444), 266(243), *284,* 359(235), *368*
Titievskaja, A. S., 290(18), *360*
Todd, P. F., 253(335), 263(163), *273, 280*
Todres, Z. V., 225(293,487), *278, 286,* 347-(204), 359(236), *367, 368*
Toibaev, B. K., 177(119), 181(119), *193*
Tomilov, A. P., 213(405), 246(121), 249-(406), 250(445), *272, 283, 284,* 341-(160), 355(160,230), *366, 368*
Tompkins, P. C., 197(446), 218(446), 223-(446), *284*
Tondeur, J. J., 344(171), 345(171), 346-(195), 348(177,209), 349(177), *366, 367*
Toome, V., 173(107), *192*
Toropova, V. P., 257(447), *284*
Torsi, G., 307(59,60), 308(60), *362*
Townshend, A., 257(171), *274*
Traini, A., 115(43), *155*
Trazza, A., 229(67), 244(68), *270*
Troepolskaia, T. V., 343(169), *366*
Trümpler, G., *192,* 198(385), 252(385), *282*

AUTHOR INDEX

Tsagareli, G. A., 203(116), 239(116), *272*
Tsukitani, Y., 264(395), *282*
Tsveniashvili, V. Sh., 225(487), *286,* 359-(236), *368*
Turcsanyi, B., 120(56), 121(56), *155*
Turian, Ia. I., 324(106,107), *363*
Tutane, I. K., 237(433), 238(433), *284*
Tza Chuan-sin, *349,* 350(213), *367*

Uehara, M., 215(448), *284*
Umemoto, K., 217(133), 228(449), 266-(133,134), *272, 284*

Vainshtein, Yu. I., 236(92), *271*
Vajda, M., 219(450), *284*
Vakulova, L. A., 213(311), 240(312), *279*
Valashek, I. E., 213(312), 244(312), *279*
Valenta, P., 223(230), 230(451), 232(452), 234(451), *276, 284*
Val'ko, A. V., 177(119), 181(119), *193*
Vanags, G. J., 237(433), 238(433), 240-(432), *284*
Vandenberghen, L., 345(177,179), 348-(177), 349(177), 352(177), 353(177), *366*
Van Heteren, W. J., 164(46), *190*
Vanlautem, N., 250(5), *267*
Vasil'eva, L. N., 185(150), *194*
Vasilieva, E. G., 346(190), *367*
Vavřička, S., 253(205), 261(243), 263-(243), 264(243), 266(245), 267(245), *275, 276,* 345(184), *360, 367*
Vazsonyi-Zilahy, A., 167(70), 168(70), *191*
Veillard-Royer, H., 237(391), 255(391), *282*
Venkatesan, V. K., 359(239), *368*
Veselovskii, B. K., 160(20), 161(20), 162-(20), *189*
Veselý, K., 113(40), *155*
Vetter, K. J., *6, 72,* 196(371), *281*
Vianello, E., 209(282), 223(69), 261(84), 262(84), *270, 271, 282*
Vierk, A. L., 160(13), *189, 192*
Vig, S. K., 130(69), *156,* 229(16), *268*
Vinogradova, E. N., 185(150,151), *194*
Vincenz-Chodkowska, A., 229(453), 232-(453), *284*
Vlček, A. A., *2,* 36(35), 41(116), *70, 72,* 342(164), 343(164), 353(217), *366, 368*
Vodzinskii, Yu. V., 198(272), 242(272), *278*

Vofsi, D., 213(6), 214(6), 250(6), *268*
Volke, J., *80,* 107(27), 110(27), 116(51,52, 55), 117(55), 118(55), *154, 155,* 214-(454), 223(230), *276, 284*
Vol'kenshtein, Yu. B., 210(307), 230(307), *279,* 342(162), *366*
Volková, V., 116(51), *155*
Voorhies, J. D., 226(456), *284*
Voronin, G. F., 158(5), *189*
Vorziatti, A. F., 164(53), *190*
Vries, W. T., d., 359(249), *369*

Wagenknecht, J. H., 254(457,468), 257-(466), *285*
Wåhlin, E., 297(23), *360*
Ward, L., 348(205), *367*
Wasa, T., 116(47,48), *155*
Watts-Tobin, R. J., 359(238), *368*
Wawzonek, S., 197(458,460,469,280), 198-(108), 199(470), 218(280,458,462,463), 219(460), 227(461), 228(461), 230(108), 232(108), 238(464), 243(467), 244(467), 248(465), 254(468), 255(459), 257(466), *271, 278, 285*
Wearing, D., 218(463), *285*
Weber, J., 306(53), 247(199), *362, 367*
Weber, P., 208(422), *283*
Weibe, F., 160(15), 162(15), *189*
Weissman, S. I., 219(348), *280*
Weller, K., 197(38), 229(30), 235(30,38), *268, 269*
Wender, I., 333(140), *365*
Westwood, J. V., *2, 69*
Wettig, K., 231(471), 232(471,472), 240-(473), *285*
Wheatley, M., 116(49), *155*
White, B. S., 218(357), *281*
White, J. C., 164(47), 166(47), *190*
Whitman, C. J., 160(18), *189*
Whitman, W. E., 231(474), *285*
Wiemann, J., 219(441), *284*
Wiesener, W., 334(141), 335(141), *335, 336*
Wiesner, K., 11(10,119), 20(10,119), *69, 72,* 116(45,49), *155, 309,* 310(71), *362*
Wiesner, W., 253(475), *285*
Wildenau, A., 253(206,222,223), *275*
Wiles, L. A., 231(474), *285*
Wilkinson, G., 198(476), 258(476), *285*
Williams, R. D., 260(199), *275*

Williams, W. G., 266(57), *270*
Willmann, A., 346(187), *367*
Wilson, C. L., 198(103), 243(103), *271*
Wilson, J. H., 168(77), 169(77), *191*
Wilson, J. M., 113(35), 130(35), *155,* 207-(338), *280*
Wilson, J. W., 56(66), *71*
Winstein, S., 219(118), 225(118), *272*
Wolf, W., 190(6), *189*
Wolff, G., 359(245), *369*
Wollish, E. G., 346(188), *367*
Woodhall, B. J., 253(295), *278*
Woods, R., 243(270), *277*
Wynne-Jones, W. F. K., 261(125), *272*

Yamada, A., *10, 40,* 41(110), 57(111), *58, 59,* 65(111), *72*
Yasukouchi, K., 227(477), *285*
Yoshida, M., 220(249), *277*
Yoshimura, C., 168(80), *191*
Yoskikawa, K., 224(137), *273*
Yuza, V. A., 73(98), *192*

Zábranský, Z., *61,* 64(40), *70*
Zahlan, A. B., 197(478), 223(478), *285*
Zahradnik, R., 218(344,479), 223(482), *280, 285*
Zaichko, L. F., 183(143), 184(145), 186-(143), *194*
Zaitsko, L. F., 186(157), *194*
Zakharov, M. S., 183(139,143), 184(145), 186(143,156,157), *193, 194*
Zebreva, A. I., 163(166), 164(43), 166(61, 67), 167(63,67), 169(61,63,67), 173(67, 108), 175(111), 177(117-119,122-126), 180(130), 181(108,118,119,122-126), 183(140), 184(146), 185(148), 186(126, 140), 187(146,162), *190, 191, 192, 193, 194*
Zelyanskaya, A. I., 173(101,104), *192*
Žežula, I., 301(61), 345(178), *362, 366*
Zhdanov, S. I., 102(11), *154,* 214(483), 219-(482), 221(281), 222(210,264,480,481, 486), 225(487), 230(373), 242(373,374, 485,486), 243(373,484,485), 244(373, 374,484,486), 247(120), 253(485), *275, 277, 285, 286,* 354(221), 355(228), 356-(236), *362, 367, 368*
Zhukovskii, G. Ya., 160(7), *189*
Zinner, H., 135(77), *156*
Zintl, E., 160(8,9), *189*
Zollinger, Hch., 227(51), *269*
Zolotovitskii, 231(350), *280*
Zuman, P., 75(1), 99(8), 100(9), 101(15), 102(16), 103(18), 104(19-21), 106(22, 23), 107(26,28), 113(18,36), 114(18), 115(43), 116(23,15,53), 118(23), 119-(23), 120(20,21,56), 121(16,17,56), 122-(19-21), 125(66), 128(67), 130(8,36,67, 69,70), 133(18), 135(76,77), 138(28), 139(23,24,28), 140(23), 148(21,76,87-89), 149(87,88), 150(87,88), *153, 154, 155, 156,* 197(493,495), 199(496), 214-(488), 215(494), 221(489), 222(491), 237(231,274,497), 238(231,397), 242-(18, 492), 244(440,490), 250(494), 257-(171), 266, 495, *268, 274, 276, 284, 286, 345,* 346(180,185), *357,* 358(233), 359-(233), *366, 367, 368*
Zuchiewecz-Zajdel, Z., 228(498), *286*
Zýka, J., 199(104), *271*
Zykov, V. I., 308(63), *362*

SUBJECT INDEX

Acetaldehyde, 103, 342, 345, 346
Acetate, buffers, 317
  complexes, 18, 27, 30, 31, 33, 35, 36, 43,
    48-50, 54, 61
Acetone, as scavenger, 236
Acetone hydrazone, 343
Acetone imine, 343
Acetonitrile, 199, 205, 216, 218, 225-228,
    240, 248, 250, 253, 257, 264-267,
    325, 348
  oxidation of, 250
Acetophenone, 129, 234, 238, 239, 244
  formation of, 99
Acetatopentamminocobalt, 353
Acetylacetonate, iron, 348
2-Acetyl-5-bromothiophene, 333
2-Acetylfuran, 236
2-Acetylthiophene, 236
Acid-base equilibria, effect of, 73-153
Acidity, effects of, 73-153, 346
Acidity functions, 82
Acids, 240
  anthraquinone-2,6-disulfonic, 229
  anthraquinone-1,5-disulfonic, 229
  benzoic, 219, 341, 358
  bromoacetic, 247, 332
  2-bromo-*n*-alkanoic, 100, 332
  2-bromobutyric, 100, 332
  α-bromocaproic, 328, 329, 332
  α-bromocarboxylic, 332
  α-bromoenanthic, 332
  bromomaleic, 247
  α-bromopalmitic, 332
  2-bromopropionic, 100
  butadiene tetracarboxylic, 247

  C-, 106, 118, 119
  cinnamic, 338
  citraconic, 138
  crotonic, 338
  derivatives of phosphoric, 250
  dibasic, 136-141
  formylbenzoic, 121
  fumaric, 138, 213, 247
  glyoxalic, 135, 136
  halogeno carboxylic, 303
  iodobenzoic, 103, 121, 356
  α-keto, 116
  maleic, 138, 143, 213, 247, 311, 312
  4-methylimidazol-2-yl-thiosulfuric, 251
  monothiobenzoic, 257
  naphthalenesulfonic, 251
  nitrobenzoic, 121, 263, 359
  nitrocarboxylic, 263
  nitronic, 248, 264
  nitrophthalic, 358
  nitrotoluic, 359
  oxalic, 141, 236
  α-oxy-β,β,β-trichloroethylphosphorous,
    325
  phenylacetic, 78
  phthalic, 141, 142, 143
    as proton donor, 265
  protonation by, 102, 103
  pyridinecarboxylic, 116
  salicyclic, 341
  sulfoxylic, 251
  terephthaladehydic, 236
  terephthalic, 141
  N-troponylamino, 345
  trans-urocanic, 143, 144

SUBJECT INDEX

Aci-nitrocompounds, *see* Acids, nitronic
Ac polarography, *see* Polarography, Ac
Acridicinium ions, 225
Acridine, 225
Acrylamide, 214
Acrylonitrile, 213, 249, 341
Addition, of amines, 114
  of hydroxyls, 113, 114
  of thiols, 114
Adiponitrile, 213
  formation, of, 249
Admittance, faradaic, 20
Adsorption, 76-78, 81, 118, 126, 139, 173, 196, 202, 204, 207, 210, 222, 232, 234, 247, 252, 253, 255-257, 260, 261, 288, 294, 296, 303-305, 308, 315-322, 325-344, 351
  elimination of, 78
  specific, 289-291, 294-296, 319
Alcohols, formation of, 198, 244
Aldehydes, 198, 230-241, 301, 345
  N-alkylpyridinium, 107, 109, 110
  aromatic, 104, 105
  hydration of, 103, 107, 108, 110, 111, 113-115, 132-134, 151, 152
  pyridine, 110, 111
  saturated, 134, 135
  α,β-unsaturated, 148-152
Alicyclics, 78
Alkali metal ions, effect of, 147, 151
Alkali metals, amalgams of, 158, 160, 169, 171
Alkaline earth metals, amalgams of, 158, 160, 161
α-Alkoxy ketones, 130, 237
Alkyl dithiocarbamates, 256
Alkyl halides, 84, 246
N-alkyliodopyridinium salts, 117
Alkyl monothiobenzoate, 257
O-alkyl oximes, 101, 102
N-alkyl-γ-piperidones, 115
N-alkylpyridinium aldehydes, 107, 109, 110
N-alkylpyridinium ions, 223
Alkyls, bulky, 77
  effects of, 325, 326, 332, 334, 335
Allyl acetate, 243
Allyl compounds, of Pd, 258
Aluminum, amalgam of, 159, 162
Aluminum chloride, effect of, 222

Amalgams, 157-189
  alkali metals, 158, 160, 169, 171
  alkaline earth metals, 158, 160, 161
  aluminum, 159, 162
  antimony, 158, 162, 166, 167, 173, 174, 177-179, 181, 182, 184
  barium, 161
  beryllium, 160
  bismuth, 158, 161, 164-166, 168, 169, 171, 172, 354
  cadmium, 159, 161, 163, 164, 167, 171, 172, 177-179, 181, 187-189
  calcium, 160
  cerium, 161
  cesium, 160, 172
  classification of, 158-159
  copper, 159, 161, 164, 165, 167, 169, 171-173, 174, 177, 179, 181, 183, 188
  electrochemistry of, 159-175
  europium, 161
  formation of, 13, 14, 42, 43, 158
  free energy of, formation of, 171
  gold, 159, 161, 177-179, 181, 182
  heat evolution, 158
  i-E curves of, 170-175
  indium, 159, 161, 164, 165, 167, 169, 171, 172, 177-179, 181-183
  lanthanum, 161
  lead, 158, 162-169, 171, 172, 181, 189
  lithium, 160, 172
  magnesium, 160
  manganese, 159, 161, 164, 165, 167, 172-174, 184
  in metal refining, 157, 159
  phosphonium, 253
  plutonium, 162
  potassium, 160, 164, 169, 172
  potentials of, 159-170, 171, 172
  rare earth metals, 158, 161
  rubidium, 172
  silver, 159, 161, 177, 179, 181, 183, 187
  sodium, 160, 164, 170, 172
  standard potential of, 158, 171
  strontium, 160
  sulfonium, 253
  supersaturated, 184, 186
  tetraalkylammonium, 253
  thallium, 159, 161, 163-165, 169-172
  tin, 158, 161, 163-168, 171, 172, 177-

388   SUBJECT INDEX

179, 181-183
uranium, 161
zinc, 158, 161, 164, 165, 167-169, 171, 172, 177-179, 181, 183, 188
Amines, 259
   addition of, 114
   aliphatic, 199
   aromatic, 199
Amino acids, 343
   catalysis by, 79
α-aminoaldehydes, 133, 134
Aminobenzaldehydes, 121
Aminochalcones, 343
α-aminoketones, 130, 227
Aminopolycarboxylates, complexes of, 30, 31, 33, 35, 53-66
Ammonia, buffer, 312-314
   complexes of, 41, 46, 49, 50, 56-58
   liquid, 267
Ammonium compounds, 252-255
n-Amyl alcohol, 295
tert-Amyl alcohol, 341
Analysis, logarithmic, 16, 76, 77, 89, 90, 100, 201, 204, 205, 304, 320, 328, 407
Angle, phase, 22, 23
Anhydrides, 240
Aniline, 341
   orientation of, 196
Anions, adsorption of, 299, 306
   effect of, 80, 81, 307, 308, 344, 348-354
   reduction of, 302, 322-325, 345
Anthracene, 218
Anthraquinones, 227-229, 347, 348
Anthrone, 236, 239
Antimony, amalgam of, 158, 162, 166, 167, 173, 174, 177-179, 181, 182, 184
   wave of, 180, 182
Aromatics, 197, 217-227
   nonbenzenoid, 222
   orientation of, 196
9-Arylaminoanthracene, 218
Association, of metals in Hg, 168
Autoinhibition, 232
Azines, 216
Azobenzene, 215
Azo bonds, 198, 214-217
Azomethine bonds, 198, 214-217
Azomethine-N-oxide, 25

Azopyridine, 216
Azopyridine dioxide, 215, 250
Azoxy compounds, 259
Azulene, 220, 221
Azulenium ions, 220, 221

Barium, amalgam of, 161
   effect of, 304, 345
Benzalaniline, 214, 215
   hydrolysis of, 214
Benzaldehydes, 122, 123, 125, 214, 231-234, 236, 238, 239, 258
p-Benzaldehyde trimethylammonium ions, 236, 253
Benzanthrone, 236
Benzene, as solvent, 258
Benzene rings, effect of, 77
Benzhydrol, 238
Benzil, 229, 348
Benzocyclobutadienoquinone, 227
Benzofurazans, 225
Benzoin, 229
Benzonitrile, 249
Benzophenone, 217, 233, 236, 345
Benzopinacol, formation of, 233
Benzoquinones, 197, 204, 227, 228
Benzoylacetone, 118, 119, 144, 145
N-benzoyllactam, 239
Benzoylpyridine, 232
ω-Benzylacetophenone, 148
Benzylamines, 253
Benzyl bromides, 247
Benzyl chlorides, 247
Benzylmercury iodide, 257
Benzyl monothiobenzoate, 257
Beryllium, amalgam of, 160
Bibenzyl, formation of, 247
Bicyclobutanes, 248
Biphenyl germanium, 258
Biphenyl silicon, 258
Bipyridine, 478
Biradical, 129, 237, 238
Bismuth, amalgam of, 158, 161, 164-166, 168, 169, 171, 172, 354
   $E_{1/2}$ of, 171, 172
   wave of, 180
Bispyridyl disulfides N-oxides, 250
Borate, buffers, 321, 345
   complexes with, 82
Borazines, 226

SUBJECT INDEX 389

Brdička test, 255
Bromide complexes, 18
Bromides, adsorption of, 289, 304, 306
  effect of, 318, 348, 350, 351, 352, 353, 354
Bromoacetate, 247
Bromobenzenes, 354, 357, 358
Bromoiodobenzenes, 357
Bromonitrobenzenes, 266
Bromopropionitriles, 247
Bromothiophene, 342
Buffers, acetate, 82, 83, 317
  adsorption of, components of, 78
  ammonia, 82, 83, 312-314
  barbital, 80, 82, 83, 108
  borate, 82, 83, 108, 321, 345
  Britton-Robinson, 82, 83
  capacity of, 81
  carbonate, 83
  choice of, 82, 83
  citrate, 83
  concentration of, 102
  dilution of, 79
  effect of, 30, 74-83, 120, 121, 143, 220, 312, 315
  formate, 83
  glycine, 83
  phosphate, 82, 83, 108, 320, 321
  simple, 83
  triethanolamine, 83
  triethylamine, 83, 108
  tris, 83
  universal, 82, 83
n-Butyl alcohol, 341
$tert$-Butyl group, effect of, 243

Cadmium, acetato complex of, 36, 54, 61
  amalgam of, 159, 161, 163, 164, 167, 171, 172, 177-179, 181, 187-189
  complexes of, 42, 46
  cyanide complexes, 46
  diffusion coefficient of, 39, 40
  $E_{1/2}$ of, 171, 172, 173
  EDTA complexes of, 61, 62, 64
  effect of, 351
  hydroxyethylethylenediaminetriacetate complexes of, 64
  nitrilotriacetate, complexes of, 53, 54, 55, 56, 64
  thiocyanate, complexes of, 29
  wave of, 181
Calcium, amalgam of, 160
  EDTA complexes of, 58, 59, 61
  effect of, 345
  salts of, 80
Capacity, differential, 292-296, 298, 318, 336, 344
  integral, 292
Carbanion-enolates, 99, 100, 106, 115, 130, 131, 135, 140, 141, 149
Carbanions, 113, 114, 239, 248, 254, 264
Carbon dioxide, 240
  effect of, 244
  as scavenger, 236
Carbonium compounds, 252-255
Carbontetrachloride, 248
Carbonyl compounds, 105, 122, 123, 125, 198, 212, 220, 230-241, 242-244, 252, 255, 301, 345
  hydration of, 113-115
  unsaturated, 198, 230, 235, 242-244
Carboxylation, 244
Catalysis, acid-base, 107, 108, 110, 111, 129, 131
  base, 134
  by cysteine-Co complex, 317-320
  by nitrobenzene, 264
  by OH$^-$, 113
  by pyridine, 312-314
  by quinine, 320, 321
  effect of, 117
  general, 107, 108, 110-112
  of cystine reduction, 255
  of H_2 evolution, 79, 243, 252, 255, 325, 343, 345, 346, 350-353
Cations, effect of, 147, 151, 217, 221, 233, 260, 264, 265, 304, 307, 308, 322, 334-340, 344-348, 354-358
  reduction of, 302
Cementation, 186-188
Cesium, adsorption of, 290
  amalgam of, 160, 172
  determination of, 345
  $E_{1/2}$ of, 172
  effect of, 323, 345, 347
Chalcones, 148-150, 343
Charge, effect of, 77, 303, 321
Charge transfer, 4-7, 14, 15, 20, 21, 23, 42, 51, 195, 196, 202-203, 207, 208, 252, 333

390 SUBJECT INDEX

complexes, 219, 227, 228, 249, 266
Chemicals, impurity in, 184
Chloramphenicol, 263
Chlorides, adsorption of, 306
  effect of, 318, 350-354
Chlorite, 345
Chloroacetanlides, 343, 344
Chlorobenzene, 347
Chlorocarbene, 248
Chloroiodobenzenes, 357
Chloronitrobenzenes, 266
Chlorophyll, 258
2-Chloropyridine, 315
6-Chloroquinoline, 224, 248
Chromanones, 107
Chromates, 344, 345
Chromium, ammonia complexes of, 41, 56-59
  complexes of, 40
  EDTA complexes of, 57-60, 65
  ethylenediamine complexes of, 41
  hydroxyethylethylenediaminetriacetate complexes of, 65
  oxalate complex of, 39
  oxidation by NO_3^-, 60
  reaction with mercury, 159, 162, 165, 166
Chronoamperometry, 359
Chronopotentiometry, 12, 13, 40, 41, 54, 55, 57, 58, 130, 204, 232, 238, 359
Cinnamaldehyde, 345
Cobalt, ammonia complexes of, 343, 353
  complexes of, 40
  cysteine complex of, 317, 318, 319, 320
  $E_{1/2}$ of, 172
  EDTA complexes of, 65
  ethylenediamine complexes of, 41
  oxalate complex of, 38, 41
  reaction with mercury, 159, 162, 172, 175
  shift of $E_{1/2}$, 304
  wave of, 180
Cobaltammines, 41, 343, 353
Coefficients, activity, 3
  α, 4-6, 16, 17, 19, 42, 48, 86, 171, 221, 239, 260, 302-306
  charge-transfer, 4-6, 16, 17, 19, 42, 48, 86, 171, 221, 239, 260, 302-306
  diffusion, 36-41
Commutator, Kalousek, 129, 201, 222, 225, 239, 258
Complexes, direct reduction of, 43
  formation of, 309
  inorganic, 1-72
  kinetics of, 42-66
    formation, 51-66
  labile, 1-72
  reduction of, 1-72
Concentration, surface, 6-8
Conformation, effect of, 221
Constants, electrochemical rate, 4, 5, 15, 171, 203
  rate, 4, 5, 54, 55, 61, 91, 102, 233, 310
  stability, 3, 7, 26-30, 30-41
Controlled-potential electrolysis, 199, 202, 232, 234, 235, 238, 244, 247, 249
Convection, 6-8
Copper, acetate complexes of, 35, 36
  amalgam of, 159, 161, 164, 165, 167, 169, 171-174, 177, 179, 181, 183, 188
  catalysis by nitrobenzene, 264
  $E_{1/2}$ of, 171, 172
  EDTA complexes of, 35, 65
  ions of, 243
  sulfatoacetato complexes of, 35, 36
  sulfate complexes of, 35, 36
  wave of, 180, 181, 183
Coumarin, 198, 341
Cozymase, 223
Current density, 6, 8, 12
Currents, anodic, 157, 171-173, 181-188, 233, 235
  catalytic, 60, 63, 79, 226, 243, 252, 253, 255, 264, 309-315, 315-322, 325, 327, 328, 342, 345
  cathodic, 180-181
  charging, 183, 288, 296, 297
  diffusion, 11, 12, 14, 15, 19, 30-36, 63, 92, 105, 207, 231, 297, 316
  faradaic, 4-6
  instantaneous, 63, 66
  kinetic, 11, 12, 19, 63, 92, 105, 207, 231, 242, 263, 309-315, 315-322, 327, 328, 342, 345, 353
  limiting, 11, 13, 15, 19, 63, 75, 83-152, 199-200, 207, 210, 211
  photo-, 236
  sign convention, 4
Current-voltage curves, *see* i-E curves

Curves, i-E, 13-16, 76, 77, 200, 203
  of amalgams, 170-175
Curve shape, approximate analysis, 13
  rigorous treatment, 13-16
Cyanide complexes, 46
ω-Cyanoacetophenone, 118, 119, 139, 140, 141
Cyano compounds, 248-249
Cyanoethyldimethylsulfonium ion, 255
Cyanomethyldimethylsulfonium ion, 254
Cyanopropyldimethylsulfonium ion, 255
3-Cyanopyridine, 223
Cyclization, 248
Cyclohexanone hydrazone, 343
Cyclohexanoneimine, 343
N-cyclohexyl acetaldoxime, 102
Cyclooctatetraene, 221, 338
N-cyclopentadienylpyridinium ion, 223
Cysteine, 255
  adsorption of, 342
  cobalt complex of, 317, 318, 319, 320
Cystine, 198, 255

Debye-Hückel, ion atmosphere, 292
Dehalogenation, 266
Dehydration, see Hydration
Deoxybenzoin, 234
Deuteration, effect of, 215
Diacetyl, 132
p-Diacetylbenzene, 89, 90, 128-130, 230, 237, 239
Dialcohol, formation of, 129
N,N'-Dialkylbipyridinium dication, 223
Dialkyldithio carbamates, 257
Diarylsulfoxides, 251
Diaziridine, 302
ω-Diazoacetophenones, 215
Diazonium compounds, 198
Dibenzoselenophen, 225
Dibenzofuran, 225
Dibenzothiophen, 225
Dibenzylcyclooctatetraene, 221
Dibenzylmercury, 234, 247
2,6-Dibromonitrobenzene, 266
Dibromosuccinate, 248
Dicarbonyl compounds, 128-132, 198
Dicarboxyacetylene, 214
Dichloroacetanilides, 343
2,6-Dichloronitrobenzene, 266
Dielectric constant, effect of, 205

Diethyldisulfide, 256
Diffuse layer, 289-292, 295, 301
Diffusion, 6, 8, 11, 12
1,3-Dihalocyclobutane, 248
Dihalogen compounds, 246
Dihydrophenazine, 225
α,β-Diketocamphorquinone, 229
α,β-Diketones, 197, 230, 235, 237
1,3-Diketones, 116, 145
Dimers, adsorption of, 222
  desorption of, 196
  formation of, 195-267
Dimethoxyethane, as solvent, 258
p-Dimethylaminoaniline, 253
p-Dimethylaminonitrobenzene, 253
p-Dimethylazobenzene, 253
Dimethylformamide, 199, 205, 213, 215, 217, 221, 223, 224, 225, 227, 228, 238-240, 244, 247-251, 253, 254, 258, 264-267, 325, 338, 339, 347, 348, 359
2,6-Dimethyliodobenzene, 355
Dimethylsulfoxide, 213, 216, 230, 256, 258
Dinitrobenzenes, 237, 253, 263, 345, 347, 348
Dioxane, 347
Dioximes, 214
Diphenylacetaldehyde, 114
9,10-Diphenylanthracene, 218
Diphenylcyclopropenone, 243
Diphenylcyclopropenylion ion, 253
Diphenyldipicrylhydrazyl, 217
Diphenyldiselenide, 256
Diphenyldisulfide, 256
Diphenylmercury, formation of, 258
Diphenylthallium, 258
Diphosphopyridine nucleotide, 223
Dipyridylethylenes, 214
Dismutation, 203, 206, 208-212, 216, 221, 229, 230, 234-236, 239, 244, 247, 249, 260, 263, 265
Dissociation constants, acid, 83-152
  equilibrium, 105, 106, 107
  polarographic, 88, 90, 91, 105
Dissociation curves, distorted, 120, 121
  polarographic, 83-152
Disulfides, 255
α,α'-Dithienylsulfide, 339, 340, 343
α,α'-Dithienylsulfoxide, 343

Dithiodiketones, 258
Dithiolium ion, 226
Di(trifluoromethyl)-*N*-hydroxylamine, 250
Ditropyl, 222
Double-layer, 252, 256, 287, 359
  effect of, 4, 36, 61
  structure of, 288-295
Drop-life, 15, 19, 232, 296

Electrocapillary curves, 296, 298
Electrodes, carbon paste, 199, 262
  curvature of, 8
  dropping amalgam, 180, 186
  dropping mercury, 8, 13-15, 39, 40, 49, 66, 75, 180, 183, 199, 288, 296, 341
  effect of material, 246
  expansion of, 8
  gold, 189
  hanging mercury drop, 183, 199, 202, 234, 262, 265, 266
  mercury streaming, 199
  platinum, 199, 218, 250, 252, 262
  rotated dropping mercury, 66
  rotating disc, 262
  stationary, 7, 8, 39, 66, 180, 183-188, 199, 202, 246, 249, 250, 252, 262, 265, 306
  stationary amalgam, 180
Electrolyte, effect of, 344-359
Electrolysis, controlled-potential, 129, 130, 199, 202, 232, 234, 235, 238, 244, 247, 249
Electron, transfer of, 333
Electron donors, 249
Electron spin resonance, 197, 198, 213, 215-217, 219, 221, 223, 224-226, 229, 233, 238-240, 249, 251, 253, 258, 261-264, 266, 267
Electrosynthesis, 75
Elimination, 247, 248, 266
Enol-keto equilibrium, 140, 141, 145
Equation, Hammett, 198, 215, 227, 233, 240, 247, 266
  Ilkovič, 14, 15, 19, 173
  Nernst, 3
Erythrose, 135
Esters, 198, 213, 240, 257, 325
Ethanol, effect on adsorption, 78
  as solvent, 81, 199, 248, 258, 329, 331, 333

Ethylbenzoyl acetate, 140, 141
Ethylbenzoyl benzoate, 106
Ethyl cinnamate, 338
Ethyl crotonate, 338
Ethylene bispyridines, 223
Ethylenediamine, complexes of, 41
Ethylenediaminetetraacetate, complexes of, 35, 57-66
Ethylenic compounds, 212-214
Ethylmethacrylate, 213
2-Ethyl-4-thiocarbamoyl pyridine, 141
Ethylxanthate, 256
Europium, amalgam of, 161
  EDTA complexes of, 61
Exchange current density, 6
Excitation, by illumination, 236

Faradaic impedance, 231, 260
Ferricyanide, 309, 323
Ferrocene, 258
Fick's Law, 6
Flash photolysis, 243
Fluorenone, 234-236, 239, 345
Fluoresceins, 229
Fluorides, effect of, 298, 299, 349
Formaldehyde, 113, 231, 233
  sulfoxylate, 251
Formate, formation of, 240
Frequency, angular, 20, 61
Fulvene, 219
Fumarates, 213
Di-2-Furoylmethane, 236

Gallium, 353
Gelatin, 253
Germanium, biphenyl of, 258
Glycine, 314
Gold, amalgam of, 159, 161, 177-179, 181, 182

Haber mechanism, 259
Halides, effect of, 350-354
α-Haloacetaldehyde, 342
α-Haloaldehydes, 133, 134, 240
Halogeno compounds, 198, 244-250, 305, 331, 355
α-Haloketones, 130
Halonitrobenzenes, 266
Halonitro compounds, 248, 263
Halopyridines, 116

SUBJECT INDEX

Hammett equation, 198, 215, 227, 233, 240, 247, 266
Helmholtz layer, 289
Helmholtz planes, 289, 290, 291, 295, 299, 300, 306, 308, 319, 336
Hemiacetal, formation of, 107, 108
Henry's Law, 326
Heptanone, 336, 338
Heredity, effect of, 342-344
Heterocyclics, 197, 217-227
  catalysis by, 79
γ-Hexachlorocyclohexane, 332
Hexafluoroacetophenone, 239
Hexafluorobenzene, 347
Hexamethylbenzene, 228, 249
Hexamethylphosphoramide, 333
Haxamminocobalt, 343
Hydration, of aldehydes, 103, 107, 108, 110, 111, 113, 114, 115, 132, 133, 134, 151, 152
  of carbonyl compounds, 113, 114, 115, 231
  of α-dicarbonyl compounds, 133
  effect of, 309
  of α-ketoacids, 116
  of pyridoxal, 107
9,9'-Hydrazoacridine, 216
Hydrazobenzene, 215
Hydrazones, 214, 343
Hydrobenzoin, 233, 234
Hydrocarbons, alternant, 206, 213, 220
  aromatic, 197, 217-227
  correlation of pK with $E_{1/2}$, 258
  polycyclic, 197, 217-227
  polynuclear, 78
  reducible, 84
Hydrogen, atomic, 208
  catalysis by, cysteine-Co complex, 317-320
    quinone, 320, 321
  catalysis of, evolution of, 243, 252, 255, 312-314, 325, 343, 345, 346, 350, 351, 352, 353
  cation effects on, 344
  effects on evolution of, 349
  evolution of, 79, 208, 232, 288, 305, 306
  overvoltage, 79
Hydrogen bonding, 205, 229, 265, 266
Hydrogenation, catalytic, 235
Hydroquinone, 197, 204, 227

  as proton donor, 215
Hydroxo complexes, 27, 46, 49, 50
α-Hydroxy aldehydes, 133, 134
1-Hydroxyanthraquinone, 347
Hydroxybenzaldehydes, 121, 236
α-Hydroxycarbonyls, 82
2-Hydroxychalcones, 107, 243
Hydroxyethylethylenediaminetriacetate complexes, 64, 65
Hydroxylamines, 127, 259, 262, 264, 265
  oxidation of, 259
Hydroxyl ions, addition of, 113, 114
  catalysis by, 103
8-Hydroxyquinoline, 224

i-E curves, 13-16, 76, 77, 200, 203
  of amalgams, 170-175
i-t curves, 201
  potentiostatic, 8-12
Illumination, effect of, 236
Imidazole, adsorption of, 78
Imidazol-2-thione, 325, 326
Impedance, Faradaic, 20-25, 61
  Warburg, 23
Indium, amalgam of, 159, 161, 164, 165, 167, 169, 171, 172, 177-179, 181-183
  chloride, 346
  $E_{1/2}$ of, 171, 172
  wave of, 181-183
Inhibitors, effect of, 204, 247, 257, 260-263, 346
Intermediates, identification of, 75, 76, 134, 195-267
Intermetallic compounds, 157-189
Inversion, surface charge, 291
Iodides, adsorption of, 289, 306
  effect of, 304, 318, 319, 348-351, 353, 354
Iodoalkanes, 247
Iodoanilines, 103, 121, 357
p-Iodoanilinium ion, 302
Iodobenzenes, 247, 348, 354, 355, 356, 357, 358
Iodo compounds, 301
  reaction with Hg, 247
Iodomethyltrialkylsilanes, 248
Iodonitrobenzenes, 266
Iodonium compounds, 198, 252-255
p-Iodophenolate ion, 302

Iodophenols, 103, 357
Iodopropionitriles, 247
Iodopyridines, 117, 118
Ionic strength, 83
Ion pairs, 37, 38, 40, 41, 74, 219, 221, 229, 233, 260, 265, 308, 348, 349, 351, 352
Iron, catalysis by, 255
  $E_{1/2}$ of, 172, 173
  reaction with mercury, 159, 162, 165, 172, 175, 186, 188
Irradiation, effect of, 197
Irreversibility, 16, 17, 29, 36, 37, 49, 74, 86, 171, 200-202, 203-205, 229, 300, 301
Isatin, 235
Isobenzpyrylium cations, 219
Isonicotinonitrile, 248

Kalousek commutator, 129, 201, 222, 225, 239, 258
Keto-enol equilibrium, 140, 141, 145
Ketol, 129
Ketones, 198, 230-241, 301, 345
  aryl, 104, 105, 119, 120, 122, 123, 125, 128-131, 144, 145-152
  phosphorylated, 240
  α-substituted, 130
  α,β-unsaturated, 107, 116, 148-152, 198
α-Ketosulfides, 130, 237
Kinetics, electrode, 300-325

Lanthanum, amalgam of, 161
  effect of, 345
Lead, acetato complexes of, 35, 36
  adsorption of, 290
  amalgam of, 158, 162-169, 171, 172, 181, 189
  $E_{1/2}$ of, 171, 172
  EDTA complexes of, 35, 65
  in excess of Cd, 189
  waves of, 181
Ligands, concentration of, 26-30, 30-33, 34, 35
  exchange of, 30-33
  number of, 42
Lithium, amalgam of, 160, 172
  $E_{1/2}$ of, 172
  effect of, 228, 238, 345, 346, 347, 348,

salts of, 78, 79, 80, 82

Magnesium, amalgam of, 160
  effect of, 323, 324
  organometallic compounds of, 258
Magnesium sulfate, effect of, 341
Maleates, 213, 311, 312
Manganese, 354
  amalgam of, 159, 161, 164, 165, 167, 172, 173, 174, 184
  $E_{1/2}$ of, 172
  EDTA complexes of, 57
  nitrotriacetate complexes of, 65
Maxima, catalytic, 253
  streaming, 102
Media, *see* Solvents
Mercaptans, 107, 256; *see also* Thiols
Mercury, compounds of, 198, 234, 243, 246, 247, 255-257
  radicals of, 254, 256, 257
  reaction with, 247
    iodo compounds, 247
    radicals, 77, 78, 243
  as solvent, 175
Metals, exchange of, 33-36
  interaction in amalgams, 157-189
  radical interaction with, 246
  solubility in mercury, 165-167, 169, 173, 174, 183, 188
Methanol, 199, 216, 228, 247, 253, 260, 267, 335
Methyl bromide, 331
Methylbutylphenacylsulfonium ion, 93, 94, 99
Methylene chloride, as solvent, 252
Methylglyoxal, 132
Methyl iodide, 331
N-Methylpyridinium ion, effect of, 336
4-Methylquinazoline, 109
Methyl vinyl ketone, 243
Minima, 322
MO calculations, 197, 198, 205, 206, 212, 214, 215, 217-219, 225-227, 240, 250, 258, 302
Monoalkyltin, 258
Monochloroacetanilide, 344

Naphthacene, 220
Naphthalene, formation of, 251

SUBJECT INDEX

Naphthoquinones, 227, 229
Nickel, acetate complex of, 32, 33, 35, 36
   compound with mercury, 159, 161, 166, 172, 175, 177, 179, 181, 184, 186
   $E_{1/2}$ of, 172, 173
   EDTA complexes of, 65, 66
   nitrilotriacetate complex of, 32, 33, 65
   peak of, 304
   o-phenylenediamine, effect on, 318, 319, 342
   pyridine complex of, 324, 325
   sulfatoacetato complexes of, 35, 36
   sulfate complexes of, 35, 36
Nicotinamide, derivatives of, 223
Ninhydrin, 231
Nitrates, effect of, 350-352
Nitriles, 248-249, 355
   unsaturated, 213
Nitrilotriacetate complexes, 31, 33, 53-56, 64, 65
Nitrite ions, formation of, 264
Nitroacetophenone, 236
Nitroalkanes, 263
Nitroanilines, 112
p-Nitroazobenzene, 263
Nitrobenzaldehyde, 236
Nitrobenzenes, 217, 262-266, 341, 347, 357
   as catalyst, 264
Nitrochalcones, 343
Nitro compounds, 198, 236, 237, 248, 259-267, 347
   aliphatic, 127
   aromatic, 127
Nitrofuran, 262, 263
Nitrofurazone, 263
Nitromethane, 319
   as solvent, 226
Nitronates, 248, 264
Nitrones, 102, 116, 215
Nitrophenols, 112, 262
N-Nitropyrazoles, 263
Nitrosobenzene, 267
Nitroso compounds, 250-252, 259-267
Nitrosohydroxyl amines, 333, 334
N-Nitroso-α-naphthylhydroxylamine, 334
p-Nitrosophenol, 113
N-Nitrosophenylhydroxylamine, 334
N-Nitroso-p-xenylhydroxylamine, 334
ω-Nitrostyrene, 263

Nitrotoluenes, 358
9-Nitrotriptycene, 266
Nuclear quadrupole resonance, 247
Nylon, manufacture of, 250

O^{17}, use of, 227
Olefins, 212-214
Onium compounds, 252-255
Organomercury compounds, 257
Organomercury radicals, 247
Organometallics, 198, 246, 257-258
   formation of, 78
Orientation, at electrode, 77, 196
Ortho effects, 358
Overvoltage, 9
Oxalates, 236
   formation of, 240
Oxalato complexes, 38, 39, 41, 46
Oxidations, 199, 216
   of metals, 309
N-Oxides, 250
Oximes, 121, 130, 138, 139
   O-alkyl, 101, 102
Oxygen, 354
   effect of, 168, 208, 259, 297
   formation of, 259
   reaction with anion radical, 267

Palladium, allyl compounds of, 258
   compound with mercury, 159, 161
   sol of, 235
Pentachlorohalobenzene, 247
Pentachloronitrobenzene, 266, 267
Pentafluorohalobenzene, 247
Pentafluoronitrobenzene, 266
Perchlorates, effect of, 348, 349, 354
Perinaphthenone, 243, 244
Periodate, catalysis by nitrobenzene, 264
Permanganate, effect of, 349
Peroxides, 240
Peroxodisulfate, 307, 308, 322, 323, 324
   catalysis by nitrobenzene, 264
   as oxidant, 267
   radical formation, 243
Perrhenate, effect of, 349
Persulfate, *see* Peroxodisulfate
pH, choice of range, 78, 79
   effects of, 73-153, 230, 231, 242, 303, 314
   at electrode, 312, 315, 319

SUBJECT INDEX

Phase exchange, 186-188
ω-Phenacylsulfonium ions, 237
Phenanthrene, 218
Phenazine, 225
Phenol, 341
  effect of, 213
  as proton donor, 219, 240
Phenothiazine, 226
Phenylacetaldehyde, 114
Phenyldiazonium ion, 252
o-Phenylenediamine, effect of, 318, 319, 342
Phenylethylsulfone, 343
Phenylhydroxylamine, 260, 267
2-Phenylindandione-1,3, 237, 238
Phenyliodobenzenes, 355
Phenylisocyanate, 215
N-Phenylisoindoline, 224
Phenylmercury, formation of, 258
Phenylmethylsulfoxide, 251
Phenylmonothiobenzoate, 257
2-Phenylpropionaldehyde, 114
3-Phenylpropionaldehyde, 151, 152, 345, 346
1-Phenyl-3-pyrazolidone, 223
Phenylvinyl ketone, 150
Phenylvinylsulfone, 343
Phosphate buffer, 320, 321
Phosphoketones, 240
Phosphonium compounds, 252-255
Phosphonium amalgam, 253
α-Phosphonium ketones, 130, 237
Phosphorus compounds, 250-252
Phosphorylated ketones, 251
Photo reactions, 66
Photoreduction, 235
Phthalimides, 125, 126, 224, 238, 240
Phthalocyanines, 258
Phthalonitrile, 249
Piazthiazoles, 225, 359
Piazselenazoles, 225, 359
Pinacolate, 238
Pinacols, desorption of, 234
  dismutation of, 234
  formation of, 198
Planarity, effect of, 221
Platinum, compound with mercury, 159, 161
Plutonium, amalgam of, 162
Polarography, Ac, 20-25, 60-62, 201, 221, 225, 262, 304
  linear sweep, 244
  oscillographic, 201, 229, 233, 234, 235, 262
  square wave, 304
Polycyanocompounds, 249
Polymerization, 242, 243, 258
  of acrylamide, 214
Polymers, 244
Polymethines, 213
Polyoxyethylene lauryl ether, 18
Porphyrin, 223
Potassium, amalgam of, 160, 164, 169, 172
  $E_{½}$ of, 172
  effect of, 345
Potentials, θ, 290, 293
  ψ, 290-293, 295-300, 301-306, 309, 312, 315, 316, 318, 319, 321-323, 328, 337, 339, 340, 342, 344, 346, 352, 354, 355
  half-wave, 14-19, 26-30, 36, 37, 42, 46, 48, 61, 73, 75-77, 82-152, 170-172, 181, 198, 206-208, 210, 211, 215, 232, 247, 250, 300, 301, 316, 322, 330, 338, 339, 340, 345
  liquid-function, 329
  of maximum adsorption, 326
  standard, 3
  of zero charge, 290, 292, 322, 359
Potentiometry, 165-167, 173, 178, 179, 181-183
Products, identification of, 75, 76, 202
Propene, formation of, 258
Propenylium ion, 243
Propionitrile, 213
  formation of, 249
Proteins, 343
Protonation, role of, 198, 230, 232, 242, 252, 261, 265, 303, 309, 325, 331
Proton donors, 73, 74, 205-207, 213, 215, 219, 220, 222, 239, 240, 255, 265, 309-312, 325
Proton transfer, 73-152, 206-208, 209, 212, 213, 217, 220, 244, 261, 263, 265, 266
Pyrazine, 223
Pyrene, 249
Pyridines, 213, 219, 222
  adsorption of, 78
  catalysis by, 312, 313, 314, 346, 352, 353

SUBJECT INDEX 397

effect of, 77
  nickel complex of, 324, 325
  protonation of, 321
  salting out, 341
  as solvent, 222, 238
Pyridine aldehydes, 110, 111
Pyridine-N-oxides, 330, 331
Pyridoxal, 107, 108, 139
Pyridoxaloxime, 138, 139
Pyridoxal-5-phosphate, 108, 139
Pyridyl alcohol, formation of, 239
Pyridoxal-5-phosphate, 108
Pyrimidine, 219, 223
Pyrylium cations, 219, 225

Quasireversibility, 26, 29
Quasiunadsorbed particles, 331-334
Quinine, adsorption of, 342
  catalysis by, 320, 321, 350-352
Quinoid intermediates, 129, 237, 238, 263, 265
Quinolines, 224
Quinoneimine, hydration of, 112
  hydrolysis of, 113
Quinones, 197, 204, 220, 227-230
  protonation of, 252

Radical anions, 73, 74, 125, 147, 150, 152, 195-267
Radical cations, 195-267
Radicals, 71, 125, 147, 150, 195-267
  reaction, with Hg, 243
    with metals, 246
  spectra of, 265
Rare earth metals, amalgams of, 158, 161
Rates, reaction, 4, 5, 15, 19, 42-66
Reaction layer, 310, 311
Reactions, acid catalyzed, 91
  base catalyzed, 91, 103
  in bulk of solution, 62-63
  catalytic, 20, 22, 60, 63, 79, 107, 226, 243, 252, 253, 255, 264, 309-315, 315-325
  chemical, effects of, 6, 8, 11, 12, 13, 19, 20, 42-66, 75, 76, 83-152, 233
  fast, 11, 12, 13, 19, 20, 21, 24, 42-66, 91, 92-153
  dark, 236
  ECE mechanism, 56, 57, 112, 113, 127, 128, 129, 130, 134, 135

electrode, 2-4
exchange, 30
following, 20, 22, 24, 25, 60
heterogeneous, see Surface
interposed, 56, 57, 112, 113, 127, 128, 129, 130, 134, 135
ligand exchange, 30-33
metal exchange, 33-36
photo-, 66, 235, 236, 243
preceding, 12, 13, 19-21, 24, 25, 61, 92-152, 309, 315, 316, 320, 325, 342
recombination, see Preceding
second order, 209, 236, 238
surface, 77, 92, 93, 101, 102, 118, 126, 140, 141, 143, 207, 210, 230-232, 261, 263, 310, 315-325, 327, 328, 331
volume, 92, 93, 101, 140, 143, 207, 230, 231, 310, 315, 321
Reactivity, 75
Reduction stepwise, 203-206
Resonance, effect of, 243, 244
Reversibility, 15-19, 26, 28, 29, 36, 42, 46, 48, 61, 73, 86, 170, 197, 200-202, 203, 204, 207, 214, 225, 233, 260, 265, 316
Ring-opening, 107
Rubidium, amalgam of, 172
$E_{\frac{1}{2}}$ of, 172
effect of, 345, 346

Salting-out, 340-342
Schiff bases, 214-217
Selenophenol, 256
Semicarbazones, 102, 103
Semiquinones, 197, 204, 227-229
Shielding, 39
Silicon, biphenyl of, 258
Silver, amalgam of, 159, 161, 177, 179, 181, 183, 187
Sodium, amalgam of, 160, 164, 169, 170, 172
$E_{\frac{1}{2}}$ of, 172
effect of, 345, 347
Solubility, 341
  of metal in mercury, 165-167, 169, 173, 174, 183, 188
Solvation, effect of, 205, 217, 219, 228, 229, 342
Solvents, adsorption of, 332

SUBJECT INDEX

aprotic, 195-267, 325
non-aqueous, 73, 74, 78, 81, 153, 195-267, 325, 331, 332
Spiro compounds, 248
Steric effects, 77, 243, 244, 246, 355, 356
Stilbenediol, 229
Stilbenequinone, 228
Stilbenes, 214
Strontium, amalgam of, 160
effect of, 80, 345
Structure, effect of, 76, 77, 198, 212, 217, 233, 240, 243, 244, 247, 258, 265, 266, 354-358
Substituent, effect of, 198, 217, 218, 229, 233, 240, 247, 265, 266, 355
Sugars, 135
Sulfate complexes, 35, 36, 40, 41
Sulfates, effect of, 350-353
Sulfatoacetato complexes, 35, 36
Sulfonium compounds, 252-255
Sulfonium amalgam, 253
Sulfonium ketones, 237
Sulfonylhydrazones, 217
Sulfur compounds, 198, 250-252, 255-257
catalysis by, 79
Sulfur dioxide, liquid, 252
Surfactants, effect of, 259, 260, 261, 262, 263
Sydnones, 104, 122

Tafel plots, corrected, 76, 306, 307, 308
Tartrato complexes, 27, 49, 50
Temperature, effect of, 321
Tetraalkylammonium salts, 74, 78-80, 239, 253, 320, 334-341, 346, 347, 349, 350, 354, 355, 357, 359
Tetraamylammonium ion, 336, 338
Tetrabromoplatinate, 336
Tetrabutylammonium ion, adsorption of, 78
Tetrabutylammonium salts, 336, 338, 341, 347, 348, 349
Tetrachloroplatinate, 336
Tetracyanoquinodimethane, 227, 228
Tetra(dimethylamino)ethylene, 213
Tetraethylammonium ion, 336, 337, 339, 347, 348
Tetrahydrofuran, 220, 239, 240
Tetra(isoamyl)ammonium ion, 336, 338
Tetraisopropylnitrobenzene, 266

Tetramethylammonium ion, 338, 347, 349, 354, 355, 356, 357
Tetraphenylcyclopentadienone, 244
Tetrapropylammonion ion, 336, 337, 338
Thallium, adsorption of, 290
amalgam of, 159, 161, 163, 164, 165, 169, 170, 171, 172
diffusion coefficient of, 39, 40
$E_{1/2}$ of, 171, 172
organometallic compounds of, 258
3-Thianaphthenone, 106
Thiocyanate, effect of, 353
Thiocyanato complexes, 29
Thioketones, 244
Thiols, addition of, 114
Thiophenol, 256
Thiosemicarbazones, 121
Thorium, reaction with mercury, 162
Time, transition, 40, 41
Tin, amalgam of, 158, 161, 163-168, 171, 172, 177-179, 181-183
$E_{1/2}$ of, 171, 172
organometallic compounds of, 258
wave of, 181, 183
Titration, amperometric, 182
Tocopherols, 227
p-Toluidine, 349
Transfer, mass, 6, 7, 23
Transition state, 77
Transition time, galvanostatic, 12, 13
Tribenzylamine, 349
Trichloroacetanilide, 343
Trichloroacetate ion, 307, 308
Trifluoromethyliodobenzenes, 357
Trimethylnitromethane, 250, 264
Trimethylphenyl radical, 218
Trimethyltin, 258
Trinitrobenzene, 266
Triphenylcarbonium ion, 252
Triphenylcyclopropenyl radical, 218
Triphenylmethyl, formation of, 252
Triphenylphosphine, 254
Triphenylphosphonium ions, 254
Triptycenequinones, 227
Trithion, 226
Tropoid ketone, 236
Tropolone, 198
Tropone, 230, 244
Tropyl alcohol, 220, 222
Tropylium ion, 100, 101, 102, 220, 221,

222, 354

Uranium, amalgam of, 161

Vitamins, E, 227
 K, 227
Voltammetry, 359
 amalgam, 157, 183-188
 anodic stripping, 157, 183-188
 cyclic, 197, 201, 218, 219, 221, 222, 225, 231, 234, 235, 239, 262, 265, 266

Wave-height, *see* Currents, limiting
Waves, anodic, *see* Currents, anodic
 equation of, 13-16
 hump-shaped, 319, 327
 polarographic, 13-16
 shape of, 75-77, 201, 204, 205, 259
Water, effect of, 74, 208, 225, 244, 265, 332
 as proton donor, 219
Water-ordering, 349

Xanthone, 231

Ylid, 99

Zinc, acetate complex of, 18, 27, 36, 43, 48, 49, 50
 amalgam of, 158, 161, 164, 165, 167-169, 171, 172, 177-179, 181, 183, 188
 ammonia complexes of, 46, 49, 50
 bromide complexes of, 18
 complexes of, 42, 46, 50
 cyanide complexes of, 46
 $E_{1/2}$ of, 171-173
 EDTA complexes of, 65
 effect of, 323, 324
 hydroxo complex of, 27, 46, 49, 50
 oxalate complexes of, 46
 peak of, 304
 reduction on amalgam, 180
 tartrate complex of, 27, 49, 50
 waves of, 180, 181, 183